T0190038

Lecture Notes in Computer Science 12490

More information about this series at http://www.springer.com/series/7409

Cesar Analide · Paulo Novais ·
David Camacho · Hujun Yin (Eds.)

Intelligent Data Engineering and Automated Learning – IDEAL 2020

21st International Conference
Guimaraes, Portugal, November 4–6, 2020
Proceedings, Part II

 Springer

Editors
Cesar Analide ⓘ
University of Minho
Braga, Portugal

Paulo Novais ⓘ
University of Minho
Braga, Portugal

David Camacho ⓘ
Technical University of Madrid
Madrid, Spain

Hujun Yin ⓘ
University of Manchester
Manchester, UK

ISSN 0302-9743   ISSN 1611-3349 (electronic)
Lecture Notes in Computer Science
ISBN 978-3-030-62364-7   ISBN 978-3-030-62365-4 (eBook)
https://doi.org/10.1007/978-3-030-62365-4

LNCS Sublibrary: SL3 – Information Systems and Applications, incl. Internet/Web, and HCI

This Springer imprint is published by the registered company Springer Nature Switzerland AG
The registered company address is: Gewerbestrasse 11, 6330 Cham, Switzerland

Preface

The International Conference on Intelligent Data Engineering and Automated Learning (IDEAL) is an annual international conference dedicated to emerging and challenging topics in intelligent data analytics and associated machine learning systems and paradigms. After the hugely successful IDEAL 2018 in Madrid, Spain, and its 20th edition in Manchester, UK, last year, IDEAL 2020 was hosted by the Synthetic Intelligence Lab (ISLab) from the ALGORITMI Center at the University of Minho, in Guimarães, the birthplace of Portugal, during November 4–6, 2020. It was also technically co-sponsored by the Portuguese Artificial Intelligence Association (APPIA) and the IEEE Computational Intelligence Society Portuguese Chapter. Due to the COVID-19 pandemic, the Organization Committee decided to hold the IDEAL 2020 completely online, to ensure the safety of the organization and the delegates of the conference.

This year marked the 21st edition of IDEAL, a conference series which has been serving an important role in data analytics and machine learning communities. The conference aims to bring researchers and practitioners together, share the latest knowledge, disseminate cutting-edge results, and forge alliances on tackling many real-world challenging problems. The core themes of the IDEAL 2020, as usual, included big data challenges, machine learning, data mining, information retrieval and management, bio-/neuro-informatics, bio-inspired models, agents and hybrid intelligent systems, real-world applications of intelligent techniques, and AI.

The IDEAL 2020 event consisted of the main track, nine special sessions, and one colocated workshop. These special sessions were:

- Special Session 1: Data Generation and Data Pre-processing in Machine Learning
- Special Session 2: Optimization and Machine Learning for Industry 4.0
- Special Session 3: Practical Applications of Deep Learning
- Special Session 4: New trends and challenges on Social Networks Analysis
- Special Session 5: Machine Learning in Automatic Control
- Special Session 6: Emerging Trends in Machine Learning
- Special Session 7: Machine Learning, Law and Legal Industry
- Special Session 8: Data Recovery Approach to Clustering and Interpretability
- Special Session 9: Automated learning for industrial applications

The colocated event was the Workshop on Machine Learning in Smart Mobility (MLSM 2020).

A total of 122 submissions were received and were then subjected to rigorous peer reviews by the members and experts of the Program Committee. Only those papers that were found to be of the highest quality and novelty were accepted and included in the proceedings. These proceedings contain 93 papers, in particular, 34 for the main track with an acceptance rate of 59%, 54 for special sessions, and 5 for the colocated workshop, were accepted and presented at IDEAL 2020.

We deeply appreciate the efforts of our invited speakers Paulo Lisboa of Liverpool John Moores University, UK; João Gama of the University of Porto, Portugal; and George Baciu of Hong Kong Polytechnic University, China; and thank them for their interesting and inspiring lectures.

We would like to thank our sponsors for their technical support. We would also like to thank everyone who invested so much time and effort in making the conference a success, particularly the Program Committee members and reviewers, the organizers of the special sessions, as well as all the authors who contributed to the conference.

A big thank you to the special sessions and workshop chairs Antonio J. Tallón-Ballesteros and Susana Nascimento for their fantastic work.

Special thanks also go to the MLSM 2020 organizers Sara Ferreira, Henrique Lopes Cardoso, and Rosaldo Rossetti.

Finally, we are also very grateful to the hard work by the local Organizing Committee at the University of Minho, Portugal, especially Pedro Oliveira, António Silva, and Bruno Fernandes for checking all the camera-ready files. The continued support, sponsorship, and collaboration from Springer LNCS are also greatly appreciated.

November 2020

Cesar Analide
Paulo Novais
David Camacho
Hujun Yin

Organization

General Chairs

Paulo Novais   University of Minho, Portugal
Cesar Analide   University of Minho, Portugal
David Camacho   Universidad Politecnica de Madrid, Spain

Honorary Chair

Hujun Yin   The University of Manchester, UK

Program Chairs

Cesar Analide   University of Minho, Portugal
Paulo Novais   University of Minho, Portugal
David Camacho   Universidad Politecnica de Madrid, Spain
Hujun Yin   The University of Manchester, UK

Steering Committee

Hujun Yin   The University of Manchester, UK
Colin Fyfe   University of the West of Scotland, UK
Guilherme Barreto   Federal University of Ceará, Brazil
Jimmy Lee   The Chinese University of Hong Kong, Hong Kong,
                         China
John Keane   The University of Manchester, UK
Jose A. Costa   Federal University of Rio Grande do Norte, Brazil
Juan Manuel Corchado   University of Salamanca, Spain
Laiwan Chan   The Chinese University of Hong Kong, Hong Kong,
                         China
Malik Magdon-Ismail   Rensselaer Polytechnic Institute, USA
Marc van Hulle   KU Leuven, Belgium
Ning Zhong   Maebashi Institute of Technology, Japan
Peter Tino   University of Birmingham, UK
Samuel Kaski   Aalto University, Finland
Vic Rayward-Smith   University of East Anglia, UK
Yiu-ming Cheung   Hong Kong Baptist University, Hong Kong, China
Zheng Rong Yang   University of Exeter, UK

Special Session and Workshop Chairs

Antonio J. Tallón-Ballesteros	University of Seville, Spain
Susana Nascimento	Universidade Nova de Lisboa, Portugal

Publicity and Liaisons Chairs

Bin Li	University of Science and Technology of China, China
Guilherme Barreto	Federal University of Ceará, Brazil
Jose A. Costa	Federal University of Rio Grande do Norte, Brazil
Yimin Wen	Guilin University of Electronic Technology, China

Local Organizing Committee

Paulo Novais

Paulo Moura Oliveira

Cesar Analide

José Luís Calvo-Rolle

José Machado

Ana Silva

António Silva

Héctor Alaiz Moretón

Bruno Fernandes

Leandro Freitas

Dalila Durães

Leonardo Nogueira Matos

Filipe Gonçalves

Marco Gomes

Francisco Marcondes

Pedro Oliveira

Fábio Silva

Program Committee

Álvaro Herrero

Ajalmar Rêgo Da Rocha Neto

Ajith Abraham

Alejandro Martín

Alexandros Tzanetos

Alfredo Cuzzocrea

Alfredo Vellido

Ana Madureira

Anabela Simões

Andre de Carvalho

Angel Arcos

Angel Arroyo

Ângelo Costa

Anna Gorawska

Antonio Fernández-Caballero

Antonio Gonzalez-Pardo

Antonio J. Tallón-Ballesteros

Antonio Neme

Armando Mendes

Bibiana Clara

Bin Li

Bogusław Cyganek

Boris Delibašić

Boris Mirkin

Bruno Baruque

Carlos A. Iglesias

Carlos Cambra

Carlos Carrascosa

Carlos Coello Coello

Carlos M. Travieso-Gonzalez

Carlos Pereira

Carlos Ramos

Carmelo J. A. Bastos Filho

Cristian Mihaescu

Cristian Ramírez-Atencia

Dalila Duraes

Daniel Urda
Dariusz Frejlichowski
Dariusz Jankowski
Diana Manjarres
Davide Carneiro
Dinu Dragan
Dongqing Wei
Dragan Simic
Edyta Brzychczy
Eiji Uchino
Emilio Carrizosa
Eneko Osaba
Ernesto Damiani
Esteban Jove
Eva Onaindia
Federico Mata
Felipe M. G. França
Fernando De La Prieta
Fernando Diaz
Fernando Nuñez
Fionn Murtagh
Florentino Fdez-Riverola
Francesco Corona
Fábio Silva
Gema Bello Orgaz
Gerard Dreyfus
Giancarlo Fortino
Gianni Vercelli
Goreti Marreiros
Grzegorz J. Nalepa
Hamido Fujita
Héctor Alaiz Moretón
Héctor Quintián
Ignacio Hidalgo
Ioannis Hatzilygeroudis
Irene Pulido
Ireneusz Czarnowski
Isaias Garcia
Ivan Silva
Izabela Rejer
J. Michael Herrmann
Jaakko Hollmén
Jason Jung
Javier Bajo
Javier Del Ser
Jean-Michel Ilie

Jerzy Grzymala-Busse
Jesus Alcala-Fdez
Jesus López
Jesús Sánchez-Oro
João Carneiro
Joaquim Filipe
Jochen Einbeck
Jose Andrades
Jose Alfredo Ferreira Costa
Jose Carlos Montoya
Jose Dorronsoro
Jose Luis Calvo-Rolle
Jose M. Molina
José Maia Neves
Jose Palma
Jose Santos
Josep Carmona
José Fco. Martínez-Trinidad
José Alberto Benítez-Andrades
José Machado
José Luis Casteleiro-Roca
José Ramón Villar
José Valente de Oliveira
João Ferreira
João Pires
Juan G. Victores
Juan Jose Flores
Juan Manuel Dodero
Juan Pavón
Leandro Coelho
Lino Figueiredo
Lourdes Borrajo
Luis Javier García Villalba
Luís Cavique
Luís Correia
Maciej Grzenda
Manuel Grana
Manuel Jesus Cobo Martin
Marcin Gorawski
Marcin Szpyrka
Marcus Gallagher
Margarida Cardoso
Martin Atzmueller
Maria Teresa Garcia-Ordas
María José Ginzo Villamayor
Matilde Santos

Mercedes Carnero

Michal Wozniak

Miguel J. Hornos

Murat Caner Testik

Ngoc-Thanh Nguyen

Pablo Chamoso

Pablo García Sánchez

Paulo Cortez

Paulo Moura Oliveira

Paulo Quaresma

Paulo Urbano

Pawel Forczmanski

Pedro Antonio Gutierrez

Pedro Castillo

Pedro Freitas

Peter Tino

Qing Tian

Radu-Emil Precup

Rafael Corchuelo

Raquel Redondo

Raul Lara-Cabrera

Raymond Wong

Raúl Cruz-Barbosa

Renato Amorim

Ricardo Aler

Ricardo Santos

Richard Allmendinger

Richar Chbeir

Robert Burduk

Roberto Carballedo

Roberto Casado-Vara

Roberto Confalonieri

Romis Attux

Rui Neves Madeira

Rushed Kanawati

Songcan Chen

Stefania Tomasiello

Stelvio Cimato

Sung-Bae Cho

Susana Nascimento

Tatsuo Nakajima

Tzai-Der Wang

Valery Naranjo

Vasile Palade

Vicent Botti

Vicente Julián

Víctor Fernández

Wei-Chiang Hong

Wenjian Luo

Xin-She Yang

Ying Tan

Additional Reviewers

Adrián Colomer

Aldo Cipriano

Alejandro Álvarez Ayllón

Alfonso González Briones

Alfonso Hernandez

Angel Panizo Lledot

Beatriz Ruiz Reina

David Garcia-Retuerta

Eloy Irigoyen

Francisco Andrade

Gabriel Garcia

Javier Huertas-Tato

Jesus Fernandez-Lozano

Joana Silva

Juan Albino Mendez

Juan José Gamboa-Montero

Julio Silva-Rodríguez

Kevin Sebastian Luck

Lorena Gonzalez Juarez

Manuel Masseno

Marcos Maroto Gómez

Mariliana Rico Carrillo

Mario Martín

Marta Arias

Pablo Falcón Oubiña

Patrícia Jerónimo

Paweł Martynowicz

Szymon Bobek

Vanessa Jiménez Serranía

Wenbin Pei

Ying Bi

Special Session on Data Generation and Data Pre-processing in Machine Learning

Organizers

Antonio J. Tallón-Ballesteros	University of Huelva, Spain
Bing Xue	Victoria University of Wellington, New Zealand
Luis Cavique	Universidade Aberta, Portugal

Special Session on Optimization and Machine Learning for Industry 4.0

Organizers

Eneko Osaba	Basque Research and Technology Alliance, Spain
Diana Manjarres	Basque Research and Technology Alliance, Spain
Javier Del Ser	University of the Basque Country, Spain

Special Session on Practical Applications of Deep Learning

Organizers

Alejandro Martín	Universidad Politécnica de Madrid, Spain
Víctor Fernández	Universidad Politécnica de Madrid, Spain

Special Session on New Trends and Challenges on Social Networks Analysis

Organizers

Gema Bello Orgaz	Universidad Politécnica de Madrid, Spain
David Camacho	Universidad Politécnica de Madrid, Spain

Special Session on Machine Learning in Automatic Control

Organizers

Matilde Santos	University Complutense de Madrid, Spain
Juan G. Victores	University Carlos III of Madrid, Spain

Special Session on Emerging Trends in Machine Learning

Organizers

Davide Carneiro	Polytechnic Institute of Porto, School of Technology and Management, Portugal
Fábio Silva	Polytechnic Institute of Porto, School of Technology and Management, Portugal

José Carlos Castillo   University Carlos III of Madrid, Spain
Héctor Alaiz Moretón   University of León, Spain

Special Session on Machine Learning, Law and Legal Industry

Organizers

Pedro Miguel Freitas   Portuguese Catholic University, Portugal
Federico Bueno de Mata   University of Salamanca, Spain

Special Session on Data Recovery Approach to Clustering and Interpretability

Organizers

Susana Nascimento   NOVA University of Lisbon, Portugal
José Valente de Oliveira   University of Algarve, Portugal
Boris Mirkin   National Research University, Russia

Special Session on Automated Learning for Industrial Applications

Organizers

José Luis Calvo-Rolle   University of A Coruña, Spain
Álvaro Herrero   University of Burgos, Spain
Roberto Casado Vara   University of Salamanca, Spain
Dragan Simić   University of Novi Sad, Serbia

Workshop on Machine Learning in Smart Mobility

Organizers

Sara Ferreira   University of Porto, Portugal
Henrique Lopes Cardoso   University of Porto, Portugal
Rosaldo Rossetti   University of Porto, Portugal

Program Committee

Achille Fonzone   Edinburgh Napier University, UK
Alberto Fernandez   Rey Juan Carlos University, Spain
Ana L. C. Bazzan   Federal University of Rio Grande do Sul, Brazil
Ana Paula Rocha   University of Porto, LIACC Portugal
Carlos A. Iglesias   Polytechnic University of Madrid, Spain
Cristina Olaverri-Monreal   Johannes Kepler University Linz, Austria
Daniel Castro Silva   University of Porto, LIACC, Portugal
Dewan Farid   United International University, Bangladesh
Eduardo Camponogara   Federal University of Santa Catarina, Brazil
Eftihia Nathanail   University of Thessaly, Greece
Fenghua Zhu   Chinese Academy of Sciences, China

Francesco Viti	University of Luxembourg, Luxembourg
Gianluca Di Flumeri	Sapienza University of Rome, Italy
Giuseppe Vizzari	University of Milano-Bicocca, Italy
Gonçalo Correia	TU Delft, The Netherlands
Hilmi Berk Celikoglu	Technical University of Istanbul, Turkey
Holger Billhardt	Rey Juan Carlos University, Spain
Joao Jacob	University of Porto, LIACC, Portugal
Josep-Maria Salanova	CERTH/HIT, Greece
Juergen Dunkel	Hochschule Hannover, Germany
Lior Limonad	IBM Research, Israel
Luís Nunes	ISCTE-IUL, Instituto de Telecomunicações, Portugal
Marin Lujak	IMT Lille, France
Mir Riyanul Islam	Mälardalen University, Sweden
Nihar Athreyas	University of Massachusetts, USA
Rui Gomes	ARMIS Group, Portugal
Soora Rasouli	Eindhoven University of Technology, The Netherlands
Tânia Fontes	University of Porto, INESC TEC, Portugal
Thiago Rúbio	University of Porto, LIACC, Portugal
Zafeiris Kokkinogenis	CEiiA, Portugal

Contents – Part II

Special Session on Practical Applications of Deep Learning

Special Session on New Trends and Challenges on Social
Networks Analysis

Special Session on Machine Learning in Automatic Control

Special Session on Emerging Trends in Machine Learning

Special Session on Machine Learning, Law and Legal Industry

Special Session on Machine Learning Algorithms for Hard Problems

Special Session on Automated Learning for Industrial Applications

Contents – Part I

Special Session on Data Generation and Data Pre-processing in Machine Learning

A Preprocessing Approach
for Class-Imbalanced Data Using SMOTE
and Belief Function Theory

Fares Grina[1]([⊠]), Zied Elouedi[1], and Eric Lefevre[2]

[1] Institut Supérieur de Gestion de Tunis, LARODEC,
Université de Tunis, Tunis, Tunisia
`grina.fares2@gmail.com`, `zied.elouedi@gmx.fr`
[2] Univ. Artois, UR 3926, Laboratoire de Génie Informatique et d'Automatique de
l'Artois (LGI2A), 62400 Béthune, France
`eric.lefevre@univ-artois.fr`

Abstract. Dealing with imbalanced datasets at the preprocessing level
is an efficient strategy used by many methods to re-balance the data
and improve classification performance. Specifically, SMOTE is a popular oversampling technique which modifies the training data by adding
artificial minority samples. However, SMOTE may create instances in
noisy and overlapping areas, far from safe regions. To tackle this issue,
we propose SMOTE-BFT, in which we use the belief function theory to
remove generated minority instances that are not in safe regions. After
applying SMOTE, each generated minority instance is represented by an
evidential membership structure, which provides detailed information
about class memberships. Rules based on the belief function theory are
then enforced to detect and remove generated instances that are in noisy
and overlapping regions. Experiments on noisy artificial datasets show
that our proposal significantly outperforms other popular oversampling
methods.

Keywords: Imbalanced data · Supervised learning · Belief function

1 Introduction

Learning from imbalanced data is one of the main challenges in machine learning.
In a binary classification problem, the imbalanced data issue occurs when one
class (the minority or positive class) is represented by a much smaller number
of instances than the other class (the majority or negative class). This problem
presents a crucial difficulty to many classifier learning algorithms that assume
a fairly balanced distribution of the classes [11]. The imbalanced data problem
can be translated to numerous real-world classification problems [13]. From an
application point of view, the correct classification of minority instances has
a greater importance than the reverse [4]. For example, in a medical diagnosis
problem, the cases affected by the disease are usually relatively rare as compared

© Springer Nature Switzerland AG 2020
C. Analide et al. (Eds.): IDEAL 2020, LNCS 12490, pp. 3–11, 2020.
https://doi.org/10.1007/978-3-030-62365-4_1

with the normal population. Assuming we have 1% of disease-affected cases, a classifier which scores correctly all negative samples (cases not affected) will get a classification accuracy of 99% even though all positive cases remain undetected. Instead of accuracy, one may use the Area under the Curve (AUC-ROC) [2], which reflects how good a model is at distinguishing between the minority and majority class.

In coping with this issue, resampling methods have been proposed to re-balance the data-set by adding new samples to the minority class (Oversampling), removing samples from the majority class (Undersampling), or both (Hybrid) [9]. Traditional replication strategies (e.g. random oversampling) [9] can cause overfitting by simply adding replicated samples to the dataset. To avoid this issue, Chawla et al. [1] suggested the Synthetic Minority Oversampling Technique (SMOTE), which adds new synthetic minority instances by interpo-lating among several minority examples that are close to each other. However, SMOTE produces synthetic samples without being aware of its surroundings which may potentially amplify the noise and overlap present in the data as illus-trated in Fig. 1, whereas, studies have shown that generated instances in safe positions improve classifier's performance [3].

Several extensions have been proposed to deal with those problems. To name a few, BorderlineSMOTE [10] was introduced to strengthen the decision bound-aries of classifiers, by replacing SMOTE's random selection of minority samples with a directed selection of examples that are close to the class border. ADASYN [12], another oversampling method which decides the number of examples to gen-erate based on density distribution. Other hybrid methods like SMOTE-RSB* [15] (based on rough set theory) and SMOTE-IPF [16] (based on iterative par-titioning filter) uses SMOTE and undersampling to clean generated minority examples and noise already present in the dataset.

In our work, we present another improvement of SMOTE based on the belief function theory. After applying SMOTE, a cleaning procedure is executed to improve SMOTE's oversampling, that is, we use the belief function theory as a way to extract information on the surroundings of each synthetic minority instance, by assigning a soft label regarding class memberships. Three rules based on belief function theory are then imposed to identify and eliminate synthetic minority instances which are not in safe regions. It is important to note that our proposal is a purely oversampling method, meaning that only generated minority instances can be removed.

The remaining of the paper is structured as follows. In Sect. 2, we introduce the belief function theory. Section 3 presents our proposal in details. In Sect. 4, we define the experimental framework and we analyze the results.

2   Belief Function Theory

Belief function theory [6, 17, 18], also known as evidence theory or the Dempster-Shafer theory is a well-founded and efficient framework for the representation and combination of a range of uncertain information. Let $\Omega = \{w_1, w_2, ..., w_M\}$

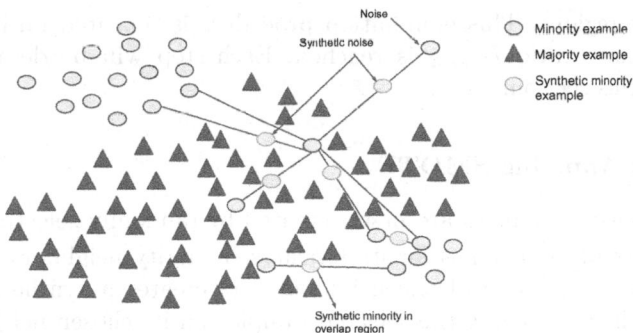

Fig. 1. SMOTE limitations (noise generation and introduction of more overlapping)

be a frame of discernment representing a finite set of M events. A basic belief assignment (BBA) represents the belief committed to the elements of 2^{Ω} by a given source of evidence is a mapping function $m : 2^{\Omega} \to [0, 1]$ such that:

$$\sum_{A \in 2^{\Omega}} m(A) = 1 \tag{1}$$

Belief and *plausibility* functions are defined by *Shafer* [17] as follows:

$$Bel(A) = \sum_{B \subseteq A} m(B) \quad and \quad Pl(A) = \sum_{B \cap A \neq \emptyset} m(B), \quad \forall \ A \in 2^{\Omega} \tag{2}$$

$Bel(A)$ represents the precise support for A and its subsets, whereas $Pl(A)$ represents the total possible support for A and its subsets. The interval $[Bel(A), Pl(A)]$ reflects the lower and upper bounds of support to A.

To combine several BBAs, *Dempster*'s rule [6] is a popular choice. Let m_1 and m_2 two BBAs defined on the same frame of discernment Ω, their combination based on *Dempster*'s rule gives the following BBA:

$$m_1 \oplus m_2(A) = \begin{cases} \dfrac{\sum\limits_{B \cap C = A} m_1(B) m_2(C)}{1 - \sum\limits_{B \cap C = \emptyset} m_1(B) m_2(C)} & for \ A \neq \emptyset \ and \ A \in 2^{\Omega}. \\ 0 & for \ A = \emptyset. \end{cases} \tag{3}$$

3   Combining SMOTE and Belief Function Theory

We focus our method on binary classification. For multi-class cases, we can apply it by decomposing the multi-class problem into two-class sub problems [5]. SMOTE-BFT consists of 3 main steps: First, we apply SMOTE to generate synthetic minority examples. Second, we compute for each generated instance an evidential soft label (BBA) using its nearest neighbors. Finally, 3 rules based on *plausibility* and *belief* functions are enforced in order to identify synthetic minority instances that are generated in noisy regions, overlapping regions, or majority class regions. This elimination procedure is repeated until a user-set minimum balance ratio Br_{min} is reached. Each step will be detailed in the remaining of this section.

3.1 Step 1: Applying SMOTE

Synthetic minority instances are firstly created by randomly selecting minority sample $\overrightarrow{a}$. Second, it searches for its k nearest minority neighbors $\overrightarrow{b}$. Finally, the algorithm chooses one of the neighbors and creates a synthetic point $\overrightarrow{s}$ anywhere on the line joining the selected sample and its chosen neighbor:

$$\overrightarrow{s} = \overrightarrow{a} + w * (\overrightarrow{b} - \overrightarrow{a}) \ , \ w \in [0, 1] \tag{4}$$

For this paper, we use the original version of SMOTE [1] to simplify the comparisons. However, one can use any other variant of SMOTE, since our proposal has a modular structure.

3.2 Step 2: Creating BBAs

To assign BBA's, we use the same evidential modeling defined by Denoeux in the Evidential KNN [8]. Let x_i be a synthetic minority instance obtained at Step 1. Each of its k neighbors (k is defined by the user) represents a piece of evidence to the evidential membership of the instance. For each neighbor x_j, a mass function m_i^j is calculated regarding the class membership of x_i as:

$$\begin{cases} m_i^j(\{w_q\} | x_j) = \alpha \phi_q(d_{ij}) \\ m_i^j(\Omega | x_j) = 1 - \alpha \phi_q(d_{ij}) \\ m_i^j(A | x_j) = 0 \qquad \forall A \in 2^{\Omega} \setminus \{\{w_q\}, \Omega\} \end{cases} \tag{5}$$

where d_{ij} denotes the euclidean distance between x_i and x_j, w_q is the class label of x_j, and α is a parameter such that $0 < \alpha < 1$. A recommended value of $\alpha = 0.95$ can be used to obtain good results on average, and a good choice for ϕ_q is $\phi_q(d) = exp(-\gamma_q d^2)$ where γ_q can be set to the inverse of the mean squared distance between training samples belonging to class w_q heuristically.

The BBAs for each neighbor x_j are then combined using the Dempster's rule defined in Eq. (3). As a result, each synthetic minority example x_i has three masses namely: $m_i(\{w_1\})$ degree of membership for the majority class, $m_i(\{w_2\})$ for the minority class and $m_i(\Omega)$ regarding both classes. Using these masses, it is now possible to compute *plausibility* and *belief* functions defined in Eq. (2).

3.3 Step 3: Eliminating Synthetic Examples

In order to perform the cleaning, we use overlap and noise rules that were introduced in [8] in addition to a misclassification rule. Each rule targets a specific type of synthetic points that are problematic to the classification task.

Overlap Threshold Rule: This situation typically arises when the synthetic minority sample is situated in a region where there is strong overlap between classes. In the belief function framework, this case is characterized by a BBA that is uniformly distributed between the two classes. As a result, the maximum plausibility $Pl_{max} = max_{\omega \in \Omega} Pl_i(\{\omega\})$ will have a relatively low value. Thus, imposing a threshold to the maximum *plausibility* will reject synthetic instances in strong overlap regions. The sample will be rejected if:

$$Pl_{max} < \beta_{Pl}, \quad \beta_{Pl} \in [0, 1] \tag{6}$$

where β_{Pl} a threshold that is set by the user. The higher this parameter is, the more synthetic instances are removed.

Noise Threshold Rule: This situation represents synthetic points which are suspected of belonging to a class which is not represented in the training set. In the belief function framework, most of the mass values will be concentrated on the whole frame of discernment Ω. As a consequence, the maximum credibility $Bel_{max} = max_{\omega \in \Omega} Bel_i(\{\omega\})$ will take on a small value. As the distance between the synthetic point and its closest neighbors goes to infinity, Bel_{max} goes to zero. Thus, the generated sample will be rejected if:

$$Bel_{max} < \beta_{Bel}, \quad \beta_{Bel} \in [0, 1] \tag{7}$$

where β_{Bel} a threshold which is reasonably fixed to the minimum value of Bel_{max} across original minority samples.

Misclassifcation Rule: This situation represents synthetic samples which are more likely to be misclassified after oversampling, meaning that they are located in the majority class region. Using BBAs, one way to make a decision about what class a sample belongs to is the maximum *plausibility*. In our situation, we want generated examples to be belonging to the minority class. Let w_1 be the majority class and w_2 be the minority one. The synthetic minority x_i example will be rejected if:

$$Pl_i(\{\omega_1\}) > Pl_i(\{\omega_2\}) \tag{8}$$

The cleaning phase is iterated over synthetic examples until the data reach a minimum balance ratio Br_{min} set by the user.

4 Experimental Study

In this section, we discuss the evaluation of our proposal in details. Section 4.1 presents the experimental setup. Section 4.2 shows the results and analysis.

4.1   Experimental Setup

To evaluate our contribution, we use synthetic imbalanced datasets with noisy and borderline examples selected from the KEEL repository [14]. All datasets are binary classification problems with combinations of two imbalance ratios (IR), 5 levels of disturbance ratios DR (reflects the amout of overlap) and three types of non-linear shapes of minority examples: *Clover*, which shapes a flower with five elliptic petals, *paw*, which shapes three elliptic subareas of minority examples, and *subclus*, which has 3 rectangles of minority instances.

Results are averaged through a ten-fold stratified cross validation to avoid inconsistencies. To compare different oversamplers, we use CART for classification and the AUC-ROC as an evaluation measure. Statistical comparisons are also performed using the Wilcoxon's signed ranks test [7] to compare the results.

We compare our proposal, SMOTE-BFT, with popular oversampling techniques: the original SMOTE [1], Borderline-SMOTE [10] and ADASYN [12]. In addition to the evaluation against no oversampling performed. For the experiments, we consider the following parameters for SMOTE-BFT: number of nearest neighbors for SMOTE and the creation of BBAs fixed to 5, minimum plausibility threshold β_{Pl} is set to 0.7, and minimum imbalance ratio Br_{min} is fixed to 0.8.

4.2   Results Analysis

Table 1 presents AUC obtained by CART on each dataset after oversampling. The best scores are marked in bold. We can observe that oversampling techniques improve the performance of the CART classifier in almost all cases.

The proposed SMOTE-BFT achieves the best AUC scores in 13 out of 30 datasets and obtains close to the best scores in the other cases. Even though CART performance worsens with higher DR, we can notice that our method performed relatively better especially when the dataset presents high percentage of DR. This shows that our method successfully identified and eliminated the noise and overlap generated by SMOTE.

In order to compare AUC results, statistical analysis was performed using the Wilcoxon's signed ranks test, which is a non-parametric pairwise test used to identify significant differences in performance of two methods [7]. In our study, we use this test to compare the performance obtained by SMOTE-BFT against other oversamplers. $R+$ represents the sum of ranks in favor of SMOTE-BFT, $R-$, the sum of ranks in favor of the other compared methods, and p-values are obtained for each comparison. As seen in Table 2, all p-values for Wilcoxon's test are lower than 0.05, which shows that SMOTE-BFT outperforms all compared methods at a significance level of $\alpha = 0.05$.

Table 1. AUC results of CART on synthetic datasets oversampled by different methods.

Datasets	NONE	SMOTE	BorderSM	ADASYN	SMOTE-BFT
$paw_{(S=600, IR=5, DR=0\%)}$	0.962	0.964	0.952	**0.972**	0.97
$paw_{(S=600, IR=5, DR=30\%)}$	0.79	0.817	**0.844**	**0.844**	0.835
$paw_{(S=600, IR=5, DR=50\%)}$	0.747	0.77	0.796	0.794	**0.798**
$paw_{(S=600, IR=5, DR=60\%)}$	0.679	0.731	0.714	0.742	**0.766**
$paw_{(S=600, IR=5, DR=70\%)}$	0.69	0.721	0.771	0.754	**0.778**
$paw_{(S=800, IR=7, DR=0\%)}$	0.94	0.952	**0.957**	0.948	0.943
$paw_{(S=800, IR=7, DR=30\%)}$	0.805	0.819	0.836	0.830	**0.837**
$paw_{(S=800, IR=7, DR=50\%)}$	0.749	0.778	0.752	0.757	**0.795**
$paw_{(S=800, IR=7, DR=60\%)}$	0.693	**0.762**	0.733	0.731	0.739
$paw_{(S=800, IR=7, DR=70\%)}$	0.634	0.721	**0.767**	0.718	0.745
$clover_{(S=600, IR=5, DR=0\%)}$	0.859	0.909	0.889	**0.91**	0.862
$clover_{(S=600, IR=5, DR=30\%)}$	0.755	0.814	**0.824**	0.817	0.823
$clover_{(S=600, IR=5, DR=50\%)}$	0.69	0.744	0.738	0.787	**0.79**
$clover_{(S=600, IR=5, DR=60\%)}$	0.696	0.728	0.705	**0.744**	**0.744**
$clover_{(S=600, IR=5, DR=70\%)}$	0.646	0.739	0.72	0.734	**0.753**
$clover_{(S=800, IR=7, DR=0\%)}$	0.845	0.905	0.869	**0.923**	0.92
$clover_{(S=800, IR=7, DR=30\%)}$	0.766	**0.836**	0.825	0.835	0.832
$clover_{(S=800, IR=7, DR=50\%)}$	0.704	0.749	0.757	**0.765**	0.763
$clover_{(S=800, IR=7, DR=60\%)}$	0.668	0.712	0.704	**0.743**	0.737
$clover_{(S=800, IR=7, DR=70\%)}$	0.659	0.727	0.725	0.724	**0.747**
$subcl_{(S=600, IR=5, DR=0\%)}$	**0.976**	0.942	0.955	0.952	0.954
$subcl_{(S=600, IR=5, DR=30\%)}$	0.822	0.811	0.82	0.801	**0.826**
$subcl_{(S=600, IR=5, DR=50\%)}$	0.715	0.728	0.734	**0.741**	0.738
$subcl_{(S=600, IR=5, DR=60\%)}$	0.671	0.728	**0.736**	0.729	0.731
$subcl_{(S=600, IR=5, DR=70\%)}$	0.688	0.746	0.718	0.71	**0.757**
$subcl_{(S=800, IR=7, DR=0\%)}$	**0.978**	0.955	0.963	0.972	0.968
$subcl_{(S=800, IR=7, DR=30\%)}$	0.793	**0.842**	0.797	0.811	0.810
$subcl_{(S=800, IR=7, DR=50\%)}$	0.702	0.753	**0.756**	0.722	0.739
$subcl_{(S=800, IR=7, DR=60\%)}$	0.702	0.724	0.742	0.739	**0.763**
$subcl_{(S=800, IR=7, DR=70\%)}$	0.605	0.716	0.699	0.704	**0.72**

Table 2. Wilcoxon's signed rank test results comparing SMOTE-BFT (R+) with other oversampling methods (R-).

Comparisons	R+	R-	P-value
SMOTE-BFT vs NONE	453.0	12.0	<0.00001
SMOTE-BFT vs SMOTE	370.0	95	0.00467
SMOTE-BFT vs BorderSM	368.0	97	0.00530
SMOTE-BFT vs ADASYN	352.5	112.5	0.02307

5   Conclusions

In this paper, we have proposed a new extension to SMOTE aiming at improving the quality of oversampling. Rules have been developed using the belief function theory to remove synthetic minority examples which are added in noisy, overlapping or majority class regions. Results from the experimental analysis show that SMOTE-BFT obtains significantly better results than compared methods, specifically on datasets with high disturbance ratios. Future work can include developing heuristic methods in order to automatically determine parameters such as the plausibility threshold β_{Pl} and the minimum balance ratio Br_{min}.

References

1. Bowyer, K.W., Chawla, N.V., Hall, L.O., Kegelmeyer, W.P.: SMOTE: synthetic minority over-sampling technique. CoRR abs/1106.1813 (2011). http://arxiv.org/abs/1106.1813
2. Bradley, A.P.: The use of the area under the ROC curve in the evaluation of machine learning algorithms. Pattern Recog. **30**(7), 1145–1159 (1997)
3. Bunkhumpornpat, C., Sinapiromsaran, K., Lursinsap, C.: Safe-Level-SMOTE: safe-level-synthetic minority over-sampling technique for handling the class imbalanced problem. In: Theeramunkong, T., Kijsirikul, B., Cercone, N., Ho, T.-B. (eds.) PAKDD 2009. LNCS (LNAI), vol. 5476, pp. 475–482. Springer, Heidelberg (2009). https://doi.org/10.1007/978-3-642-01307-2_43
4. Chawla, N., Japkowicz, N., Kołcz, A.: Editorial: special issue on learning from imbalanced data sets. SIGKDD Explorations **6**(1), 1–6 (2004)
5. Chen, K., Lu, B.L., Kwok, J.T.: Efficient classification of multi-label and imbalanced data using min-max modular classifiers. In: The 2006 IEEE International Joint Conference on Neural Network Proceedings, pp. 1770–1775. IEEE (2006)
6. Dempster, A.P.: A generalization of bayesian inference. J. R. Stat. Soc. Ser. B (Methodol.) **30**(2), 205–232 (1968)
7. Demšar, J.: Statistical comparisons of classifiers over multiple data sets. J. Mach. Learn. Res. **7**(1), 1–30 (2006)
8. Denoeux, T.: A k-nearest neighbor classification rule based on dempster-shafer theory. systems, man and cybernetics. IEEE Trans. **219**, 804–813 (1995)
9. Haixiang, G., Yijing, L., Shang, J., Mingyun, G., Yuanyue, H., Bing, G.: Learning from class-imbalanced data: Review of methods and applications. Expert Syst. Appl. **73**, 220–239 (2017)
10. Han, H., Wang, W.-Y., Mao, B.-H.: Borderline-smote: a new over-sampling method in imbalanced data sets learning. In: Huang, D.-S., Zhang, X.-P., Huang, G.-B. (eds.) ICIC 2005. LNCS, vol. 3644, pp. 878–887. Springer, Heidelberg (2005). https://doi.org/10.1007/11538059_91
11. He, H., Garcia, E.A.: Learning from imbalanced data. IEEE Trans. Knowl. Data Eng. **21**(9), 1263–1284 (2009)
12. He, H., Bai, Y., Garcia, E.A., Li, S.: Adasyn: Adaptive synthetic sampling approach for imbalanced learning. In: 2008 IEEE International Joint Conference on Neural Networks, pp. 1322–1328. IEEE (2008)
13. Krawczyk, B.: Learning from imbalanced data: open challenges and future directions. Progress Artif. Intell. **5**(4), 221–232 (2016). https://doi.org/10.1007/s13748-016-0094-0
14. Napierała, K., Stefanowski, J., Wilk, S.: Learning from Imbalanced data in presence of noisy and borderline examples. In: Szczuka, M., Kryszkiewicz, M., Ramanna, S., Jensen, R., Hu, Q. (eds.) RSCTC 2010. LNCS (LNAI), vol. 6086, pp. 158–167. Springer, Heidelberg (2010). https://doi.org/10.1007/978-3-642-13529-3_18

15. Ramentol, E., Caballero, Y., Bello, R., Herrera, F.: SMOTE-RSB *: a hybrid pre-processing approach based on oversampling and undersampling for high imbalanced data-sets using SMOTE and rough sets theory. Knowl. Inf. Syst. **33**(2), 245–265 (2012)
16. Sáez, J.A., Luengo, J., Stefanowski, J., Herrera, F.: SMOTE-IPF: addressing the noisy and borderline examples problem in imbalanced classification by a re-sampling method with filtering. Inf. Sci. **291**, 184–203 (2015)
17. Shafer, G.: A Mathematical Theory of Evidence. Princeton University Press, New Jersey (1976)
18. Smets, P.: The transferable belief model for quantified belief representation. In: Smets, P. (ed.) Quantified Representation of Uncertainty and Imprecision. HDRUMS, vol. 1, pp. 267–301. Springer, Dordrecht (1998). https://doi.org/10.1007/978-94-017-1735-9_9

Multi-agent Based Manifold Denoising

Mohammad Mohammadi$^{(\boxtimes)}$ ⓘ and Kerstin Bunte ⓘ

Faculty of Science and Engineering, University of Groningen,
Groningen, The Netherlands
m.mohammadi@rug.nl, kerstin.bunte@googlemail.com

Abstract. Manifold learning plays a central role in many Machine Learning (ML) methods where it assumes information lies on a low-dimensional manifold, but the presence of high dimensional noise may defect their performance. In this contribution, we propose a novel (swarm) algorithm to suppress the noise of manifolds of potentially varying dimensionalities. Inspired by colonial insects this method employs multiple agents with different strategies moving through the data space in parallel. During this process, they use local information to reconstruct the manifolds and then move data objects close to them. Moreover, principles of evolutionary game theory are used to encourage agents to select better strategies and hence optimize the hyper-parameters automatically. While other denoising techniques can be seen as single-agent approaches, the new algorithm is a multi-agent approach which makes it more flexible and suitable for scenarios including multiple manifolds. In the experiments, we simulate several situations from a simple manifold with a specific noise level, to more complex manifolds where there are variations on the density, noise level or dimensionalities. Furthermore, we demonstrate the improvement of the proposed algorithm for the performance of the Parzen Window (PW) density estimator.

Keywords: Manifold learning · Swarm intelligence · Multi-agent systems · Evolutionary game theory · Replicator dynamics · Parzen window

1 Introduction

Developments in modern sensor technology allow to gather increasing amounts of high dimensional data [8]. With growing dimension the curse of dimensionality arises as the lack of data points in data spaces exhibiting several undesired properties. It causes for example problems such as addressing distances and noise. A common assumption is that for many real problems the data points spread along a lower dimensional topological structure, called manifold [20]. The goal of many ML methods is therefore to obtain a good representation of it. However, in practice data are often disrupted by noise during data collection, storage and processing. The presence of such errors can degrade the output of ML techniques [20]. Furthermore, cleaning manifolds is valuable in applications

© Springer Nature Switzerland AG 2020
C. Analide et al. (Eds.): IDEAL 2020, LNCS 12490, pp. 12–24, 2020.
https://doi.org/10.1007/978-3-030-62365-4_2

such as 3D mesh smoothing in the computer graphics community [18]. Hence, multiple denoising methods have emerged as preprocessing strategy to suppress the noise and therefore increasing the performance of subsequent ML methods.

Many strategies to denoise manifolds have been proposed, which are typically applied as a preprocessing step on data sets to reduce the noise before solving the ML task. For instance, Manifold Denoising (MD) [9], Mean Shift (MS) [23] and Manifold Blurring Mean Shift (MBMS) [20] decrease the noise, iteratively, via pushing data points towards the underlying manifold. Unlike these iterative algorithms, Locally Linear Denoising (LLD) [8] optimizes a cost function fulfilling two objectives: a) preserving local linear structures and b) a smooth manifold. However, its computation and memory costs make LLD undesirable for big data sets. In general, the sucess of these methods highly depends on using proper values for their hyper-parameters which often require information of the manifold and the noise itself. Thus it is highly desirable for a manifold denoising algorithm to make fewer presumptions about the data and the nature of noise.

Swarm intelligence (SI) is a self-organized system of multiple agents performing individual actions leading to intelligent collective behaviour. A fascinating example of SI is the well-known and studied ant colony. The ants' strategy to find food sources has motivated many optimization methods searching for optima of complicated non-convex cost functions [2,5,6,17]. Moreover, some clustering techniques are inspired by the ability of their biological counterparts to cluster corpses and sort larvae [3,4,13,15,16,19]. They modeled this capability by two simple rules called picking up and dropping down, where several ants walk in the data space, pick up data objects and then drop them down somewhere else. Although no central authority controls them, their local interactions recover underlying clusters by moving similar objects close to each other. While typically less efficient than classical clustering methods, swarm strategies proved captivating and beneficial through their flexibility, simplicity, and robustness.

In this contribution, we propose a multi-agent evolutionary algorithm to denoise manifolds. In the new algorithm, agents with different capabilities walk in the data space and move samples close to manifolds. While the previous works need to be tuned with appropriate hyper-parameters, which require prior knowledge of the manifold and noise, we use the Evolutionary Game Theory (EGT) to find the best hyper-parameters automatically. Thus, in contrast to others, agents pick different strategies and then EGT is used to adapt the population in order to promote better strategies. Using multiple agents equipped with different capabilities makes the novel formulation suitable for multi-manifold scenarios where manifolds may have priorly unknown with variation in dimensionality and noise.

This paper is organized as follows: we first detail the problem setting and introduce the MBMS [20], which is the most related work to our algorithm, and EGT (Sect. 2). In Sect. 3 we explain our novel algorithm. The performance is then demonstrated and compared with empirical analysis by means of four example scenarios in Sect. 4. Finally, we conclude in Sect. 5.

2   Related Work

2.1   Problem Setting

We assume that the data is distributed along a d-dimensional smooth manifold embedded in a D dimensional space, with $D > d$. Let N samples $\{\vec{y}_i \in \mathbb{R}^D : i = 1, \ldots, N\}$ be randomly selected from the manifold. Due to noise ϵ the observations $\vec{x}_i$ do not lie exactly on the manifold, such that $\vec{x}_i = \vec{y}_i + \epsilon_i$. The goal of denoising is to recover $\vec{y}_i$ from noisy samples $\vec{x}_i$. If we assume the underlying manifold is linear and the noise ϵ follows a Gaussian distribution, then Principal Component Analysis (PCA) can be used to denoise the manifold. From PCA, we obtain the parameter set $\{(\lambda_i, \vec{u}_i)\}_{i=1}^D$, where λ_i and $\vec{u}_i$ represent the i-th eigenvalue and its corresponding eigenvector. Without loss of generality we assume the eigenvalues are sorted in descending order and re-scaled, such that $\sum_{i=1}^D \lambda_i = 1$. Defining $\mathcal{U}_i = \vec{u}_i \vec{u}_i^T$ the covariance matrix C can be re-written as:

$$C = \sum_{i=1}^D \lambda_i \mathcal{U}_i = S_1 \mathcal{U}_1 + \frac{1}{2} S_2 (\mathcal{U}_1 + \mathcal{U}_2) + \cdots + \frac{1}{D} S_D (\mathcal{U}_1 + \cdots + \mathcal{U}_D) \quad (1)$$

where $S_i = i \times (\lambda_i - \lambda_{i+1})$ $(\forall i = 1, \cdots, D - 1)$ and $S_D = D \cdot \lambda_D$. As suggested in [14,21] one can estimate the intrinsic dimensionality of the manifold by:

$$\widehat{d} = \arg \max_i S_i \ . \quad (2)$$

Then the first $\widehat{d}$ eigenvectors $\{u_1, \cdots, u_{\widehat{d}}\}$ can be used to recover the linear manifold and $\vec{x}_i$ is denoised via projecting on the reconstructed space.

2.2   Manifold Blurring Mean Shift (MBMS)

The primary goal of any denoising algorithm is to repress noise while simultaneously keeping the important details. One way to achieve it is to move the data points towards the manifold. Thus, for a given sample $\vec{x}_i$ the update is: $\vec{x}_i^{\text{new}} = \vec{x}_i^{\text{old}} + \delta \vec{x}_i$, where $\delta \vec{x}_i$ represents the direction of the motion. Since the position of samples on the manifold reveals important details, the motion should not take place along the manifold [20]. Thus, MBMS restricts motions to be parallel to the manifold normals. Although the manifold may be non-linear, it can be locally estimated by linear patches using eigenvectors derived by PCA. Therefore, the update term is:

$$\delta \vec{x}_i = (I - UU^T)(\vec{\mu} - \vec{x}_i) \quad (3)$$

where U's columns contain the first d eigenvectors and $\vec{\mu}$ is the kernel average of $\vec{x}_i$'s neighbours. Our novel strategy keeps the idea of limiting the motion to be parallel to manifold normals. However, to avoid fixing the dimensionality d of the manifold beforehand it is estimated locally as integral part of our strategy.

2.3   Evolutionary Game Theory (EGT)

Game theory studies strategic interaction such as competition or cooperation of players. While classical game theory assumes that all players make rational decisions EGT replaces it by biologically inspired notions, such as natural selection [10]. Let a population contain n strategies (or types) E_1 to E_n with frequencies p_1 to p_n, and $\vec{p} = (p_1, \cdots, p_n)$ denote the distribution of strategies within the population. Let f_i represent the fitness of strategy i which reflects its reproductive success. If the size of the population is very large, then its evolution can be modelled as a differential equation. For a given strategy i the rate of change in its frequency (i.e. $\frac{\dot{p_i}}{p_i}$) measures its evolutionary success. Following Darwin's doctrine the replicator dynamics expresses evolutionary success as the difference between the fitness of the strategy and the average fitness of the population:

$$\frac{\dot{p_i}}{p_i} = f_i(\vec{p}) - \bar{f}(\vec{p}) \ . \tag{4}$$

Thus, if a strategy has a higher fitness than the average value, its population share increases and vice versa.

3   A Multi-agent Paradigm for Manifold Denoising

In this contribution we combine aspects of EGT with principles of the above mentioned SI clustering techniques resulting in a novel strategy for manifold denoising. The basic idea is to release several ant-like agents in the data space, which can, such as their biological counterparts, pick up data samples and bring them closer to the manifold. To achieve this goal we need to define individual behaviour comprising of the following objectives:

O1: an approximation strategy to recognize the manifold structure,
O2: a good policy to keep agents close to the manifold,
O3: and an update rule bringing samples close in proximity of the respective manifold (decreasing the noise).

In the following subsections we explain how we address the above challenges, describing the strategy for one agent first followed by the generalization to several agents. While in the first case an agent with a specific capability is introduced, the second case takes advantage of having several agents with different abilities (or strategies). For the latter we use EGT to automatically determine the winning strategies and therefore avoiding the necessity to fix them beforehand.

3.1   Manifold Alignment Aware Agent

While a linear manifold can be approximated by PCA it is challenging to estimate non-linear or multiple manifolds. Hence, non-linear manifolds are commonly estimated by local linear patches, namely given a data point $\vec{x}$ a manifold in $\vec{x}$'s neighbourhood can be approximated by the tangent space at $\vec{x}$. A way to estimate tangent spaces is to recover them by the eigenvectors derived from PCA. Given a radius r the set of $\vec{x}$'s neighbours is defined as $\mathcal{N}_x = \{\vec{x}_i : \|\vec{x}_i - \vec{x}\| < r\}$. Using PCA on $\mathcal{N}_x$ one can use (2) to approximate the dimensionality of the manifold $\hat{d}$ and then recover it by the first eigenvectors $\{u_1, \cdots, u_{\hat{d}}\}$. Thus, a non-linear manifold is described by a set of (overlapping) linear patches.

When the manifold is approximated the denoising can be translated to bringing samples closer to it. Thus, our aim is to define a random walk which reinforces an agent to walk through the data cloud and bring data objects closer to the underlying manifold. Let an agent, with a specific view range r, be on a data point $\vec{x}$. It uses local PCA to approximate the manifold and its intrinsic dimensionality. Hence, the distance of samples to the manifold is estimated by:

$$\Delta_i = \|(I - UU^T)(\vec{\mu} - \vec{x}_i)\| \tag{5}$$

where $\|.\|$ is the Euclidean norm and $\vec{\mu} = \frac{1}{|\mathcal{N}|} \sum_{k:\vec{x}_k \in \mathcal{N}_{\vec{x}}} \vec{x}_k$. Then for every $\vec{x}_i \in \mathcal{N}_{\vec{x}}$ we define the following weight values:

$$w(\vec{x}, \vec{x}_i) = \begin{cases} 1 - \frac{\Delta_i}{b} & \text{if } \Delta_i \leq b \\ 0 & \text{if } \Delta_i > b. \end{cases} \tag{6}$$

where b is selected such that only $p = 50\%$ of neighbors have non-zero weights. Next the agent selects its next destination by the following probability:

$$P_{\text{rw}}(\vec{x}, \vec{x}_i) = \frac{w(\vec{x}, \vec{x}_i)}{\sum_{j=1}^{k} w(\vec{x}, \vec{x}_j)} \tag{7}$$

Thus, using only local information the manifold is approximated and then agents are encouraged to stay close to it fulfilling objectives O1 and O2.

During the random walk agents pick up and later drop down some of the data points in their neighbourhood aiming to reduce the noise of the manifold. To fulfill objective O3, in every step an agent randomly selects a data point $\vec{x}_i$ for picking up with the following probability:

$$P_{\text{pick}}(\vec{x}_i) = \frac{1 - w(\vec{x}, \vec{x}_i)}{\sum_{j=1}^{k} \left(1 - w(\vec{x}, \vec{x}_j)\right)} \cdot \tag{8}$$

Hence, the further $\vec{x}_i$ is from the tangent space the more probable it is to be picked up. Then the main goal of the agent is to drop it down closer to the manifold. Motivated by MBMS, it is achieved by the following rule:

$$\vec{x}_i^{\text{new}} = \vec{x}_i^{\text{old}} + \eta(I - UU^T)(\vec{\mu} - \vec{x}_i^{\text{old}}) , \tag{9}$$

where $\eta > 0$ is a value which controls how much change an agent can do in a single step. Moreover, to prevent unnecessay change, we introduce a threshold τ, such that agents only change the neighbourhood if the mean distance of neighbours

---

Algorithm 1: One Agent

Input: data set, radius r, step size η, number of steps N_s, τ;
Output: denoised dataset (S), number of changes in the environment (n),
number of visited samples in different dimensionalities $(\vec{\xi})$;
Function OneAgent($dataset$, r, η, N_s, τ):

   | $\vec{\xi} = \vec{0}_{D \times 1}$, $n = 0$;
   | Randomly select a node $\vec{x}$ as an initial point for the k-th agent
   | $\vec{x}_c = \vec{x}$;
   | **for** $s = 1$ to N_s **do**
      | find $\vec{x}_c$'s neighbours $\mathcal{N}$;
      | $(\mu, \lambda, U) \leftarrow \text{PCA}(\mathcal{N})$;
      | approximate the intrinsic dimensionality $\widehat{d}$ in $\mathcal{N}$ by (2);
      | $\vec{\xi}[\widehat{d}] = \vec{\xi}[\widehat{d}] + 1$;
      | **if** $\sum_{i=d+1}^{D} \lambda_i > \tau$ **then**
         | $n = n + 1$;
         | select a neighbour to pick up by Eq. (8);
         | bring the selected neighbour closer to the manifold using Eq. (9);
      | **end**
      | use (7) to select the next place $\vec{x}_i$ and set $\vec{x}_c = \vec{x}_i$;
   | **end**
   return

---

to the tangent space (quantified by $\sum_{i=d+1}^{D} \lambda_i$) is bigger than τ. In other words, a neighbourhood only changes when the effect of noise and curvature is large enough. Thus, we can decrease the chance of over-smoothing and shrinkage.

In summary, an agent is released on the data set performing a random walk following the local preference (7). During this process it picks up a data object according to (8) and in the following carry the sample closer to the manifold using Eq. (9). The pseudo code for the method is given in Algorithm 1. In addition to denoising the data set, the agent also counts the number of changes it does and the number of samples it visits in any dimensionality. These outputs provide useful information about the manifold.

3.2 Evolutionary Manifold Alignment Aware Agents (EM3A)

As it is shown in [11] the performance of PCA and the methods above highly depend on the neighbourhood size r. While high curvatures encourage the use of smaller radii, it should be large enough to suppress the effect of noise and contain enough samples [12]. Hence, one needs extra information over the data in order to pick a proper value for r. In the lack of such extra information and for increased flexibility, we propose an optimization combining SI strategies with concepts of EGT, to automatically adapt such parameters.

EGT studies how a population of strategies evolves, where the population share of better strategies increases over time. Thus, if we define r as a strategy,

EGT can be used to find appropriate values for it. Since r is a continuous variable, we assume $r \in [R_{\min}, R_{\max}]$, and discretize by division into m smaller intervals $([r_i, r_{i+1}), 0 \le i < m)$ where each interval is viewed as a strategy. Let each agent randomly select a strategy following a population share distribution $\vec{p}^{(t)} = [p_k^{(t)}]_{m \times 1}$, where $p_k^{(t)}$ denotes the population share of the k-th strategy in the current generation. Given an agent with strategy k, r is chosen from $[r_k, r_{k+1})$ uniformly. Next, a queen (or master branch) sends a copy of the data set to all agents, for them to separately do the random walk as explained in Algorithm 1. After finishing the walk with N_s steps, each agent sends back the output to the queen. The queen updates the data set taking the average over all copies received from the agents and it computes the fitness of the strategies. Let S_k denote a set of agents with strategy k, then $\sum_{i \in S_k} n_i$ denotes the number of times they change the data set. The fitness is defined as:

$$f_k = \frac{\sum_{i \in S_k} n_i}{|S_k| \cdot N_s} \tag{10}$$

where $|S_k|$ is the number of agents with strategy k. The queen uses the replicator equation to adapt the population share of each strategies for the next generation:

$$p_k^{(t+1)} = p_k^{(t)} + p_k^{(t)}(f_k^{(t)} - \bar{f}^{(t)}) \ . \tag{11}$$

In the next generation agents use the distribution $\vec{p}^{(t+1)} = \left[p_k^{(t+1)} \right]$ to select their own strategy. The resulting method is detailed in Algorithm 2. Note that among the parameters in EM3A the most important one is η which controls the size of changes. Since agents have different view range r the term $(\vec{x}_i - \vec{\mu})$ in Eq. (9) is in different scales for different r. Therefore, we re-define $\eta = \frac{\rho}{\sqrt{r}}$ with constant ρ. The selection of ρ has a direct effect on the smoothness of manifolds, and it should be $\rho \in (0, 1]$. Moreover, the number of steps N_s and agents N_a should be big enough to ensure all parts of the data set are reachable. One can select it such that $N_s \times N_a > c \cdot N$ for a constant c.

3.3 Complexity Analysis

Every agent should perform two operations: a) detecting its neighbours and b) applying PCA on the neighbours. The neighbourhood search needs to compute the pairwise distances of the N samples of the data set, resulting in a complexity $O(N^2)$. Approximate neighbour search strategies can reduce the computation, for example using a k-d tree reduces the complexity to $O(N \log N)$ [1]. In addition to the neighbour search the agents perform PCA which is done by Singular Value Decomposition. This complexity depends on the specific implementation and in the worst case it scales cubic with the dimensionality of the data and quadratic with the number of samples in the neighbourhood $|\mathcal{N}_{\vec{x}}|$. Again approximate strategies can reduce the computational complexity [7]. Moreover, the execution of the EM3A is parallelizable via distributing agents among several processors. In the following section, we show the acceleration of the process.

Algorithm 2: EM3A

Input: data set, number of strategies m and their boundaries $(R_{\min}, R_{\max})$, step size η, number of agents N_a, number of steps N_s, stopping criteria: N_{iter};

Output: denoised data set (S^{new}), fitness values $\vec{f}$,;

Function EM3A($data, r, \eta, N_s$):

   $S^{\mathrm{new}} = data$;

   $\vec{p}^{(0)} = \frac{1}{m}\vec{1}_{D \times 1}$;

   while $t = 1$ *to* N_{iter} **do**

      $\vec{n} = \vec{0}_{D \times 1}, \vec{s}^{(t)} = \vec{0}_{m \times 1}, \vec{\xi}^{(t)} = \vec{0}_{D \times 1}$;

      for $i = 1$ *to* N_a **do**

         following $\vec{p}^{(t)}$, select its strategy (say k), and $\vec{s}^{(t)}[k] = \vec{s}^{(t)}[k] + 1$;

         select a neighborhood size $r \in [r_{k-1}, r_k)$;

         $[S_i, n_i, \vec{\xi}_i] = \mathbf{OneAgent}(S^{\mathrm{new}}, r, \eta, N_s)$;

         $\vec{n}[k] = \vec{n}[k] + n_i$;

      end

      $S^{\mathrm{new}} = \frac{1}{N_a}\sum_{i=1}^{N_a} S_i, \vec{\xi}^{(t)} = \sum_i \vec{\xi}_i$;

      compute $f_k^{(t)}$ and $\vec{p}^{(t+1)}$ by Eq. (10) and (11), respectively;

   end

 return

---

4   Experiments

In this section, we demonstrate and investigate several aspects of the proposed method on several data sets and application scenarios, including:

– handling variation in density, noise, and dimensionality and
– improving the result of density estimators.

In subsection 4.1, we use four synthetic datasets to simulate different situations, and we use mean square error (MSE) to evaluate outputs (see Table 1). Section 4.2 demonstrates the effect of EM3A used as preprocessing step before density estimation, exemplified by Parzen Window (PW) (see Table 2).

4.1   Denoising Manifolds

To see the strengths and weaknesses of different denoising algorithms, we use three synthetic data sets[1]: a) an s-shaped manifold, b) a circular manifold where the noise and density along a one dimensional manifold is varied, and c) a combination of multiple manifolds of different dimensions. Here, we set $R_{\min} = 0$, $R_{\max} = 2$ and $m = 10$ with $\tau = 0.01$. To ensure $N_a \times N_s > 2N$, we set $N_s = 100$ and $N_a = 100$. For the simple manifold (a) we assume $\rho = 1$ and for the more complex ones (b,c) we choose a smaller value of $\rho = 0.5$.

---

[1] Code and supplementary material: https://git.lwp.rug.nl/m.mohammadi/em3a.

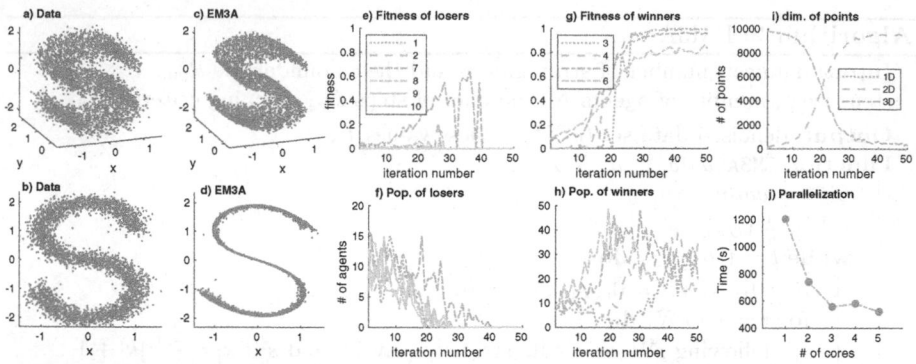

Fig. 1. Two views of the data set (a,b) and the results of EM3A (c,d). The fitness and population of losing (e,f) and winning (g,h) strategies. (i) the number of visited points with different dimensionalities. (j) computation time with number of processors used.

Simple Manifold. To investigate how EGT helps to achieve proper values for r, we extract 4000 data points from the s-curve manifold exhibiting Gaussian noise $\mathcal{N}(0, 0.15)$. The panels c) and d) in Fig. 1 display an example output of EM3A, where it successfully decreases noise, especially in less curved regions. Panels (e, f) show the fitness values and population sizes of the losing strategies for every iteration, while panels (g,h) report the corresponding values for the winning strategies. One may divide the strategies into three categories. The first group, including strategies 1 and 2, contains agents with low view range r where noise has a major effect on the eigenvalues. Hence, agents with these strategies rarely observe the low-dimensional manifold and their fitness values become small and their population decreases over time. On the other hand, the second category strategies 7–10, includes agents with large view range. The major obstacle for them in observing 2D manifold is curvature. Particularly, high curvature regions prevents them to detect the manifold so their populations continuously reduces as well. Finally, the winning strategies 3–5 involve agents with medium view range where r is big enough to suppress the effect of noise and small enough not to be defected by curvature. Thus, they survive while the rest disappears. Moreover, the algorithm provides information over the dimension of the manifold as learned in the iterations (panel i). At first, since the noise level is high, most agents see a 3D object. Since the noise level decreases over time and agents pick better strategies, agents observe the 2D manifold more often. Panel j) exemplifies the achieved speed up for this experiment if agents are distributed among several processors. Because of the transfer time adding more than 3 cores does not accelerate the computation further for this experiment. Note that optimizing parallelization is possible but not focus of this contribution.

Manifold with Varying Noise. Figure 2 (a) displays a circular manifold which contains a sharp change in its density, and is disrupted by Gaussian noise with varying $\sigma \in (0, 0.3)$ along the manifold. To apply denoising techniques, we set

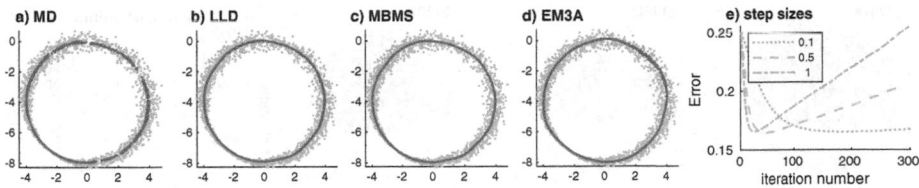

Fig. 2. Gray points: a 1D manifolds varying in density and noise level. Red points: the denoised manifold, using: a) MD, b) LLD, c) MBM and d) EM3A. e) evolution of the MSE (to the ground truth) per iteration for values of $\rho \in \{0.1, 0.5, 1\}$.

their hyper-parameters as follow: 1) MD: $\delta t = 0.01, k = 200, N_{\text{iter}} = 200$. 2) LLD: $k = 200$, $\lambda = 0.01$, $\sigma = 0.02$, $d = 0.01$. 3) MBMS: $k = 50$, $\sigma = \infty$, $N_{\text{iter}} = 10$. The Fig. 2 shows their outputs and Table 1 reports MSE. To report the MSE for EM3A, we run it 5 times. Since MBMS and EM3A limits the motion of samples to be parallel to the normal vectors they outperform the others. EM3A can handle variations in density and noise of the manifold very well, while not needing to make presumptions about the manifold beforehand. The effect of EM3A's parameter ρ (and η) on the reconstruction error is shown in Fig. 2e. Smaller values lead to more stable results but need more iterations to run. Thus, the selection of $\rho \in (0, 1]$ builds a trade-off between the iterations necessary and the stability of the result, and it needs to be small for more complex structures.

Multi-manifold Example. Often real data sets contain several manifolds with different dimensionalities. For demonstration, we generate manifolds forming a basket-like structure shown in Fig. 3 (a). The top manifold contains 3600 data points uniformly sampled from a hemisphere, and the bottom manifold includes 1000 data points generated from a semicircle. Then, Gaussian noise with $\sigma = 0.3$ is added. Since LLD and MBMS needs to know the intrinsic dimensionality d, they can not be used directly on a data set containing multiple manifolds. In order to compare MBMS we cheat and separate the two manifolds to be able to apply it. On the other hand MD and EM3A do not need to know d and can be used in a realistic scenario. As seen in Fig. 3b MD depicts some undesired effects: it creates artificial holes in the hemisphere structure and does not completely clean the 1D basket handle Although MBMS provides comparable clean manifolds as EM3A, it needs to denoise them separately. EM3A outperforms others without any preprocessing (see Fig. 3c) and additionally provides information about dimensions of the manifolds as visualized in panel e.

4.2   Manifold Denoising as a Preprocessing Step

Density Estimation. One major problem in machine learning is to approximate the underlying probability density function (pdf) of a data set. If there is

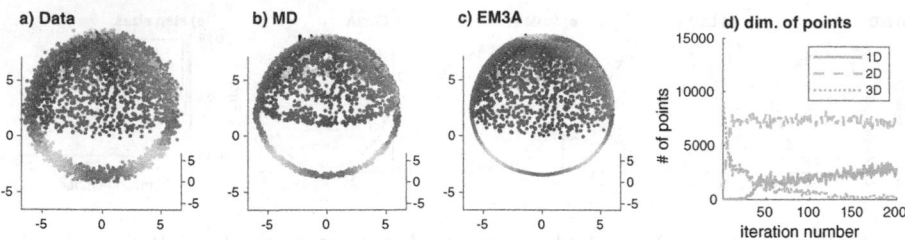

Fig. 3. comparison of the original noisy basket data (panel a) to the results after applying b) MD: $\delta t = 0.01, k = 50, N_{\text{iter}} = 150$. c) MBMS: $k = 50, \sigma = \infty, N_{\text{iter}} = 5$. d) EM3A. e) the number of visited samples in given dimensionalities per iteration.

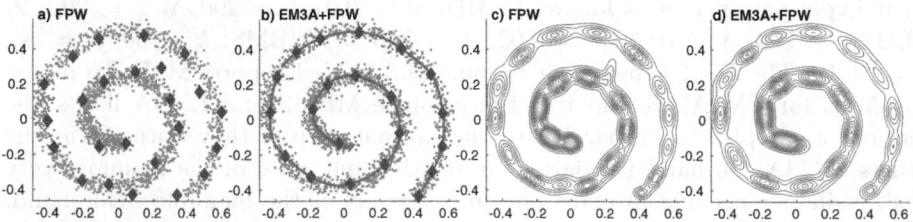

Fig. 4. a) Fast Parzen Window (FPW), b) EM3A+FPW, and contour curves for the likelihood function of FPW and EM3A+FPW in panel c) and d) respectively.

no prior knowledge about the form of the pdf, the Parzen Window (PW), a well-known non-parametric density estimation technique, is appealing. However, for larger data sets, this method becomes computationally expensive and strategies have been proposed to reduce its burden. For instance, Fast Parzen Windows (FPW) [22] decreases the amount of computation by two steps:

1. partition the data set via hyper-balls with fixed radii r and
2. fit a Gaussian distribution on each ball, separately.

In order to find the centres, it randomly selects a sample as a centre and then removes all its neighbours closer than r and continues this process on the remaining samples until only centres remain. Although it provides a sparse representation for the pdf, its performance can be degraded by noise. We demonstrate the benefit of denoising using EM3A as preprocessing and perform FPW subsequently on the denoised data set to find the centres. In order to show its impact we use a 2D data set aligned along a spiral structure extracting 2000 samples from the distribution $x_1 = 0.04t \sin(t) + \epsilon_1$, $x_2 = 0.04t \cos(t) + \epsilon_2$, with t distributed uniformly over the interval $(3, 15)$, and ϵ_i being Gaussian with $\mathcal{N}(0, 0.03)$. Here, we set $R_{\max} = 0.2$ and $\rho = 0.1$ (to keep $\eta = \frac{\rho}{\sqrt{r}}$ small). Table 2 shows the mean result for 20 independent runs. While the denoising improves the performance of FPW significantly, it also leads to a sparser representation of the pdf.

Table 1. Average distances to the ground truth

Dataset	Noisy	Denoising Methods			
		MD	LLD	MBMS	EM3A
Circle	0.2498	0.2108	0.1782	0.1636	0.1648 (0.0005)
Basket	0.4782	0.4014	–	0.3763	0.3675 (0.0002)

Table 2. Density estimation

	FPW	EM3A+FPW
Log lik	429 (185)	1220 (49)
Centres	25 (0.9)	21.6 (1)

5 Conclusion and Future Work

The presence of noise may degrade the results of many algorithms, and data cleaning can improve the performance of subsequently applied ML algorithms. Inspired by swarm intelligence our method uses colonies of agents with different properties to decrease noise and concepts from evolutionary game theory to control their populations for automatic hyper-parameter optimization. In contrary to our proposed strategy, related manifold denoising techniques, such as LLD and MBMS, require information about the manifolds and noise beforehand. Since every agent in the new algorithm just uses local information for their actions it has several advantages: it is flexible, parallelizable and applicable in different scenarios including multi-manifold cases. In future work we will perform a more detailed theoretical and empirical analysis of the hyperparameters and demonstrate the novel EM3A method on large scale astronomical simulations where it is desired but computationally demanding to obtain a probabilistic model describing the distribution of particles.

Acknowledgments. This project has received financial support from the European Union's Horizon 2020 research and innovation program under the Marie Sklodowska-Curie grant agreement No. 721463 to the SUNDIAL network.

References

1. Bentley, J.L.: Multidimensional binary search trees used for associative searching. Commun. ACM **18**(9), 509–517 (1975)
2. Blum, C., Roli, A., Dorigo, M.: Hc-aco: The hyper-cube framework for ant colony optimization. Proceedings of MIC, vol. 2, pp. 399–403 (2001)
3. Chu, Shu-Chuan., Roddick, John F., Su, Che-Jen, Pan, Jeng-Shyang: Constrained ant colony optimization for data clustering. In: Zhang, Chengqi, W. Guesgen, Hans, Yeap, Wai-Kiang (eds.) PRICAI 2004. LNCS (LNAI), vol. 3157, pp. 534–543. Springer, Heidelberg (2004). https://doi.org/10.1007/978-3-540-28633-2_57
4. Deneubourg, J.L., Goss, S., Franks, N., Sendova-Franks, A., Detrain, C., Chrétien, L.: The dynamics of collective sorting robot-like ants and ant-like robots. In: Proceedings of the First International Conference on Simulation of Adaptive Behavior on from Animals to Animats, pp. 356–363 (1991)
5. Dorigo, M., Maniezzo, V., Colorni, A.: Positive feedback as a search strategy. Technical Report. 91–016, Politecnico di Milano, Italy (1991)
6. Dorigo, M., Maniezzo, V., Colorni, A., et al.: Ant system: optimization by a colony of cooperating agents. IEEE Trans. Syst. Man Cybern. Part B: Cybern **26**(1), 29–41 (1996)

7. Golub, G.H., Van Loan, C.F.: Matrix computations, vol. 3. JHU press (2012)
8. Gong, D., Sha, F., Medioni, G.: Locally linear denoising on image manifolds. In: Proceedings of the 13th International Conference on AI and Stats, pp. 265–272 (2010)
9. Hein, M., Maier, M.: Manifold denoising. In: Advances in Neural Information Processing Systems, pp. 561–568 (2007)
10. Hofbauer, J., Sigmund, K., et al.: Evolutionary Games and Population Dynamics. Cambridge University Press, Cambridge (1998)
11. Kaslovsky, D.N., Meyer, F.G.: Non-asymptotic analysis of tangent space perturbation. Inf. Infer. J. IMA 3(2), 134–187 (2014)
12. Little, A.V., Maggioni, M., Rosasco, L.: Multiscale geometric methods for estimating intrinsic dimension. Proc. SampTA, 4(2) (2011)
13. Lumer, E.D., Faieta, B.: Diversity and adaptation in populations of clustering ants. In: Proceedings of the 3rd International Conference on Simulation of Adaptive Behavior: from Animals to Animats 3, pp. 501–508. MIT Press (1994)
14. Mordohai, P., Medioni, G.G.: Unsupervised dimensionality estimation and manifold learning in high-dimensional spaces by tensor voting. In: IJCAI, pp. 798–803 (2005)
15. Runkler, T.A.: Ant colony optimization of clustering models. Int. J. Intell. Syst. 20(12), 1233–1251 (2005)
16. Shelokar, P., Jayaraman, V.K., Kulkarni, B.D.: An ant colony approach for clustering. Anal. Chim. Acta 509(2), 187–195 (2004)
17. Stützle, T., Hoos, H.H.: Max-min ant system. Future Gen. Comput. Syst. 16(8), 889–914 (2000)
18. Taubin, G.: A signal processing approach to fair surface design. In: Proceedings of the 22nd Conference on Computer graphics and interactive techniques, pp. 351–358 (1995)
19. Tsai, C.F., Tsai, C.W., Wu, H.C., Yang, T.: ACODF: a novel data clustering approach for data mining in large databases. J. SS 73(1), 133–145 (2004)
20. Wang, W., Carreira-Perpinán, M.A.: Manifold blurring mean shift algorithms for manifold denoising. In: 2010 IEEE CVPR, pp. 1759–1766. IEEE (2010)
21. Wang, X., Tiňo, P., Fardal, M.A.: Multiple manifolds learning framework based on hierarchical mixture density model. In: ECML PKDD, pp. 566–581 (2008)
22. Wang, X., Tino, P., Fardal, M.A., Raychaudhury, S., Babul, A.: Fast parzen window density estimator. In: 2009 IJCNN, pp. 3267–3274. IEEE (2009)
23. Xiang, Y., Chen, Y.C.: Statistical inference using mean shift denoising (2016). arXiv preprint arXiv:1610.03927

A Novel Evaluation Metric for Synthetic Data Generation

Andrea Galloni[1](✉) , Imre Lendák[1,3], and Tomáš Horváth[1,2]

[1] Faculty of Informatics, Department of Data Science and Engineering,
ELTE – Eötvös Loránd University, Budapest, Hungary
andrea.galloni@inf.elte.hu
[2] Faculty of Science, Institute of Computer Science, Pavol Jozef Šafárik University,
Košice, Slovakia
Tomas.Horvath@upjs.sk
[3] Faculty of Technical Sciences, University of Novi Sad, Novi Sad, Serbia
lendak@uns.ac.rs

Abstract. Differentially private algorithmic synthetic data generation (SDG) solutions take input datasets D_p consisting of sensitive, private data and generate synthetic data D_s with similar qualities. The importance of such solutions is increasing both because more and more people realize how much data is collected about them and used in machine learning contexts, as well as a consequence of newly introduced data privacy regulations, e.g. the EU's General Data Protection Regulation (GDPR). We aim to develop a novel and composite SDG evaluation metric which takes into account macro-statistical dataset similarities and data utility in machine learning tasks against privacy boundaries of the synthetic data. We formalize the mathematical foundations for quantitatively measuring both the statistical similarities and the data utility of synthetic data. We use two well-known datasets containing (potentially) personally identifiable information as inputs (D_p) and existing SDG algorithms PrivBayes and DPGroupFields to generate synthetic data (D_s) based on them. We then test our evaluation metric for different values of privacy budget ϵ. Based on our experiments we conclude that the proposed composite evaluation metric is appropriate for quantitatively measuring the quality of synthetic data generated by different SDG solutions and possesses an expected sensitivity to various privacy budget values.

Keywords: Synthetic data generation · Differential privacy · Evaluation metrics

1 Introduction

The task of synthetic data generation (SDG) tackled in this paper is, given a private dataset $D_p \subset X_1 \times X_2 \times \cdots \times X_m$ of n_p rows and m attributes, to generate a synthetic dataset D_s which has the same number and type of attributes $X_1, X_2, \cdots, X_m$ as D_p and a pre-defined number n_s of rows. The attributes (columns) $X_i = (x_{i_1}, x_{i_2}, \ldots, x_{i_n})$ refer to n-dimensional vectors having numeric,

© Springer Nature Switzerland AG 2020
C. Analide et al. (Eds.): IDEAL 2020, LNCS 12490, pp. 25–34, 2020.
https://doi.org/10.1007/978-3-030-62365-4_3

ordinal or nominal values n being equal either to n_p or n_s. The goal is to generate D_s such that it keeps the main statistics and utility of D_p while preserves the individual privacy of its objects (rows). Privacy is especially important when the private data-set holds personally identifiable information (e.g. medical information) and the SDG process must guarantee that it will not be possible to identify any person based on analyzing only the publicly available generated D_s. A well-designed SDG solution generates data with similar statistical characteristics to the private data, e.g. maintains the correlations between inter-related attributes. Synthetic data is often generated as a substitute for the private data to be used in machine learning tasks. Generated data with high utility can be used to train a model, which in turn is expected to have adequate (classification or regression) performance when fed with the private data-set.

A consistent and comprehensive methodology or score for quantitatively measuring and evaluating the quality of the results of SDG is still missing, since at the moment of writing existing techniques only rely on macro-statistics and Machine Learning performances separately. Any such measure would ideally take into consideration more factors, such as *privacy-degree*, *macro-statistics* and *data utility*.

2 Related Work

Differential privacy is a mathematical tool for quantifying and bounding privacy loss [8]. Within the context of statistical disclosure control and cryptography it provides an accurate statistical information about a population while protecting the privacy of each individual within it. A differentially private algorithm holds a series of constraints which are exploited in order to publish information about a statistical database, these constraints limit and quantifies the disclosure of private information of records whose information is present within the database. The most common and widely used approaches to achieve differential privacy are the *Laplace Mechanism* [8] and the *Exponential Mechanism* [15].

Several differentially private generative approaches were proposed by the scientific community, where lately the most relevant models are utilizing Bayesian networks [12] or Copula functions [18]. These are the following[1]: Privelet+ [20]; PSD (Private Spatial Decomposition) [6]; Filter Priority [7]; P-HP method [1]; PrivateERM [21]; PrivGene [23] using genetic algorithms; PrivBayes [22] utilizing Bayesian networks which was extended with a more complete framework in [17] and [11] with a web interface; DPCopula [13] focusing on Gaussian Copula functions as the Synthetic Data Vault (to date not differentially private) [16]; or an improved and parallelized approach using Copula functions [2].

Different authors use different evaluation metrics for assessing the quality of their SDG solutions. In [22] authors evaluate their results making use of α-*way marginals* derived from query counts of subsets of attributes introduced in [4] measuring the accuracy of each marginal by computing the total variation distance [19] between noisy marginals and the original marginals. Furthermore

---

[1] Since this paper is focusing on evaluation, most popular SDG approaches are just listed here since their detailed description is out of the scope of this paper.

authors trained multiple SVM classifiers over several attributes of the synthetic dataset, where each classifier predicts one attribute in the data based on all other attributes, in this case the metric used is misclassification rate compared against other differentially private generative algorithms such as [21] and [23] which also apply k-means clustering. In [2] authors evaluate the generated data on all one-way marginal and two-way positive conjunction queries and three-way positive conjunction queries. In [13] authors evaluate the utility of the synthetic data generated by DPCopula answering random range-count queries and compare results against other methods.

In [10] (the most similar contribution) authors provide an evaluation of synthetic datasets under utility perspectives related to machine learning, leaving room for improvements. The authors compare two different approaches including differentially private algorithms which are evaluated under machine learning performance tasks over a single target attribute. Our work wants to provide a more comprehensive approach considering also macro-statistics within our evaluation metric and propose it as a standard methodology.

Based on these, is possible to state that at the moment of writing there is no comprehensive and standard evaluation methodology which evaluates multiple characteristics of the synthetic data against original private data in a unified score. We argue that any synthetic data generation (SDG) solution should possess at least the following three characteristics: i) guarantee a measurable amount of privacy, ii) resemble original data macro-statistics such that correlations, ranges, etc. and iii) maintain data utility for real scenarios usage, meaning that the performance of machine learning algorithms should be similar on both the generated and the original data. A good SDG methodology should maintain a good trade-off between privacy and utility.

3 Proposed Solution

Our metric G_ϵ is a composition of several indicators. Indeed, in the context of SDG there are several aspects, which have to be considered, such that:

- *privacy guarantee* (ϵ);
- the *macro-statistics* between attributes: significant correlation among attributes X_i has to be maintained;
- *data utility* in terms of machine learning performances: we would like to have similar classification performances in terms of accuracy when deploying the same algorithm over the private data-set and the original one.

Privacy Guarantee

The privacy guarantee is a parameter to quantify the privacy budget and it is direct consequence of ϵ-*differentially private* mechanism definition introduced in [8] and well formalized in [9]. Namely epsilon is a guaranteed boundary of privacy loss.

Definition 1 (Differential Privacy (DP) [8]**).** *A randomized algorithm $\mathcal{M}$ with domain $\mathbb{N}^{|X|}$ is (ϵ)-differentially private if for all $S \subseteq$ Range ($\mathcal{M}$) and for all $x, y \in \mathbb{N}^{|X|}$ such that $\|x - y\|_1 \leq 1$*

$$\Pr[\mathcal{M}(D_p) \in \mathcal{S}] \leq \exp(\varepsilon) \Pr[\mathcal{M}(D_s) \in \mathcal{S}]$$

where the probability space is over the coin flips of the mechanism $\mathcal{M}$ we say that $\mathcal{M}$ is ϵ-differentially private; thus the following privacy boundary is guaranteed:

$$log(\Pr[\mathcal{M}(D_p) \in \mathcal{S}]) - log(\Pr[\mathcal{M}(D_s) \in \mathcal{S}]) \leq \epsilon \qquad (1)$$

The term ϵ represents the so called *privacy budget* and it is a parameter of any differentially private mechanism. As its value decreases the more privacy is guaranteed through the injection of random noise when learning the probability distributions of the data-points; it is the direct measure of privacy boundary within the context of differential privacy. Thus for achieving a fair and consistent evaluation of SDG methods ϵ has to hold the same fixed value among those mechanisms to be compared at time of model creation and/or data generation.[2]

Macro-statistics

In [3], authors provide a new and practical correlation coefficient ϕ_k. It is based on several refinements to Pearson's hypothesis test of independence of two variables which works consistently between categorical, ordinal and interval variables. It also captures non-linear dependency. Moreover, it reverts to the Pearson correlation coefficient in case of a bi-variate normal input distribution. These are useful features when studying the correlation between variables with mixed types. Particular emphasis is paid to the proper evaluation of statistical significance of correlations and to the interpretation of variable relationships in a contingency table, in particular in case of low statistics samples and significant dependencies.

The proposed overall macro-statistics measure μ between D_s and D_p, both having m attributes $X_1, X_2, \ldots, X_m$ will be computed as

$$\mu(D_s, D_p) = \frac{\|\phi_k(D_s) - \phi_k(D_p)\|_2}{m(m-1)/2} \qquad (2)$$

Data Utility

D_s is intended to be used mostly for analytic purposes on which various machine learning (ML) tasks might be performed.

Since, in the time of generation of D_s we can not be sure which of the attributes from $X_1, X_2, \ldots, X_m$ would serve as labels in the future, we consider m different prediction tasks on D_s and D_p, respectively. The corresponding models are denoted as $M_{X_1,D}, M_{X_2,D}, \ldots, M_{X_m,D}$. Here, $M_{X_i,D}$, with $1 \leq i \leq m$, denotes a ML model learned/optimized using the data D (D_p or D_s) using the attributes $X_1, \ldots, X_{i-1}, X_{i+1}, \ldots, X_m$ to predict the attribute X_i. Within our experimental framework the machine learning task is classification over categorical attributes.

---

[2] The important role of ϵ in DP justifies its presence as subscript of G in our evaluation metric definition since we evaluate G at varying of ϵ.

It is important that different classes/types of ML algorithms should be used due to their different biases. Thus, for a more generic model, we allow K different ML models deployed over D, i.e. the synthetic (D_s) and the private (D_p) dataset. We denote these models by $M^1_{X_i,D}, M^2_{X_i,D}, \ldots, M^K_{X_i,D}$, where $1 \leq i \leq m$.

Let $acc(M_{X_i,D})$ denote the performance of $M_{X_i,D}$ measured on D $(D_p$ or $D_s)$. acc can be an arbitrary accuracy measure such that miss-classification rate or AUC, to name a few. For a more generic model we allow L different accuracy measures which will be denoted as $acc^1(M^k_{X_i,D}), acc^2(M^k_{X_i,D}), \ldots, acc^L(M^k_{X_i,D})$, where $1 \leq i \leq m$ and $1 \leq k \leq K$.

The proposed overall data utility measure δ between D_s and D_p will be computed as

$$\delta(D_s, D_p) = \frac{1}{mKL} \sum_{i=1}^{m} \sum_{k=1}^{K} \sum_{l=1}^{L} \|acc^l(M^k_{X_i,D_s}) - acc^l(M^k_{X_i,D_p})\|_2 \qquad (3)$$

The Combined Metric

Our proposed formula for a combined evaluation metric considering privacy-guarantee, macro-statistics and data utility is defined as

$$G_\epsilon = \alpha\mu(D_s, D_p) + \beta\delta(D_s, D_p) \qquad (4)$$

where α and β are weights allowing the user to define the importance of micro-statistics similarity and data utility similarity, while ϵ represents the value fed to the SDG algorithm while implementing DP while building the model.

4   Experimental Setup and Results

In order to evaluate our method we've deployed *PrivBayes* developed in *Python3.7* [22] based on *Bayesian Networks* and *DPGroupFields* mentioned in [5] and among the winners of *Differential Privacy Synthetic Data Challenge 2019* developed in *Java* and based on histogram sampling techniques (both algorithms are publicly available on *github.com*). Since we expect a wider application in real use-cases of DP SDG techniques by the industry, the selection of the two datasets is contextual to the most probable fields of application: indeed both our datasets do contain information about individuals in two main areas: *healthcare data* and *census data*. Namely *diabetes* is a well known dataset holding 8 real attributes and a categorical-binary one (for classification) for a total of 9 attributes and holding 768 records and the *adults* data-set composed by 14 numerical and categorical attributes (majority) and a categorical-binary one (for classification). It is important to mention that at generation time the same number of records of the original databases have been generated, namely: $n_s == n_p$ for all experiments. All the algorithms have been run using a computer equiped with an Intel *CPU i7-7500U@2.70 GHz* and *16 GB RAM DDR4*. All the ML tasks have been deployed using Python's *Scikit-Learn 0.22.2*. For a matter of brevity we're going to show the most relevant results.

For sake of simplicity, we have used the following settings: $K = 1$ and $L = 1$ in Eq. (3), $\alpha = 1$ and $\beta = 1$ in Eq. (4). While for what concerns the values of the privacy budget ϵ (parameter of the SDG routines at generation time) the following values have been selected: (2.0, 1.5, 1.0, 0.9, 0.8, 0.7, 0.6, 0.5, 0.4, 0.3, 0.2, 0.1, 0.05, 0.01). Those values have been chosen taking into consideration the public literature describing experiments and this value interval represent the most common setup, namely we've extended the ranges proposed within [22] and [13]. Our K task is going to be classification over all the categorical attributes ($K = 1$) of each dataset solved through the well known SVM algorithm with *RBF kernel* and $C = 1$ (regularization parameter) and $\gamma = 1/(m * variance(X))$ over the categorical attributes (default value of the SKLearn SVC classifier). In case of categorical target attributes. For what concerns L in our settings we set its value is 1 and is going to be *miss-classification rate* over categorical attributes.

4.1 Macro-statistics μ

In our first experiment our goal was to measure the effects of the chosen privacy budget on the macro statistics $\phi(D_s)$ against $\phi(D_p)$. Given the random nature at the root of the generative algorithms for each value of ϵ we've generated three different synthetic D_s for each input datasets both making use of PrivBayes and DPFieldGroups, then we've computed the average values of μ for each value of ϵ as plotted in Fig. 1. As expected the distance in terms of macro statistics defined in Eq. 2 grows as the magnitude of the matrices of the difference among correlation coefficients defined as ϕ in 2 grows. Generally, as expected, we might observe that δ tends to grow at decreasing of ϵ (differential privacy budget), indeed at decreasing ϵ more noise is introduced within the learned model thus correlation coefficients tends to differ more and more between the original dataset and the synthetic ones. In Fig. 1 it is possible to observe the behaviour of the δ term over ϵ against two datasets (Adults and Diabetes). It is possible to note that for Diabetes the computed values of μ appear more unstable if compared to the ones produced by the Adults dataset, This behavior is due to the splitting of continuous values performed by PrivBayes. Still the value range looks acceptable thus in this case the values of δ will play a decisive role when computing G_ϵ (at same values of α and β).

4.2 Data Utility δ

Also the δ factor as expected holds a similar behavior as μ but with a slightly different magnitude this characteristic justifies the presence of the two constants α and β in Eq. 4. Also in this case over our experimental setup we've run three times the data generation algorithm *PrivBayes* and plotted the average of the measures of δ for each value of ϵ. The values of δ are the result of repeating ten times the same machine learning task (classification in our setup) over each attribute, randomly selecting training and testing records with a ratio of 0.2 for testing. As expected the measure in terms of data utility defined in Eq. 3 tends to grow. Generally, it is possible to observe that values of δ tends to grow at

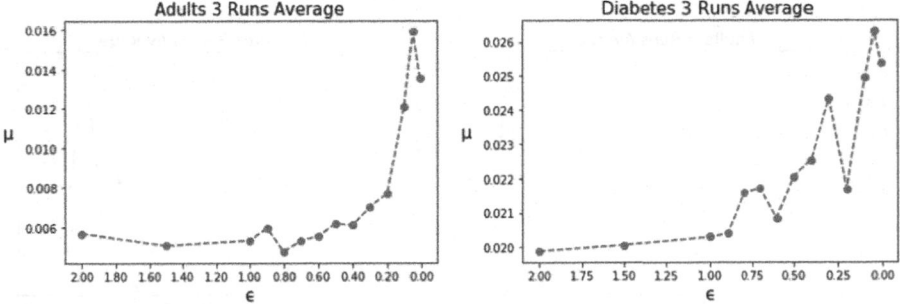

Fig. 1. Values of μ against the two datasets Adults and Diabetes using PrivBayes synthetic data generation method.

decreasing of ϵ (differential privacy budget), indeed at decreasing ϵ more noise is introduced and performances in terms of accuracy tends to differ. In Fig. 2 it is possible to observe the behaviour of the δ term over ϵ against two datasets (Adults and Diabetes). It is important to note that within this setup we can observe that values of δ over ϵ tend to be more stable holding a more clear trend in the case of *adults* (where the majority of the attributes is categorical) when compared to the *Diabetes* graph in which the majority of attributes is numerical. This outcome is due to the splitting of the continuous values within the Bayesian Networks model. This observation represents an important/insight for the user, indeed our metric could suggest the scientist to alter (augment) only the number of splits for a continuous or several continuous variables *obtaining better scoring results without altering the privacy budget magnitude* ϵ. In this case the values of δ grow *"faster"* than μ for the same dataset, thus this factor could dominate smaller values of ϵ.

4.3   Composite Measure G

Finally we calculated the values of composite measure G defined by Eq. 4 for $\alpha = 1$ and $\beta = 1$. We find that the evaluation method provides coherent results. Figure 4 shows results of our experiments and G_ϵ for the Adults dataset.

Within this setup we can note that at the same privacy budget ϵ PrivBayes clearly better preserves data-utility (lower values of G) when compared to DPFieldGroups. This is due to the fact that histogram sampling (DPField-Groups) performs worse when applied over dataset holding a double digit number of attributes. Thus, we state that our method reflects and confirms earlier literature results [13,14,22]. Figure 3, instead, shows results of G_ϵ regarding just PrivBayes since DPFieldGroups has not been designed to handle attributes holding floating numbers (not integers). Also in this case the effect of continuous attributes splitting has a noticeable impact on the stability of the score.

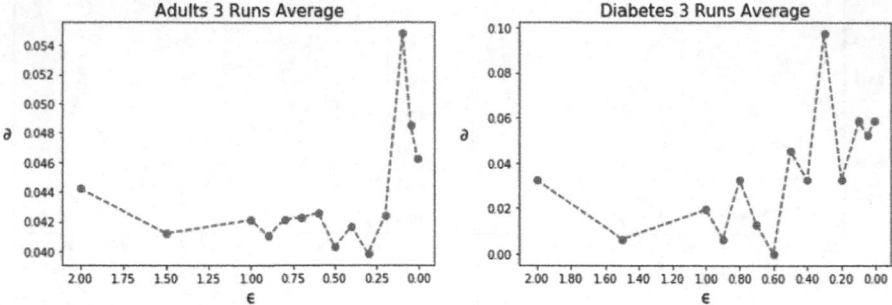

Fig. 2. Values of δ against the two datasets Adults and Diabetes using PrivBayes for both as synthetic data generation method. Values of δ tend to grow *"faster"* due to the nature of PrivBayes which splits continuous intervals.

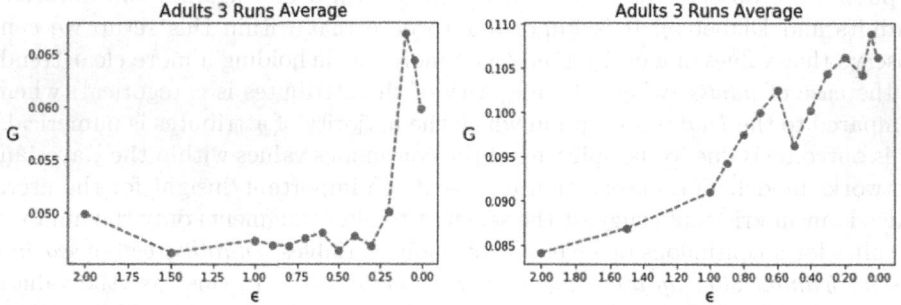

Fig. 3. Values of G_ϵ over the two algorithms PrivBayes (left) and DPFieldGroups (right) deployed on Adults dataset. Within this setup PrivBayes clearly keeps a better data utility over varying ϵ (lower G).

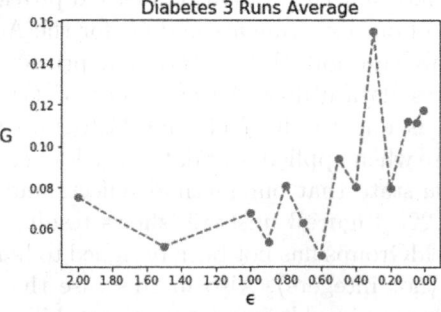

Fig. 4. Values of G_ϵ over Diabetes dataset using PrivBayes as generative algorithm.

5   Conclusions and Future Work

We propose a novel composite and comprehensive evaluation metric G_ϵ for quantitatively measuring synthetic data generation solutions. The metric takes into account dataset similarities and data utility, against privacy budget ϵ, between synthetically generated data and the original data. We test the introduced evaluation metric against two datasets comparing two different differentially private synthetic data generation algorithms. The results are consistent with literature and will open the path for further investigation and possibly it will be used as a base-model and standardized methodology to assess and evaluate the trade-off between privacy and data utility within the context of differentially private synthetic data generation.

As for the next steps there is room for testing this evaluation metric against several different DP SDG techniques, possibly find an empirical way to set the best values of α and β based on the objective of the synthetic data generation task and the properties of the dataset.

Acknowledgments. This research was co-funded by EIT Digital Industrial Doctorate and Ericsson Hungary. Project no. ED_18-1-2019-0030 (Application domain specific highly reliable IT solutions subprogramme) has been implemented with the support provided from the National Research, Development and Innovation Fund of Hungary, financed under the Thematic Excellence Programme funding scheme. We are thankful to Gian Marco Canneori for the fruitful discussions leading to the final mathematical foundations presented in this paper.

References

1. Acs, G., Castelluccia, C., Chen, R.: Differentially private histogram publishing through lossy compression. In: 2012 IEEE 12th International Conference on Data Mining, pp. 1–10. IEEE (2012)
2. Asghar, H.J., Ding, M., Rakotoarivelo, T., Mrabet, S., Kaafar, D.: Differentially private release of datasets using Gaussian copula. J. Priv. Confidentiality **10**(2) June 2020
3. Baak, M., Koopman, R., Snoek, H., Klous, S.: A new correlation coefficient between categorical, ordinal and interval variables with pearson characteristics. Comput. Stat. Data Anal. **152**, 107043 (2020)
4. Barak, B., Chaudhuri, K., Dwork, C., Kale, S., McSherry, F., Talwar, K.: Privacy, accuracy, and consistency too: a holistic solution to contingency table release. In: Proceedings of the Twenty-Sixth ACM SIGMOD-SIGACT-SIGART Symposium on Principles of Database Systems, pp. 273–282 (2007)
5. Bowen, C.M., Snoke, J.: Comparative study of differentially private synthetic data algorithms and evaluation standards (2019). arXiv preprint arXiv:1911.12704
6. Cormode, G., Procopiuc, C., Srivastava, D., Shen, E., Yu, T.: Differentially private spatial decompositions. In: 2012 IEEE 28th International Conference on Data Engineering, pp. 20–31. IEEE (2012)
7. Cormode, G., Procopiuc, C., Srivastava, D., Tran, T.T.: Differentially private summaries for sparse data. In: Proceedings of the 15th International Conference on Database Theory, pp. 299–311 (2012)

8. Dwork, C., McSherry, F., Nissim, K., Smith, A.: Calibrating noise to sensitivity in private data analysis. In: Halevi, S., Rabin, T. (eds.) TCC 2006. LNCS, vol. 3876, pp. 265–284. Springer, Heidelberg (2006). https://doi.org/10.1007/11681878_14
9. Dwork, C., Roth, A., et al.: The algorithmic foundations of differential privacy. Found. Trends Theor. Comput.Sci. **9**(3–4), 211–407 (2014)
10. Hittmeir, M., Ekelhart, A., Mayer, R.: On the utility of synthetic data: an empirical evaluation on machine learning tasks. In: Proceedings of the 14th International Conference on Availability, Reliability and Security, pp. 1–6 (2019)
11. Howe, B., Stoyanovich, J., Ping, H., Herman, B., Gee, M.: Synthetic data for social good (2017). arXiv preprint arXiv:1710.08874
12. Koller, D., Friedman, N.: Probabilistic Graphical Models: Principles and Techniques. MIT press, Cambridge (2009)
13. Li, H., Xiong, L., Jiang, X.: Differentially private synthesization of multi-dimensional data using copula functions. In: Advances in Database Technology: Proceedings. International Conference on Extending Database Technology, vol. 2014, p. 475. NIH Public Access (2014)
14. Li, H., Xiong, L., Zhang, L., Jiang, X.: Dpsynthesizer: differentially private data synthesizer for privacy preserving data sharing. In: Proceedings of the VLDB Endowment International Conference on Very Large Data Bases, vol. 7, p. 1677. NIH Public Access (2014)
15. McSherry, F., Talwar, K.: Mechanism design via differential privacy. In: 48th Annual IEEE Symposium on Foundations of Computer Science (FOCS'07), pp. 94–103. IEEE (2007)
16. Patki, N., Wedge, R., Veeramachaneni, K.: The synthetic data vault. In: 2016 IEEE International Conference on Data Science and Advanced Analytics (DSAA). pp. 399–410, IEEE (2016)
17. Ping, H., Stoyanovich, J., Howe, B.: Datasynthesizer: privacy-preserving synthetic datasets. In: Proceedings of the 29th International Conference on Scientific and Statistical Database Management, pp. 1–5 (2017)
18. Sklar, A.: mfonctions de répartition à n dimensions et leurs marges, n publ. Inst. Statist. Univ. Paris **8**, 229–231 (1959)
19. Tsybakov, A.B.: Introduction to Nonparametric Estimation. Springer Science & Business Media, Berlin (2008)
20. Xiao, X., Wang, G., Gehrke, J.: Differential privacy via wavelet transforms. IEEE Trans. Knowl. Data Eng. **23**(8), 1200–1214 (2010)
21. Zhang, J., Zheng, K., Mou, W., Wang, L.: Efficient private ERM for smooth objectives. In: Proceedings of the 26th International Joint Conference on Artificial Intelligence, IJCAI 2017, pp. 3922–3928. AAAI Press (2017)
22. Zhang, J., Cormode, G., Procopiuc, C.M., Srivastava, D., Xiao, X.: Privbayes: private data release via bayesian networks. ACM Trans. Data. Syst. (TODS) **42**(4), 1–41 (2017)
23. Zhang, J., Xiao, X., Yang, Y., Zhang, Z., Winslett, M.: Privgene: differentially private model fitting using genetic algorithms. In: Proceedings of the 2013 ACM SIGMOD International Conference on Management of Data, pp. 665–676 (2013)

Data Pre-processing and Data Generation in the Student Flow Case Study

Luís Cavique[1,2(✉)] [iD], Paulo Pombinho[2] [iD], Antonio J. Tallón-Ballesteros[3] [iD], and Luís Correia[2] [iD]

[1] Universidade Aberta, Lisbon, Portugal
luis.cavique@uab.pt
[2] MAS-BioISI, FCUL, Lisbon, Portugal
pmimatos@fc.ul.pt, luis.correia@ciencias.ulisboa.pt
[3] University of Huelva, Huelva, Spain
antonio.tallon.diesia@zimbra.uhu.es

Abstract. Education covers a range of sectors from kindergarten to higher education. In the education system, each grade has three possible outcomes: dropout, retention and pass to the next grade. In this work, we study the data from the Department of Statistics of Education and Science (DGEEC) of the Education Ministry. DGEEC maintains those outcomes for each school year, therefore, this study seeks a longitudinal view based on student flow. The document reports the data pre-processing, a stochastic model based on the pre-processed data and a data generation process that uses the previous model.

Keywords: Data pre-processing · Data generation · Student flow · Stochastic model

1 Introduction

Models of the student flow throughout the educational system can be very important for planning infrastructure resources, providing the distribution of human resources in the labor market, and identifying problems across the educational system, as well as to propose specific actions to overcome them.

In a broader definition the student flow is characterized by a number of features that changes over time. Two time-variables should be distinguished: curricular year (or grade), and the school (or academic) year. The school year is usually made up of a couple of consecutive years, for instance 2019–2020. To simplify the nomenclature, sometimes in this work we use the year of enrollment, so for the given example the value of 2019.

In the grade-by-grade student flow models, each grade has three possible outcomes: dropout, retention and pass to the next grade. The cohort definition is a group of individuals who share specific characteristics, usually in the same period, such as year of birth. Some studies use the concept of cohort and calculate the survival rate of the cohort.

The goal of this paper is to present the student flow approach in an educational system, by detailing the model and proposing techniques. This work is part of a wider project, the student flow modeling in the Portuguese educational system, with the acronym ModEst.

C. Analide et al. (Eds.): IDEAL 2020, LNCS 12490, pp. 35–43, 2020.
https://doi.org/10.1007/978-3-030-62365-4_4

In our study, the Public Administration organism is the Department of Statistics of Education and Science (DGEEC), of the Portuguese Education Ministry. DGEEC has a vast amount of data regarding 2 million students per year on the Portuguese school system, from pre-scholar to doctoral programs, with annual in/out flow of more than 0.5 million students. The data provided by DGEEC have been anonymized in order to meet the requirements of the General Data Protection Regulation (GDPR).

In this work, we only used the data from the 1st to the 12th grade, leaving the study of higher education for future work. A stochastic model, based on the Markov chain process, is applied in order to generate new data volumes.

The procedure used in this work is developed in R language and can be summarized in three steps: data pre-processing, modeling and data generation. In step (i) the data is pre-processed and the structure of the problem is formalized, in Step (ii) a stochastic model is generated, based on the pre-processed data, that explains the past, and in step (iii), data generation in the prediction phase, we forecast the future number of students in the system. The proposed procedure can be presented in the following data pipeline: data pre-processing → stochastic model → data generation.

The paper is organized in five additional sections. In Sect. 2, background information is presented. Section 3 details the data pre-processing for the student flow case study. The stochastic model is introduced in Sect. 4. Section 5 reports the data generation and the prediction of the number of students. Finally, in Sect. 6 conclusions are drawn.

2   Background Information

2.1   Student Flow

Education covers a range of sectors from kindergarten, basic (1st to 9th grade) and secondary school (10th to 12th grade, referred as High School in some countries), to higher education, from level 0 to level 8 of the International Standard Classification of Education, ISCED 2011 classification [4].

The education sector can be seen as a series of components where each student follows a pathway, which meets his/her own aspirations.

Many examples of the application of mathematical models to education planning exist since the late 1960s [7]. In this state-of-the-art in student flow problem, we identified different approaches such as Key Performance Indicators (KPI), Visualization, Markov models and What-if simulation. These approaches are ordered by the usual sequence of data analysis.

In Science Education two KPI are usually established in student flow: dropout and failure or retention [5]. The knowledge and prediction of dropout and retention in the student flow is highly relevant since these KPI directly influence the performance of the education system.

The visual representation of student's march to graduation can be analogous to Napoleon's march to Moscow created by Charles J. Minard in late 1800 s [3]. The visual-analytics capabilities are relevant for two reasons. First, these tools allow a rapid exploration of large data sets, providing the ability to easily detect trends and anomalies in data related to student progress. Second, the ability to supply visual proof of student progress grounded in facts, rather than speculation or supposition.

In business intelligence, what-if analysis fills this gap between data mining and decision making, by enabling users to simulate and inspect the behavior of a complex system under some given hypotheses. What-if dynamic simulation model allows to analyze the impact of potential changes in core curriculum policy, prerequisite structure, and staffing capacity to be tested prior to implementation [8]. Discrete Event Simulation is more adaptable to real-world applications than pure a Markov model. It accommodates more easily the complexities and interdependencies of the many components involved in the system [6].

2.2 Markov Chains

Stochastic processes and, in particular, Markov chains are very useful tools to deal with uncertainty [1, 2].

A stochastic process consists of a set of indexed random variables X(t), for example the number of students at the beginning of the school year t. The time parameter t can take discrete values, $t = \{0, 1, 2,\ldots, n\}$, or take continuous values, i.e. $t \geq 0$. Variable X can assume a set of states, $S = \{1, 2,\ldots, m\}$. A stochastic process is said to have a finite state space, if $|S|$ is finite, and the stochastic process has a space of continuous states, otherwise.

The Markov chain is a special case of stochastic process, with the following propriety: if the process is currently in state i, then it will occupy state j in the next period with probability P(i, j), i.e. $\text{Prob}(s_{t+1} = j \mid s_t = i) = P(i, j)$, where P(i, j) is a parameter fixed for each pair of states $(i, j) \in S \times S$.

An important property in Markov chains is the stationarity of the process. The stochastic process X is said to be stationary when its behavior is independent of time and the system is in a steady state. In a stationary process, as t becomes very large, X^t converges to the probability vector X such that $X = X.P$.

3 Data Pre-processing

The available dataset contains the enrollment information of pre-primary, primary, secondary and post-secondary education levels, from level 0 to 5 of ISCED 2011 [4], of a period of 10 school years, from 2008–2009 to 2017–2018. The dataset deals with around 17,000,000 instances, which corresponds to an average of 1,700,000 students per school year, which supports several enrolments of around 5,500,000 individual students during 10 school years.

In the data pre-processing the ETL (extract, transform and loading) process was applied to the dataset. During the cleaning phase the classifications were uniformed, the duplicate student registrations were removed, and missing data were also removed.

In this section, firstly, we detail possible transitions from a generic grade in order to find the outcomes of the student flow. And secondly, we identify the most relevant dimensions of DGEEC dataset.

To better explain the developed work, we introduce the concept in three steps: the data collection, where the data is prepared and the attributes selected, the selection of dropout and loyalty rules with a decision tree algorithm, and the creation of loyalty actions to prevent dropout.

3.1 Grade Transitions

The sequence of grades can be seen as a sequence of states in a state transition system with specific transitions, as shown in Fig. 1. The three possible outcomes, or output transitions, are dropout, retention or pass to the next grade. On the other hand, the input transitions are retention in the same grade, pass from the previous grade and ingress of new students coming from outside the system.

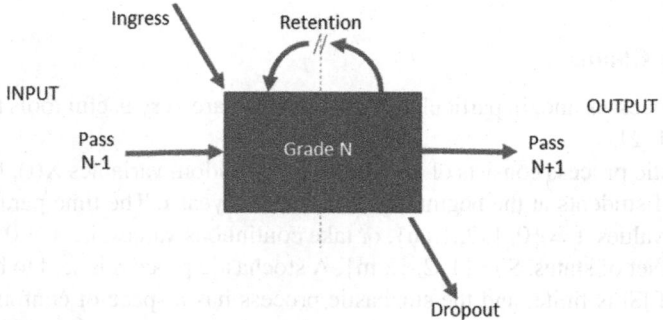

Fig. 1. A state (concerning a grade) with input and output state transitions

In the ModEst project, given the anonymous identification of the student, the school year, and his current grade, we can identify his transition state:

- Dropout: a student drops out if there is no information about him/her in the next school year;
- Retention: in the following school year, a student is retained if he/she is in the same grade;
- Pass: in the following school year, a student passes if he/she is enrolled in a higher grade;
- Ingress: a student ingresses if there is no information about him/her in the previous school year.

The information of the outcome of each student can be easily implement in SQL using sub-queries, as show in Fig. 2.

3.2 Data Dimensions

For each enrolled student, there is a set of information associated that can be classified into dimensions. The fact table of ModEst Project has a set of six dimensions which includes:

- School year of enrollment (e.g. 2008–2009);
- Curricular year (or grade) in which the student enrolls (e.g. 10th grade);
- Modality of the course (e.g. regular or professional education);
- Nature of course (e.g. public or private education);

```
-- Update Dropout
UPDATE factTable AS F1
SET outcome = "dropout"
WHERE F1.student NOT IN (SELECT F2.student
                         FROM factTable AS F2
                         WHERE F1.schoolYear+1 = F2.schoolYear);
-- Update Retention
UPDATE factTable AS F1
SET outcome = "retention"
WHERE EXISTS (SELECT *
              FROM  factTable AS F2
              WHERE F1.student = F2.student
              AND F1.schoolYear+1 = F2.schoolYear
              AND F1.grade = F2.grade);
-- Update Pass
UPDATE factTable AS F1
SET outcome = "pass"
WHERE EXISTS  (SELECT *
              FROM factTable AS F2
              WHERE F1.student = F2.student
              AND F1.grade+1 = F2.grade
              AND F1.schoolYear+1 = F2.schoolYear);
```

Fig. 2. SQL code to update outcome information of each student

- Geographic information at the level of 'nomenclature of territorial units for statistics' NUTS-II and NUTS-III (e.g. Lisbon Metropolitan Area);
- Outcome of the enrollment: dropout, retention or passing grade.

The student data was anonymized with a specific identifier. Each line of the fact table corresponds to a student enrolled in only one school year, with the additional information of the grade, the course Modality, the course Nature, the NUTS and the outcome. The education level (primary education and secondary education) aggregates the dimensions of grade and course modality. Fact table and respective dimensions are shown in Fig. 3.

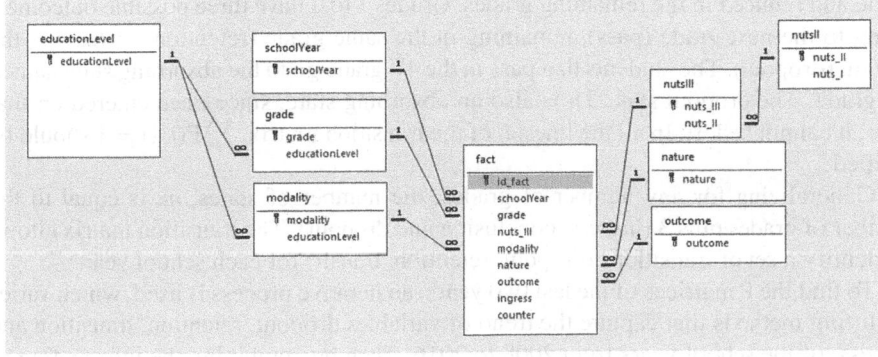

Fig. 3. Fact table and student dimensions

In the fact table the number of ingress can also be found. For each combination of the dimensions we have a counter with the number of students in a fact table with 85,000 instances.

As it is usual in projects with real data, data cleaning procedures were performed. In particular, the dimension grade was normalized from 0 (kindergarten) to 13th (post-secondary) for the different course modalities.

4 Stochastic Model

As previously defined, in a stochastic model with m states and n time periods, the variable $X = [x_1, x_2,..., x_m]$ in the time period $n + 1$ is given by $X^{n+1} = X^n.P$. In this section, we define the structure of the transition matrix with the respective m states.

Given the number of students, for each school year, and for each grade, we obtain the number of ingresses, dropouts, failures and transitions, as the sample shown in Table 1. With this we aim to define the structure of the transition matrix.

Table 1. Sample of the input data of the transition matrix

Year	In/outcome	Grade1	Grade2	Grade3	Grade4
20XX	ingress	98,627	10,382	10,382	10,382
20XX	dropout	7,135	10,112	9,823	11,466
20XX	retention	1,019	6,741	3,274	4,586
20XX	transit	93,774	95,499	96,043	98,605

In our study the student flow presents an open-loop system, with an input flow, corresponding to the ingresses of new students, and two output flows, which correspond to dropout students and completion of studies.

Figure 4 shows the transition diagram of the open-loop Markov chain, of a sequence of school years in the years 2000, with four grades. The state ingress, I, is larger in the 1st grade and reduced in the remaining grades. Grades 1 to 4 have three possible outcomes: going to the next grade (pass), remaining in the same grade (retention) or leaving the system (dropout). The students that pass in the 4th grade go to the absorbing state named 5th grade. The dropout state, D, is also an absorbing state, since once entered on that state, it cannot be left. In all the lines, i, of the transition matrix, $\sum P(i, j) = 1$ should be verified.

Generalizing for any number of grades, the number of states, m, is equal to the number of grades plus 3 (ingress, conclusion and dropout). The transition matrix allows to identify a set of transitions (dropout, retention, transit) for each school year.

To find the P matrices of the last two years, an iterative process is used, which varies the fitting methods that capture the trend of variables dropout, retention, transition and ingress, of the school years from 2008 to 2016. Then by combining the information of the two P matrices, the transition matrix that best predicts the last two years is found.

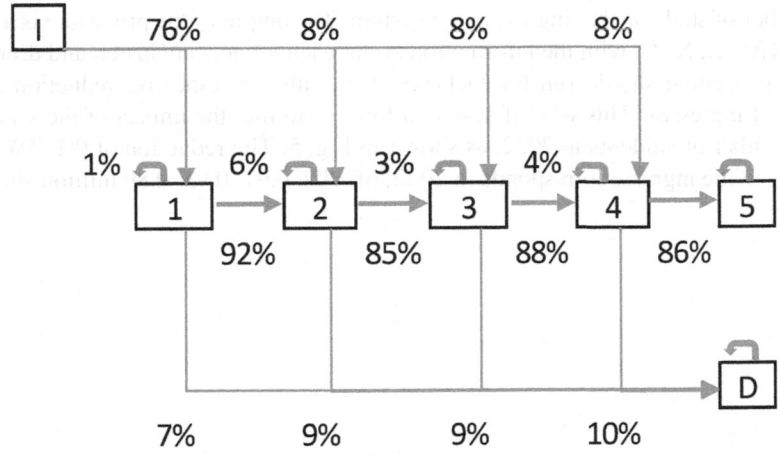

Fig. 4. Transition diagram

Using Y to express the prediction and X the real data, the prediction for the year n + 1 is given by $Y^{n+1} = X^n.P$. The error of the prediction, for a given year n, is calculated by the expression $error^n = \sum(abs(X^n - Y^n))/\sum X^n$, where Y^n corresponds to the prediction and X^n to the real data. In this study, the following error percentages were found:

- to predict 2015, error2015 $= 4.93\%$,
- to predict 2016, error2016 $= 6.22\%$.

The error of the predictions of the school years 2015 and 2016 is about 5%, i.e., ensuring an average accuracy of 95%.

5 Data Generation and Prediction

As already mentioned, the prediction of the next time period is given by $X^{n+1} = X^n.P$ where P is the probability matrix described in the previous section. Based on the previous stochastic model, data can be generated in order to predict the number of students. In this section we develop a what-if approach, with four scenarios, to predict X in the year 2022, i.e., X^{2022}.

The stochastic variable X must be compatible with P, having the following structure X = [ingress, grade1, grade2,..., grade12, conclusion, dropout]. The number of ingresses is a new variable that changes each year, and should be inserted into the stochastic model. The other variables of X are given by the previous year.

The reduction of the moving average of the number of ingresses in the last 6 school years varies from 10% to almost 30%. In this work, we create four conservative scenarios of the number of ingresses with the reduction of 0%, 5%, 10% and 15%.

In the procedure to predict the number of students, the input is data from school year 2016, X^{2016}, the transition matrix P, and a new vector Ingress with the information of the number of students that ingress in the system. The output of the procedure is a set of vectors $\{X^i,\ldots, X^{i+k}\}$ with the information about each grade, conclusion and dropout.

The procedure should run for each scenario with the respective reduction of the number of ingresses. This what-if system allows us to find the impact of the scenarios in the number of students in 2022, as shown in Fig. 5. The reduction of 0%, 5%, 10% and 15% in the ingress, corresponds in 2022, of 1.02, 0.97, 0.93, 0.89 million students, respectively.

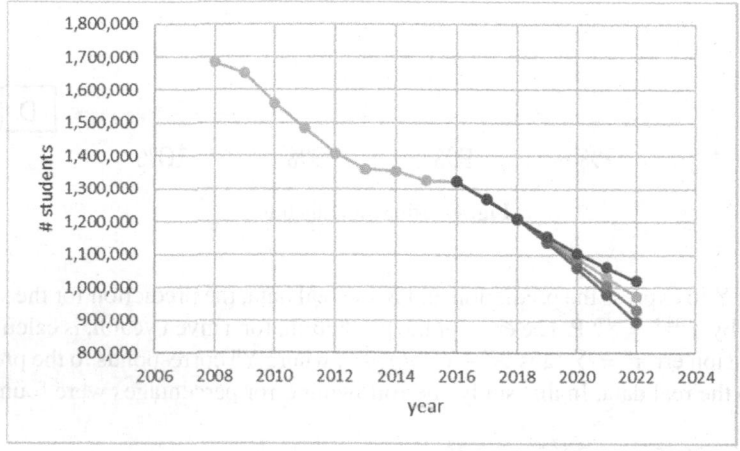

Fig. 5. Impact of the scenarios in the number of students in 2022

6   Conclusions and Future Work

In this work, we focused on the data from the 1st to the 12th grade, provided by the DGEEC dataset. The document reports the data pre-processing and the stochastic model that allows the data generation.

Data pre-processing included the ETL, the definition of the grade/state transition and the data warehouse with one fact table and six dimensions.

The stochastic model includes the formulation of the problem and the model tuning: (i) formulation: define m and the transition matrix P for each school year; (ii) stochastic modelling: given the data of time periods n from 2008 to 2016, find the transition matrix P that best predicts the last two years. In the model tuning we obtained an accuracy around 95% for the years 2015 and 2016.

Data was generated in order to predict the number of students, from 2017 to 2022. In the prediction phase, three of four scenarios indicate that the number of students from grade 1st to 12th, in the enrolment year of 2022, would be less than one million.

Two major contributions can be highlighted related to the student flow case study. The first one corresponds to the dichotomy between the annual surveys and the view of the data flow, i.e., DGEEC validates the annual data collected from the schools and ModEst considers a longitudinal view of the student flow obtained in the pre-processing phase.

Given the stochastic model, the second contribution concerns the variety of possibilities of data generation, in order to overcome the limitations imposed by just 10 years of real data.

Since this work is part of a wider project, for future work we point out the following topics:

- to apply the tools developed to predict the number of students using filters on the dimensions (modality, nature, NUTS);
- to visualize the student flow with cohorts using Sankey diagrams, among others.

Acknowledgments. The authors would like to thank the FCT Projects of Scientific Research and Technological Development in Data Science and Artificial Intelligence in Public Administration, 2018–2022 (DSAIPA/DS/0039/2018), for its support. LCav, PP and LCor also acknowledge support by UID/MULTI/04046/2103 center grant from FCT, Portugal (to BioISI).

References

1. Bronson, R., Naadimuthu, G.: Schaum's Outline of Operations Research, 2nd edn. McGraw-Hill, New York (1997)
2. Grinstead, C.M., Snell, J.L.: Introduction to probability. American Mathematical Society, Providence (1997)
3. Heileman, G.L., Babbitt, T.H., Abdallah, C.T.: Visualizing student flows: busting myths about student movement and success, change. Mag. High. Learn. **47**(3), 30–39 (2015)
4. ISCED: International Standard Classification of Education. UNESCO Institute for Statistics, Montreal, Quebec, Canada (2011)
5. Junior, P.L., Silveira, F.L., Ostermann, F.: Survival analysis applied to student ow in undergraduate physics courses: an example from a Brazilian university. Revista Brasileira de Ensino de Física **34**(1), 1403 (2012)
6. Fiallos A.X.: Ochoa Discrete event simulation for student flow in academic study periods, Twelfth Latin American Conference on Learning Technologies (LACLO). IEEE, Argentina (2017)
7. Lovell, C.C.: Student Flow Models: A Review and Conceptualization. National Center for Education Research and Development, Washington, D.C. (1971)
8. Saltzman, R.M., Roeder, T.M.: Simulating student flow through a college of business for policy and structural change analysis. J. Oper. Res. Soc. **63**(4), 511–523 (2012)

Enhanced Credit Prediction
Using Artificial Data

Peter Mitic[1,2(✉)] and James Cooper[3]

[1] Santander UK, 2 Triton Square, Regent's Place, London NW1 3AN, UK
[2] Department Computer Science, UCL, Gower Street, London WC1E 6BT, UK
p.mitic@ucl.ac.uk
[3] Santander US, 75 State St, Boston, MA 02109, USA
jsc42@cantab.net

Abstract. Analysing credit data using a neural network has hitherto proved to be very resilient to attempts to improve success rates in prediction. We present a technique using simulated data which results in a marginal improvement in success rate. The empirical probability distribution for each feature of the training data is determined, and random samples are drawn from those distributions. The result is termed 'artificial' data. It is then possible to generate equal volumes of data for each of the binary outcomes (default or not), thereby alleviating a class imbalance classification problem. The simulation method uses a copula (to preserve the correlation structure of the original data) and optimal feature weighting to give acceptable results. The results indicate that overall percentage success rates for the more common outcome only are improved, but there is a more significant improvement in the AUC metric. The significance of this result in the context of assessing credit worthiness is discussed.

Keywords: Artificial data · Copula · Importance weight · Neural network · Lorenz curve

1 Introduction

Recent advances in medical diagnoses using artificial intelligence (AI) have been remarkably successful. See, for example, Chabon, [5], Awan [2], Yala [19], and McKinney [13]. Overall classification success rates (i.e., total number of correct predictions divided by total number of predictions) exceeding 90% are common, with AUC values as high as 0.9.

Attempts to apply neural network technology to credit data have not hitherto proved to be as successful as the widely-used logistic regression methods that most lenders employ. Using the same technology (a neural network implemented in *Tensorflow*) that Google employed for the Chabon [5] study, it was difficult to achieve success rates of more than 74%. An attempt to explain this result was made in [14]. It appears that the indicators used when assessing credit

© Springer Nature Switzerland AG 2020
C. Analide et al. (Eds.): IDEAL 2020, LNCS 12490, pp. 44–53, 2020.
https://doi.org/10.1007/978-3-030-62365-4_5

worthiness, or combinations of them, are not strong pointers to future success in repayment.

In this paper we attempt to improve on the results reported in [14] using a variant of the *Probabilistic Novelty Detection* technique (hereinafter referred to as *PND*) developed by Clifton et al. [6] to generate *artificial data*. The literature review below summarizes the principal drivers for this paper. Following that, our method of deriving and using *artificial data* is described. Result comparisons are then made, and explanations are offered.

In this paper, the most common outcome (i.e., the one with the most instances) will be referred to as the *major* outcome, and the least common outcome will be referred to as the *minor* outcome.

2   Literature Review: Credit Risk and Artificial Data

In this review we concentrate on the application of novelty detection methods and to the assessment of credit worthiness. In doing so we provide some specific details of our previous work which provide a basis on which to improve.

AI with Credit Data: Previous Research
Earlier application of AI technology to credit data have yielded mediocre results compared to the recent medical successes already mentioned. For example, Louzada [12] quotes mean success rates of 77.7% for German credit data and 88.1% for Australian credit data (see [8]). Those figures mask success rates for the major and minor outcomes. Although 'better' results have been reported ([11]: AUC = 0.915 and [1]: AUC = 0.975), we suspect that either the data set used contains some behavioural indicator of default, or that loans in the dataset are only for 'select' customers who have a high probability of non-default.

Summary of the Metric Framework for Data Concentration
In [14] the first author explores whether the relative lack of success in using artificial neural networks to model credit risk may arise from inherent structures in the data. Three metrics (*Copula*, *Hypersphere* and *k-Neighbours*) are used to measure the 'shape' of the data. They are combined in a metric $\hat{H}$. It was observed that a high value of $\hat{H}$ implies that either the data are too noisy or that they provide insufficient predictive information to train a neural network.

The richness and complexity of the data comes from having different paths to success (or failure), which implies that there is little room to improve on initial results. Effectively, data corresponding to the major and minor outcomes appear to be almost coincident.

Summary of Probabilistic Novelty Detection
An general overview of Novelty Detection methods is given in [16]. This review concentrates on a specific example from that paper: the *PND* method of Clifton et al. [6]. It is designed to cope with situations where the instances of the minor outcome are extremely rare, or even non-existent. Original data is used to define a hypersphere of radius r, defined by the centroid of the real data. The data set for the major outcome is assumed to exist within a hypersphere of radius $2r$

and the minor class is assumed to exist outside that radius (i.e., the minor class comprises outliers). The artificial data are used as a training set and the original data are used as the test set for an ensuing *AI* process.

Two sets of *PND* application results, both derived using SVM, are reported. Both show a clear separation between artificial data for the major and minor outcomes. Summarizing:

1. Combustion monitoring: AUC ~ 0.81–0.96
2. Patient vital sign monitoring. AUC ~ 0.9, indicated by ROC curves.

We have found that the *PND* method resulted in a deterioration of our previous results when applied to credit data. We suggest reasons in Sect. 4.3.

The statistical outlier detection method in [18] adopts a different approach. Outliers (equivalent to the 'minor outcome' set in the Clifton method) are determined by first dividing a training set into as many partitions as there are classes. An instance of a feature is considered an outlier if any feature value is more than three times the inter-quartile range from the third quartile for feature values in each partition. The German data set ([8]) mentioned in Sect. 2 was analyzed in this way, and it was found that less than 10% of that data set could be considered as an outlier. The significance of this result is discussed in Sect. 6.

Subsequent research advanced the *PND* method further. Gorokhov et al. [10] applied a convolutional neural networks to extract features from text data by sequentially filtering features from training and test sets (AUC = 0.92). Pidhorskyi et al. [15] use a *generative-PND* method to compute the density function of image data on a training set, and generate samples from it (AUC $\geq$ 0.98 using the MNIST data). Two further studies adopt the same general approach: Rad et al. [17] (mobility assessment, AUC $\in$ (0.65, 0.95)), and Contreras et al. [7] (robotics, 77% of predictions exceeded 90% accuracy). Bhattacharjee et al. [3] treats data that cannot be classified with confidence as 'novelties' (image classification, AUC $\in$ (0.77, 0.90))

3   Methodology: Artificial Data Generation and Use

We have found that the algorithm presented in [6] did not produce satisfactory results. Reasons are suggested in Sect. 4. Therefore we have developed an alternative, the overall strategy for which is summarized the following algorithm. The step numbers correspond to the steps in Fig. 1.

1. Partition the original data into training and test sets, D_{train} and D_{test} respectively (Step A).
2. Partition D_{train} into two subsets $D_{train,0}$ and $D_{train,1}$ according to the binary outcomes 0 and 1 respectively (Step A).
3. Generate artificial data $D_{art,0}$ and $D_{art,1}$ from the subsets $D_{train,0}$ and $D_{train,1}$ respectively (Steps B and C).

4. Combine $D_{art,0}$ and $D_{art,1}$ to form a single artificial data set D_{art} (Step D).
5. Use D_{art} for training and D_{test} for testing.

The steps above are summarized in Fig. 1. The source of importance weights is discussed in Step B.1 of the detailed algorithm. (Sect. 3.1). The numbers in black roundals refer to the steps in Sect. 3.1.

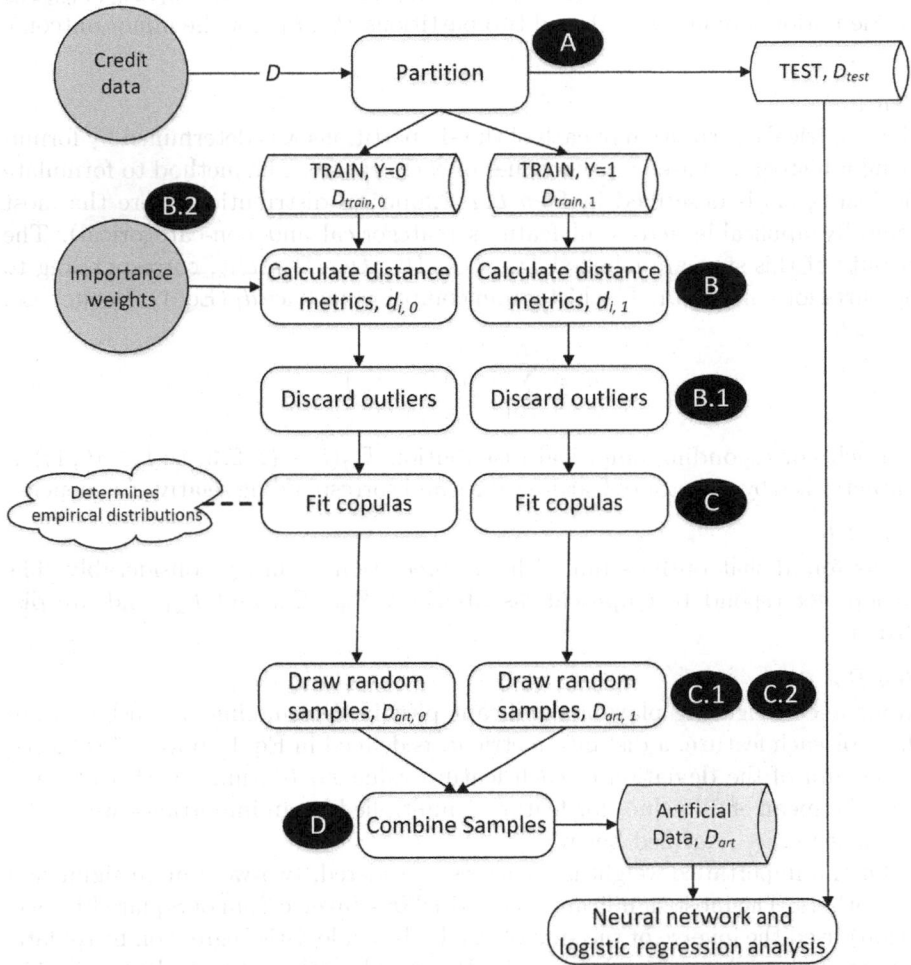

Fig. 1. Artificial Data generation algorithm. The step numbers in black roundals refer to sub-sections in 3.1.

3.1   Artificial Data Algorithm: Details

The details of our algorithm to generate artificial data are summarized in the steps that follow. The starting point is a dataset comprising N feature columns labeled $X_1, X_2, ...X_N$. The outcome column is labeled Y and takes values zero for the major outcome (correct prediction-credit pass) and one for the minor outcome (incorrect prediction-credit fail). The data are imbalanced: the number of major outcomes is approximately $1/3$ of all outcomes.

Step A
Partition the original data D such that there is sufficient data in each partition to model the empirical data accurately. In our case, four partitions $P_{01}, P_{02}, P_{03}, P_{04}$ for the major outcome $Y = 0$, and two partitions P_{11}, P_{12} for the minor outcome $Y = 1$.

Step B
The empirical distribution of each of the six partitions was determined by formulating a histogram based on the values of each feature. The method to formulate the histogram is described in *Step B.1*. Empirical distributions were the most generally applicable across all features (categorical and non-categorical). The outputs of this step are labelled $D_{01}, D_{02}, D_{03}, D_{04}, D_{11}, D_{12}$, corresponding to the partitions in *Step A*. The histogram comprises metrics d_i (Eq. 1). In our case, $N = 22$.

$$d_i = \sum_{j=1}^{N} w_j (M_j - x_{ij})^2 \tag{1}$$

Each corresponding empirical distribution $E_{ij}(i \in (1,2,3,4); j \in (0,1))$ is characterized by a vector of feature values and corresponding relative frequencies.

Step B.1
It was found that outliers diminish the prediction accuracy considerably. The outliers correspond to empirical distributions E_{40}, E_{30} and E_{21} and are discarded.

Step B.2
Importance weighting plays a significant part in determining the distributions D_{ij}. For each feature, a distance metric d_i is defined in Eq. 1 above. This metric is the sum of the deviation of each feature value x_{ij} (datum i with feature j) from the mean of all values for feature j, multiplied by an importance weight for that feature, w_j (clarified below).

Of the importance weighting schemes considered, two were more significant than others. The most significant (termed *ISSE* - Inverse Sum of Squared Expectation) used the inverse of the sum of residuals of a logistic regression fit to data. Importance weights derived using the Boruta algorithm also worked well. The *ISSE* importance weights, w_i are calculated from Eq. 2, which summarises the *ISSE* calculation for a logistic regression function ρ acting on each of N features in the training data $x_i[Train]$ and outcome y_i, with a logistic regression prediction function $Pred$ which takes test data $x_i[Test]$ as an additional argument.

$$w_i = \left(\sum_{j=1}^{N} Pred\Big(\rho(x_i[Train], y_i) - x_i[Test]\Big)^2 \right)^{-1} \tag{2}$$

Step C
Fit a copula C_{ij} to pseudo observations of each partition P_{ij}. The copula preserves the dependency structure of the features of the original data. The *Normal* and *Frank* copulas proved to be optimal.

Step C.1
Uniformly distributed random samples U_{ij} were extracted from each copula C_{ij}. The sample size was set for each partition so as to be sufficient to generate enough artificial data to use in a neural network and to produce approximately the same number of '$Y = 0$' cases as '$Y = 1$' cases. It was found that using partitions P_{12}, P_{03}, P_{04} resulted in diminished results, and sample sizes of 1 were allocated to these sets.

Step C.2
The random samples U_{ij} were transformed to the appropriate empirical distributions D_{ij} using inverse empirical distribution function transformations.

Step D
The outputs of the previous step were combined columnwise. This combination constitutes the artificial data.

4   Results

4.1   Data and Implementation

The data set used was the data set labelled *INT* in [14]. It comprises 8202 records: 2690 records for the minor outcome $Y = 1$ (credit fail), and 5512 for the minor outcome $Y = 1$ (credit pass). Each record had $N = 22$ features, each normalized to $[0,1]$, and a binary decision flag Y. Calculations were done using R on an i7 processor with 16 MB RAM. We are grateful for the *Tensorflow* neural network code supplied by Chollet and Allaire in [4].

4.2   Copula and Importance Weighting Results

In order to choose an importance weighting scheme for *Step B* of the Artificial Data algorithm, the overall algorithm at the start of Sect. 3 was run with the most generally applicable copula (the *Normal* copula), cycling through a range of importance weighting schemes. Repeated trials showed that *ISSE* importance weighting (see Sect. 3.1, *Step B.1*) was optimal (AUC = 0.865), and produced particularly stable results. The *Boruta* method was almost as good (AUC = 0.845). The AUC without importance weighting was 0.649, so is not a viable option. Other weighting schemes tested were *Principal Components, Pseudo-R^2, Recursive Feature Elimination, Log-Likelihood ratio, Random Forest, Logistic Regression* and *LVQ*.

Given the optimal *ISSE* choice, the copulas tested were *Normal, Student-t, Joe, Clayton, Gumbel* and *Frank*. There was very little variation between them, and the *Frank* copula was optimal (AUC = 0.871). The Frank copula stresses outlier and near-origin data more than the others, which may explain its optimality.

4.3 Results Using Artificial Data

Table 1 shows a comparison of neural network and logistic regression results with original data only, with data derived from the *PND* method [6], and with data derived from our *Artificial Data* method. The mean and standard deviation results for 25 runs using each method are shown.

Table 1. Neural network and logistic regression results (Mean, SD), using the Artificial data method (see note 1), the Probabilistic Novelty Detection method (see note 2), and with original data exclusively (see note 3).

Method	*Metric mean*	Original data	*PND*	Artificial data
NN	% Success	(65.91, 6.88)	(95.89, 0.86)	(77.26, 4.89)
NN	% Success Major	(65.83, 7.01)	(98.5, 0.98)	(76.74, 6.89)
NN	% Success Minor	(74.07, 7.25)	(5.18, 3.98)	(77.27, 5.18)
NN	Gini	(0.63, 0.01)	(0.23, 0.07)	(0.72, 0.03)
NN	AUC	(0.82, 0.01)	(0.61, 0.04)	(0.86, 0.01)
LR	% Success	(72.02, 1.54)	(97.11, 0.06)	(55.43, 4.8)
LR	% Success Major	(72.13, 1.71)	(99.87, 0.1)	(54.58, 5.02)
LR	% Success Minor	(69.6, 5.19)	(1.11, 1.79)	(85.04, 3.46)
LR	Gini	(0.52, 0.04)	(0.24, 0.03)	(0.65, 0.04)
LR	AUC	(0.76, 0.02)	(0.62, 0.02)	(0.83, 0.02)

Note 1: Artificial Data. Frank copula, *ISSE* importance weighting, 2000 major outcome data, 5000 minor outcome data. 25 runs, each $\sim$10 min

Note 2: PND, with parameters defined in [6] da $= 0.25$, dn $= 0.01$, $r_a = 3r$. 2000 records generated, 10 runs, each $\sim$5 h

Note 3: Results with original data only, from [14]. LR training sets were obtained by random sampling.

The results in Table 1 indicate that using *Artificial Data* gives an improvement on the results derived using original data only. In particular, the balance between % success for the major and minor outcomes is preserved using a neural network. If logistic regression is used instead, it is not. In contrast, there is a marked deterioration of results using *PND*. We suggest that the reason is some or all of the following points.

- The dependency structure of the original data is not preserved.
- It is assumed that the minor outcome corresponds to outliers, as defined by the hypersphere. That is unlikely to be the case for credit data.
- There is no clear way to tune the model parameters.
- There is an over-dependence on uniformly-distributed data. Only a few credit data feature distributions resemble uniform distributions.

In contrast, our *Artificial Data* set is specifically designed to preserve the dependency structure of the original data, and models individual features for the major and minor outcomes as closely as possible.

5   Discussion: Analysis of the Lorenz Curve

We now consider an alternative approach, in the context of credit risk, to measuring 'success' by AUC or % of correct predictions. Lorenz curves are a useful tool to measure, in the context of credit risk, the proportion of predictive success in the binary outcomes $Y = 0$ and $Y = 1$. More often they are used to quantify economic inequality: proportion of income against proportion of population. See a recent discussion in [9].

A Lorenz curve is a plot, parameterised by threshold, of modelled propensity against % minor outcome class included up to a given threshold (horizontal) and % major outcome class included up that threshold (vertical). Lorenz curves are well established for visualizing the ability of a model to rank order by likelihood of default. A perfect rank ordering would start at the origin, rise vertically as it works through the major class, and then horizontally across the top of the unit square. A perfect rank ordering would start at the origin rise vertically as it works through the major class and then horizontally across the top of the unit square (note, there is no requirement for the propensity cut off to be 0.5). The power of such a model is given by gini (=2*AUC-1). Gini values lie between -1 and 1, with 0 representing random selection and negatives a reverse ordering). Modelling is typically geared towards maximizing the gini, because of a broad relationship between gini and the capital a bank needs to hold for credit risk exposure. Figure 2 shows a hypothetical Lorenz curve.

However, the practical use of this model is often focused on a particular region. For illustration:

- Always lend to people with a predicted default probability less than 1%,
- Never lend to people with a predicted default probability more than 5%.

So in terms of decisioning, the area of the curve near $p = 3\%$ might be critical. It shows how different the population between $p = 3\%$ and $p + \Delta = 4\%$ looks compared to the population between $p - \Delta = 2\%$ and $p = 3\%$. In that way we get a sense of the performance of the model at the decision boundary. This may be thought off as the difference in gradient of line segments as shown in Fig. 2. The flatter the gradient the higher the local density of defaults per non-defaulter. Although not considered in this paper, understanding the effect near the decision boundary would be required for implementation. The point of such analysis would be to reduce the incidence of false-negatives, which cause far more harm to a bank than false-positives.

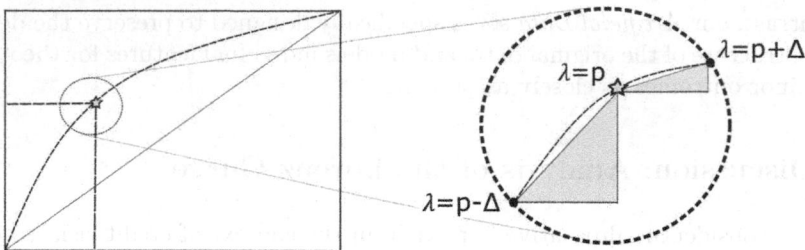

Fig. 2. Lorenz curve illustration, showing a gradient discontinuity near a typical decision boundary. The axes are explained in the associated text.

6 Conclusion

In this paper we have attempted to improve upon a previous result obtained when applying neural network technology to credit data. Using *Artificial Data* has made it possible to improve the previous result marginally, in terms of both AUC and success rates. Correct predictions of the minor outcome (credit fail) is a major factor in credit analysis, since every defaulted loan requires multiple non-defaulted loans to compensate for any shortfall incurred. Therefore a valuable theme to pursue is to improve on the minor outcome success rate without compromising the major outcome, using the idea suggested in Sect. 5.

In Sect. 2 a method of outlier detection ([18]) was noted. The particular case of the German credit data ([8]) has a bearing on the results of this paper. In that case, less than 10% of instances were classified as outliers. We consider, following analysis using the Novelty Detection method of [6], that there are similarities between the data used in our analysis and the German credit data. Specifically, outliers cannot be used to generate artificial data, because outliers are sparse. The subsets corresponding to 'credit fail' and 'credit pass' are almost coincident. Outliers comprise a mixture of the two subsets.

References

1. Addo, P.M., Guegan, D., Hassani, B.: Credit risk analysis using machine and deep learning models. Risks **6**, 38 (2018). https://doi.org/10.3390/risks6020038
2. Awan, R., Koohbanani, N.A., Shaban, M., Lisowska, A., Rajpoot, N.: Context-aware learning using transferable features for classification of breast cancer histology. In: Campilho, A., Karray, F., ter Haar Romeny, B. (eds.) ICIAR 2018. LNCS, vol. 10882. Springer, Cham (2018). https://doi.org/10.1007/978-3-319-93000-8
3. Bhattacharjee, S., Mandal, D., Biswas, S.: Multi-class novelty detection using mix-up technique. In: Proceedings WACV2020 (2020). https://doi.org/10.1109/WACV45572.2020.9093303
4. Chollet, F., Allaire, J.J.: Deep Learning with R. Manning, NY (2018)
5. Chabon, J.J., et al.: Integrating genomic features for non-invasive early lung cancer detection. Nature **580**, 245–251 (2020). https://doi.org/10.1038/s41586-020-2140-0

6. Clifton, L., Clifton, D.A., Zhang, Y., Watkinson, P., Tarassenko, L., Yin, H.: Probabilistic novelty detection with support vector machines. IEEE Trans. Reliab. **63**(2), 455–467 (2014)

7. Contreras-Cruz, M.A., Ramirez-Paredes, J.P., Hernandez-Belmonte, U.H., Ayala-Ramirez, V.: Vision-based novelty detection. Sensors **19**, 2965 (2019). https://doi.org/10.3390/s19132965

8. Dua, D., Graff, C.: Statlog Data: UCI Machine Learning Repository Irvine CA (2019). http://archive.ics.uci.edu/ml

9. Costa, R.N., Perez-Duarte, S.: Not all inequality measures were created equal. European Central Bank Statistics Paper Series 31 (2019). https://www.ecb.europa.eu//pub/pdf/scpsps/ecb.sps31269c917f9f.en.pdf

10. Gorokhov, O., Petrovskiy, M., Mashechkin, I.: Convolutional neural networks for unsupervised anomaly detection in text data. In: Yin, H., et al. (eds.) IDEAL 2017. LNCS, vol. 10585, pp. 500–507. Springer, Cham (2017). https://doi.org/10.1007/978-3-319-68935-7_54

11. Kvamme, H., Sellereite, N., Aas, K., Sjursen, S.: Predicting mortgage default using convolutional works. Expert Syst. Appl. **102**, 207–217 (2018). https://doi.org/10.1016/j.eswa.2018.02.029

12. Louzada, F., Ara, A., Fernandes, G.B.: Classification methods applied to credit scoring. Surv. Oper. Res. Manage. Sci. **21**(2), 117–134 (2016). https://doi.org/10.1016/j.sorms.2016.10.001

13. McKinney, S.M., et al.: International evaluation of an AI system for breast cancer screening. Nature **577**, 89–94 (2020). https://doi.org/10.1038/s41586-019-1799-6

14. Mitic, P.: A metric framework for quantifying data concentration. In: Yin, H., Camacho, D., Tino, P., Tallón-Ballesteros, A.J., Menezes, R., Allmendinger, R. (eds.) IDEAL 2019. LNCS, vol. 11872, pp. 181–190. Springer, Cham (2019). https://doi.org/10.1007/978-3-030-33617-2_20

15. Pidhorskyi, S., Almohsen, R., Adjeroh, D.A., Doretto, G.: Generative probabilistic novelty detection with adversarial autoencoders. In: Proceedings NIPS'2018. Montreal Canada, pp. 6823–6834 (2018). https://doi.org/10.5555/3327757.3327787

16. Pimentel, M., Clifton, D.A., Clifton, L., Tarassenko, L.: A review of novelty detection. Signal Process. **99**, 215–249 (2014)

17. Rad, N.M., van Laarhoven, T., Furlanello, C., Marchiori, E.: Novelty detection using deep normative modeling. Sensors **18**, 3533 (2018). https://doi.org/10.3390/s18103533

18. Tallon-Ballesteros, A.J., Riquelme, J.C.: Deleting or keeping outliers for classifier training? In: Sixth World Congress on Nature and Biologically Inspired Computing (NaBIC), pp. 281–286. IEEE (2014)

19. Yala, A., Lehman, C., Schuster, T., Portnoi, T., Barzilav, R.: A deep learning mammography-based model for improved breast cancer risk prediction. Radiol. Online **292**, 60 (2019). https://doi.org/10.1148/radiol.2019182716

Data Generation Using Gene Expression Generator

Zakarya Farou[1(✉)] 🆔, Noureddine Mouhoub[2], and Tomáš Horváth[1,3] 🆔

[1] Department of Data Science and Engineering, Telekom Innovation Laboratories, Faculty of Informatics, ELTE–Eötvös Loránd University, Pázmány Péter sétány 1/C, 1117 Budapest, Hungary
{zakaryafarou,tomas.horvath}@inf.elte.hu
[2] Science and Technology Campus, Bordeaux Computer Science Laboratory-LaBRI, University of Bordeaux, 351 Cours de la Liberátion, 33400 Talence, France
noureddine.mouhoub@labri.fr
[3] Institute of Computer Science, Pavol Jozef Šafárik University, Jesenná 5, 040 01 Košice, Slovakia
http://t-labs.elte.hu/

Abstract. Generative adversarial networks (GANs) could be used efficiently for image and video generation when labeled training data is available in bulk. In general, building a good machine learning model requires a reasonable amount of labeled training data. However, there are areas such as the biomedical field where the creation of such a dataset is time consuming and requires expert knowledge. Thus, the aim is to use data augmentation techniques as an alternative to data collection to improve data classification. This paper presents the use of a modified version of a GAN called Gene Expression Generator (GEG) to augment the available data samples. The proposed approach was used to generate synthetic data for binary biomedical datasets to train existing supervised machine learning approaches. Experimental results show that the use of GEG for data augmentation with a modified version of leave one out cross-validation (LOOCV) increases the performance of classification accuracy.

Keywords: Data generation · Generative adversarial networks · Gene expression data · Cancer classification

1 Introduction

Automated diagnosis of oncology diseases such as colon cancer is extremely complex and requires special attention. Gene expression profiling is widely adopted to learn and analyze the conditions of cells and their response to various conditions, which is useful in the pathogenesis of diseases. Gene expression micro arrays can be used for diagnostic purposes as well as for insights into biology. *Gene Expression Data* (GED) is a high dimensional data with a high number of features that indicate gene levels, but with very few records. Usually, a satisfying number of measurements are available for each tumor sample, with each measurement belonging to a particular gene.

© Springer Nature Switzerland AG 2020
C. Analide et al. (Eds.): IDEAL 2020, LNCS 12490, pp. 54–65, 2020.
https://doi.org/10.1007/978-3-030-62365-4_6

Obtaining accurate results while training a classifier becomes difficult due to the lack of available, varied and meaningful data. It is therefore recommended that additional samples be added to train the classifier more efficiently. Increasing the original data, i.e., generating additional samples from the existing ones, for the imbalanced class would avoid overfitting situations and improve the classification process.

1.1 Motivation

Building a satisfying Machine Learning (ML) model usually requires a large number of samples. In the biomedical field, the collection of such data is very expensive and time-consuming. Experts must store, examine and annotate the recorded data (which can be either an image, information or clinical tests) to obtain a clean, meaningful and useful dataset. Recently, Deep Neural Networks (DNNs) [9] has made many improvements in ML, especially when massive datasets are available. The popularity of DNNs has led to their use also in cases where only a small number of samples are available, and simpler ML techniques, requiring less computational effort, would be considerably better or even better than DNNs. However, for a very limited number of data samples, it is difficult to train any type of ML model with reasonable performance. Thus, this work aims to highlight that even the performance of simple classification techniques can be improved by providing the appropriate augmentation technique.

In the classification of cell dysplasia (cancer data), the sensitivity of the data should be maintained. This means that when new instances are generated using data augmentation techniques, it is important to ensure that the generated data is close to the original data, not only in terms of values but also in terms of data semantics. Generative Adversarial Networks (GANs) [6], established in recent years, have attracted the attention of researchers (Fig. 1). Several variants of GANs have been proposed to generate high-quality synthetic data, where they

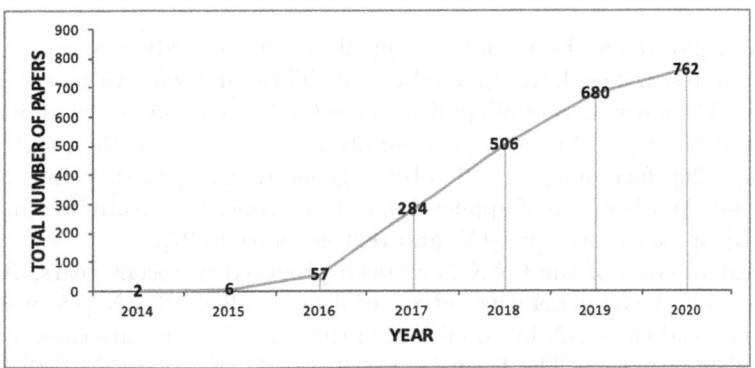

Fig. 1. Cumulative number of GAN-related paper publications/journals per year since its introduction in 2014

have been used for data augmentation in cases where traditional data augmentation methods do not yield good results. Thus, the feasibility of using GANs as a data augmentation technique to improve the performance of classifiers in GED classification is worth exploring. To our knowledge, the use of GANs in relation to cancer classification by gene expression has not yet been performed.

1.2 Contributions

As the collection of large amounts of medical data is costly and difficult to acquire due to certain privacy constraints, data generation is an alternative to data collection, especially when synthetic data can be generated from a small number of existing samples. To this end, this work provides several contributions.

First, a modified version of the GAN called *Gene Expression Generator* (GEG) is proposed. GEG learns the GED distribution and tries to synthesize new samples that are consistent with the original data. The GEG discriminator uses the Wasserstein distance to reflect the similarity between the original and synthetic data distributions, which allows for greater stability during training. GEG also uses data restriction, i.e., the discriminator is fed only with data belonging to a single class, which allows avoiding unnecessary steps leading to labeled generated data after GEG training. Although in this work GEG was used only for binary classification, it would also work for multi-class classification problems.

Next, the paper introduces a modified version of leave one out cross-validation (LOOCV) by not merging the synthetic data with the original data but using the generated instances only for the training of the classifiers. By using the generated data as an extension of the training samples, testing the model performance with the original data only and improving the results, more attention is paid to data sensitivity. The generated data helps the model to better understand the data without interfering with the original data.

2   Related Work

The GAN, introduced by Goodfellow et al. [6], has quickly received increased interest and researchers have explored its capabilities in a wide variety of applications [28]. The most successful application of GAN is computer vision, including image translation [8], image super-resolution [10], image synthesis [29], video generation [26], face aging [2], 3D object generation [23] or detection of small objects [30]. Another area of application of GAN concerns natural language processing [5], speech processing [13], and text generation [27].

Several variants of the GAN have been proposed in recent years. A conditional version of the generative adversarial nets called CGAN [18] was introduced to extend the GAN by conditioning the generator and discriminator with additional information. The Deep Convolutional GAN (DCGAN) [20] has certain architectural constraints and is powerful while dealing with unsupervised learning problems. Another variant, called GRAN [7] was proposed to generate images with recurrent adversarial networks. MGAN [12] is a Markovian GAN, a technique for training generative neural networks for efficient texture synthesis. GAN-CLS [21] demonstrated the ability to generate plausible images of birds and flowers using simple text descriptions. An alternative to the GAN called VIGAN [22], an extended version of CycleGAN [31], handled missing data imputations from the MNIST.

A distributed adversarial network called DAN was proposed in [11] where the adversarial training relies on the entire sample as a unit and not on its sample points. Unlike researchers who usually use thousands of images, Marchesi et al. [16] investigated the possibility of applying GAN to produce high-quality megapixel images using a limited amount of data. Lu et al. [15] suggested a new approach called Bi-GAN which uses two generators dedicated solely to generating synthetic data from Gaussian noise, two evaluators and a discriminator for data classification. Later, Wang et al. [28] suggested working with a set of generators against a single discriminator. Finally, Ian Goodfellow introduced a new robust and simple to train method called latent adversarial generator (LAG) [3], which generates high-resolution images using latent spaces.

The review of the GAN and its areas of application reveals that most researchers are focusing on image/video data generation, while there are still some unexplored areas such as biomedical field, which is a delicate, sensitive and costly area (in terms of data acquisition). Recently, there have been some attempts to use GAN for genetic data [17]. Due to the relatively small number of researchers working on the GAN for genetic data, this work aims to examine the GAN alongside the GED.

3   Proposed Gene Expression Generator

Generative adversarial networks (GANs) are deep neural networks based on game theory [6]. They are considered as intelligent and creative machines. Figure 2 shows that a GAN has two principal components: a generator G and a discriminator D (red boxes); G and D are neural networks. The output of G i.e., x' where $x' = G(z)$ is directly linked to the input of D next to the original data x. G produces synthetic instances x' starting from random Gaussian noise z such that x' and x are consistent. D checks the closeness of the synthetic data to the original data and returns a probability between 0 and 1 for each instance created. In the basic GAN, a cross-entropy loss is used to estimate the error between predicted/actual label of D output and the actual labels.

3.1   GEG Architecture

The gene expression generator, i.e., GEG, is a variant of GAN that attempts to reproduce the probability distribution of gene expression. GEG uses original samples from a single class as input to generate more artificial instances referring to the same class label. The Wasserstein distance WD is used as a loss function to reflect the similarity between the distribution of the original and synthetic data generated by GEG. When a discriminator uses the Wasserstein loss function, it

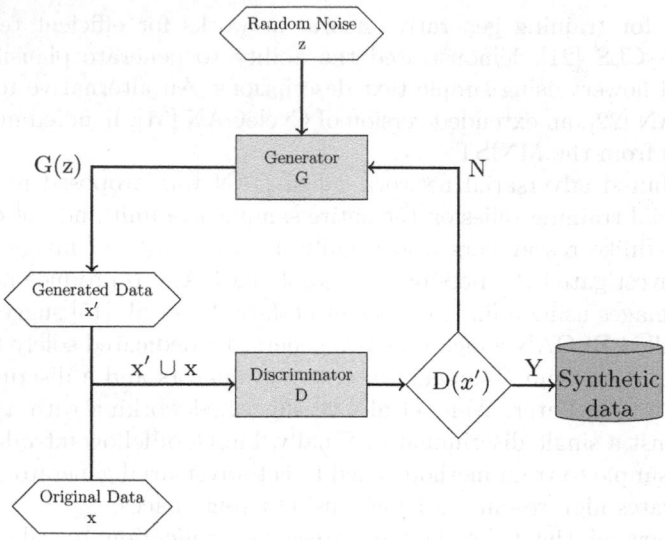

Fig. 2. Process of creating synthetic data using GAN

does not try to classify the instances by giving them a class label (i.e., synthetic or real instance), but attaches a number to each instance so that high values are used for the original data and low values for the synthetic data. However, in simple GANs, the discriminator gives a probability p for each instance, where $p \in [0..1]$ and, based on a threshold (e.g. 0.5), it can classify the instances either as artificial (negative) or original (positive) instances.

A GAN can use a unique loss function θ for both G and D. Therefore, one of them (G or D) should use $-\theta$ (the same loss function differing only by the sign). In this case, GEG uses different loss functions. It uses a generator loss $- \mathbf{D(x')}$ where G tries to maximize this function; in other words it tries to maximize the output of the discriminator for its negative instances. Furthermore, it uses a discriminator loss $\mathbf{D(x') - D(x)}$ where D tries to maximize WD defined as the difference between $D(x)$ and $D(x')$; in other words it tries to maximize the difference between its output on positive instances and its output on negative instances[1].

The training data of D belongs to two groups, the original data being denoted by x and the synthetic data by x' which is generated by G. During the training of D, x and x' are used as positive and negative instances respectively. It is important to note that during the training of D, G is semi-suspended, i.e., its weights are kept constant, but at the same time G remains to generate new instances to feed D with more training data. The stopping criterion happens when the critic cannot distinguish between original and synthetic samples; the output of

---

[1] x denotes original (positive) instances, x' denotes synthetic (negative) instances, z is a random Gaussian noise. $D(x)$ and $D(x')$ are discriminator's outputs for original and synthetic instances respectively, and $G(z)$ is the generator's output.

the critic for negative samples is as high as for positive samples, making WD very close to 0. When GEG training is complete, an unlabeled synthetic dataset will be available. For all data belonging to this set, a class label is automatically added due to the way GEG is trained (class-based data restriction). This greatly facilitates the process of identifying class label for the synthetic data by avoiding the additional steps of classifying the unlabeled instances with a semi-supervised or supervised approach.

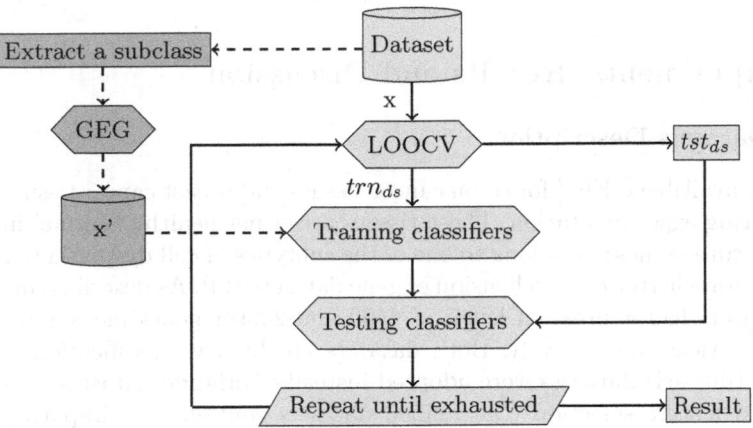

Fig. 3. Proposed LOOCV training diagram

3.2 Proposed LOOCV Training Diagram

Before starting the classification process, it is essential to pre-process the data. Since GED features (genes) have different ranges, this can influence the classification accuracy, resulting in misclassification of the data. The normalization step is integrated to avoid such a situation. Normalization is the process of scaling the values of numerical columns into a single range such as [0, 1] that will ensure that balanced weights are allocated to each feature. Therefore, normalization minimizes the training error, thus demonstrating the classification accuracy. The gene-level of each feature is normalized using Eq. 1:

$$new_{value} = \frac{actual_{value} - val_{min}}{val_{max} - val_{min}} \qquad (1)$$

where, val_{max} and val_{min} are the maximal and minimal original value of a given gene respectively. new_{value} is the normalized expression level. After normalization, all genes will be included between [0, 1]. As this is very small and sensitive data, the leave one out cross-validation (LOOCV) is suitable for validating model performance. LOOCV is a cross-validation technique where N data samples are sliced into two sets: a training set trn_{ds} containing $N-1$ samples and a test set tst_{ds} containing a single sample. A classifier was trained with

trn_{ds}, and the constructed model was tested with tst_{ds} by adopting any performance measurement (such as accuracy). The procedure was repeated N times, so that all instances are used once in trn_{ds} and once in tst_{ds}, but never again in both sets simultaneously. Figure 3 shows the traditional LOOCV training process (with the gray color). Additionally, it is suggested to improve the training of classifiers by adding more instances to trn_{ds} (purple color in Fig. 3). These instances are generated using GEG to deepen the learning methods and improve the classification performance only on the original data. This means that the proposed models are only tested with the original GED.

4   Experiments, Results and Discussion

4.1   Datasets Description

Publicly available GED [2] for colon cancer tissues and breast cancer tissues [4] was used during experimentation. These tissues are either healthy (normal histological structure tissues) or belong to one of the subtypes of cell dysplasia (cancerous tissues), for a better comprehension of gene datasets [14]. As described in Table 1, the datasets had expression levels of 6500 and 24,481 genes measured from 62 and 99 patients respectively. Both datasets are binary classification datasets. Refined (filtered) datasets were adopted instead of original datasets as an alternative to feature selection (the refined datasets contain only important genes that have a high impact factor for GED classification).

Table 1. Description of datasets used for the experiments

Dataset	Nbr of original genes	Nbr of used genes	Classes distribution
Colon cancer	6500	2000	40/22
Breast cancer	24481	4997	44/34

- **Colon Cancer Dataset** is composed of 40 different types of dysplasia colon tumors and 22 normal cell and histological structure of tissue samples from an Affymetrix oligonucleotide array of 6500 genes. A clustering algorithm revealed wide consistent patterns that suggest a high degree of organization underlying gene expression in these tissues [1]. The filtered dataset has 2000 gene expressions (instead of the original 6500 genes).
- **Breast Cancer Dataset** consists of 99 tumor samples from breast cancer patients with an initial gene count of 24,481 genes which decreased to 7,650 genes in [24]. Based on the estrogen receptor (ER), cancer tumors can be divided into two subgroups. Subjects with $ER+$ tumors have a better survival rate than those with $ER-$ because patients with $ER+$ tumors can benefit from anti-estrogens, such as tamoxifen. Since breast cancer is highly correlated

---

[2] DNA microarray data: https://homes.di.unimi.it/~valentini/DATA/MICROARRAY-DATA/.

with ER, only those genes that were linked to ER are considered, so about 5,000 genes were retained. Of the 99 patients, only 78 were considered. Of these 78 patients, 34 had a poor prognosis and were therefore labeled as $ER-$ patients, while 44 had a good prognosis, making them $ER+$ patients.

4.2   Used Methods

The baseline results (simple classifiers without data augmentation) were compared to classifiers using GEG as a data augmentation technique to study the effect of GEG on classification accuracy. The classifiers used are Support Vector Machine (SVM), K-nearest Neighbors (KNN) and Decision Trees (DT) [25]. Supervised learning methods were applied as cancer datasets are labeled. The Sklearn library [19] was used for the implementation of the algorithms, with default hyper-parameters; where, SVM with a linear kernel and $C = 1$, KNN with $K = 5$, and DT with $max\_depth = 2$. It's important to note that for all experiments performed with GEG, the synthetic samples were only used during the training stage as additional training data and not as test data. Table 2 provides details of the samples generated by class.

Table 2. Number of samples generated per class for each dataset using GEG

Dataset	Originals C1/C2	Generated C1/C2	Total samples
Colon cancer	40/22	10/28	100
Breast cancer	44/34	6/16	100

4.3   Evaluation Metric

The classification accuracy was adopted as an evaluation metric to assess the performance of GEG. Accuracy is a good metric in this case because for each dataset, every classifier is trained with a balanced dataset (original data + generated data by a data augmentation technique). Accuracy is defined as the ratio between the total number of correctly classified instances and the total number of instances. Formally, accuracy can be calculated using Eq. 2:

$$Accuracy = \frac{TP + TN}{Total} \tag{2}$$

where, True Positive (TP) (resp. True Negative (TN)) are correctly classified positive (resp. negative) samples by the classifier; False Positive (FP) (resp. False Negative (FN)) are incorrectly classified negative samples as positive (resp. positive samples as negative); and Total ($Total = TP + FP + FN + TN$) is the total number of samples.

In ML, the dataset is usually divided into a training set and a test set. The cross-validation was used to give a more accurate estimate of a model's performance. Figure 3 shows the LOOCV training used for the experiments, where the result of each iteration is the accuracy. The comparative study between the proposed models and the baseline was carried out using the following formula:

$$Model\ accuracy\ \% = 100 \times \frac{1}{N} \times \sum_{i=1}^{N} A_i \qquad (3)$$

where, N is the number of samples and A_i is the calculated accuracy of iteration i, multiplied by 100 to reflect model accuracy in %.

4.4   Classification Results

Several classification tests were conducted using different classifiers with different training approaches. In each case, only original data (OD) was used without data augmentation and synthetic data generation using the proposed GEG (proposed approach).

Table 3. Summary of experimental results based on classification accuracy

Dataset	Method used	SVM	KNN	DT
Colon	OD	87.10	82.26	75.81
cancer	GEG	**88.71**	**83.87**	**83.87**
Breast	OD	60.26	51.28	57.69
cancer	GEG	**75.64**	**57.69**	**73.08**

The results in bold in Table 3 show the used data augmentation, i.e., GEG improved the classification accuracy compared to the baseline (original data only). Based on the good results obtained, GEG can be considered as a promising data augmentation technique to improve the classification accuracy. The results for each dataset can be summarized as follows:

- **Colon Cancer Dataset:** despite using three classifiers, GEG produced the best results in terms of classification accuracy. The highest result was achieved using GEG as a data augmentation technique and SVM as a classifier. GEG was able to achieve 88.71% accuracy and 55 out 62 as the number of correctly classified instances. The SVM results are due to the linearity of the data, which means that the data classes are linearly separable, justifying the good initial results obtained without the usage of data augmentation. Therefore, the use of GEG gave a slightly higher result, meaning that the support vectors used by the SVM were adjusted (due to the use of synthetic GEG data).
- **Breast Cancer Dataset:** for this dataset, the combination of GEG and SVM was again a success. It improved the classification accuracy by more than 15% to reach the value of 75.64% and 59 out of 78 as a number of instances correctly classified compared to the baseline using only SVM. SVM correctly classified only 47 out of 78 instances. Similar to the colon cancer dataset, the

improvement in the results was due to the adjustment of the support vectors, and thus the optimal hyper-plane when synthetic GEG-generated data was used during the training process.

5 Conclusion

The collection of medical data for cancer detection is costly and difficult to obtain due to privacy constraints. Since the available data show a disproportionate ratio between the number of available instances and the number of features, and since GED analysis uses only a small number of available samples, this could lead to inappropriate classification results. To this end, the use of sophisticated data augmentation approaches such as GANs with appropriate hyper-parameters could be very beneficial in this application area. As an alternative to data collection, it is suggested to generate synthetic samples and increase the amount of training data.

However, these generated instances must be very consistent with the original instances to obtain good results. The results of the classification accuracy of the duality between GEG and simple supervised learning methods are promising. The application of GAN to other datasets and its comparison with other data augmentation techniques remains to be done. Nevertheless, as a first step, it can be noted that GAN can be successfully used to produce synthetic samples that are in harmony with real samples, and not only for images, videos and text, but also for gene expression data.

Acknowledgments. This work was supported by the Telekom Innovation Laboratories (T-Labs) and the Research and Development unit of Deutsche Telekom.
The authors would like to express their deepest gratitude towards **Tsegaye Misikir Tashu** for his advice, valuable feedback, proofreading and assistance in overcoming technical problems.
  Project no. ED_18-1-2019-0030 (Application domain specific highly reliable IT solutions subprogramme) has been implemented with the support provided from the National Research, Development and Innovation Fund of Hungary, financed under the Thematic Excellence Programme funding scheme.

References

1. Alon, U., et al.: Broad patterns of gene expression revealed by clustering analysis of tumor and normal colon tissues probed by oligonucleotide arrays. Proc. Nat. Acad. Sci. **96**(12), 6745–6750 (1999)
2. Antipov, G., Baccouche, M., Dugelay, J.L.: Face aging with conditional generative adversarial networks. In: 2017 IEEE International Conference on Image Processing (ICIP), pp. 2089–2093. IEEE (2017)
3. Berthelot, D., Milanfar, P., Goodfellow, I.: Creating high resolution images with a latent adversarial generator (2020). arXiv preprint arXiv:2003.02365
4. Buza, K.: Classification of gene expression data: a hubness-aware semi-supervised approach. Comput. Methods Prog. Biomed. **127**, 105–113 (2016)

5. Damian, A., Piciu, L., Turlea, S., Tapus, N.: Advanced customer activity prediction based on deep hierarchic encoder-decoders. In: 2019 22nd International Conference on Control Systems and Computer Science (CSCS), pp. 403–409. IEEE (2019)
6. Goodfellow, I., Pouget-Abadie, J., Mirza, M., Xu, B., Warde-Farley, D., Ozair, S.: Generative adversarial nets. In: Advances in Neural Information Processing Systems, pp. 2672–2680 (2014)
7. Im, D.J., Kim, C.D., Jiang, H., Memisevic, R.: Generating images with recurrent adversarial networks (2016). arXiv preprint arXiv:1602.05110
8. Isola, P., Zhu, J.Y., Zhou, T., Efros, A.A.: Image-to-image translation with conditional adversarial networks. In: Proceedings of the IEEE Conference on Computer Vision and Ppattern Recognition, pp. 1125–1134 (2017)
9. Krizhevsky, A., Sutskever, I., Hinton, G.E.: Imagenet classification with deep convolutional neural networks. In: Advances in Neural Information Processing Systems, pp. 1097–1105 (2012)
10. Ledig, C., et al.: Photo-realistic single image super-resolution using a generative adversarial network. In: Proceedings of the IEEE Conference on Computer Vision and Pattern Recognition, pp. 4681–4690 (2017)
11. Li, C., Alvarez-Melis, D., Xu, K., Jegelka, S., Sra, S.: Distributional adversarial networks (2017). arXiv preprint arXiv:1706.09549
12. Li, C., Wand, M.: Precomputed real-time texture synthesis with markovian generative adversarial networks. In: Leibe, B., Matas, J., Sebe, N., Welling, M. (eds.) ECCV 2016. LNCS, vol. 9907, pp. 702–716. Springer, Cham (2016). https://doi.org/10.1007/978-3-319-46487-9_43
13. Li, J., Monroe, W., Shi, T., Jean, S., Ritter, A., Jurafsky, D.: Adversarial learning for neural dialogue generation. In: EMNLP (2017)
14. Lin, W.J., Chen, J.J.: Class-imbalanced classifiers for high-dimensional data. Briefings Bioinf. **14**(1), 13–26 (2013)
15. Lu, Y., Kakillioglu, B., Velipasalar, S.: Autonomously and simultaneously refining deep neural network parameters by a bi-generative adversarial network aided genetic algorithm (2018). arXiv preprint arXiv:1809.10244
16. Marchesi, M.: Megapixel size image creation using generative adversarial networks (2017). arXiv preprint arXiv:1706.00082
17. Marouf, M., et al.: Realistic in silico generation and augmentation of single cell RNA-seq data using generative adversarial neural networks. bioRxiv, p. 390153 (2018)
18. Mirza, M., Osindero, S.: Conditional generative adversarial nets (2014). arXiv preprint arXiv:1411.1784
19. Pedregosa, F., et al.: Scikit-learn: machine learning in Python. J. Mach. Learn. Res. **12**, 2825–2830 (2011)
20. Radford, A., Metz, L., Chintala, S.: Unsupervised representation learning with deep convolutional generative adversarial networks (2015). arXiv preprint arXiv:1511.06434
21. Reed, S., Akata, Z., Yan, X., Logeswaran, L., Schiele, B., Lee, H.: Generative adversarial text to image synthesis (2016). arXiv preprint arXiv:1605.05396
22. Shang, C., Palmer, A., Sun, J., Chen, K.S., Lu, J., Bi, J.: Vigan: missing view imputation with generative adversarial networks. In: 2017 IEEE International Conference on Big Data (Big Data), pp. 766–775. IEEE (2017)
23. Smith, E.J., Meger, D.: Improved adversarial systems for 3d object generation and reconstruction. In: Conference on Robot Learning, pp. 87–96 (2017)

24. Sotiriou, C., et al.: Breast cancer classification and prognosis based on gene expression profiles from a population-based study. Proc. Nat. Acad. Sci. **100**(18), 10393–10398 (2003)
25. Taan, A., Farou, Z.: Supervised learning methods for skin segmentation classification (2020). https://doi.org/10.13140/RG.2.2.12444.51843/2
26. Vondrick, C., Pirsiavash, H., Torralba, A.: Generating videos with scene dynamics. In: Advances in Neural Information Processing Systems, pp. 613–621 (2016)
27. Wang, H., Qin, Z., Wan, T.: Text generation based on generative adversarial nets with latent variables. In: Phung, D., Tseng, V.S., Webb, G.I., Ho, B., Ganji, M., Rashidi, Lida (eds.) PAKDD 2018. LNCS (LNAI), vol. 10938, pp. 92–103. Springer, Cham (2018). https://doi.org/10.1007/978-3-319-93037-4_8
28. Wang, Z., She, Q., Ward, T.E.: Generative adversarial networks: a survey and taxonomy (2019). arXiv preprint arXiv:1906.01529
29. Zhang, H.: Generative Adversarial Networks for Image Synthesis. Ph.D. thesis, Rutgers The State University of New Jersey-New Brunswick and University of Medicine and Dentistry of New Jersey (2019)
30. Zhang, Y., Bai, Y., Ding, M., Ghanem, B.: Multi-task generative adversarial network for detecting small objects in the wild. Int. J. Comput. Vis. **128**, 1–19 (2020). https://doi.org/10.1007/s11263-020-01301-6
31. Zhu, J.Y., Park, T., Isola, P., Efros, A.A.: Unpaired image-to-image translation using cycle-consistent adversarial networks. In: Proceedings of the IEEE International Conference on Computer Vision, pp. 2223–2232 (2017)

Stabilization of Dataset Matrix Form for Classification Dataset Generation and Algorithm Selection

Ilya Sahipov, Alexey Zabashta$^{(\boxtimes)}$, and Andrey Filchenkov

Machine Learning Lab, ITMO University, Kronverksky Prospekt 49, bldg. A,
St. Petersburg 197101, Russia
{azabashta,afilchenkov}@itmo.ru

Abstract. Datasets for the classification task are usually encoded by a matrix of numbers, the order of rows and columns does not matter. Swapping any two objects or features in it does not change the hidden target function and performance of the machine learning algorithms train of the dataset. However, in the dataset generation problem solution such symmetry is an obstacle. In this paper, we study several methods of the inverse transformation of classification dataset aiming to break the symmetry. We experimented with it in the meta-learning problems of datasets generation and algorithm selection which were solved by conditional generative adversarial nets with convolutional networks.

Keywords: Machine learning · Meta-learning · Generative adversarial nets · Dataset generation

1 Introduction

Meta-learning is a field of machine learning, objects of which are datasets [3,5]. One of the problems solved by meta-learning is the algorithm selection problem [2,16]. Predictions are made on how different algorithms will work with different datasets by treating datasets as meta-objects. These meta-objects are described by meta-features. This description can be used for detecting the dependence of the machine learning algorithms performance on datasets.

Once treating datasets as meta-objects, the problem of increasing the size and variety of such collection arises, which consequently lead to the problem of dataset generation that is to generate datasets with specific properties [21].

Typically, datasets for a classification task are encoded by a matrix of numbers. The order of rows and columns in such a matrix does not matter. One of the most efficient algorithms allowing to work with matrices as objects are convolutional neural networks (CNNs) [8]. Initially, they were designed for image processing [9]. Despite the fact that images are often encoded as a matrix of numbers, as are datasets, if you rearrange the rows or columns in it, you will obtain a different image (actually you will corrupt it). This is due to the convolution networks are aimed to work with a signal localized in a set of closely connected pixels. The application of CNNs for processing dataset matrix form requires additional steps for reducing this variation in equivalent representations of the same dataset.

In this paper, we study methods that uniform datasets for the classification task by rearranging objects and features in it. Such an operation can be applied to any order-dependent algorithm. It will also create patterns in the matrix of datasets that will simplify the work of convolution networks with them.

The rest of the paper is organized as follows. In Sect. 2 the problem of dataset symmetry and several stabilization methods to solve it are described. Section 3 contains a description and results of experiments with different stabilization methods.

2   Dataset Stabilization

2.1   Dataset Symmetry

In this paper, we study the symmetry of datasets for the classification task with respect to permutations of objects or features. Conventional machine learning algorithms are independent on the order of objects or features. Although in practice, algorithms may work slightly differently depending on the order of the features [17]. For the algorithms that may depend on the order of the objects such as stochastic gradient descend, it is recommended to shuffle that. Thus, since traditional machine learning algorithms are invariant to permutation of the order of objects and features in the dataset, meta-learning algorithms must also take into account this symmetry.

A dataset is considered as a set of two-dimensional matrices, objects of one class will belong to one matrix. In this representation, there is no special column representing the class feature, but not all possible transposition within each matrix can be applied. Only transpositions of objects within the matrix of one class or synchronously transposition of features in all matrices are allowed. An example of transposition is shown in Fig. 1.

Let us define **stabilization** as the process of transforming a dataset into a canonical form. Formally, each dataset creates an equivalence class [11] with respect to permutations of rows and columns, the task of stabilization is choosing a representative member of the equivalence class. The representative member will be called **stable**.

2.2   Dataset Normalization

Classification algorithm performance may depend on the statistics of the features. For example, when the nearest neighbor algorithm calculates distances, it is sensitive to the variance of features, because features with large variance are more likely to affect the resulting distance. To solve this problem, normalization methods are used [1]. The most popular normalization methods are: reducing the minimum value of each feature to zero, and the maximum to unity; reducing

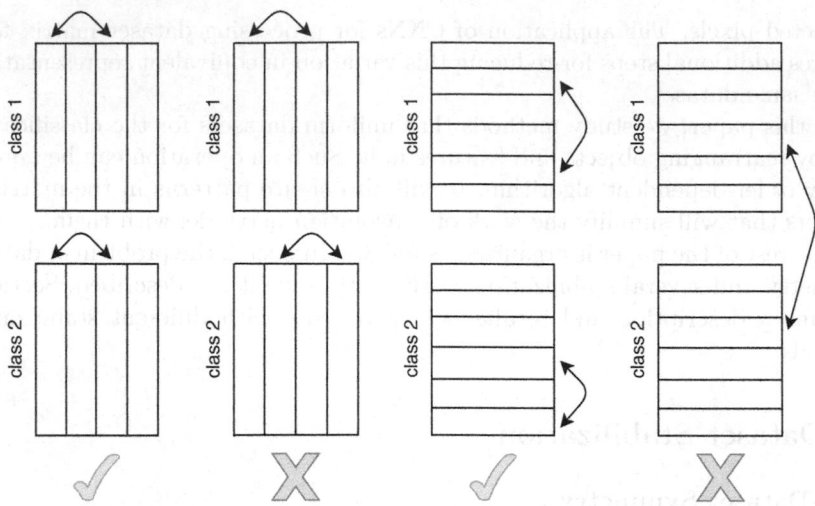

Fig. 1. Example of two correct (first and third) and two incorrect (second and fourth) transpositions of rows (objects) and columns (features) in a dataset.

the mean value of each feature to zero, and the variance to unity. In this paper, all stabilization methods will use the second normalization method (min-max normalization) as the initial step.

The basic method (**BASE**) does not change the order of rows (objects) or columns (features) and perform only normalization.

2.3   Diagonalization

The diagonalization method (**DIAG**) was proposed in paper [7]. It is based on column maximum potential, which is defined as $v_h\left(\xi_j^t\right) = Top_h\left(\xi_j^t\right)/h - (Sum\left(\xi_j^t\right) - Top_h(\xi_j^t))/(n_t - h)$ where ξ_j^t is the values in the j-th column of submatrix x^t (objects with the same class t), $Top_h(\xi_j^t)$ is the sum of h largest elements in the column ξ_j^t, and $Sum(\xi_j^t)$ is the sum of all elements in the column ξ_j^t.

The stabilization algorithm built on the described column maximum potential is as follows:

1. Mark all rows and columns as unfixed.
2. Evaluate maximum potential for each unfixed j-th column as $v_h\left(\xi_j\right) = \max_t v_h\left(\xi_j^t\right)$.
3. Move the column that has reached its maximum potential to the end of the series of fixed columns (or to beginning if there are unfixed columns) and mark it as fixed.
4. Sort all unfixed rows in descending order by the values of the moved column.

5. Mark the first $\left\lfloor \frac{n_t}{m} \right\rfloor$ unfixed rows as fixed in each submatrix t, where n_t is the number of objects and m is the number of objects.
6. Return to the second step if there are still unfixed columns left, otherwise stop.

2.4   Minimize Difference

The method of minimizing the difference of neighboring elements is based on the idea that a dataset is stable if its neighboring elements are most similar to each other. The similarity criterion for neighboring elements can be expressed as: $\phi(x, r, c) = \sum_{i \in [1;n)} \sum_{j \in [1;m]} (x_{r[i],c[j]} - x_{r[i+1],c[j]})^2 + \sum_{i \in [1;n]} \sum_{j \in [1;m)} (x_{r[i],c[j]} - x_{r[i],c[j+1]})^2$, where x is the submatrix with objects of the same class, n is the number of objects in the submatrix, m is the number of features, r is the row permutation in the submatrix, and c is the permutation of columns. Thus, the task of the algorithm is to find the minimum $\min_{r_1,r_2,\ldots,r_k,c} \phi(x^1, r_1, c) + \ldots + \phi(x^k, r_k, c)$ among all possible permutations of $r_1, r_2, \ldots, r_k, c$, where x^t is a submatrix with objects of the class t.

Since the sum itself is invariant to permutation, $\phi(x, r, c)$ can be simplified to: $\sum_{i \in [1;n)} \sum_{j \in [1;m]} (x_{r[i],j} - x_{r[i+1],j})^2 + \sum_{i \in [1;n]} \sum_{j \in [1;m)} (x_{i,c[j]} - x_{i,c[j+1]})^2$. As seen, the left side of the sum depends only on the permutation of the rows r, and the right side on the permutation of the columns c. Therefore, the problem can be reduced to several independent solutions of the Travelling Salesperson problem (TSP), in which the path is minimized along objects, the distance between which is calculated through the difference of features. Similarly, the problem can be solved for features.

It is proposed to solve this problem with the following heuristics:

– Greedy Search (**MING**): each time selects the closest object to the current among the set of objects that have not yet been considered. A vector consisting entirely of zeros is used as the first object [12].
– Search by Hill climbing (**MINH**): sequential search of transpositions of two elements for the current permutation, replacing the current solution with a more optimal one. The search begins with a random permutation. Since only the relative order is important, the cyclic shift of the permutation with the minimum pairwise distance is selected as the solution [4].

2.5   Sort by Correlation

The method of sorting by correlation [18] with the class is based on some metric of the relationship between the feature values and the class. As such a metric in this paper we use:

– The absolute value of Pearson correlation coefficient (**CORP**). For two vectors a and b it is equal to $corr(a, b) = \frac{\sum (a_i - \bar{a}) \cdot (b_i - \bar{b})}{\sqrt{\sum (a_i - \bar{a})^2 \cdot \sum (b_i - \bar{b})^2}}$.

– The absolute value of Spearman's rank correlation coefficient (**CORS**). It is equal to the Pearson correlation coefficient calculated from the rank vectors of the values of the corresponding vectors.

Theoretically, it can be used any relationship metric for the feature selection algorithm.

The stabilization algorithm built on the described metric is as follows:

1. For each j-th feature, the value ρ_j of the calculated metric of its relationship with the class is remembered.
2. The coefficient ω_i is calculated for the vector of each object x_i according to the formula $\omega_i = \sum x_{i,j} \cdot \rho_j$.
3. All objects are ordered inside the corresponding submatrices of the same class in ascending order ω_i.
4. All items are sorted in ascending order by ρ_j.

The key difference between this method and the previous ones is that information about the class is used to organize objects and features. Therefore, this method cannot be directly applied to the datasets for the clustering task. However, it can be modified if you use some row or column statistics (average, maximum, etc.) as the values to be sorted.

3   Experiments

3.1   Dataset Description

For the experiments, 976 real-world datasets were collected from the OpenML [19]. They were processed as follows:

– If the dataset contained more than 2 classes, it was divided into all possible combinations of datasets with 2 classes.
– If the dataset contained less than 16 features or less than 64 objects of the same class, it was eliminated.
– If the data set contained more than 16 features, then only features selected by *SelectFromModel* algorithm from *scikit-learn* [15] library remained in it.
– If the dataset contained more than 64 objects of the same class, then only 64 randomly selected objects of that class were left.

As a result, 9911 data sets were obtained. They were divided into training and test sets. There were 8000 objects in the training set and 1911 objects in the test set. For each new experiment with each described stabilization method all datasets in both training and test set were stabilized by it.

3.2   Time Comparison

The Table 1 shows the theoretical and actual running time of different stabilization methods on the described collection of real-world datasets.

Table 1. Comparison of theoretical and actual running time of stabilization methods. For theoretical time n is a number of objects, and m is a number of features, and $C = 20000$ is a maximum number of iteration. The Intel Core i5-5350U CPU was used for the wall-clock time calculation.

Stabilization method	Theoretical time	Wall-clock time (sec.)
BASE	$n \cdot m$	10
DIAG	$n \cdot m^2$	56
CORP	$n \cdot m + n \cdot \log n + m \cdot \log m$	40
CORS	$n \cdot m + n \cdot \log n + m \cdot \log m$	44
TSPG	$n \cdot m^2 + n^2 \cdot m$	94
TSPH	$n \cdot m \cdot C$	3118

3.3   LM-GAN Description

The proposed stabilization approaches were tested with LM-GAN architecture [7] because it simultaneously solves two tasks of meta-learning: the algorithm prediction problem and the dataset generation problem.

LM-GAN is a modification of Conditional Generative Adversarial Nets [10], which is a modification of the traditional Generative Adversarial Nets (GAN) [6]. It is implemented on *Pytorch* library [14]. Its generator generates a dataset given concatenate vector of noise and meta-features. Also the same vector of meta-features used in the discriminator that predicts a source of a given dataset (real or generated) and the best classifier for it. In turn, the generator tries to fool the discriminator by generating a dataset that would look like a real one.

LM-GAN is based on tabular dataset representation [13,20], so a convolutional network is used as the discriminator. As mentioned earlier, the convolutions do not take into account the symmetry of the input dataset. Therefore, the stabilization process should greatly affect LM-GAN work.

LM-GAN was trained as in the original paper. Each experiment was repeated 3 times. Confidence intervals were calculated from experiments.

3.4   The Algorithm Prediction Task

For the algorithm prediction task, the discriminator from LM-GAN was trained to predict the best classification algorithm for a given dataset from the following set: *kNN, Naïve Bayes, Decision tree*. Truly the best algorithm was determined by the 3-fold cross-validation F-score. All random states were fixed to 0, and the rest of the parameters were set by default.

The relative performance of the classification algorithms was encoded by a vector of ones and zeros. '1' indicates that the corresponding algorithm was the best. A vector could contain several '1' at once if several algorithms turned out to be the best simultaneously.

The mean squared error (MSE) was used as a loss function for training. Accuracy was used as the performance metric for testing. To determine the class, an arg-maximum was extracted from the predicted vector and it was checked that the corresponding position in the true vector contains '1'.

3.5   The Dataset Generation Task

For the dataset generation task, the generator from LM-GAN was tasted by generating datasets with meta-features that were taken from datasets of the test set. This ensured that the dataset with the required meta-features exists.

The generator from LM-GAN was not directly trained to generation datasets by meta-features. But since the discriminator also received meta-features as input in addition to the dataset, the generator needed to generate a dataset that would correspond to the given meta-features to fool the discriminator. The generator did not see stabilized datasets, but it was trained with respect to the derivative of the discriminator, which used stabilized datasets. Thus, stabilization should also affect the generator performance.

After the generator generates a new dataset for a given target vector of meta-features, a new meta-features were extracted from this new dataset. To compare the loss of different approaches to stabilization, MSE was used between the target meta-features and the obtained meta-features.

4   Results

4.1   The Algorithm Prediction Result

Figure 2 shows a graph of dependency of the algorithm prediction accuracy of the discriminator from LM-GAN on the number of training epochs for the different stabilization approaches.

Table 2 shows the results of comparing the discriminator from LM-GAN with standard meta-classifications algorithms that received information about the dataset only from their meta-features.

As can be seen from the results, similar methods showed similar performance. Correlation-based stabilization methods show better performance, but only after 35-th epoch. The LM-GAN discriminator that used full information about a dataset along with meta-features outperforms methods that used only meta-features.

4.2   The Dataset Generation Result

Figure 3 shows a graph of dependency of the dataset generation loss of the generator from LM-GAN on the number of training epochs for the different stabilization approaches.

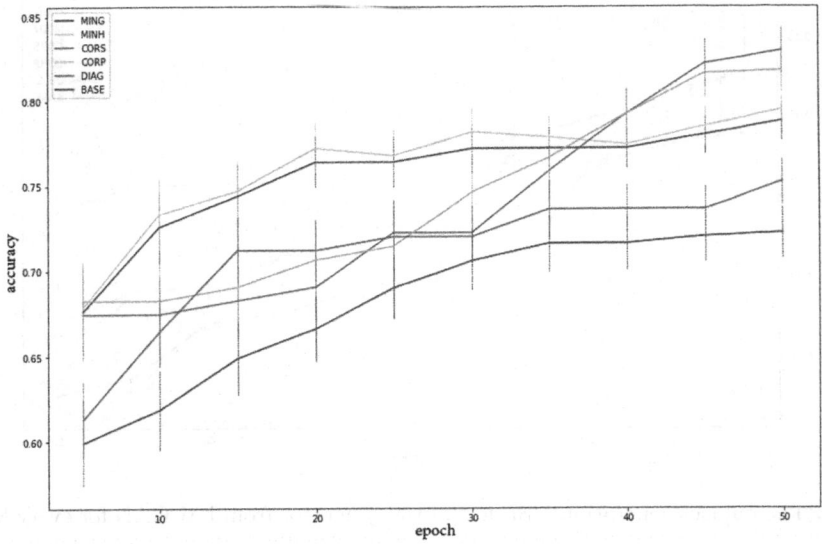

Fig. 2. The algorithm prediction accuracy of the discriminator from LM-GAN for every fifth epoch.

Table 2. Comparison of algorithm prediction accuracy for different meta-classifiers: *LMGAN-D* is the discriminator from LM-GAN, which was trained on differently stabilized datasets, *DT* is a Decision Tree, *kNN* is a k-Nearest Neighbors algorithm, *MLP* is a Multilayer Perceptron.

Meta-classifier	Average accuracy
LMGAN-D BASE	0.722 ± 0.015
LMGAN-D DIAG	0.752 ± 0.013
LMGAN-D CORP	0.817 ± 0.013
LMGAN-D CORS	$\mathbf{0.829 \pm 0.013}$
LMGAN-D TSPH	0.794 ± 0.008
LMGAN-D TSPG	0.787 ± 0.012
DT	0.650 ± 0.000
kNN	0.674 ± 0.000
MLP	0.694 ± 0.001

As can be seen, stabilization methods the are based on the minimization of differences are the best for the data generation task, in contrast to the previous one. The performance of the correlation-based stabilization methods was similar to the diagonal maximization method.

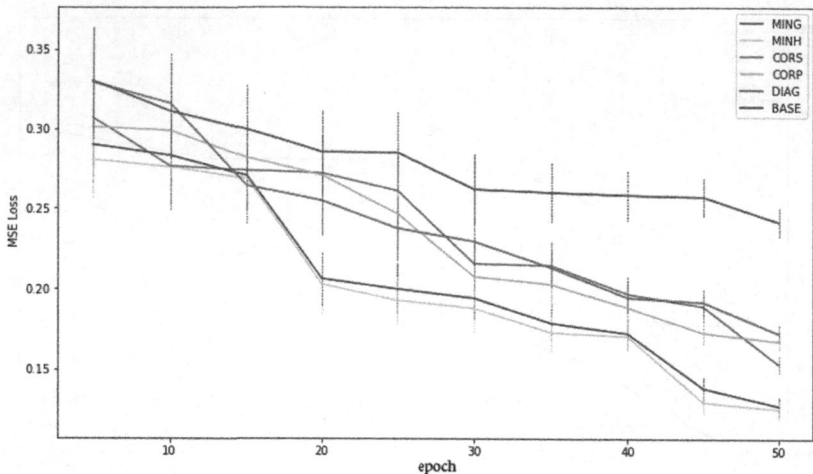

Fig. 3. The dataset generation MSE loss of the generator from LM-GAN for every fifth epoch.

5   Conclusion

In this paper, we study several stabilization methods that unified datasets for a classification task by rearranging their row (objects) and columns (features). We experimentally test these methods with LM-GAN architecture in two meta-learning tasks, namely algorithm prediction and dataset generation. The best performance in the algorithm prediction task showed the discriminator from LM-GAN which was trained on stabilized datasets by correlation-based methods. The stabilization methods based on the minimization of differences were best in the dataset generation task.

Thus, it was shown that the use of stabilization methods for datasets can improve the performance of algorithms that do not take into account the symmetry of datasets with respect to permutations of their rows and columns.

Acknowledgments. The research was financially supported by The Government of the Russian Federation, Grant 08-08, the Russian Science Foundation, Agreement 17-71-30029, and RFBR, project number 19-37-90165.

References

1. Al Shalabi, L., Shaaban, Z., Kasasbeh, B.: Data mining: a preprocessing engine. J. Comput. Sci. **2**(9), 735–739 (2006)
2. Alexandros, K., Melanie, H.: Model selection via meta-learning: a comparative study. Int. J. Artif. Intell. Tools **10**(04), 525–554 (2001)
3. Brazdil, P., Carrier, C.G., Soares, C., Vilalta, R.: Metalearning: Applications to Data Mining. Springer Science & Business Media, New York (2008)

4. Damghanijazi, E., Mazidi, A.: Meta-heuristic approaches for solving travelling salesman problem. Int. J. Adv. Res. Comput. Sci. **8**(5), 19 (2017)
5. Giraud-Carrier, C.: Metalearning-a tutorial. In: Tutorial at the 7th International Conference on Machine Learning and Applications (ICMLA), San Diego, California, USA (2008)
6. Goodfellow, I., et al.: Generative adversarial nets. In: Advances in Neural Information Processing Systems, pp. 2672–2680 (2014)
7. Kachalsky, I., Zabashta, A., Filchenkov, A., Korneev, G.: Generating datasets for classification task and predicting best classifiers with conditional generative adversarial networks. In: Proceedings of the 2019 3rd International Conference on Advances in Artificial Intelligence, pp. 97–101 (2019)
8. Khan, A., Sohail, A., Zahoora, U., Qureshi, A.S.: A survey of the recent architectures of deep convolutional neural networks. Artif. Intell. Rev. **53**(8), 5455–5516 (2020). https://doi.org/10.1007/s10462-020-09825-6
9. LeCun, Y., Bengio, Y., et al.: Convolutional networks for images, speech, and time series. Handb. Brain Theor. Neural Networks **3361**(10), 1995 (1995)
10. Mirza, M., Osindero, S.: Conditional generative adversarial nets (2014). arXiv preprint arXiv:1411.1784
11. Morash, R.P.: Bridge to Abstract Mathematics: Mathematical Proof and Structures. McGraw-Hill College, New York (1991)
12. Nilsson, C.: Heuristics for the traveling salesman problem. Linkoping Univ. **38**, 00085–9 (2003)
13. Park, N., Mohammadi, M., Gorde, K., Jajodia, S., Park, H., Kim, Y.: Data synthesis based on generative adversarial networks. Proceedings of the VLDB Endowment, vol. 11, p. 10 (2018)
14. Paszke, A., et al.: Pytorch: an imperative style, high-performance deep learning library. In: Advances in Neural Information Processing Systems, pp. 8026–8037 (2019)
15. Pedregosa, F., et al.: Scikit-learn: machine learning in python. J. Mach. Learn. Res. **12**, 2825–2830 (2011)
16. Rice, J.R.: The algorithm selection problem. Adv. Comput. **15**, 65–118 (1976). Elsevier
17. Tallón-Ballesteros, A.J., Fong, S., Leal-Díaz, R.: Does the order of attributes play an important role in classification? In: Pérez García, H., Sánchez González, L., Castejón Limas, M., Quintián Pardo, H., Corchado Rodríguez, E. (eds.) HAIS 2019. LNCS (LNAI), vol. 11734, pp. 370–380. Springer, Cham (2019). https://doi.org/10.1007/978-3-030-29859-3_32
18. Uurtio, V., Monteiro, J.M., Kandola, J., Shawe-Taylor, J., Fernandez-Reyes, D., Rousu, J.: A tutorial on canonical correlation methods. ACM Comput. Surv. (CSUR) **50**(6), 1–33 (2017)
19. Vanschoren, J., van Rijn, J.N., Bischl, B., Torgo, L.: Openml: networked science in machine learning. SIGKDD Explor. **15**(2), 49–60 (2013)
20. Xu, L., Skoularidou, M., Cuesta-Infante, A., Veeramachaneni, K.: Modeling tabular data using conditional GAN. In: Advances in Neural Information Processing Systems, pp. 7335–7345 (2019)
21. Zabashta, A., Filchenkov, A.: Active dataset generation for meta-learning system quality improvement. In: Yin, H., Camacho, D., Tino, P., Tallón-Ballesteros, A.J., Menezes, R., Allmendinger, R. (eds.) IDEAL 2019. LNCS, vol. 11871, pp. 394–401. Springer, Cham (2019). https://doi.org/10.1007/978-3-030-33607-3_43

Special Session on Optimization and Machine Learning for Industry 4.0

Distributed Coordination
of Heterogeneous Robotic Swarms
Using Stochastic Diffusion Search

Eneko Osaba[1]([✉]), Javier Del Ser[1,2], Xabier Jubeto[2], Andrés Iglesias[3,4],
Iztok Fister Jr.[5], Akemi Gálvez[3,4], and Iztok Fister[5]

[1] TECNALIA, Basque Research & Technology Alliance (BRTA), 48160 Derio, Spain
eneko.osaba@tecnalia.com
[2] University of the Basque Country (UPV/EHU), 48013 Bilbao, Spain
[3] Universidad de Cantabria, 39005 Santander, Spain
[4] Toho University, Funabashi, Japan
[5] Maribor University, Maribor, Slovenia

Abstract. The term Swarm Robotics collectively refers to a population of robotic devices that efficiently undertakes diverse tasks in a collaborative way by virtue of computational intelligence techniques. This
paradigm has given rise to a profitable stream of contributions in recent
years, all sharing a clear consensus on the performance benefits derived
from the increased exploration capabilities offered by Swarm Robotics.
This manuscript falls within this topic: specifically, it gravitates on an
heterogeneous Swarm Robotics system that relies on Stochastic Diffusion
Search (SDS) as the coordination heuristics for the exploration, location
and delimitation of areas scattered over the area in which robots are
deployed. The swarm is composed by agents of diverse kind, which can
be ground robots or flying devices. These agents communicate to each
other and cooperate towards the accomplishment of the exploration tasks
comprising the mission of the overall swarm. Furthermore, maps contain
several obstacles and dangers, implying that in order to enter a specific
area, robots should meet certain conditions. Experiments are conducted
over three different maps and three implemented solving approaches.
Conclusions are drawn from the obtained results, confirming that i) SDS
allows for a lightweight, heuristic mechanism for the coordination of the
robots; and ii) the most efficient swarming approach is the one comprising a heterogeneity of ground and aerial robots.

Keywords: Swarm Robotics · Stochastic Diffusion Search · Swarm
Intelligence · Unmanned Aerial Vehicles · Robotics

1 Introduction

Since its inception, the so-called Swarm Intelligence paradigm (SI [1]) has
attracted a great deal of attention from the community working on Operations

© Springer Nature Switzerland AG 2020
C. Analide et al. (Eds.): IDEAL 2020, LNCS 12490, pp. 79–91, 2020.
https://doi.org/10.1007/978-3-030-62365-4_8

Research and Computational Intelligence. In a nutshell, SI harnesses complex collective behaviors conducted by decentralized and self-organized systems, usually comprised by a spatially scattered population of agents or individuals, in order to efficiently undertake a number of (possibly interrelated) tasks. These individuals interact with each other and with the environment through the application of simple and heterogeneous local behavioral rules, leading to an inherently effective, robust and flexible swarm, capable of coping with complex environments. From the algorithmic perspective, this paradigm widely embraces the efficiency of the behavioral patterns observed in nature, which are often emulated and implemented in the robots composing the swarm. As such, animals' behaviors, social and political behaviors, or physical processes are phenomena inspiring coordination and actuation mechanisms in SI [2].

A myriad of successful applications has been reported in the literature that rely, to an extent, on SI concepts. Among them, a particular research area protrudes in recent years: Swarm Robotics (SR, [3]). Specifically, SR denotes the application of SI approaches to environments in which agents represent robotic devices. Thus, the main objective of SR is to analyze how a swarm composed by simple robotic tools can communicate, coordinate and collectively complete complex missions/tasks, which would be not possible to accomplish otherwise [4]. As can be inferred from the related literature, approaches under the umbrella of SR have been successfully used for a wide variety of real-world problems, including agricultural foraging and seeding [5] or supervision [6]. Arguably, purposes for which SR can unleash their full potential relate mostly to exploration: localization [7], disaster rescue missions [8], or scenery mapping [9,10].

The importance related to SR for exploration tasks stems from the fact that most environments represent an impediment to the realization by humans due to their inaccessibility or the risk of the tasks therein defined. This is why the ultimate goal of SR is to prevent people from mapping vast spaces, or places where natural disasters, large landslides, radioactive leaks, or other dangerous events have occurred. In the recent literature, several works have focused on the application of SR to exploration scenarios. Alfeo et al. [11], for example, tackle the problem of discovering static hidden targets in non-homogeneous environments by using a swarm of small Unmanned Aerial Vehicles (UAVs). Innocente and Grasso show in [12] that SR can collaboratively battle against the spread of wildfires. Another extreme scenario is tackled in [13], which relates to different levels of radioactive and chemical leakage from drums in a nuclear storage facility. Furthermore, in [14] a Bat Algorithm is introduced for guiding a swarm of small robots in their exploration of a closed environment towards reaching a fixed objective location. In [10], exploratory SR are also utilized along with *trophallaxis* (i.e., energy sharing) as one of the key ideas for efficient SR scouting. A similar work is published in [9], where the mission to be performed by the swarm is the 3D mapping of the environment over which it is deployed.

This work takes a step over the above state-of-the-art in what refers to the algorithmic solution utilized for SR coordination. Specifically, we herein present an alternative heuristic that uses, as inspiration, the Stochastic Diffusion Search heuristic (SDS [15]). SDS was introduced in 1989 as a population-based pattern matching method. Interestingly, SDS can be used to address optimization problems where the objective can be decomposed into different components that can be evaluated separately [16]. This is the main motivation for the adoption of SDS as the heuristic engine for SR-based exploration missions, in which each agent in the swarm may have a hypothesis about each component of the objective. Throughout cooperation and communication, the entire swarm arrives at the most optimal final hypothesis, namely, the completion of the exploration mission. Despite the upsurge of literature on SR for exploration purposes, to the best of the authors' knowledge the application of SDS to this end has not been addressed before. With the intention of accommodating this manuscript to the extension limits, we refer the reader to [17] for details on the mathematical foundations of SDS, and to [18] for recent applications leveraging this algorithm.

As will be later detailed, the mission under consideration in this work is the exploration, location and delimitation of areas scattered over a large area of land. Several obstacles and dangerous areas will be established so as to enforce certain conditions on robots to enter them safely. Three different coordination schemes have been developed to decide the waypoints that should be followed by robots at every point in time. The first one corresponds to random decisions, and it serves as baseline for comparing the other two alternatives. The second approach, labeled as GROUND-SDS, performs exploration tasks using only ground robots coordinated with SDS heuristics. Finally, the third designed scheme is coined as UAV-SDS, and assumes an heterogeneous swarm composed by UAVs and ground robots. The latter are in charge of performing the exploration tasks, whereas UAVs are deployed to coordinate communications between nearby ground robots. We compare the performance of these three SR schemes over three different synthetically generated scenarios, each with a different number of areas to locate and explore.

The rest of this paper is structured as follows. First, Sect. 2 delves on the formulation of the problem under study, whereas Sect. 3 describes the proposed coordination strategies. Section 4 presents the simulation setup and discusses the obtained results. Finally, Sect. 5 concludes the paper and outlines some research lines rooted on our reported findings.

2   System Model and Problem Statement

As has been mentioned in the introduction, the problem under consideration models an exploration mission to be autonomously performed by a swarm of N robots represented by the set $\mathcal{R} = \{R_1, R_2, \ldots, R_N\}$, each of a specific type $T_n^R \in \{1, 2, \ldots, \tau^R\}$ (with τ^R denoting the number of different robot types in the swarm). Without loss of generality we assume that the scenario to be explored by the swarm is square with dimensions $D \times D$ [squared units of distance], which comprises M delimited areas scattered all over its surface that have to be explored by the swarm. Such areas, which we hereafter denote as $\mathcal{A} = \{A_1, A_2, \ldots, A_M\}$, are characterized by a set $T_m^A \subseteq \{1, \ldots, \tau^R\}$ indicating which robot types are admitted for the exploration of the area at hand. In other words, types in $\mathcal{T}_m^A$ dictate the requirements that a robot R_n should meet in order to be qualified to explore area A_m. This qualification of robots to examine a certain area simulates real-world scenarios in which it is often the case that robots have to be equipped with sensors, actuators and other devices aimed at handling radioactive leaks, probing contaminated lakes or circumvent physical obstacles. Similarly, robots are also categorized into types towards modeling which areas they can analyze.

In order to record the information gathered during the mission in a computationally efficient manner, each robot is endowed with a grid (*scan map*) that corresponds to a section of the map to be discovered. Using this scan map, agents are able to simplify the data they capture. This process is particularly important, since devices have a limited computing capacity, which makes it almost impossible to deal with all the data available in the environment.

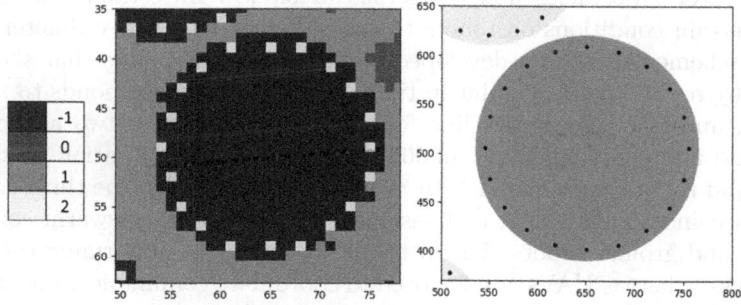

Fig. 1. Example of a grid (left) constructed from a synthetic map (right).

Figure 1 shows an area of the map (right) that is translated to an exploration grid (left). We can see in this plot that robots translate the map into numerical values that refer to the content of the map. This numbering is as follows: −1 represents a zone, 0 indicates that this part has not yet been examined, 1 refers to an explored part which does not corresponds to an objective area, and 2 are the points that delimit the area. Following this nomenclature, the grid of all agents is initialized to 0 as per their absolutely uncertainty about what can be found in the scenario to be explored.

For properly interacting with the environment, agents implementing the intelligent functionalities of robotic devices must incorporate observation and detection mechanisms. A realistic assumption is that sensors installed in robots cannot observe the complete map, but rather a vision cone with maximum range d and angle α centered on its line of sight. Consequently, agents are only able to see what falls within their cone of vision, including the areas to be explored. We depict in Fig. 2 the process performed to determine the vision cone of each agent. First, the vision cone is created using the above-mentioned angle α and a d range. After that, and depending on what is in that place on the map, one value or another is saved in the grid (−1 or 1). This process is conducted after a preliminary check about the restrictions of the area (T_m^A) and the type of robot (T_n^R). If $T_n^R \in T_m^A$, the agent overwrites the square corresponding to its location in its scan map, and continues exploring.

Fig. 2. Determination of the vision cone of a robot at a certain instant in time: cone conformation (left); identification of elements in line of sight (center); qualification compliance check and boundary points determination (right).

With all this, the objective of the mission is to delimit all areas of the map under consideration, i.e. to determine the boundaries of these areas without assuming any prior knowledge of their location when deploying the swarm. Accordingly, a *task* of this problem in the context of swarm robotics is to locate all points that enclose an area A_m. In addition, we seek coordination policies within the robotic swarm that allow completing this mission as quickly as possible. Mathematically, we denote as $\{\mathbf{a}_m^p\}_{p=1}^{P_m}$ the set of boundary points of area A_m, with $\mathbf{a}_m^p = (a_m^{p,x}, a_m^{p,y})$. Furthermore, we let $t(\mathbf{a}_m^p)$ denote the time – relative to the start of the mission – at which point a_m^p has been first observed by a qualified robot R_n fulfilling $T_n^R \in T_m^A$. The problem under consideration can be formulated as:

$$\min_{\mathbf{W}_1,\ldots,\mathbf{W}_N} \left[\max \left\{ t(\mathbf{a}_1^1), ..., t(\mathbf{a}_1^{P_1}), t(\mathbf{a}_2^1), ..., t(\mathbf{a}_2^{P_2}), \ldots, t(\mathbf{a}_M^1), ..., t(\mathbf{a}_M^{P_M}) \right\} \right], \quad (1)$$

where $t(\mathbf{a}_m^p) = \infty$ if point $\mathbf{a}_m^p$ has not been visited by any robot, and $\mathbf{W}_n = \{\mathbf{w}_n^1, \mathbf{w}_n^2, \ldots, \mathbf{w}_n^{\beta_n}\}$ represents the β_n waypoints (i.e. locations over the map) established by the coordination mechanism for robot R_n during the mission. We note that optimization variables comprise both the number of waypoints β_n and the waypoints themselves $\mathbf{w}_n^b = (w_n^{b,x}, w_n^{b,y})$. It is in the design of different distributed coordination strategies where differences emerge when solving the above problem, as discussed in the next section.

3   Proposed Coordination Strategies

It has been previously mentioned that a possible heuristic to tackle the problem in Expression (1) is to randomly decide the set of waypoints (number and coordinates) to be followed by the robots in the swarm, expecting that, at some time since the mission start, the boundary points of all areas in the scenario at hand are visited at least once by a qualified robot. However, this strategy is highly inefficient as it neglects the potential of the swarm to communicate and coordinate the decisions taken locally by every robot. There lies indeed the power of SR methods: distributed coordination of local decisions towards the emergence of collective intelligence.

We leverage this concept by designing two different distributed coordination strategies. The first one, Ground-SDS, assumes a swarm fully composed by ground robotic devices, which communicate with other within a range D_R to obtain as much information as possible from their counterparts. Given the decentralized nature of the robotic swarm, it may occur that certain robots are left isolated for relatively long periods of time, thus exchanging/receiving no knowledge with the rest of robots. The second approach, UAV-SDS, consists of a heterogeneous swarm composed by ground and aerial agents (UAVs). On one hand, ground devices are responsible for the exploration of the map, locate and delimit areas in $\mathcal{A}$. On the other hand, UAVs are in charge of establishing communication links between ground robots with coverage $D_{UAV} \leq D_R$, ensuring that all agents have updated information and not isolated.

In both strategies, decisions made locally by every robot in the swarm rely on SDS. This algorithm has demonstrated a great efficiency in problems in which the overall objective can be split into local sub-objectives, which can be evaluated separately. When extrapolating this feature to the problem at hand, each agent (robot) of the swarm has a hypothesis about the waypoint to be reached. Specifically, a hypothesis is a specific point on the exploration grid of a robot, in which it is assumed that an unexplored area could be found. First, the agent tries to communicate with a near agent. After that, it checks is any compatible area to delimit is present in its cone of vision. If so, the robot is directed towards the zone for properly delimiting it. Otherwise, it searches over its scan grid for a point marked as unexplored, and establishes it as the new hypothesis. Thanks to the cooperation and communication with the rest of individuals, local hypotheses are refined until the entire swarm eventually arrives at the final solution, namely, the delimitation of all areas in the scenario.

As in any other optimization metaheuristic, SDS is divided in different phases. The first one is the initialization of each individual. This first step is followed by an iterative process composed by the analysis of the hypotheses and the diffusion mechanism, in which agents communicate between them. Algorithm finishes when a convergence criterion is reached. In the specific case of Ground-SDS and UAV-SDS, first the final objective is divided into several hypotheses, which are shared among the agents considering the coverage constraints as per D_R and D_{UAV}. In the second phase, each robot evaluates its hypothesis and checks the fulfillment of its partial hypothesis. In the third phase, individuals communicate between them comparing the outcomes obtained, ruling out invalid solutions. Lastly, when the convergence is reached, the algorithm presents the joint hypothesis of all the agents, building the final solution.

Going deeper, it is interesting to highlight that both Ground-SDS and UAV-SDS are internally composed by two algorithms. The first one controls the movements that each agent conducts on its own, whereas the second algorithm controls the cooperation among swarm agents. Figure 3 depicts the decision workflow of the algorithm in charge of the individual behavior of the agents, which is followed by both Ground-SDS and UAV-SDS strategies.

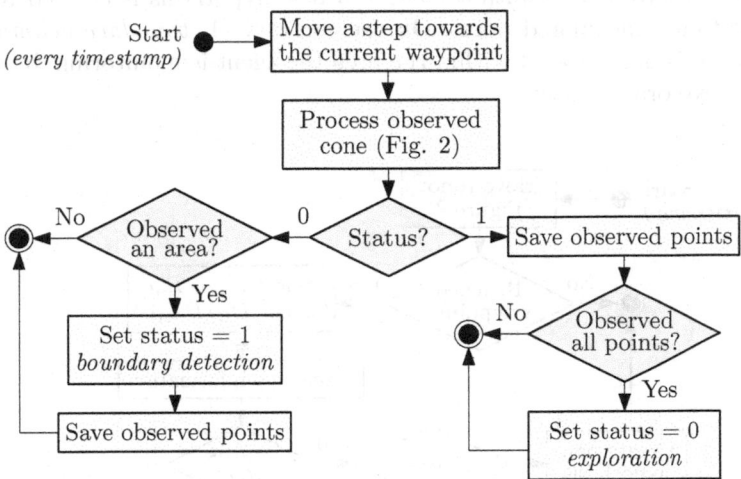

Fig. 3. Workflow determining the movement of robots inside the swarm.

The above procedure is executed by every robot at each timestamp, and starts with the agent moving in the direction of its hypothesis. Then, the robot stores the information inside its vision cone, ignoring possible points that could belong to an area of interest. The next step in this procedure depends on the status of the agent. On the one hand, and if the status is 0 (*exploration mode*, set as default), the robot determines if it has entered an area. If so, the robot modifies its status to 1 (*boundary detection mode*) whenever it is compatible with the type enforced by the observed area (i.e., if $T_n^R \in \mathcal{T}_m^A$). On the other hand, if the agent is already in *boundary detection mode* (status $= 1$), it first stores all boundary points of the observed area falling inside its cone of vision. This procedure is repeated in subsequent timestamps, until the entire area is delimited. Then, the status of the robot returns to its default value (0, *exploration mode*).

As remarked previously, Ground-SDS and UAV-SDS resort to the same movement mechanism depicted in Fig. 3. Thus, the difference between them lies in the algorithm that governs the communication and coordination of the agents. Figure 4 shows the workflow developed for this purpose. It should be highlighted that the difference in Ground-SDS and UAV-SDS in this procedure is the *device-to-device communication* step. At this moment, it is also interesting to mention that to carry out the communications, the methodology proposed in SDS is followed, determining that agents communicate with a single robot at each timestamp. Furthermore, if a robot is in its diffusion phase, and it cannot find any agent within its coverage range, it makes a new hypothesis considering only its own data. This way, communication algorithm is also run by each agent at each timestamp. This method starts running the algorithm depicted in Fig. 3. Then, in the case the agent has reached its hypothesis, it tries to find another robot for sharing the gathered information. Next, the robot makes a decision based on its status. If equal to 0 (*exploration model*), a new hypothesis is created to direct the robot to an unexplored point. On the contrary (1, *boundary delimiting*), a new hypothesis is generated within the area the agent is monitoring, in order to finish this exploration task.

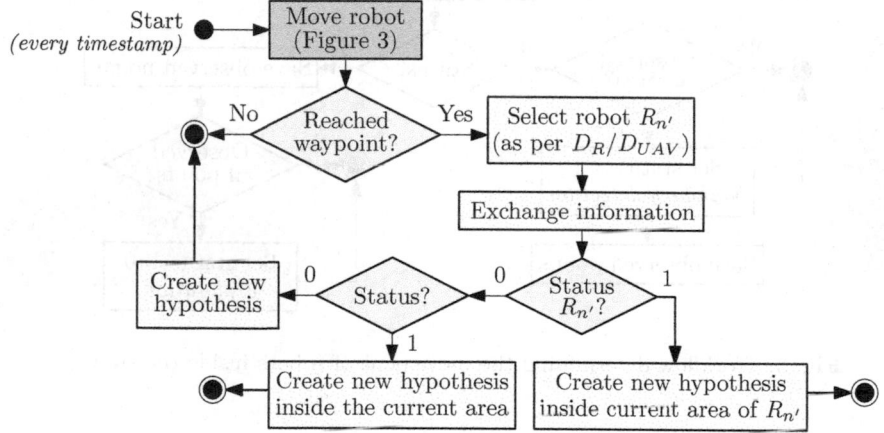

Fig. 4. Workflow defining the communication between devices.

For `Ground-SDS`, the sharing of knowledge is conducted between ground robots. On the other hand, in `UAV-SDS`, this communication is produced between ground robots and flying agents. First, UAVs create at each timestamp a map with the positions of all the ground robots. Then they use these maps to decide where they should place themselves to maximize the number of agents covered (i.e. non-isolated). Finally, UAV agents communicate also to each other to have updated map information and refine further their navigational decisions.

4   Experiments and Results

The performance of the proposed distributed coordination strategies for robotic swarms has been assessed in two different ways. On the one hand, comparative plots have been produced to ascertain the effectiveness of each approach when carrying out different exploration missions. In these cases, the objective is to examine the number of boundary points that each strategy is able to explore within a given number of timestamps. On the other hand, we also inspect the exploration grids, in which agents record the distribution and geographical location of the areas to locate. With these graphics we intend to determine how much information each agent collects with respect to the whole environment.

Three different maps (MAP_1_10, MAP_2_15, and MAP_3_20) have been synthetically generated for measuring the efficiency of each implemented strategy. The size of each map is the same ($D \times D = 1000 \times 1000$), differing in terms of the number of areas $|\mathcal{A}|$ to be explored. Specifically, in the label MAP_X_Y utilized for referring to the maps, Y indicates the amount of areas M to be identified and delimited. Maps are shown in Fig. 5, wherein each circle in a map represent an area A_m. Colors in which these areas are filled refer to the cardinality of their type set $\mathcal{T}_m^A$. Specifically, in areas colored in blue only one type of robot is admitted (e.g. $|\mathcal{T}_m^A| = 1$). In red are depicted those areas that support two types of robots (corr. $|\mathcal{T}_m^A| = 2$), whereas green areas accept the three types of robots considered in the simulations ($|\mathcal{T}_m^A| = 3$). All the $N = 25$ robots depart from the lower left corner of the scenario, among which 15 are ground robots and the remaining 10 are UAVs. Communication ranges are set to $D_R = 50$ and $D_{UAV} = 250$ units of distance.

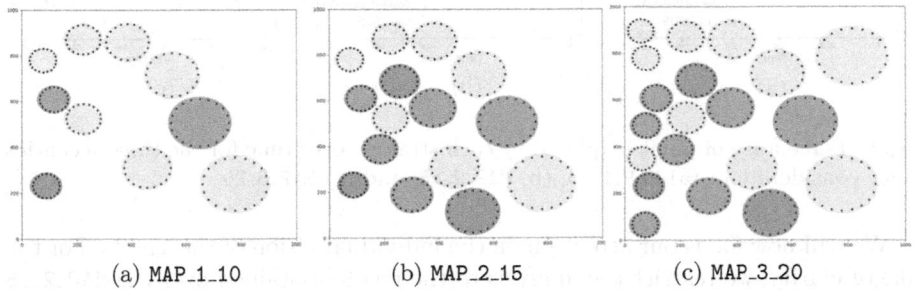

(a) MAP_1_10   (b) MAP_2_15   (c) MAP_3_20

Fig. 5. Representation of the three maps used in the experimentation. Blue: one type of robots supported. Red: two kinds of robots admitted. Green: three types supported. (Color figure online)

In order to account for the statistical significance of the obtained results, every coordination strategy has been run 20 times for each map, using a termination criterion of 900 timestamps. For each run, the number of points a_m^p found at each timestamp is recorded in percentage with respect to the total of points $\sum_{m=1}^{M} P_m$. After these runs, first, second and third quartiles of the results have been reported. We represent these outcomes in Fig. 6. For the sake of understandability, solid lines correspond to the second quartile (i.e., the median). Shaded areas depict the inter-quartile range (IQR). It should be also pointed out that the red color corresponds to the algorithm based on random decisions, blue lines to GROUND-SDS and green to UAV-SDS.

The results in Fig. 6 reveal that the designed strategies behave as expected: both GROUND-SDS and UAV-SDS significantly improve the convergence of a purely random strategy. Despite not surprising, this observation is valuable as a first evidence of the improved convergence yielded by our designed strategies. A second conclusion that can be drawn from these plots is that in general, GROUND-SDS features the highest dispersion among the compared schemes. The reason behind this large variance hinges on its high dependence on the device-to-device communications established between robots along the execution. This second remark emphasizes on the critical importance of the communication scheme for low-range swarm robotics, as occurs in the modeled GROUND-SDS scenario. In fact, the results for UAV-SDS reinforce this statement, as the use of UAVs reduce their dispersion. This statistical robustness brought by heterogeneous robotic swarms is one of the main advantages of UAV-SDS. This is so since each ground agent counts with a maximized amount of information discovered, being able to perform more efficient decisions.

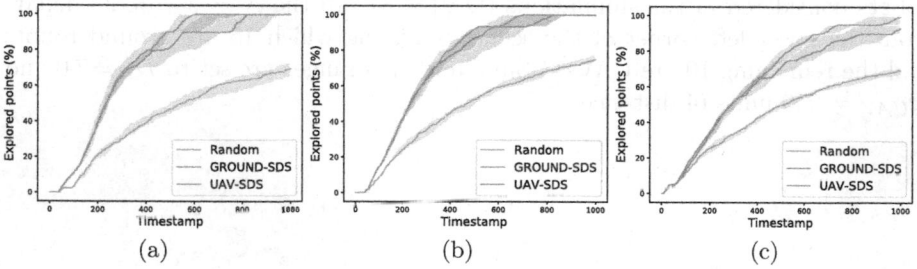

Fig. 6. Percentage of points explored by each strategy over time for the three scenarios under consideration: (a) MAP_1_10; (b) MAP_2_15; and (c) MAP_3_20.

We will now focus our attention on the individual vision of the agents. For the sake of clarity, we restrict the analysis to the results obtained over the MAP_2_15 scenario. Thus, we represent in separated figures the performance of GROUND-SDS (Figs. 7.a and 7.b) and UAV-SDS (Figs. 7.c and 7.d), leaving aside the results of the strategy based on random decisions. We depict the scan map of a single agent (the one with the best performance) and the overlay of all the scan maps of the entire swarm.

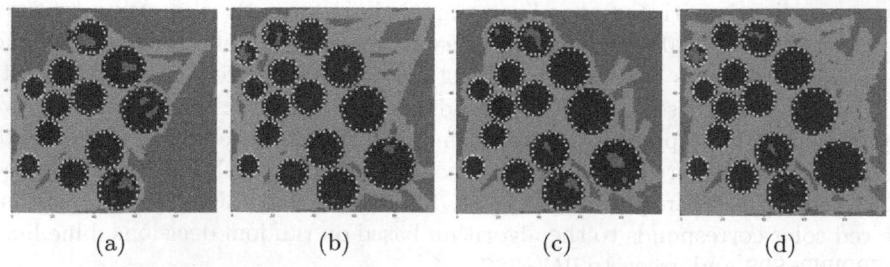

Fig. 7. Scan maps of MAP_2_15 corresponding to (a) GROUND-SDS, single robot; (b) GROUND-SDS, entire swarm; (c) UAV-SDS, single robot; (d) UAV-SDS, entire swarm.

Analyzing the performance of the single device in Fig. 7.a, we observe that most of the areas have been identified (12 out of 15) and fully delimited. Furthermore, the superposition of the scan maps of all agents (Fig. 7.b) shows all the areas of the map and all the points that delimit them. On other hand, Fig. 7.c evinces that the best robot has practically the same information than the entire swarm (Fig. 7.d), clearly materializing one of the main advantages of this strategy: only one agent from the whole swarm can collect enough information from the rest of the swarm to furnish a fairly realistic approximation of the arrangement of the zones on the map. At the same time, by having more region explored and stored, the algorithms to determine new hypotheses are much more effective. On a closing note, this second experimentation is conclusive about the advantages of using a hybrid swarm that combines ground robots and UAVs.

5 Conclusions and Future Work

This manuscript has focused on Swarm Robotics for the exploration, location and delimitation of areas scattered over a large area of land. Specifically, the main contribution of this work is the development of novel distributed coordination strategies for robotic swarms based on Stochastic Diffusion Search. We have implemented three different approaches: an algorithm based on random decisions, a system composed by ground robots (GROUND-SDS) and an heterogeneous swarm composed by UAVs and ground robots (UAV-SDS). The performance of these strategies has been assessed over three synthetically generated maps with varying complexity. The discussed results highlight the improved behavior of heterogeneous robotic swarms when endowed with i) an extended range for device-to-device communications; and ii) the coordination of local hypotheses enabled by SDS.

Several research lines are envisioned for the near future towards gaining further insight beyond the conclusions reported in this work. We plan to introduce areas with different shapes and volumes, so that more realistic terrains can be modeled. Furthermore, efforts will be invested to include collision mechanics in the movement of robots across the environment, both with physical obstacles and among agents. Finally, we plan to perform an exhaustive performance comparison to other distributed coordination mechanisms such as reinforcement learning, and other nature-inspired methods [19]. In particular, we will quantify whether this family of learning techniques attains better levels of generalization than simpler heuristics as the one proposed in this paper.

Acknowledgments. Eneko Osaba and Javier Del Ser would like to thank the Basque Government for its support through the EMAITEK and ELKARTEK (Elkarbot project) programs. Javier Del Ser also receives support from the Consolidated Research Group MATHMODE (IT1294-19) granted by the Department of Education of this institution.

References

1. Kennedy, J.: Swarm intelligence. In: Handbook of Nature-Inspired and Innovative Computing, pp. 187–219. Springer (2006)
2. Del Ser, J., et al.: Bio-inspired computation: where we stand and what's next. Swarm Evol. Comput. **48**, 220–250 (2019)
3. Beni, G.: From swarm intelligence to swarm robotics. In: Şahin, E., Spears, W.M. (eds.) SR 2004. LNCS, vol. 3342, pp. 1–9. Springer, Heidelberg (2005). https://doi.org/10.1007/978-3-540-30552-1_1
4. Osaba, E., Del Ser, J., Iglesias, A., Yang, X.-S.: Soft computing for swarm robotics: New trends and applications, p. 101049 (2020)
5. Albani, D., IJsselmuiden, J., Haken, R., Trianni, V.: Monitoring and mapping with robot swarms for agricultural applications. In: IEEE International Conference on Advanced Video and Signal Based Surveillance (AVSS), pp. 1–6 (2017)
6. Lewkowicz, M.A., Agarwal, R., Chakraborty, N.: Distributed algorithm for selecting leaders for supervisory robotic swarm control. In: IEEE International Symposium on Multi-Robot and Multi-Agent Systems (MRS), pp. 112–118 (2019)
7. de Sá, A.O., Nedjah, N., de Macedo Mourelle, L.: Distributed efficient localization in swarm robotic systems using swarm intelligence algorithms. Neurocomputing **172**, 322–336 (2016)
8. Couceiro, M.S.: An overview of swarm robotics for search and rescue applications. In: Artificial Intelligence: Concepts, Methodologies, Tools, and Applications. IGI Global, pp. 1522–1561 (2017)
9. Carrillo, M., Sánchez-Cubillo, J., Osaba, E., Bilbao, M.N., Del Ser, J.: Trophallaxis, low-power vision sensors and multi-objective heuristics for 3D scene reconstruction using swarm robotics. In: Kaufmann, P., Castillo, P.A. (eds.) EvoApplications 2019. LNCS, vol. 11454, pp. 599–615. Springer, Cham (2019). https://doi.org/10.1007/978-3-030-16692-2_40
10. Carrillo, M., et al.: A bio-inspired approach for collaborative exploration with mobile battery recharging in swarm robotics. In: Korošec, P., Melab, N., Talbi, E.-G. (eds.) BIOMA 2018. LNCS, vol. 10835, pp. 75–87. Springer, Cham (2018). https://doi.org/10.1007/978-3-319-91641-5_7
11. Alfeo, A.L., Cimino, M.G., De Francesco, N., Lega, M., Vaglini, G.: Design and simulation of the emergent behavior of small drones swarming for distributed target localization. J. Comput. Sci. **29**, 19–33 (2018)
12. Innocente, M.S., Grasso, P.: Self-organising swarms of firefighting drones: harnessing the power of collective intelligence in decentralised multi-robot systems. J. Comput. Sci. **34**, 80–101 (2019)
13. Huang, X., Arvin, F., West, C., Watson, S., Lennox, B.: Exploration in extreme environments with swarm robotic system. In: IEEE International Conference on Mechatronics (ICM), vol. 1, pp. 193–198 (2019)
14. Suárez, P., Iglesias, A.: Bat algorithm for coordinated exploration in swarm robotics. In: International Conference on Harmony Search Algorithm, pp. 134–144 (2017). https://doi.org/10.1007/978-981-10-3728-3_14
15. Bishop, J.: Stochastic searching networks. In: IEEE International Conference on Artificial Neural Networks (Conference Publication No. 313), pp. 329–331 (1989)
16. Nasuto, S.J., et al.: Time complexity analysis of the stochastic diffusion search. In: International ICSC/IFAC Symposium on Neural Computation, pp. 260–266 (1998)
17. Al-Rifaie, M.M., Bishop, J.M.: Stochastic diffusion search review. Paladyn J. Behav. Robot. **4**(3), 155–173 (2013)

18. Yuan, H., Gu, X., Lai, R., Wen, Z.: Global optimization with orthogonality constraints via stochastic diffusion on manifold. J. Sci. Comput. **80**(2), 1139–1170 (2019)

19. Precup, R.-E., David, R.-C., Petriu, E.M., Szedlak-Stinean, A.-I., Bojan-Dragos, C.-A.: Grey wolf optimizer-based approach to the tuning of pi-fuzzy controllers with a reduced process parametric sensitivity. IFAC-PapersOnLine **49**(5), 55–60 (2016)

An Intelligent Procedure
for the Methodology of Energy
Consumption in Industrial Environments

Izaskun Mendia[1(⊠)], Sergio Gil-Lopez[1(⊠)], Javier Del Ser[1,2(⊠)],
Iñaki Grau[3(⊠)], Adelaida Lejarazu[1(⊠)], Erik Maqueda[1(⊠)],
and Eugenio Perea[1(⊠)]

[1] TECNALIA, Basque Research and Technology Alliance (BRTA),
Mikeletegi Pasealekua 2, 20009 Donostia-San Sebastián, Spain
{izaskun.mendia,sergio.gil,javier.delser,adelaida.lejarazu,
erik.maqueda,eugenio.perea}@tecnalia.com
[2] University of the Basque Country (UPV/EHU), 48013 Bilbao, Spain
[3] Gestamp, 28014 Madrid, Spain
igrau@gestamp.com

Abstract. The concern of the industrial sector about the increase of
energy costs has stimulated the development of new strategies for the
effective management of energy consumption in industrial setups. Along
with this growth, the irruption and continuous development of digital
technologies have generated increasingly complex industrial ecosystems.
These ecosystems are supported by a large number of variables and
procedures for the operation and control of industrial processes and
assets. This heterogeneous technological scenario has made industries
difficult to manage by traditional means. In this context, the disruptive potential of cyber physical systems is beginning to be considered
in the automation and improvement of industrial services. Particularly,
intelligent data-driven approaches relying on the combination of Energy
Management Systems (EMS), Manufacturing Execution Systems (MES),
Internet of Things (IoT) and Data Analytics provide the intelligence
needed to optimally operate these complex industrial environments. The
work presented in this manuscript contributes to the definition of the
aforementioned intelligent data-driven approaches, defining a systematic,
intelligent procedure for the energy efficiency diagnosis and improvement of industrial plants. This data-based diagnostic procedure hinges
on the analysis of data collected from industrial plants, aimed at minimizing energy costs through the continuous assessment of the production-
consumption ratio of the plant (i.e. energy per piece or kg produced). The
proposed methodology aims to support managers and energy-efficiency
technicians to minimize the plant's energy consumption without affecting
the production and therefore, increase its competitiveness. The data used
in the design of this methodology are real data from a company dedicated
to the design and manufacture of automotive components and one of the
main manufacturers in the automotive sector worldwide. The present
methodology is under the pending patent application EU19382002.4-120.

© Springer Nature Switzerland AG 2020
C. Analide et al. (Eds.): IDEAL 2020, LNCS 12490, pp. 92–103, 2020.
https://doi.org/10.1007/978-3-030-62365-4_9

Keywords: Energy efficiency · Smart manufacturing · Intelligent systems · Industry 4.0 · Big data · Cyber physical systems

1   Introduction

The progressive increase in energy costs is a growing concern in the industrial sector, with electricity prices for non-household consumers achieving unprecedented values in 2019 (i.e. more than 0.15 EUR per kWh (including taxes) for the European Union in 2019 [4]). Besides this rising trend, another problem stems when comparing the cost of electricity to the value of industrial production, which is starting to have an impact on the competitiveness of manufacturing companies as their production costs become higher over time. Although renewable energies are becoming increasingly important to counteract this issue, the vulnerability derived from the consumption of fossil fuels and the dependence on oil-exporting countries in recent years has spawned the search and development of alternative strategies for the efficient management of energy consumption in the industrial sector.

In this context, the concept of energy efficiency involves the efficient allocation of the amount of energy required to produce products and services with a given set of industrial assets. Improvements in energy efficiency are generally achieved through the adoption of new technologies, upgrades in the production chain to make it more efficient, or through the application of commonly accepted methods to reduce energy losses. Beyond the industrial sectors, there are many motivations for businesses to improve their energy efficiency, the main one is that by keeping the energy use to its minimum, electricity costs can be reduced, without affecting their production entailing larger economic savings. This profitability holds as long as energy savings offset any additional costs of implementing an energy-efficient technology. Reducing energy use is also considered essential for the global reduction of greenhouse gas and due to regulations like ISO 50.001.

Energy efficiency lies at the core of Industry 4.0 [7], which refers to a new industrial paradigm that defines the transition from *traditional* – i.e., based on industrial machinery – to the concept of digital manufacturing [10]. The concept of Industry 4.0 covers a range of industrial developments, including Cyberphysical Systems (CPS), Internet of Things (IoT), Internet of Services (IoS), Robotics, Big Data, Cloud Manufacturing and Augmented Reality [11]. Most of these technologies under the Industry 4.0 influence already existed years ago. However, their successful deployment over industrial environments has not been a reality until such technologies have evolved mature enough and have encountered a rich digital, interconnected industrial ecosystem, as provided by massive monitoring sensors, manufacturing databases and the end-to-end traceability of products and services. This digital transformation process is driven by an abrupt increase in the volume of data, the power of computer systems and connectivity, and their storage capacity that has increased at an exponential pace in recent decades [15]. Industrial machines can now operate with each other in a symbiotic way. As a result, the development and improvement of each technology contributes to the advance of the rest of technologies. These changes allow the different industrial sectors to adapt, evolve and create synergies to become stronger and more competitive [1]. Having greater interconnection capacity, greater adaptability and greater speed of information exchange has a huge competitive advantage potential for all types of industries, either for their internal functioning or for the services they offer to their customers for their products and services [14].

Modelling of industrial and manufacturing processes is very important in an environment where production processes must be digitally supported by new technologies [16]. Raw data do not provide significant value for decision making in a cyber-physical system, unless these data are effectively processed and analyzed [8]. In order to analyze massive amounts of data generated by both IoT applications and existing ICT systems, data science and analysis techniques must be developed and employed [2,3,13]. It is necessary to collect, analyze and optimize all details of a manufacturing process related to the desired process results. The real goal of Industry 4.0 is to create a seamless integration of processes to the intelligent cyberphysical factory [6,17].

Energy efficiency is not an exception in the need of data-based modelling pipelines suited for manufacturing processes noted above. The contribution of the present work is framed within this statement: specifically, we define a procedure for the diagnosis of energy (in)efficiencies in industrial plants. Our proposal repouds to machine learning algorithms that allow examining the production-consumption ratio (i.e. energy per piece or kg produced), internalizing the operation with a global vision of the plant, process and machine, and inferring abnormal or unknown operational patterns from the collected data. The ultimate aim of the procedure is to develop new methodologies endowed with Machine Learning functionalities that guarantee a sustained industrial production while minimizing energy costs. To the best of our knowledge, in the current literature there is no prior work dealing with global solutions that integrate the data at plant, process and machine level. Effectively, existing approaches are considered at plant level due to lack of measures at lower levels, or they only optimize the individual behavior of certain machines, without considering their integration and interaction with the rest of the plant machinery. A case study will be presented and discussed in order to support the novelty of our procedure with empirical evidence based on real data obtained from several industrial plants.

2 Background and Motivation

Classical industrial control systems are not able to adequately cope with the essential problems of today's connected world in industrial environments, mainly due to issues with data types, information modelling and the relationships between data providers and control systems. This article proposes an intelligent procedure that, based on the interpretation of historical data, analyzes different behaviors with the aim of supporting decision making processes. In order to carry out this methodology of energy consumption in an industrial plant, the proposed approach includes advanced data-based models capable of inferring knowledge not only at the plant level, but also at the machine level. In the state of the art, there are no global solutions that integrate data at plant, process and machine levels. Instead, they are either considered at the plant level (due to lack of measures at lower levels), or they only optimize the individual behavior of certain machines, without considering their integration and interaction with the rest of the plant. The use of the proposed solution yields an overall vision of the system data, disregarding the level at which they are produced. This approach introduces the concept of plant, line, process, machine and piece's reference models for a better adaptation to the needs of modern industrial applications.

In the field of manufacturing, information technologies must be integrated with process control engineering. To do this, systems must be interoperable, which requires the exchange of information throughout the production process. On one hand, an EMS (Energy Management System) is used to collect data on the production equipment energy consumption. EMS systems provide in-depth knowledge of the energy consumption mode of the industrial plant, either through high-level measurements of its transformers and/or electrical connections, or through individual measurements of the equipment installed in the plant. On the other hand, an MES (Manufacturing Execution System) is used for production control, being its ultimate goal to increase the efficiency of the production plant (OEE, Overall Equipment Effectiveness) and thereby, to reduce costs and improve productivity. Thus, while EMS is focused on the energy control of the plant, an MES system is focused on the information of the production itself.

From an energy point of view, the industrial plant can be considered as a collection of loads grouped at different submetering levels. Thanks to the deployment of IoT, equipment like meters and gateways enable the measurement of a variety of physical magnitudes, as well as the connection of different industrial assets, enabling the collection of data at diverse submetering levels. Throughout the analysis of patterns emerging from production data, energy behaviors can be interpreted. Energy inefficiencies in a production process can also be detected, and the root cause for those inefficiencies can be identified. Therefore, our methodology allows to discern whether the discovered energy behaviors are due to production changes.

The challenge of this intelligent procedure, proposed in this article, is not only to merge the energy consumption and production data, but also to analyze the information collected by these subsystems. This information is provided to the plant manager through a set of tools that allow him to implement the necessary actions to maximize the energy efficiency of the industrial ecosystem thanks to the supervision, management of alarms and reports, consultation of indicators, control panels and so on.

3   Proposed Methodology

As has been mentioned in preceding sections, a new methodology is proposed for the analysis and integral evaluation of energy consumption in industrial plants.

Figure 1 illustrates this methodology, which responds to a typical CPS architecture [9]. Each of the processing layers are explained in the following sections::

3.1 Physical-Energy Layer

The physical layer is responsible, through sensors of different nature, of capturing information over time. We assume that data captured by sensors over time are deployed over different assets of the industrial plant. In particular, each of such sensors captures and conveys energy measurements [kWh] to the MES module, yielding a collection of daily energy consumption curves $\mathbf{x}^{s,d} \doteq [x_t^{s,d}]_{t=1}^{T^s}$, wherein $s \in \{1, \ldots, S\}$ indicates the index of the sensor, and $d \in \{1, \ldots, D^s\}$ denotes the day index out of a total of D^s days for sensor s. We assume that curves are registered at a rate of 1 measurement every Δ^s units of time, which yields the exact number of daily measurements T^s collected for sensor s.

The set of consumption loads are divided into (i) productive loads, those loads associated with production processes (i.e. hot-stamping, cold-stamping). (ii) auxiliary loads, those loads not directly associated with production (i.e. lighting, air compressors), and (iii) unmonitored loads (processes not associated with any energy measurement equipment). Preferably, all loads should have the same temporal discretization, for which a time homogenization function is implemented for the sake of an uniformized sampling of loads. Separating energy measurements of different nature allows not processing them together, which would cause uncertainty when detecting energy inefficiencies. An energy consumption curve $\mathbf{x}^{s,d}$ represents a profile of energy consumption measured at a rate of 1 sample every Δ^s seconds, where the superindex d denotes a specific energy consumption curve.

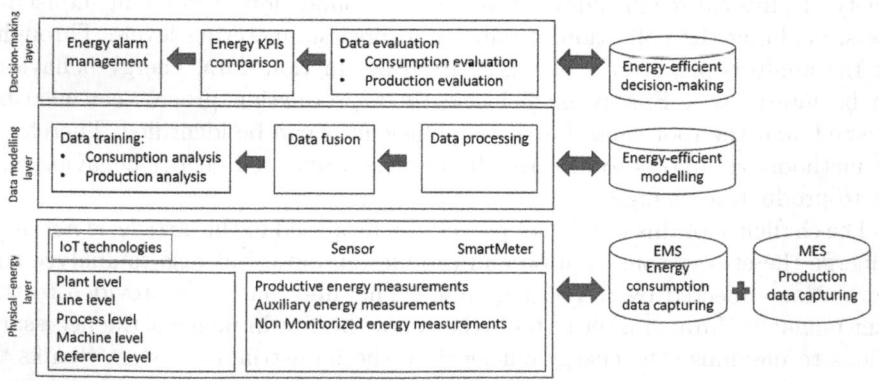

Fig. 1. Architecture for the proposed methodology for energy consumption monitoring and evaluation of industrial plants.

As mentioned above, energy consumption measurements are collected from the loads at four different levels: plant level, production line level, process level and machine level. Production level measurement sensors, or information captured by MES and measured in kg, allow production to be represented as a daily value aggregated at a plant level. From the information of energy consumption and production, this methodology allows to evaluate and consider other type of indirect metrics. For example, if the principle of energy conservation is met, the value of the unmonitored loads can be computed as the difference between the sum of the total loads at the transformer level, and the sum of the total monitored plant curves.

3.2 Data Modelling Layer

Basically the Data Modelling Layer is responsible for merging the information on energy consumption and production, and for identifying the normal mode in which the plant operates, based on similar behavioral patterns inferred from historical information. Initially, this information is divided into several groups of different behaviors. Loads from the same group have similar shapes and consumption levels, and therefore are declared to imprint similar effects on the energy consumption/production ratio. The grouping of loads and their relation with the daily production are the most important concepts of this methodology. The data layer is divided into the following modules:

Data Processing. For each of the consumption curves $\mathbf{x}^{s,d}$ of the plant, this module verifies first their quality in terms of data completion. In case of missing measurements in any of the load curves (due to sensor connection failure or any other reason), the missing entries are imputed whenever the time interval of missing data does not exceed a predefined threshold. Different techniques for missing time series data imputation are used to interpolate missing points in the energy consumption curves. For example, missing data can be imputed with the mean value at the involved time instants of the energy consumption curves belonging to a same cluster or pattern (which is the output of the module explained in Subsect. 3.2). This example, however, comes at no loss in generality for the proposed methodology: other imputation strategies can be chosen.

Data Fusion. This module homogenizes data collected from EMS and MES in terms of information content in time, and transforms it into useful information for subsequent modeling. The module also comprises a generic database for integrating such data at all levels (plant, line, process, machine).

Data Clustering. Once energy consumption curves have been collected for the S sensors over the plant, this module utilizes a clustering algorithm to infer K daily energy consumption patterns $\mathcal{C}_k^s$, where $k = 1,\ldots,K$. Each energy consumption pattern $\mathcal{C}_k^s$ represents a subset of the daily energy consumption curves of sensor s grouped together according to a similarity metric. Let us denote as $x_t^{s,d}$ the energy consumption measurement collected at time slot

$t \in \{1, \ldots, T^s\}$ for sensor s and day $d \in \{1, \ldots, D^s\}$, where T^s denotes the number of daily slots, and D^s accounts for the number of daily consumption traces for sensor s. The representative $\mathbf{c}_k^s = [c_{t,k}^s]_{t=1}^{T^s}$ of each energy consumption pattern $\mathcal{C}_k^s$ is calculated as the mean value of each component $x_t^{s,d}$ of each daily energy consumption curve $\mathbf{x}^{s,d} \in \mathcal{C}_k^s$ belonging to that pattern:

$$c_{t,k}^s = \frac{1}{|\mathcal{C}_k^s|} \sum_{d:\mathbf{x}^{s,d} \in \mathcal{C}_k} x_t^{s,d}, \tag{1}$$

where $|\mathcal{C}_k^s|$ denotes cardinality of a set. Thus, an energy consumption pattern represents an energy behavior that characterizes a set of similar daily energy consumption curves.

The inference of the above consumption patterns is done by applying a clustering technique. Without loss of generality we resort to the well-known K-Means clustering algorithm. Considering the notation introduced in this section, the clustering algorithm used to compute these energy consumption patterns implements an iterative process for minimizing the distance between elements forming a cluster and its representative:

$$\arg\min_{\mathcal{C}_1,\ldots,\mathcal{C}_K} \sum_{k=1}^{K} \sum_{d:\mathbf{x}^{s,d} \in \mathcal{C}_k} \left\| \mathbf{x}^{s,d} - \mathbf{c}_k^s \right\|^2, \tag{2}$$

where $\|\mathbf{x}^{s,d} - \mathbf{c}_k^s\|$ denotes the squared Frobenius norm. It is well-known that for the K-Means algorithm, the number K of clusters to be sought is a parameter that must be set beforehand. Many criteria can be followed for this purpose [5,12]; in our system we adopt the so-called Elbow method, which draws the value of K from the point of maximum deflection in the representation of the variance between groups divided by the total variance of the selection from the set of curves.

Once the K patterns have been computed, for a new energy consumption curve obtained from sensor s, its degree of similarity/dissimilarity with respect to current patterns is evaluated by comparing the new energy consumption curve with the K patterns already defined. Depending on whether the new energy consumption curve is close to one of such patterns in terms of a threshold imposed on the similarity measure, it is declared that the new energy consumption curve belongs to that pattern. The new energy consumption curve is associated to the group of curves represented by the pattern, and the pattern is updated taking into account the new curve. The treatment of the new curve comprises the following steps:

1. When the new curve fails to match any of the prevailing set of energy consumption patterns, we retrieve from the MES production data corresponding to the same time span of the energy consumption curve under analysis.
2. Considering this information, two quantities are computed: 1) the total energy consumed by the asset monitored by sensor s over the time span of the curve under analysis; and 2) the production rate of the asset over the same period of time.

3. A comparison is made to historical data in terms of the relationship between the above two parameters, yielding a quantitative tool to assess the energy efficiency of monitored asset as per its contribution to the overall productivity of the plant.

Data Regression. As stated above, the analysis performed every time a new energy consumption curve is obtained from data captured from the industrial plant, is explained based on this statistical inferred behavior. Once several K patterns are obtained during a training stage using data collected for a certain time horizon, every time a new energy consumption curve is obtained, for example on a daily basis, it is determined to which pattern the new curve belongs. This is done by comparing the new curve with the K patterns. A new energy consumption curve belongs to a certain pattern if the distance to the centroid of that certain pattern is smaller than the distance to any other centroid (centroid of any other pattern) and the distance to it is smaller than the maximum distance from the rest of energy consumption curves set in the training stage. The new energy consumption curve is compared with the day of maximum energy consumption.

Figure 2 represents an example of pattern calculation performed with the above mentioned procedure from a collection of energy consumption curves. The energy consumption curves were defined by electricity consumption data collected at the electric distribution grid connection point of an industrial plant, where daily energy consumption patterns were measured every $\Delta^s = 15$ min and measured in kWh. We note that further details on the particularities of the use case cannot be provided for confidentiality reasons. The plot clearly shows three different energy patterns, whose corresponding energy consumption curves are perfectly discriminable from each other. In each pattern, the representative is computed as per Expression (1).

Departing from these centroids, we continue with the exemplifying case by depicting in Fig. 3 the production (in kg) versus the aggregated energy consumption (in kWh) corresponding to the real-world industrial asset under consideration. In this graphical representation, there are as many points as historic days are used in the analysis for obtaining energy consumption patterns. The continuous line is a straight line obtained by means of an adjustment for least squares. This line permits to represent the relationship between energy and production.

In the case of an adjustment for least squares, this adjustment is characterized by a χ^2 statistical test. This statistic is used to determine whether a new daily consumption trace belongs to the distribution spanned by the historical consumption curves. If this hypothesis fails to hold, a production alarm may be triggered. This involves starting with all days as candidate variables, testing the deletion of each day using a chosen model criterion, deleting the day whose loss gives the most statistically insignificant deterioration of the model fit, and repeating this process until no further days can be deleted without a statistically significant loss of fit. In this way it is possible to understand the contribution of each day.

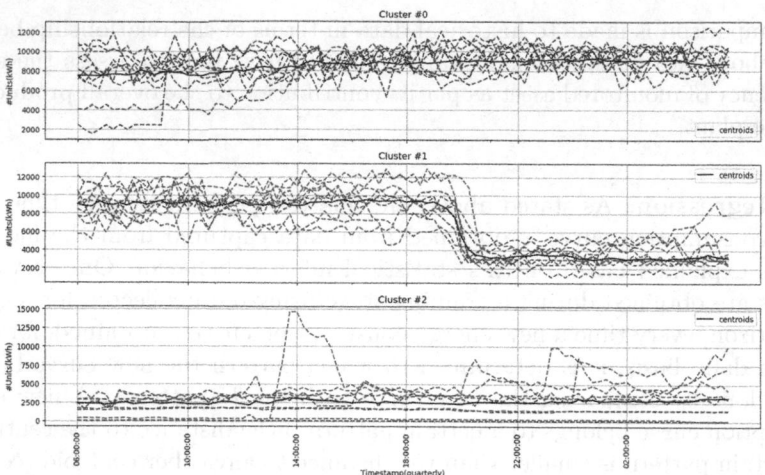

Fig. 2. Consumption curves and inferred representative patterns of a monitored asset of a real-world industrial plant.

Thus, if for example production data is available every 24 h, the aggregated energy consumption for 24 h is obtained for each curve forming each pattern. Later, in the evaluation stage, during for example daily execution of the methodology by comparing the relationship of the energy consumption versus production with the relation of all the curves forming the K patterns, it is possible to detect energy efficiency alarms. In other words, if an increase in energy consumption is caused by an increase in production, no alarm should be triggered. On the contrary, if an increase in energy consumption is not associated to an increase in production, an alarm should be triggered. Therefore, production is extremely important in order to prevent false alarms. The goal is to detect intensive energy consumption, higher than expected, not associated with production increases.

3.3   Decision-Making Layer

This layer of the proposed methodology resorts to visualization analytics towards explaining the outcome of the Machine Learning algorithm so as to make it more understandable to plant managers. The ultimate aim is to render the whole logic of the methodology transparent to the manager, and to allow expediting the best possible operational choice. The functionalities of this decision-making layer include the identification of anomalous trends, automatic notification of control alarms, centralization of forensic data on consumption and production, definition of control Key Performance Indicators (KPIs), and the generation of customized reports about production processes, including visual comparisons between processes of the plant.

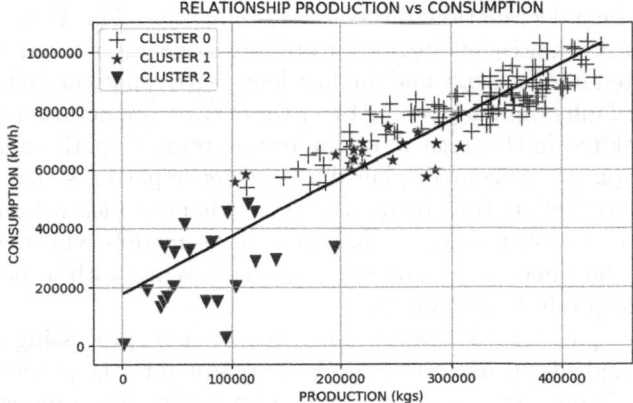

Fig. 3. Relationship between production and energy consumption of the monitored asset.

4 Conclusions and Outlook

Within the industrial sector, process and auxiliary equipment is a source of energy expenditure that results in a variable operating cost in the income statement of the industry. For a plant with an energy expenditure of 1M euros/year, the company, a manufacturer of automotive parts, spends in the order of 100 KEuros to 250 KEuros per year. It is a priority for the industry to have optimized equipment not only in regard to its operation, but also in terms of its energy consumption. The optimized management of this equipment, as well as the detection of inefficiencies in its operation and the consequent reprogramming of its operation entails great energy savings (it can reach up to 5% of the total plant consumption in certain sectors - based on Gestamp's EE experts).

 This work has presented a simple and effective methodology that not only allows for an energy-efficient management of production processes and equipment, but also provides a solution to automate the detection of energy inefficiencies at the plant level through the inspection of production-consumption ratio. In fact, this is methodology is being implemented in several real plants around the world. It implements descriptive analytic methodologies (i.e., clustering and regression techniques). The proposed methodology is based on the study of real data from different production plants and different processes. The production processes are planned/controlled (EMS) and their production monitored/quantified by external systems (MES). In the same way and under the ISO 50.001 standard, more and more energy management systems are implemented, which measure consumption at plant, process and machine level.

 We envisage several promising research directions rooted on this work:

- The methodology herein described is focused on the descriptive analysis of plant operation. In the future, these descriptive models should be able to predict future moments in order to anticipate the occurrence of certain events that reflect on a degradation of their energy efficiency. This is crucial in order to establish preemptive strategies for optimizing energy-efficient systems.

- The presented methodology and further descriptive capacities to be developed will prevent faults from happening by virtue of early predictions and proactive decision making. In this regard, an interesting research path aims at making production plants resilient to plausible occurrence patterns that might have never occurred before, thereby hindering any chance to learn from them. For this purpose, we plan to investigate generative models which, in addition, might help plant managers understand unseen insights within the complexity of real-world production systems.
- New technologies for distributed and efficient data processing such as fog computing and edge computing can be incorporated to the proposed methodology towards deploying part of the data processing and mining on sensors themselves. These paradigms, however, bring about issues related to latency, reliability and the distribution of algorithmic components over the network. A closer look will be taken at how to redesign each layer to make their modules compliant with Fog and Edge computing, as well as new functionalities emerging therefrom (e.g. correlation between alarms triggered at different sensors).

Acknowledgments. This work has received funding support from the HAZITEK program of the Basque Government (Spain) through the NAIA (Ref. ZL-2017/00701) research grants. It is also appreciate the deference of the company GESTAMP, especially to Iñaki Grau, to provide data from several of its plants. Finally, Javier Del Ser acknowledged funding support from the Consolidated Research Group MATHMODE (IT1294-19), granted by the Department of Education of the Basque Government, as well as by ELKARTEK and EMAITEK programs of this same institution.

References

1. Antsaklis, P.: Goals and challenges in cyber-physical systems research editorial of the editor in chief. IEEE Trans. Autom. Control **59**(12), 3117–3119 (2014)
2. Cheng, Y., Chen, K., Sun, H., Zhang, Y., Tao, F.: Data and knowledge mining with big data towards smart production. J. Ind. Inf. Integr. **9**, 1–13 (2018)
3. Diez-Olivan, A., Del Ser, J., Galar, D., Sierra, B.: Data fusion and machine learning for industrial prognosis: trends and perspectives towards industry 4.0. Inf. Fusion, **50**, 92–111 (2019)
4. EUROSTAT: Electricity prices for non-household consumers (2020). https://appsso.eurostat.ec.europa.eu/nui/show.do?dataset=nrg_pc_205. Accessed 2 April 2020
5. Hamerly, G., Elkan, C.: Learning the k in k-means. In: Advances in Neural Information Processing systems, pp. 281–288 (2004)
6. Ivanov, D., Dolgui, A., Sokolov, B., Werner, F., Ivanova, M.: A dynamic model and an algorithm for short-term supply chain scheduling in the smart factory industry 4.0. Int. J. Product. Res. **54**(2), 36–402 (2016)
7. Lasi, H., Fettke, P., Kemper, H.G., Feld, T.: Hoffmann, M.: Industry 4.0. Bus. Inf. Syst. Eng. **6**(4), 239–242 (2014)

8. Li, H.X., Xu, L.D.: Feature space theory-a mathematical foundation for data mining. Knowl. Based Syst. **14**(5–6), 253–257 (2001)
9. Ma, S., Zhang, Y., Lv, J., Yang, H., Wu, J.: Energy-cyber-physical system enabled management for energy-intensive manufacturing industries. J. Clean. Product. **226**, 892–903 (2019)
10. Oztemel, E., Gursev, S.: Literature review of industry 4.0 and related technologies. J. Intell. Manufact. **31**(1), 127–182 (2020)
11. Pereira, A., Romero, F.: A review of the meanings and the implications of the industry 4.0 concept. Proc. Manufact. **13**, 1206–1214 (2017)
12. Pham, D.T., Dimov, S.S., Nguyen, C.D.: Selection of k in k-means clustering. Proc. Inst. Mech. Eng. Part C: J. Mech. Eng. Sci. **219**(1), 103–119 (2005)
13. Sheng, G., Zhao, X., Zhang, H., Lv, Z., Song, H.: Mathematical models for simulating coded digital communication: a comprehensive tutorial by big data analytics in cyber-physical systems. IEEE Access **4**, 9018–9026 (2016)
14. Weyer, S., Schmitt, M., Ohmer, M., Gorecky, D.: Towards industry 4.0-standardization as the crucial challenge for highly modular, multi-vendor production systems. Ifac-Papersonline, **48**(3), 579–584 (2015)
15. Wielki, J.: The impact of the internet of things concept development on changes in the operations of modern enterprises. Polish J. Manage. Stud. **15** (2017)
16. Xu, L.D., Xu, E.L., Li, L.: Industry 4.0: state of the art and future trends. Int. J. Product. Res. **56**(8), 2941–2962 (2018)
17. Zhou, K., Liu, T., Zhou, L.: Industry 4.0: towards future industrial opportunities and challenges. In: 2015 12th International conference on fuzzy systems and knowledge discovery (FSKD), pp. 2147–2152. IEEE (2015)

Unified Performance Measure for Binary Classification Problems

Ana R. Redondo[✉], Jorge Navarro, Rubén R. Fernández[✉],
Isaac Martín de Diego, Javier M. Moguerza, and Juan José Fernández-Muñoz

Rey Juan Carlos University, c/ Tulipán s/n, 28933 Móstoles, Spain
{anaisabel.rodriguez,ruben.rodriguez,isaac.martin,javier.moguerza,
juanjose.fernandez}@urjc.es, j.navarro.2016@alumnos.urjc.es
http://www.datasciencelab.es

Abstract. Different performance measures are used to inspect, compare and evaluate the behaviour of classifiers in *Machine Learning (ML)*. ML researchers employ one or several of these performance measures in their classification studies to report their success. However, no widespread consensus has been reached on a unified chosen measure. In this work, we introduce a reliable and informative measure, the *Unified Performance Measure (UPM)*, by modifying the F_1-score to avoid its undesired behaviour in imbalanced classification problems. *UPM* is compared with alternative performance measures, like the F_1-score or Accuracy, in both simulated confusion matrices and real datasets. The proposed measure outperforms the alternatives, providing a promising new research line.

Keywords: Classification models · Confusion matrices · F_1-score · Accuracy · Matthews correlation coefficient · Imbalanced data · Binary classification.

1 Introduction

Classification models are an important part of *Machine Learning (ML)* whose objective is to make a perfect division in different classes that are previously defined. There are several types of classification problems: binary, multi-class, multi-labelled and hierarchical [9]. This paper will focus on binary classification problems. The purpose of binary classification is to split the given observations into two mutually exclusive classes. A typical example of such problems is to identify the presence or absence of a disease in an individual.

In order to evaluate the performance of a *ML* classification model, different measures exist. The choice of a proper measure is relevant, since it will be used to select or discard classification models, and to evaluate if the selected model meets the requirements. In general, these measures are obtained from a confusion matrix, which faces the observed and the predicted classes for a set of observations. Accuracy is one of the most common performance measures. It represents the ratio of correctly predicted instances. However, in many classification problems, alternative measures such as the F_1-score are more appropriate.

© Springer Nature Switzerland AG 2020
C. Analide et al. (Eds.): IDEAL 2020, LNCS 12490, pp. 104–112, 2020.
https://doi.org/10.1007/978-3-030-62365-4_10

Table 1. Confusion matrix for binary classification.

		Observed	
		-1	$+1$
Predicted	-1	TN	FN
	$+1$	FP	TP

In this paper, a new binary classification measure, the *UPM*, is proposed. It is obtained from the combination of several metrics that are based on a confusion matrix. A comparison with alternative metrics is proposed to show the benefits of the new measure.

The rest of the paper is structured as follows. Section 2 introduces the most common metrics that are calculated based on a confusion matrix. The proposed measure is described in Sect. 3. Experiments using simulated and real datasets are detailed in Sect. 4. Finally, Sect. 5 concludes and provides further research lines.

2   State of the Art

In the context of a binary classification problem, the confusion matrix is one of the best and most intuitive representations for optimally evaluating the results of a classification model [1]. The confusion matrix is not a metric as such, but most of them are calculated from it. It faces the observed values and the values predicted by a *ML* classification model (see Table 1).

The components of a confusion matrix are:

- *True Positives* (TP): the observed $+1$ cases also predicted as $+1$.
- *True Negatives* (TN): the observed -1 cases also predicted as -1.
- *False Positives* (FP): the observed -1 cases predicted as $+1$ (type I error).
- *False Negatives* (FN): the observed $+1$ cases predicted as -1 (type II error).

Notice that the relative importance of the two types of errors (FP and FN) depends on the problem under consideration. For instance, in a cancer detection study, a relevant question to consider is which is the most costly type of mis-classification: a classifier which predicts cancer when the person does not have cancer (FP) or a classifier which predicts no-cancer when the person has cancer (FN). There is no single answer to this question, and it depends on the problem and the consequences of the misclassification.

Probably the most intuitive measure of a classifier success is its Accuracy [3], which represents the ratio between the correctly predicted instances and all the instances in the dataset:

$$Accuracy = \frac{TP + TN}{TP + TN + FP + FN} \tag{1}$$

Table 2 shows several metrics that are defined based on the confusion matrix. The Sensitivity, also known as Recall or *True Positive Rate* (*TPR*), can be

Table 2. Definition of classification performance metrics from a confusion matrix.

Symbol	Metric	Defined as
TPR	Sensitivity/Recall	$\frac{TP}{TP+FN}$
TNR	Specificity	$\frac{TN}{TN+FP}$
PPV	Precision	$\frac{TP}{TP+FP}$
NPV	Negative Predictive Value	$\frac{TN}{TN+FN}$

understood as the probability that an observed +1 is classified as +1 by the *ML* model. The Specificity, also known as *True Negative Rate* (*TNR*), is the proportion of −1 cases that are correctly predicted as such. The Precision, also known as *Positive Predictive Value* (*PPV*), can be viewed as the probability of success when a case is classified as +1. Finally, the *Negative Predictive Value* (*NPV*) is the proportion of −1 cases classified as −1 by the *ML* model. The optimal value for all these measures is 1.00. However, there is usually a trade-off between *PPV* and *TPR*, and consequently between *TNR* and *NPV*. That is, when one measure increases, the other decreases. To manage this trade-off, measures that are a combination of these performance measures are defined. Given that the harmonic mean is more intuitive than the arithmetic mean when computing a mean of ratios, the F_1^+ (usually called F_1-score [8]) is defined as the harmonic mean of Precision and Recall.

$$F_1^+ = 2 \cdot \frac{Precision \cdot Recall}{Precision + Recall} = \frac{2 \cdot TP}{2 \cdot TP + FP + FN} \qquad (2)$$

Notice that F_1^+ gives the same weight to Precision and Recall. However, in some real problems, one of them is preferred: a high success rate when predicting a positive instance (Precision), or covering more positive instances (Recall). To control this trade-off, the general formula for F^+ is defined as follows:

$$F_\beta^+ = (1 + \beta^2) \cdot \frac{Precision \cdot Recall}{\beta^2 \cdot Precision + Recall}, \qquad (3)$$

where the positive parameter β is such that Recall is considered β times as important as Precision is. Two commonly used values are $\beta = 2$ (F_2^+), where Recall is twice as important as Precision, and $\beta = 0.5$ ($F_{0.5}^+$), where Recall is half as important as Precision.

Even though F_1^+ is widely employed in statistics, it can be misleading, since it does not consider the component TN of the confusion matrix in its final score computation. That is, no relevance is given to the ratio of −1 correctly classified by the *ML* model. Furthermore, imbalanced classes are a common problem in *ML* classification where there is a disproportionate ratio of observations in each class [4]. In such case, Precision, Recall, Accuracy and F_1^+ are not informative performance measures for evaluating *ML* models.

Matthews Correlation Coefficient (MCC) [5] is a performance measure that uses all the components of a confusion matrix, and it is defined as follows:

$$MCC = \frac{TP \cdot TN - FP \cdot FN}{\sqrt{(TP + FP) \cdot (TP + FN) \cdot (TN + FP) \cdot (TN + FN)}} \qquad (4)$$

MCC is obtained as a geometric mean taking values in $[-1, 1]$. In this case, 1.00 means that both classes are perfectly classified; 0.00 means no better than a random prediction; -1.00 means a total disagreement between the observed and the predicted classes. Notice that the criterion behind *MCC* is intuitive and straightforward: to get a high-quality score, the classifier has to make correct predictions in most of the negative and positive instances, independently of their ratios in the overall dataset.

3   Unified Performance Measure

On the basis of the F_1^+ structure, it is possible to define the F_1^- that is the harmonic mean of the Specificity and Negative Predictive Value (see Table 2):

$$F_1^- = 2 \cdot \frac{NegativePredictiveValue \cdot Specificity}{NegativePredictiveValue + Specificity} = \frac{2 \cdot TN}{2 \cdot TN + FP + FN} \qquad (5)$$

F_1^- is a trade-off between the success of predicting an observation as -1 and the ratio of right predictions in the negative class. As F_1^+, F_1^- is defined in $[0, 1]$ and it can be generalized to F_β^-.

Thus, F_1^+ and F_1^- reflect different, though related, aspects of classification tasks. The first focuses on the positive class, whereas the second on the negative class. In this paper, we propose a new performance measure, the *Unified Performance Measure (UPM)*, which considers all the elements in a confusion matrix. It is defined as the harmonic mean of the F_1^+ and F_1^- measures:

$$UPM = 2 \cdot \frac{F_1^+ \cdot F_1^-}{F_1^+ + F_1^-} \qquad (6)$$

It is straightforward to express *UPM* as the harmonic mean of the four measures in Table 2:

$$UPM = \frac{4}{\frac{1}{Precision} + \frac{1}{Recall} + \frac{1}{Specificity} + \frac{1}{NegativePredictiveValue}} \qquad (7)$$

Notice that *UPM* is the product of the metrics divided by the sum of the products for each combination of three elements of these metrics.

$$UPM = 4 \cdot \frac{PPV \cdot TPR \cdot TNR \cdot NPV}{PPV \cdot TPR \cdot NPV + PPV \cdot TPR \cdot TNR + NPV \cdot TNR \cdot PPV + NPV \cdot TNR \cdot TPR} \qquad (8)$$

In addition, *UPM* can be calculated directly from the confusion matrix:

$$UPM = \frac{4 \cdot TP \cdot TN}{4 \cdot TP \cdot TN + (TP + TN) \cdot (FP + FN)} = \frac{1}{1 + \frac{(TP+TN) \cdot (FP+FN)}{4 \cdot TP \cdot TN}} \tag{9}$$

This new measure is defined in the range [0,1]. When the *UPM* value is equal to 1.00, it means perfect agreement. If it is equal to 0.00, the measure indicates that at least one of the metrics in Table 2 equals 0.00. This means that (at least) one of the two classes does not have any observation correctly classified. When *UPM* is equal to 0.50, it indicates randomness. It is expected that *UPM* performs properly for imbalanced classification problems, since it is built by using information regarding the performance of a classifier on both classes.

4 Experiments

In order to evaluate the relative performance of *UPM* regarding several alternative measures, two different scenarios have been proposed. First, a battery of simulated confusion matrices is considered. In the second scenario, two real datasets are used. The proposed *UPM* is compared, both on interpretability and stability, with Accuracy, different options of the F-score, *MCC* and the metrics in Table 2: Recall, Specificity, Precision and Negative Predictive Value.

4.1 Simulated Confusion Matrices

To evaluate the proposed measure, several simulated confusion matrices are considered. These matrices, reported in Table 3, contain all the information needed to calculate the previously mentioned performance measures.

Table 3. Balanced and imbalanced confusion matrices.

a)	Obs. −1	Obs. +1		b)	Obs. −1	Obs. +1		c)	Obs. −1	Obs. +1		d)	Obs. −1	Obs. +1		e)	Obs. −1	Obs. +1
Pred. −1	5	94		Pred. −1	4	94		Pred. −1	1	0		Pred. −1	1	1		Pred. −1	1	5
Pred. +1	0	1		Pred. +1	1	1		Pred. +1	94	5		Pred. +1	94	4		Pred. +1	4	90

f)	Obs. −1	Obs. +1		g)	Obs. −1	Obs. +1		h)	Obs. −1	Obs. +1		i)	Obs. −1	Obs. +1		j)	Obs. −1	Obs. +1
Pred. −1	90	4		Pred. −1	45	5		Pred. −1	90	1		Pred. −1	25	25		Pred. −1	28	22
Pred. +1	5	1		Pred. +1	5	45		Pred. +1	2	7		Pred. +1	25	25		Pred. +1	22	28

The simulated confusion matrices include balanced and imbalanced classification problems (both with high and low performance). The matrices a), b), c),

d), e), f), and h) correspond to imbalanced classification problems. The matrices
g), i) and j) are examples of balanced classification problems. In cases a), b), c)
and d), the prediction is very poor for the predominant class. In cases e) and f),
the prediction is very poor for the non-predominant class. Examples g) and h)
present a high global performance in both classes for balanced and imbalanced
datasets, respectively. Finally, examples i) and j) correspond to low performance
in both classes. Table 4 shows the results of the metrics for all these cases.

Table 4. Performance measure values in the simulated confusion matrices.

Metrics	Confusion matrices									
	a)	b)	c)	d)	e)	f)	g)	h)	i)	j)
Recall	0.01	0.01	1.00	0.80	0.94	0.20	0.90	0.87	0.50	0.56
Specificity	1.00	0.80	0.01	0.01	0.20	0.94	0.90	0.98	0.50	0.56
Precision	1.00	0.50	0.05	0.04	0.95	0.16	0.90	0.78	0.50	0.56
NPV	0.05	0.04	1.00	0.50	0.16	0.95	0.90	0.99	0.50	0.56
Accuracy	0.06	0.05	0.06	0.05	0.91	0.91	0.90	0.97	0.50	0.56
$F_{0.5}^{+}$	0.05	0.05	0.06	0.05	0.95	0.17	0.90	0.79	0.50	0.56
$F_{0.5}^{-}$	0.06	0.05	0.05	0.05	0.17	0.95	0.90	0.99	0.50	0.56
F_{1}^{+}	0.02	0.02	0.09	0.07	0.95	0.18	0.90	0.82	0.50	0.56
F_{1}^{-}	0.09	0.07	0.02	0.02	0.18	0.95	0.90	0.98	0.50	0.56
F_{2}^{+}	0.01	0.01	0.21	0.17	0.95	0.19	0.90	0.85	0.50	0.56
F_{2}^{-}	0.21	0.17	0.01	0.01	0.19	0.95	0.90	0.98	0.50	0.56
MCC	0.02	−0.29	0.02	−0.29	0.13	0.13	0.80	0.80	0.00	0.12
UPM	0.03	0.03	0.03	0.03	0.30	0.30	0.90	0.90	0.50	0.56

In the confusion matrices a) and b), the Precision and Specificity are very
high, and the Recall and NPV are very low. On the other hand, in the matrices
c) and d), the Precision and Specificity are very low, and the Recall and NPV
are very high. These cases present low Accuracy. In all these cases, the proposed measure, UPM, achieves a similar low value. Notice that a similar value
is expected since there is only a change in the classification of one observation
between a) and b), and their equivalent confusion matrices c) and d) with the
classes swapped.

In the confusion matrix e), the Precision and Recall are very high, and the
Specificity and NPV are very low. Thus, F^{+} achieves high values, since the
performance is good for the positive class. However, F^{-} achieves low values since
the performance is poor for the negative class. In the confusion matrix f), the
Precision and Recall are very low, and the Specificity and NPV are very high. In
this case, F^{+} achieves low values, since the performance is poor for the positive
class. However, F^{-} achieves high values since the performance is good for the
negative class. Both cases e) and f) are examples of high performance for the

predominant class but very poor for the non-predominant class. Notice that the confusion matrix e) is equal to the confusion matrix f) with the classes swapped. In these two cases, *UPM* achieves a value close to 0.50. Thus, *UPM* provides the same value regardless of the definition of the positive and negative classes. The confusion matrices g) and h) are examples of high-performance prediction models. All the considered measures, including *UPM*, are high in both cases. The confusion matrices f) and g) correspond to very similar values of Accuracy, which makes sense for the latter since it classifies both classes correctly, however, for f) the negative class is classified almost entirely correctly, but for the positive class, it is classified incorrectly. The confusion matrices i) and j) are examples of random predictions. *UPM* is close to 0.50 in both cases.

It should be noted that when the classification performs well, all the metrics perform well too. Therefore, *UPM* and *MCC* are the only measures that do not depend on which class is predominant. This poses an advantage over F^+, F^- and Accuracy which are not invariant to swapping the classes. According to this research, both *UPM* and *MCC* are better for imbalanced data. Both *UPM* and *MCC* take into account true and false positives and negatives. By examining the confusion matrices a) and b), it can be seen that *MCC* returns a different value for similar cases when only one observation is classified differently. Moreover, *MCC* returns similar values for the confusion matrices i) (random), a) (high Precision and low Recall), and c) (low Precision and high Recall). On the other hand, *UPM* presents low values when at least one of the involved metrics is low. In the random confusion matrix, its value is 0.50. Given the low performance on the non-predominant class, *UPM* achieves values lower than 0.50 for the confusion matrices e) and f). However, *MCC* achieves higher values for the confusion matrices e) and f) (0.13 in both cases) than for the random confusion matrix i) (0.00). In addition, the balanced confusion matrix j) has a *MCC* value similar to the imbalanced confusion matrices e) and f) that are almost wrongly classifying an entire class. In this case, *UPM* obtains a higher value in the confusion matrix j).

4.2   Real Data

The performance of the *UPM* is evaluated on two real datasets. First, an imbalanced dataset in the tourism domain is considered [6]. The goal is to determine if a client would recommend a hotel based on a questionnaire filled upon departure. This dataset has been divided into two sets: training set (70%) to fit a logistic regression model and testing set (30%). The second dataset was provided by the Digitanimal project [2]. The aim is to detect events that might signal calving in cattle. This dataset contains a time series for each animal over a period of 40 days. The first 30 days are used to fit the method proposed in [7], whereas the last 10 days are used as the testing set.

For each real dataset, several values for $\varepsilon \in [0, 1]$ are used as different classification thresholds, generating confusion matrices for the testing set. Then, the previously presented metrics are calculated for each confusion matrix.

Figure 1 presents a comparison of the metrics for each dataset and threshold. It can be seen that in graphical terms, *UPM* is more centred and similar to the basic metrics compared to *MCC*, which indicates that it is more stable. Then, the Euclidean distance between both *UPM* and *MCC*, and each of the basic metrics in Table 2, is shown in Table 5. The average distance between *UPM* and the four basic metrics is 1.46 ± 0.35, and 0.70 ± 0.26 for the tourism and cattle datasets, respectively. Regarding *MCC*, this average distance is 1.65 ± 0.67, and 1.82 ± 0.67 for each dataset. Notice that the means and standard deviations are lower for *UPM*. This supports the previous findings in the graphical analysis.

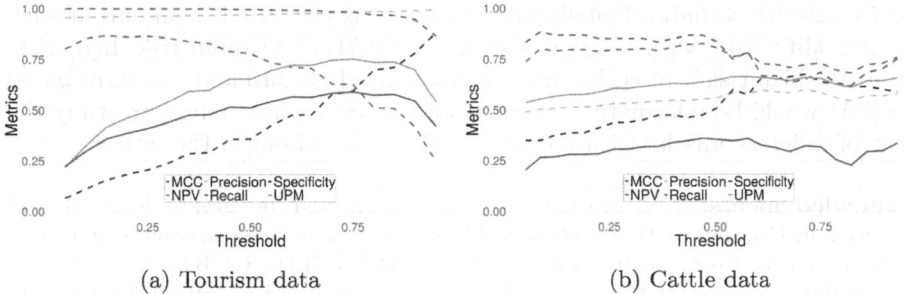

(a) Tourism data   (b) Cattle data

Fig. 1. Comparison of *UPM* and *MCC* with the basic metrics, regarding different values of ε for each real dataset.

Table 5. Euclidean distance between both *UPM* and *MCC*, and each of the basic metrics.

Dataset	Metrics	Metrics			
		Precision	Recall	Specificity	NPV
Tourism	UPM	1.72	1.79	1.06	1.26
	MCC	2.17	2.23	0.81	1.40
Cattle	UPM	0.36	0.99	0.66	0.79
	MCC	1.27	2.45	1.21	2.35

5   Conclusions

Performance measures are critical instruments for assessing and expressing the success of a binary classification study. Although many measures are available in the literature, only a few are reported in the conducted studies. In addition, there is no consensus on a unique optimal measure. For instance, Accuracy, F_1^+ and F_1^- can generate misleading results on imbalanced datasets because they fail to consider the ratio between positive and negative elements.

In this study, a unified performance measure (UPM) that combines the information from other performance measures has been proposed. It has been compared with MCC, Accuracy, F_1^+, F_1^- and other basic metrics. In the imbalanced classification problems, the new measure UPM improves the stability of MCC and veracity of Accuracy, F_1^+ and F_1^-. The evidence from this study suggests the use of UPM for both imbalanced and balanced data.

Future work will focus on generalizing the use of UPM for multi-class problems. An extension of UPM to evaluate the performance of a multi-class classifier on binary classification problems will be considered. In many binary real problems, the most informative output is not a binary output but an output with more than two options. For instance, consider the case of life-or-death prediction in a health scenario. Usually, the researcher is interested in an output such as "mortality risk" with categories such as: "unlikely, medium risk, high risk, extremely high risk". In such a case, a generalized performance measure based on UPM would be informative. Furthermore, a sensitivity analysis to study the effect of different misclassification costs will be carried out in the future.

Acknowledgments. This research has been supported by grants from Madrid Autonomous Community (Ref: IND2018/TIC-9665) and the Spanish Science and Innovation, under the Retos-Colaboración program: SABERMED (Ref: RTC-2017-6253-1); and the Retos-Investigación program:MODAS-IN (reference: RTI-2018-094269-B-I00). Special thanks to MISC International S.L and HOTELS QUALITY.

References

1. Canbek, G., Sagiroglu, S., Temizel, T.T., Baykal, N.: Binary classification performance measures/metrics: a comprehensive visualized roadmap to gain new insights. In: 2017 International Conference on Computer Science and Engineering (UBMK), pp. 821–826. IEEE (2017)
2. Digitanimal (2020). https://www.digitanimal.com/
3. Goodall, D.W.: The distribution of the matching coefficient. Biometrics **9**, 647–656 (1967)
4. He, H., Garcia, E.A.: Learning from imbalanced data. IEEE Trans. Knowl. Data Eng. **21**(9), 1263–1284 (2009)
5. Matthews, B.W.: Comparison of the predicted and observed secondary structure of t4 phage lysozyme. Biochimica et Biophysica Acta (BBA)-Protein Struct. **405**(2), 442–451 (1975)
6. Muñoz, J., Moguerza, J., Martín Duque, C., Bruna, D.: A study on the effect of imbalanced data in tourism recommendation models. Int. J. Qual. Serv. Sci. ahead-of-print, 08 2019. https://doi.org/10.1108/IJQSS-05-2018-0050
7. Navarro, J., Diego, I.M.d., Fernández-Isabel, A., Ortega, F.: Fusion of GPS and accelerometer information for anomalous trajectories detection. In: Proceedings of the 2019 the 5th International Conference on e-Society, e-Learning and e-Technologies, pp. 52–57 (2019)
8. Sasaki, Y., et al.: The truth of the f-measure, vol. 2007 (2007)
9. Sokolova, M., Lapalme, G.: A systematic analysis of performance measures for classification tasks. Inf. Process. Manage. **45**(4), 427–437 (2009)

Data Augmentation for Industrial Prognosis Using Generative Adversarial Networks

Patxi Ortego[1,2]($\boxtimes$), Alberto Diez-Olivan[1,2], Javier Del Ser[1,3], and Basilio Sierra[2]

[1] TECNALIA, Basque Research and Technology Alliance (BRTA),
20009 Donostia-San Sebastián, Spain
patxi.ortego@tecnalia.com
[2] Department of Computer Sciences and Artificial Intelligence, University
of the Basque Country (UPV/EHU), 20018 Donostia-San Sebastián, Spain
[3] Department of Communications Engineering, University of the Basque
Country (UPV/EHU), 48013 Bilbao, Spain

Abstract. The Industry 4.0 revolution allows monitoring and intelligent processing of big amounts of data. When monitoring certain assets, very few data is found for operation under faulty conditions because the cost of not operating properly is unacceptable and thus preventive strategies are put in practice. Because machine learning algorithms are data exhaustive, synthetic data can be created for these cases. Deep learning techniques have been proven to work very well for these cases. Generative Adversarial Networks (GANs) have been deployed in numerous applications with data augmentation objectives, but not so much for balancing unidimensional series with few data. In this paper, a GAN is applied in order to augment data for assets operating under faulty conditions. The proposed method is validated on a real industrial case, yielding promising results with respect to the case with no strategy for class imbalance whatsoever.

Keywords: Generative Adversarial Networks · Data augmentation · Imbalanced data · Deep learning

1 Introduction

Machine learning techniques in general have gained popularity recently in any data intensive field as a consequence of the digitization of our era. The traditional manufacturing and industrial processes in particular, are immersed in the Industry 4.0 revolution. The digitization of the industry has led to significant changes including the application of the latest advances in information technology. Nevertheless, this Industry has its particularities and a wide range of different necessities depending on the application.

A common issue when applying data analytic techniques to problems related to Industry 4.0 is the lack of some particular class data, i.e., the class imbalance

C. Analide et al. (Eds.): IDEAL 2020, LNCS 12490, pp. 113–122, 2020.
https://doi.org/10.1007/978-3-030-62365-4_11

problem. In installation processes, for instance, the majority of available data belongs to zero defects products for safety requirements (e.g., aircraft manufacturing). In some cases in fact, there is almost no data, or even none at all, of incorrectly installed operations. Models trained with imbalanced datasets can lead to underperforming and/or biased models [9].

Different approaches can be found in the literature to tackle this issue, such as [11] that proposes the use of instanced-based machine learning models on the existing data, without any oversampling or undersampling. Nevertheless, for the modeling of complex behaviors (which is usually the case for most industrial assets) where Deep Learning models are applied, oversampling the minority class does not only minimize the effects of class imbalance, but can also improve the algorithm's performance as they are data intensive algorithms. Given the ample consensus on the paramount role that machine learning methods play nowadays in industrial applications [4] and the high incidence of the class imbalance problem in this domain, there is a clear need for assessing the performance of different strategies to deal with this issue in real-world problems.

This being said, there are different methods and techniques in the literature to deal with imbalanced datasets, from the simplest ones, random undersampling and oversampling, to more complex ones like ADASYN [8] or SMOTE [2]. For binary classification problems, outlier detection can be a feasible option. Samples belonging to the minority class are considered outliers. Nevertheless, methods grounded on this assumption are limited to binary problems, and it is difficult to determine the exact behavior represented by each sample detected as outlier, which is one of the main issues in industrial processes [1].

Nevertheless, with the recent upsurge of Deep Learning models, techniques such Generative Adversarial Networks (GANs) have gained popularity for tackling the class imbalance problem. Actually, synthetic data generation by means of GANs has been demonstrated to outperform SMOTE and ADASYN (and its variants) in different contributions throughout the literature [14,20]. GANs were first introduced in [6]. They can learn deep representations by deriving backpropagation signals through two networks (the Generator and the Discriminator), which are trained simultaneously by using a two-player minimax game. Different GAN surveys and reviews can be found in the literature [3,7,17,19]. GANs have also been applied for Data Augmentation, especially for image data [13,18].

The goal of this work aligns with the current momentum around GAN-based strategies for class imbalance. Specifically, we generate synthetic data to balance an imbalanced problem in a real industrial scenario composed of one dimensional data. Previous research in this field can be found in [16], where the authors propose an auxiliary classifier GAN(ACGAN)-based framework to generate one dimensional data from mechanical sensor signals. Also related is the work in [12], which resorts to GANs to generate raw one dimensional sound data. Likewise, [10] proposes a GAN based approach for anomaly detection on imbalanced one-dimensional data. In this paper we build upon this prior work to propose a GAN-based approach for creating synthetic data to oversample the minority class in order to balance the dataset, which involves a real industrial scenario. To evaluate and measure the improvement, baseline models are trained and evaluated with and without the augmented dataset. The obtained results are conclusive, confirming that the use of GANs for balancing data related to correctly and incorrectly installed operations in aircraft manufacturing processes allows obtaining higher accuracy in classification tasks.

The remainder of the manuscript is structured as follows: Section 2 describes the proposed GAN model data augmentation. Section 3 describes the experimental setup and discusses on the obtained results. Finally, Sect. 4 ends the work by underscoring the main conclusions of this work and outlining future research inspired by our findings.

2   Proposed GAN Model for Class Imbalance

The GAN architecture proposed in this paper is described in this section. The Generator is composed of an input layer with 100 neurons (the size of the latent vector dimension). The hidden layers are three fully connected layers with 128, 256 and 512 neurons respectively, a dropout rate of 0.3 and *Leaky ReLU* activation functions. The output layer neurons are the same as the signal length (230 as described in Sect. 3.2 and a *tanh* activation function.

The Discriminator is composed of an input layer with neurons matching the signal length, which is equal to 230 samples. It has 3 hidden layers of 512, 256 and 128 neurons, respectively. At the Discriminator network we set a dropout rate of 0.3 and *Leaky ReLU* activation functions. The output layer has a single neuron and a *sigmoid* activation function. Figure 1 illustrates both models in an schematic way.

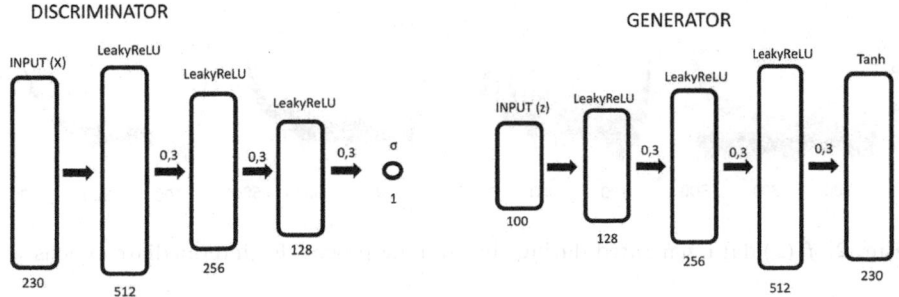

Fig. 1. Discriminator and Generator networks utilized in our scenario.

The training process consists of two steps that are repeated. First, the Discriminator is trained while the Generator is frozen (if it is the first epoch, the parameters of the Generator are initialized at random as per a standard Glorot strategy). Using the latent vector generator and real data, the Discriminator is trained for one epoch. During the next epoch. the Generator is trained by using the Discriminator's loss. ADAM optimizer ($\beta_1 = 0.5$ and $\beta_2 = 0.999$) is adopted for model training with learning rate 0.0002 for both the Discriminator and the Generator. *Binary cross-entropy* is used as the loss function.

3   Experimental Evaluation

The evaluation of the proposed approach is done in two different steps. First, the designed GAN is validated with four defined analytical functions towards guaranteeing that it learns efficiently the distribution of one-dimensional signals. The second step consists of testing the method with real data produced in an real-world industrial case.

3.1   Step 1: GAN Applied to Defined Functions

The advantages of testing the approach with user-defined analytical functions is that the synthetic data quality can be easily evaluated. It is first evaluated by simple *visual inspection* in a similar way to the case of synthetically generated images. The distribution of data is known and can be plotted, so a qualitative evaluation can be easily performed. The functions in use are the following ones:

1. $f_1(x) = x^{10} - x^7 + x^3 + 0.01$ for $x \in \mathbb{R}[-1, 1]$
2. $f_2(x) = x^{10} + x^7 + x^3 + 0.01$ for $x \in \mathbb{R}[0, 50]$
3. $f_3(x) = x^{10} - x^8$ for $x \in \mathbb{R}[-1, 1]$
4. $f_4(x) = \sin(x)$ for $x \in \mathbb{R}[0, 50]$

Each function is evaluated in 500 evenly distributed points, and Gaussian noise is added to create different series. They are normalized between 0 and 1.

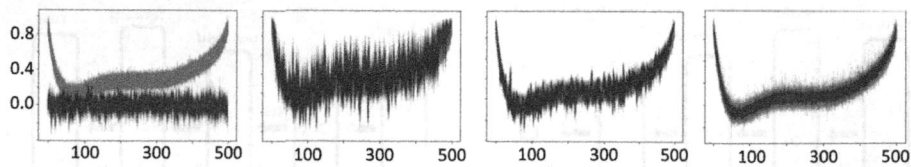

Fig. 2. $f_1(x)$ data generated during the training process for intermediate epochs.

Figure 2 illustrates the training process for some intermediate epochs for data generation using $f_1(x)$. It can be observed that, at the beginning, the Generator output is mostly noise centered around 0 (due to the fact that the activation function is tanh()). As the learning process evolves, the characterization process improves to the point of learning to model the underlying distribution faithfully. A deeper analysis can be performed to analyze the convergence properties of the designed GAN model. However, we have omitted it since the interest of this first experimental step is to verify that the GAN works as expected. In light of the results in Fig. 2 and those corresponding to the rest of analytical functions, our GAN is able to model them efficiently, paving the way towards the application of the developed GAN to a more challenging real-world case.

3.2   Step 2: GAN on Real-World Industrial Data

The approach has been tested with real data belonging to a real industrial case. The problem to solve is described in detail in [15] and [5]. In essence, the problem involves the analysis and modeling of condition monitoring data acquired from a blind fastener installation process related to the aeronautical industry. A fastener is a mechanical device to assemble two or more components together. The installation process is monitored to obtain a torque rotation signal that characterizes the operation. In this particular case, it is a blind fastener because the fastener is only visible from the side it is installed. This makes the labeling process (*correctly installed* or *incorrectly installed*) time consuming, hence the benefit of obtaining realistic synthetic data belonging to the real distribution for the training of the models.

A torque-rotation signal describes the evolution along the time of the fastener installation. The raw signals are preprocessed by aligning them in the rotation axis (equivalent to time dimension) to the highest point by cross-correlation and then normalized between 0 and 1 and in both dimensions, head diameter and height, aiming at minimizing the variation effect in process conditions.

Each sequence consists of 11,587 samples which are downsampled to 230 (one out of fifty, approximately). The downsampling process makes physical sense because of the dynamics of the signal,. Depending on particularities of the process, the installation can be classified in different classes depending on the reason for failure. Therefore, the *correctly installed* and *incorrectly installed* classes can be further developed into other subclasses depending on the installation characteristic. With the goal of having an imbalanced data set, only one of these subclasses belonging to the *incorrectly installed* class is used. There are 149 *correctly installed* samples and 30 *incorrectly installed*, roughly a 0.2 imbalance ratio. The data is shown in Fig. 3.

To evaluate the quality of the synthetic data, a first evaluation is done by visual inspection. Even though this method is not accurate and identifying signals can not be done by eyesight, a signal that clearly does not belong to any of the underlying distributions can be spotted and therefore it is a good starting point to ascertain whether the GAN training process converges. Figure 4 shows some intermediate stages of the training process. Similarly to the prior experimental phase, the proposed GAN model seems to converge towards modeling the real data of the scenario under analysis.

The further quantitative evaluation is performed by training different models and evaluating the model with and without synthetic data. A stratified k-fold cross-validation is used with $k=5$. The test samples only consist of real samples. Each model is trained twice, one with only real data and another with both real data and synthetic data (augmented scenario where training data is balanced).

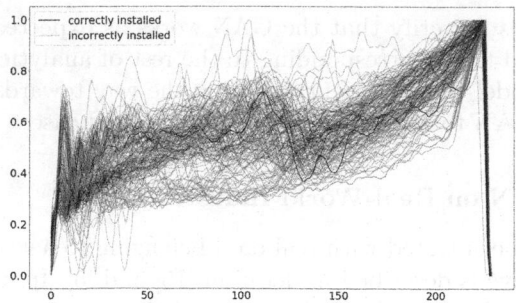

Fig. 3. Real data corresponding to *correctly installed* and *incorrectly installed* classes.

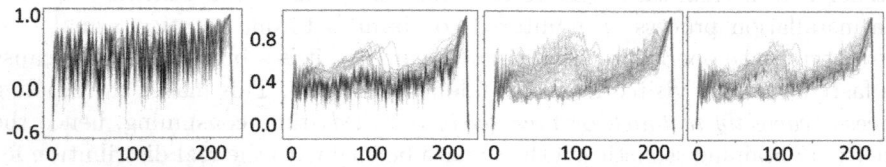

Fig. 4. Real data training process for some intermediate epochs.

The synthetic data belongs to the minority class, and the number of samples is the difference between the majority class and minority class for each particular fold, so the data becomes completely balanced ($n_{synthetic} = n_{majority} - n_{minority}$ where n denotes number of samples). The utilized models and their fixed hyper-parameters are described in Table 1. Table 2 denotes the optimized values found by exhaustive grid search for some critical hyper-parameters of the models. This search process was driven by the maximization of the $F1$ score.

Table 1. Fixed parameters for each of the considered classification models.

Model	Parameters
Random Forest (RF)	Criterion = gini, max_depth = 50
Logistic Regression (LR)	Penalty = 'l2'
Support Vector Classifier (SVC)	Kernel = 'linear', Penalty = 'squared l2'
Gaussian Naive Bayes classifier (GNB)	None
k-Nearest Neighbors (KNN)	Leaf_size = 30, p = 2

Results of this performance study are shown in Table 3 in the form of *F1* score and *Cohen's kappa coefficient*. As it can be seen in that table, the models that are trained with synthetic data generally obtain better results than those with only real data. Gaussian Naive Bayes algorithm outperformed using only real data in comparison to its augmented case, whereas regarding the other algorithms the augmented case always worked better. Logistic Regression is the

Table 2. Parameters search for each classification model over different data partitions.

Model (parameters)	Real and synthetic data					Only real data				
	K = 0	K = 1	K = 2	K = 3	K = 4	K = 0	K = 1	K = 2	K = 3	K = 4
RF (estimators)	13	7	14	14	6	15	5	18	18	45
LR (C)	8.2	1.6	3.8	1.1	2.6	2.9	7	8.6	5.5	3.8
SVC (C)	0.1	0.2	0.1	0.1	0.1	0.3	0.4	0.7	0.2	0.5
KNN (neighbors)	7	4	17	3	3	3	3	3	3	11

Table 3. Results obtained for the baseline models with and without GAN-generated synthetic data for class imbalance. Best results are highlighted in bold. Note that *aug* is the case for training with both synthetic and real data.

K	LR		RF		SVC		GNB		KNN	
	aug	real	aug	real	aug	real	aug	real	aug	real
0	1.00	0.80	0.67	0.73	1.00	0.91	0.57	0.80	1.00	0.80
1	1.00	1.00	0.75	0.62	1.00	1.00	0.71	1.00	1.00	1.00
2	1.00	1.00	0.83	0.67	1.00	1.00	0.83	0.91	0.91	0.91
3	1.00	0.91	0.92	0.91	1.00	0.91	0.91	0.91	0.91	0.91
4	1.00	1.00	0.92	0.91	1.00	1.00	0.71	1.00	1.00	0.80
MEAN	**1.00**	0.77	**0.82**	0.77	**1.00**	0.96	0.75	**0.92**	**0.96**	0.88
STD	**0.00**	0.11	**0.10**	0.12	**0.00**	0.04	0.12	**0.07**	**0.04**	0.08

(A) F1-score

K	LR		RF		SVC		GNB		KNN	
	aug	real	aug	real	aug	real	aug	real	aug	real
0	1.00	0.77	0.60	0.68	0.89	0.89	0.47	0.77	1.00	0.77
1	1.00	1.00	0.68	0.53	1.00	1.00	0.63	1.00	1.00	1.00
2	1.00	1.00	0.80	0.63	1.00	1.00	0.80	0.89	0.89	0.89
3	1.00	0.89	0.91	0.89	0.89	0.89	0.89	0.89	0.89	0.89
4	1.00	1.00	0.91	0.89	1.00	1.00	0.62	1.00	1.00	0.75
MEAN	**1.00**	0.72	**0.78**	0.72	0.96	0.96	0.68	**0.91**	**0.96**	0.86
STD	**0.00**	0.12	**0.12**	0.15	0.05	0.05	0.15	**0.09**	**0.05**	0.09

(B) Kappa score

algorithm that best performs using augmented data, obtaining the best overall scores. It must be noted that the test set is composed of exclusively real data that none of the models (neither the signal generator nor the baseline classifiers) have utilized during training.

Finally, Fig. 5 shows the diversity of the synthetic data instances belonging to the *incorrectly installed* class. In particular we depict the mean of real and synthetic data, as well as the standard deviation of the latter. As elucidated in this plot, the generated data fits the underlying distribution of the few real instances belonging to this minority class, which is the goal of the GAN approach and provides evidence of its suitability for the real-world scenario at hand.

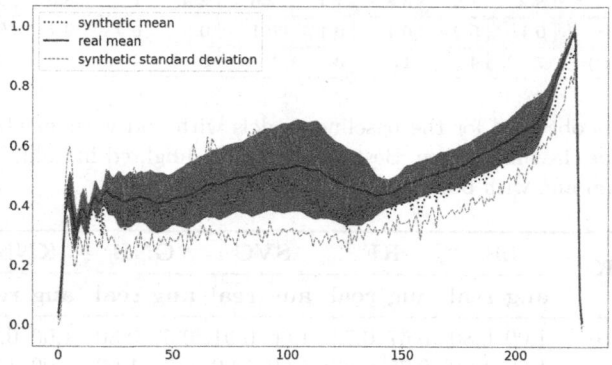

Fig. 5. Real and synthetic data mean and standard deviation for the *incorrectly installed* class.

4   Conclusions and Future Work

This article has elaborated on the implementation of a methodology for minority class synthetic data generation and its validation on real data. Specifically, GANs have been used for this purpose in a blind fastener installation process, which is a essential part of an aircraft manufacturing plant. The majority data (149 samples) belongs to the *correctly installed* class, whereas the minority data (30 samples) to the *incorrectly installed* class. Baseline models have been trained with both synthetic data and only real data. The models trained with augmented data obtain better results than the models trained with only real data, except for the Gaussian Naive Bayes. Logistic Regression obtained the best overall result when trained with augmented data. These results indicate that the provision of synthetically generated data improves the model at the training stage. Nevertheless, as there are few minority class data samples (not only compared to the majority class but overall), an in-depth study could include using common knowledge from all classes.

These results are a promising start for further research in this topic. As explained in Sect. 3.2, each raw signal consists of 11,587 points, out of which only 230 samples are used. As mentioned before, this downsizing strategy makes sense from a physical perspective because of the dynamics of the process and also because there are not enough samples to properly use all data. Future work could include more sophisticated downsampling methods to this end, potentially focusing on those time instants where more signal variability is detected. Further along this line, more complex neural networks can be designed for synthetic data generation. Finally, a framework to quantify the quality of GAN-based generated data is needed, so that the approach can be extrapolated to other real-world industrial cases where this particularity is of utmost importance (e.g. precision machining).

Acknowledgments. This project was supported by the Spanish Centro para el Desarrollo Tecnologico Industrial (CDTI, Ministry of Science and Innovation) through the "Red Cervera" Programme (AI4ES project), as well as by the Basque Government through EMAITEK and ELKARTEK (ref. KK-2020/00049) funding grants. J. Del Ser also acknowledges support from the Department of Education of the Basque Government (Consolidated Research Group MATHMODE, IT1294-19).

References

1. Beyan, C., Fisher, R.: Classifying imbalanced data sets using similarity based hierarchical decomposition. Pattern Recogn. **48**(5), 1653–1672 (2015)
2. Chawla, N.V., Bowyer, K.W., Hall, L.O., Kegelmeyer, W.P.: SMOTE: synthetic minority over-sampling technique. J. Artif. Intell. Res. **16**(9), 321–357 (2002)
3. Creswell, A., White, T., Dumoulin, V., Arulkumaran, K., Sengupta, B., Bharath, A.A.: Generative adversarial networks: an overview. IEEE Signal Process. Mag. **35**(1), 53–65 (2018)
4. Diez-Olivan, A., Del Ser, J., Galar, D., Sierra, B.: Data fusion and machine learning for industrial prognosis: trends and perspectives towards industry 4.0. Inf. Fus. **50**, 92–111 (2019)
5. Diez-Olivan, A., Penalva, M., Veiga, F., Deitert, L., Sanz, R., Sierra, B.: Kernel density-based pattern classification in blind fasteners installation. In: Martínez de Pisón, F.J., Urraca, R., Quintián, H., Corchado, E. (eds.) HAIS 2017. LNCS (LNAI), vol. 10334, pp. 195–206. Springer, Cham (2017). https://doi.org/10.1007/978-3-319-59650-1_17
6. Goodfellow, I., et al.: Generative adversarial nets. In: Ghahramani, Z., Welling, M., Cortes, C., Lawrence, N.D., Weinberger, K.Q. (eds.) Advances in Neural Information Processing Systems, vol. 27, pp. 2672–2680. Curran Associates, Inc. (2014). http://papers.nips.cc/paper/5423-generative-adversarial-nets.pdf
7. Gui, J., Sun, Z., Wen, Y., Tao, D., Ye, J.: A review on generative adversarial networks: Algorithms, theory, and applications (2020)
8. He, H., Bai, Y., Garcia, E.A., Li, S.: ADASYN: adaptive synthetic sampling approach for imbalanced learning. In: IEEE International Joint Conference on Neural Networks, pp. 1322–1328 (2008)
9. He, H., Garcia, E.A.: Learning from imbalanced data. IEEE Trans. Knowl. Data Eng. **21**(9), 1263–1284 (2009)
10. Jiang, W., Hong, Y., Zhou, B., He, X., Cheng, C.: A GAN-based anomaly detection approach for imbalanced industrial time series. IEEE Access **7**, 143608–143619 (2019)
11. Lee, T., Lee, K.B., Kim, C.O.: Performance of machine learning algorithms for class-imbalanced process fault detection problems. IEEE Trans. Semicond. Manuf. **29**(4), 436–445 (2016)
12. Madhu, A., Kumaraswamy, S.: Data augmentation using generative adversarial network for environmental sound classification. In: European Signal Processing Conference, pp. 1–5 (2019)

13. Mehta, K., Kobti, Z., Pfaff, K., Fox, S.: Data augmentation using CA evolved GANs. In: IEEE Symposium on Computers and Communications, pp. 1087–1092 (2019)
14. Oh, J.H., Hong, J.Y., Baek, J.G.: Oversampling method using outlier detectable generative adversarial network. Expert Syst. Appl. **133**, 1–8 (2019)
15. Ortego, P., Diez-Olivan, A., Del Ser, J., Veiga, F., Penalva, M., Sierra, B.: Evolutionary LSTM-FCN networks for pattern classification in industrial processes. Swarm Evol. Comput. **54**, 100650 (2020). https://doi.org/10.1016/j.swevo.2020.100650
16. Shao, S., Wang, P., Yan, R.: Generative adversarial networks for data augmentation in machine fault diagnosis. Comput. Ind. **106**, 85–93 (2019)
17. Sharma, A., Jindal, N., Thakur, A.: Comparison on generative adversarial networks - a study. In: International Conference on Secure Cyber Computing and Communication, pp. 391–396 (2018)
18. Shorten, C., Khoshgoftaar, T.M.: A survey on image data augmentation for deep learning. J. Big Data **6**(1), 60 (2019)
19. Wang, K., Gou, C., Duan, Y., Lin, Y., Zheng, X., Wang, F.: Generative adversarial networks: introduction and outlook. IEEE/CAA J. Automatica Sinica **4**(4), 588–598 (2017)
20. Xie, Y., Zhang, T.: Imbalanced learning for fault diagnosis problem of rotating machinery based on generative adversarial networks. In: 2018 37th Chinese Control Conference (CCC), pp. 6017–6022, July 2018

A Comparison of Evolutionary Multi-objective Optimization Algorithms Applied to Antenna Design

Pedro B. Santos[1]([✉]), Marcello C. Melo[2], Everaldo Faustino Jr.[1],
Arismar Cerqueira S. Jr.[2], and Carmelo J. A. Bastos-Filho[1]

[1] University of Pernambuco - POLI-UPE, Recife, PE, Brazil
{pjbls,efsj,carmelofilho}@ecomp.poli.br
[2] Lab. WOCA, National Institute of Telecommunication - INATEL,
Santa Rita do Sapucaí, MG, Brazil
{marcellocaldano,arismar}@inatel.br

Abstract. In complex engineering problems, it is common to deal with more than two output target variables, making it challenging to obtain the best trade-offs among all output variables. Multi-objective optimization algorithms are promising candidates for providing Pareto Fronts that describe these possibilities. Particularly in antenna design, the input variables are geometrical elements associated with the antenna type. On the other hand, the output variables are the desirable performance indicators, such as resonance frequency, bandwidth, and gain. This paper aims to use several state-of-the-art multi-objective evolutionary algorithms and study the underlying mechanics of their operators to understand how we can optimally choose the antenna design parameters. Moreover, we propose an entire pipeline to automate this task, which is based on main phases: performing simulations using six multi-objective evolutionary algorithms, analyzing the convergence, Pareto front approximation, and quality indicators. Numerical results demonstrate the *OMOPSO* is a potential approach for the two evaluated studies of cases on antenna design.

Keywords: Evolutionary algorithms · Multi-objective optimization · Antenna · Device design

1 Introduction

Communication systems have critical importance in the modern world. A key protagonist in most communication systems is the antenna, which works as a transmitter and receiver. The antenna design must be conducted in an optimal way to ensure performance according to the application requirements. With the expansion of 5G and the beginning of discussions for 6G, new methodologies must be applied to develop new high-performance antennas for this era. A significant challenge in antenna design is that simulation software used requires

© Springer Nature Switzerland AG 2020
C. Analide et al. (Eds.): IDEAL 2020, LNCS 12490, pp. 123–134, 2020.
https://doi.org/10.1007/978-3-030-62365-4_12

high computational power and has a long-running time. HFSS is one of the most robust commercial software for electromagnetic simulations. That makes it difficult to efficiently apply optimization techniques to find the best parameters to build the antenna.

Advances in machine learning and artificial intelligence allow us to use a surrogate model in the form of high accuracy regression models to build a black box that mimetics the behavior of such a high computational cost software. With a trained regression model, we can use the surrogate model to work as an objective function in optimization algorithms. These algorithms are faster and require less computational power when compared with electromagnetic software. It also allows the creation of a well designed automated pipeline for this task and other similar engineering problems.

Solving problems with more than one objective function requires multi-objective algorithms. The goal is to find the best *trade-offs* among those objectives since the idea of a unique global best is lost in multi-objective optimization problems. In last years, methodologies were proposed to use multi-objective algorithms and surrogate models as a pipeline to be applied in the project and design of antennas [2,6,7]. Thus, no further investigation was made on the impact of the usage of different multi-objective algorithms, neither how the various underlying aspects of those algorithms impact on the entire process.

This paper aims to investigate which algorithms are more suitable for the project and design of antennas. The rest of this paper is structured as follows: Section 2 presents multi-objective optimization concepts; also, several algorithms are tested. We evaluate algorithms from different families such as population-dominance based such as *NSGA-II* and *SPEA2*, decomposition-based such as *MOEA/D*, quality-indicator based such as *IBEA* and PSO-derived algorithms such as *OMOPSO* and *SMPSO*. Quality indicators are also reviewed and how they are used to evaluate those algorithms. Section 3 is focused on the challenges of antenna design, where it is also presented the two study cases, which are the focus of this work. Section 4 presents the results of the proposed methodology. Finally, Sect. 5 reports the conclusion of this work, also presenting future directions for research.

2   Multi-objective Optimization

A Multi-Objective Optimization Problem (*MOOP*) differs from the traditional Scalar-Objective Optimization Problem (*SOOP*) by having a vectorial output space, with at least two dimensions. The main difference between *SOOPs* and *MOOPs* is that the first has a unique optimal solution, and the second presents a set of best *trade-off* solutions. Thus, the idea of a global minimum differs from *SOOPs*. A *MOOP* can be stated as follows:

$$\underset{x}{\text{minimize}} \quad F(x) = (f_1(x), f_2(x), \ldots, f_k(x)) : \mathbb{R}^d \longrightarrow \mathbb{R}^k$$
$$\text{s.t.} \qquad x \in X, \tag{1}$$

where k is the number objectives, and the input vector x is a d-dimensional vector from a euclidean space. Most of *MOEAs* deploy some concepts in their implementations. We state these definitions as follows.

Definition 1. *Pareto Dominance. Let $u = (u_1, \ldots, u_k)$ and $v = (v_1, \ldots, v_k)$ be two equal sized d-dimensional vectors. It is said that u dominates v, denoted by $u \preceq v$, if and only if $u_i \leq v_i \ \forall i$ and $\exists i$ s.t. $u_i < v_i$.*

Definition 2. *Pareto Optimal Set. From the definition of Pareto Dominance, the definition of Pareto Optimal Set, denoted by P, is straightforward:*

$$P = \{\, u \in \Omega \mid \nexists v \in \Omega \ s.t. \ F(v) \preceq F(u) \,\}.$$

In other words, P is the set of all decision vectors composed of non-dominated solution vectors in the objective space.

Definition 3. *Pareto Front. is the projection of the Pareto Optimal Set in the objective space. Thus, the Pareto Front, denoted as PF, can be stated as:*

$$PF = \{\, F(u) \mid u \in P \,\}.$$

2.1   Multi-objective Evolutionary Algorithms

MOEAs are population-based algorithms. They can find an approximation for P and PF sets in a single run. One of the most significant challenges for most of the *MOEAs* is to balance exploration and exploitation. The first concerns the search of the entire decision space and is related to the global aspect of the search process. The other regards to a refined search in promising areas of the decision space and is associated with the local search.

In this paper, the goal is to find a set of general-purpose *MOEAs* among the state-of-the-art algorithms, take a look at their mechanics and provide a comparison among them as far as the task of project and design of antennas is concerned. Many specific implementations were proposed in recent years [6,10], and several surveys in *MOEAs* and quality indicators were also published in recent years [1,3,9,11,14]. Some critics were made about these very problem-oriented implementations due to the lack of comparison with other *MOEAs* and the lack of applications in different real-world problems. In this paper, we have grouped a set of six algorithms from different families, such as population-dominance based such as *NSGA-II* and *SPEA2*, decomposition-based such as *MOEA/D*, quality-indicator based such as *IBEA* and PSO-derived algorithms such as *OMOPSO* and *SMPSO*. These algorithms are among the state-of-the-art approaches and well-know *MOEAs*. They are very often present in recent surveys and were widely used in many real-world engineering applications. A discussion of the structure and characteristics of the algorithm is given as follows.

SPEA2 was first proposed by Zitzler and Thiele in [20]. The *SPEA2* algorithm is an improved version of its original predecessor, the *SPEA* algorithm. It uses an external archive, where it keeps the non-dominated solutions found so far. According to [4], the *SPEA2* differs from its predecessor in three main points, which are, a fine-grained fitness assignment strategy, the nearest neighbor density estimation to guide the search and enhanced archive truncation method.

NSGA-II was first proposed in [5], as a generic non-explicit building block applied to MOOPs [4]. Its general purpose design allowed the development of several *MOEAs* that uses *NSGA-II* as an inner mechanism. In *NSGA-II*, the population of individuals competes against each other through an elitist mechanism. It ranks and sorts each individual according to their non-dominated level and makes usage of several genetic operators to generate diversity. Both *NSGA-II* and *SPEA2* became baselines for comparing new *MOEAs*, due to their large popularity and efficiency.

OMOPSO is a multi-objective variation of the original *PSO* algorithm proposed by Kennedy and Eberhart in [8]. Since then, several MOEAs were proposed [13]. Two main challenges in multi-objective *PSO* variants are: (1) how to select the leader from the archive to update the particles and (2) how to keep diversity to avoid premature sub-optimal convergence. *OMOPSO* was first proposed in [16], using crowding distance as an additional mechanism to save the non-dominated leaders on the archive through each iteration. It uses uniform and non-uniform mutation to maintain diversity in the swarm to balance between exploration and exploitation.

SMPSO was an improved multi-objective *PSO* based on the *OMOPSO* algorithm. It was first proposed in [12], and it has two main differences from its predecessor: (1) the velocity constriction mechanism, which changes the interval of C_1 and C_2 from the original *PSO* formula, and (2) the mutation operator, which adds polynomial mutation together with the uniform and non-uniform mutation. Although the changes seem small, it allowed the *SMPSO* to increase the speed of convergence and an improvement on the quality of the solutions on the final Pareto Front when compared with *OMOPSO*.

MOEA/D was first proposed in [17]. The major *MOEAs* treat the *MOOPs* as a whole. Dominance-based and quality indicators guided approaches have well-known limitations. Their main struggle is when dealing with high dimensional objective space, they have difficulties maintaining diversity, and soon the entire population becomes non-dominated. *MOEAs* based on decomposition are a recent trend on algorithmic framework design that aims to decompose the *MOOP* in several *SOOPs*, associating each solution to a subproblem and sharing information with their neighbors. Each subproblem keeps in memory the unique solution used to generate new ones with the use of genetic operators and their neighbors' help. If the new solution is better than the previous one, then the *SOOPs* is updated.

IBEA was one of the first indicator-based approaches, first presented in [19]. Quality-Indicator guided *MOEAs* to tackle *MOOPs* have become popular in recent years. Several indicators were proposed and used to measure the quality of the final Pareto Front [9]. Furthermore, the usage of such measurements to solve *MOOPs* arose naturally. *IBEA* was designed to use an arbitrary indicator to compare pairs of candidates' solutions. The quality indicator is used in the fitness assignment step and to guide the population improvement towards the metric. Quality-Indicators are also a way to insert user preference in the search process.

2.2 Quality Indicators for MOEAs

Many *MOEAs* have been developed in recent years, thus the need to compare and measure their performances arise in a natural fashion [1,3,9,11,14]. Many papers have proposed quality indicators to address the three main aspects of a *MOEA*: (1) Accuracy, which addresses how close to the true Pareto Front an algorithm has reached, (2) Cardinality, which addresses how many solutions the algorithm was able to found through the optimization process and (3) Diversity, which is how well-spread the solutions are on the Pareto Front approximation. In [14], a survey was performed to address the most popular and the most efficient quality indicators used to measure the performance of *MOEAs*. The chosen metrics were Hypervolume (HV), Spacing, and Spreading.

Hypervolume is also known as hyper-area or S metric [21], which is a unary metric that quantifies how much of the objective space was covered by the Pareto Front. It requires the definition of a reference point, which is used to compute the hypervolume. This point is selected far from the Pareto Front's points and is a point that is dominated by all other points in the Pareto Front. The normalized HV metric is within $[0, 1]$. It considers all three aspects of *MOEAs*, and it can be used to determine when the Pareto Front of a *MOEA* is better than another. The larger the value of the HV metric, the better.

Spacing was first proposed in [15] by Schott. The goal is not to measure convergence, but the actual spread of the solutions on the Pareto Found approximation found by a *MOEA*. It addresses only the diversity aspect of a *MOEA*. It takes into account the distance between a solution and its closest neighbor. It is simple and very straightforward to compute, as can be seen in Eq. 2.

$$S(A) = \sqrt{\frac{1}{|A| - 1} \sum_{i=1}^{|A|} (\bar{d} - d_i)^2} \tag{2}$$

where $\bar{d}$ is the average of all d_i and d_i is the Manhattan distance between solution i and its closest neighbor.

Spreading was first introduced in [18]. It can be seen as an extension of the Spacing metric and it is also a very straightforward metric. It incorporates extreme points to its equation, which is equivalent to calculating Spacing as if the extreme points were the part of the Pareto Found approximation found by a *MOEA*. It provides additional information about the spread of the solution when compared with Spacing, but it is very dependent on the extreme points [9,11]. The formula can be seen in Eq. 3.

$$\Delta(A) = \frac{\sum_{k=1}^{K} d_{ek} + \sum_{i=1}^{|A|} |\bar{d} - d_i|}{\sum_{k=1}^{K} d_{ek} + |A|\bar{d}}, \tag{3}$$

where $\bar{d}$ and d_i are computed in the same way as Spacing, and d_{ek} is the distance between the extreme solution on the kth objective and the closest point in the Pareto Front approximation.

3   Antenna Design

The process of antenna design is a challenging task. The traditional method, based on full-wave electromagnetic simulations, is usually time-consuming and demands high computational cost due to the parameter sweeps and re-running process. This issue limits their applications in multi-parameters antenna design problems. In this Section, a multi-objective optimization framework combining evolutionary algorithms and a back-propagation neural networks (BPNN) antenna surrogate model is applied to antenna development.

The surrogate model replaces the full-wave EM simulation for evaluating the individual fitness value. The fitness function is related to antenna parameters' performance, such as scattering parameters, resonance frequency, bandwidth, among others. With the surrogate model constructed, evolutionary optimization algorithms are used to tune the connection between the parameters for obtaining the optimal solution O_{op}. Next, the O_{op} response is applied to construct the antenna. For the sake of validation, the method is applied in a printed-dipole and a Quasi-Yagi antenna. Figure 1 shows the antennas, and the results are presented in Sect. 4.

The printed-dipole comprises a director element, excited by a matched port of 50 Ω. The dielectric substrate is a $100 \times 25 \times 1.6$ mm FR-4 ($\varepsilon_r = 4.4$ and $\tan \delta = 0.02$). The variables used are $v = [L, t, f, BW]^T$, all in mm. When constructing the BPNN dipole surrogate model, Matlab rand function is first adopted to obtain the input set $M = [L, t]^T \in R^n$ in the antenna design space, and the finite element method (FEM) within the ANSYS HFSS commercial software is used to obtain the high fidelity response set $N = [f, BW]^T$ of these samples. Then, the set of 171 samples of M and N were used as input/output pairs to construct the antenna surrogate model based on BPNN. Next, the *MOEA* is used to tune the connection parameters to maximize the antenna bandwidth and minimize the error in the resonance frequency centered at 3.5 GHz at the same time.

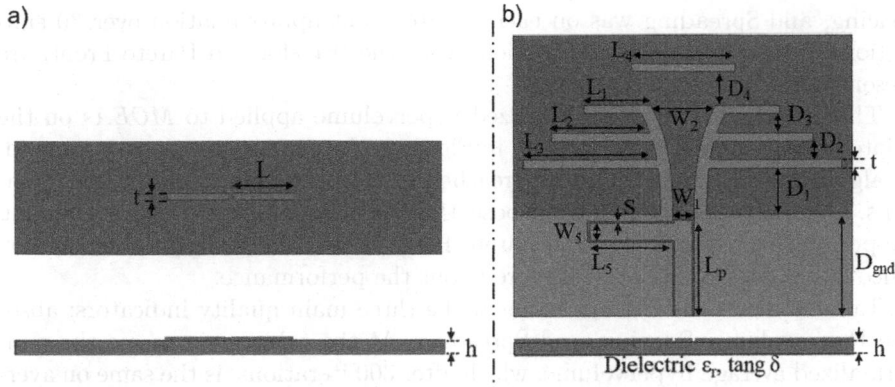

Fig. 1. a) Printed-dipole antennas structure, top and side view. b) Quasi-yagi antenna structure, top and side view.

The Quasi-Yagi antenna was designed to operate over three different bandwidths centered at 1.95, 2.6, and 3.5 GHz. Its overall size is 50×60 mm with a radiator made of cooper. It comprises of a coplanar waveguide to stripline (CPW-to-CPS) transition, L_5 and L_p, a simple ground plane as reflector, three driver dipole elements, L_1, L_2, L_3, with different lengths and a parasite element, L_4. The set of variables used are $g = [L_1, L_2, L_3, L_4, L_5, L_p, f_1, f_2, f_3]^T$, in which the f_n represents the reflection coefficient. Initially, the BPNN antenna surrogate model is built using the Latin Hypercube Samplings (LHS) feature of the Design of Experiments (DOE) available in the ANSYS HFSS software. The LHS generates the input sample set, $J = [L_1, L_2, L_3, L_4, L_5, L_p]^T$, in a pseudo-random way and the FEM from HFSS is used to obtain the high fidelity response containing the output set $K = [f_1, f_2, f_3]^T$. Then, the 1719 samples were set up to set up the surrogate model based on BPNN of the Quasi-Yagi antenna. Next, a *MOEA* is used to tune the connection parameters in order to minimize the antenna reflection coefficient. All input dimensions are in mm.

4 Results

This section concerns a discussion on the obtained results of the two proposed antenna types, namely printed-dipole and Quasi-Yagi. As explained in Sect. 3, the surrogate model was trained and used to formulate the *MOOP* for each antenna. Furthermore, for both antenna models, all *MOEAs* described in Sect. 2.1 were run over 30 simulations, the Pareto Front was transformed to the original output variables so a *DM* could make real use of it. It is essential to notice that the converge analysis with the usage of HV quality indicator and analysis of Spread and Spacing was made on all algorithms in the original Pareto Front for both antenna models.

For each experiment, we performed a convergence analysis with the Hypervolume indicator over 30 simulations. Furthermore, a separate analysis of *HV*,

Spacing, and Spreading was on each Pareto Front approximation over 30 simulations. The convergence curve, along with the transformed Pareto Front, are presented for both experiments.

The convergence of the normalized hypervolume applied to *MOEAs* on the Printed-Dipole antenna can be seen in Fig. 3. As far as convergence is concerned, all algorithms, but *MOEA/D* has reached a stationary curve before 200 iterations. It shows that the general-purpose *MOEAs* did not struggle with converging the population towards the true Pareto Front. Thus, analyzing just the convergence curve does not tell much more about the performance.

Table 1 shows the average value of the three main quality indicators: absolute Hypervolume, Spacing, and Spreading. At the table, we can see the non-normalized average hypervolume, which after 300 iterations, is the same on average for all algorithms, with changes being noticed only after four decimal places. Thus, just hypervolume is not enough to distinguish the performance of the *MOEAs*. As far as Spacing and Spreading, as smaller the value of those quality indicators, the better. Furthermore, by looking at Table 1 we can see that both *OMOPSO* and *SPEA2* had the smallest values for the present quality indicator.

Table 1. Quality Indicators applied to the Printed-Dipole Antenna *MOOP*.

	SMPSO	OMOPSO	NSGA-II	SPEA2	MOEA/D	IBEA
Hypervolume	0.095(±0.001)	0.095(±0.001)	0.095(±0.001)	0.095(±0.001)	0.095(±0.001)	0.095(±0.001)
Spacing	0.073(±0.044)	**0.000(±0.000)**	0.017(±0.004)	**0.007(±0.005)**	0.123(±0.121)	0.026(±0.003)
Spreading	1.552(±0.066)	**0.000(±0.000)**	0.553(±0.202)	**0.216(±0.083)**	1.120(±0.277)	0.468(±0.047)

The reason why both *OMOPSO* and *SPEA2* were mentioned as the best *MOEAs* when analyzing Spread and Spacing is because, as it can be observed in Fig. 2, all *MOEAs* but *OMOPSO* have a fixed-sized archive. After applying the crowding distance on the final Pareto Front approximation, the Printed-Dipole antenna leads to having only one point in the output.

A similar analysis can be made with the Quasi-Yagi antenna. Figure 5 shows the convergence curve of the normalized hypervolume applied to all mentioned *MOEAs*. Since this is a three-objective optimization problem, it becomes visual that the algorithms begin to struggle a little with the convergence aspect. The *MOEA/D* algorithm seems to be the one who had the most difficulty with convergence, not presenting a stationary behavior even around 600 iterations. The *NSGA-II* also seems to have a small struggle, presenting a little but noticeable stationary-oscillatory behavior. The four other *MOEAs* seem to have converged well, presenting a stationary behavior around 100–200 iterations.

Table 2 depicts the absolute hypervolume, spacing, and spreading applied to all *MOEAs*. The absolute hypervolume covers the region of the objective space covered by the Pareto Front approximation. Thus, higher values for this metric present better performance and robustness.

The *OMOPSO* presented the best values for absolute hypervolume, showing the smallest values for the Spacing metric. In the case of Spreading, the *SPEA2*

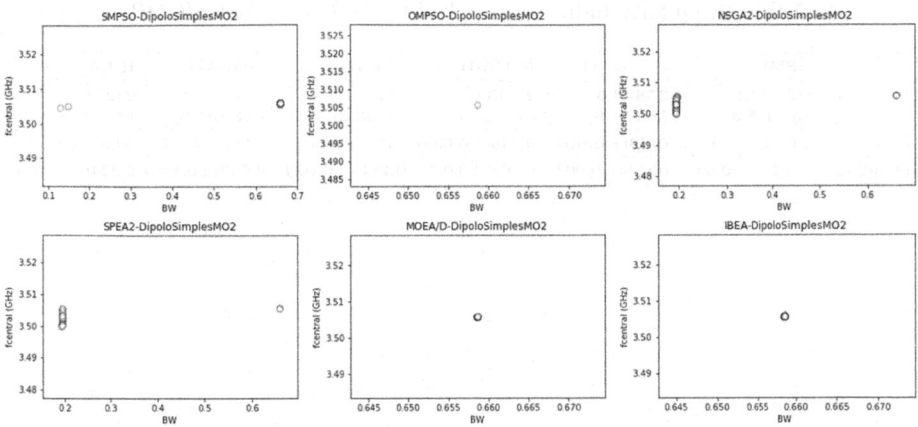

Fig. 2. Transformed Pareto Fronts for Printed-Dipole MOOP.

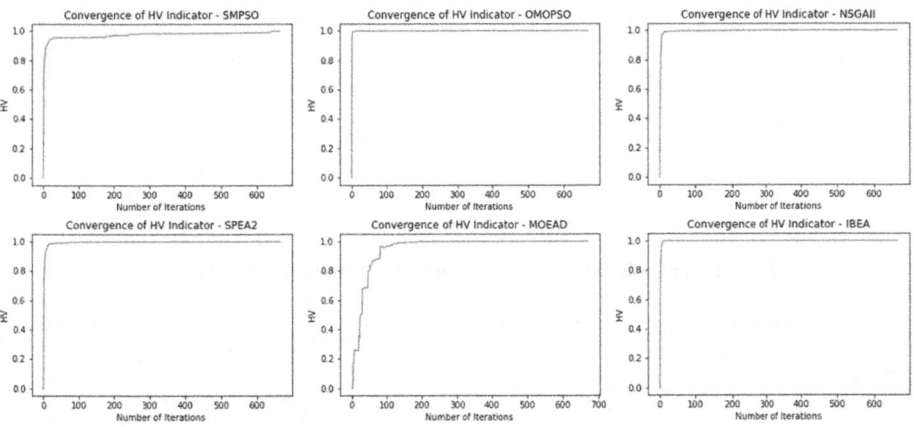

Fig. 3. Normalized HV Convergence for Printed-Dipole MOOP.

gave the lowest value. It can be observed in Fig. 4, that those numbers can be visually observed on the Pareto Front approximation. The *SPEA2* has fewer number of solutions, but they are well spread on the objective space. In the case of *OMOPSO*, we have a higher number of solutions, on average, well-spread on the objective space. The other *MOEAs* seems to have struggled, having solutions clustered in specific locations of the objective space, given fewer options for a *DM* (Decision Maker) to analyze.

Table 2. Quality Indicators applied to the Quasi-Yagi *MOOP*.

	SMPSO	OMOPSO	NSGA-II	SPEA2	MOEA/D	IBEA
Hypervolume	24732.1 (±155.3)	**25838.5** (**±154.9**)	21613.6 (±1051.3)	23461.7 (±216.2)	11025.254 (±3696.2)	24218.8 (±414.1)
Spacing	0.022(±0.02)	**0.09(±0.001)**	0.046(±0.009)	0.033(±0.008)	0.094(±0.029)	0.027(±0.005)
Spreading	0.435(±0.038)	0.485(±0.007)	0.461(±0.052)	**0.263(±0.024)**	0.652(±0.130)	0.324(±0.057)

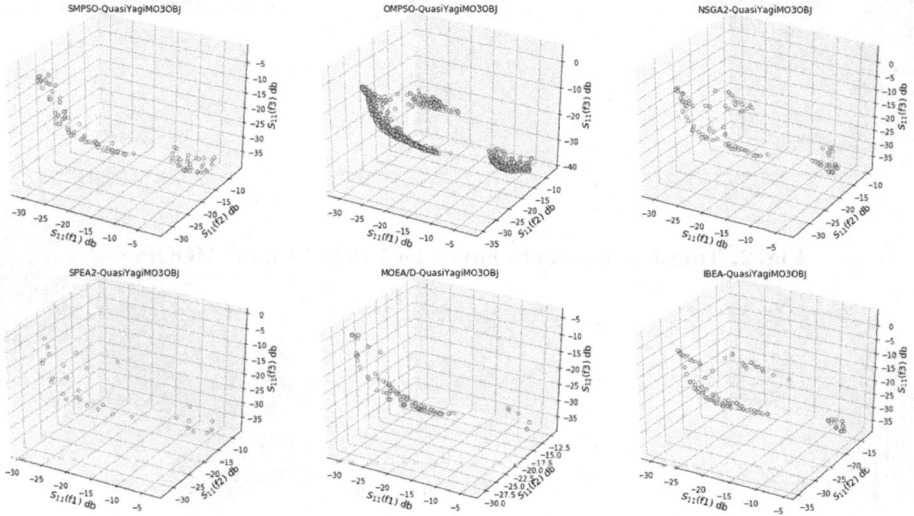

Fig. 4. Transformed Pareto fronts for Quasi-Yagi MOOP.

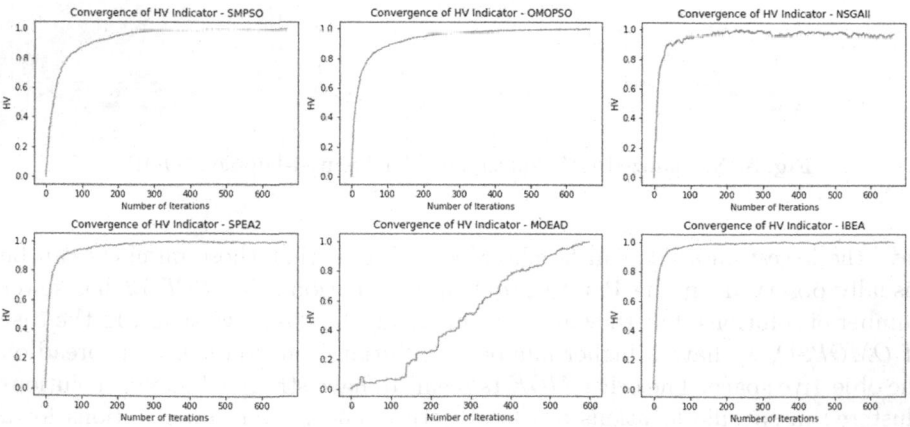

Fig. 5. Normalized HV Convergence for Quasi-Yagi MOOP.

5   Conclusion

We present the application of general-purpose Multi-Objective Evolutionary Algorithms in antenna design. Given six widely-used *MOEAs*, we performed tests to two antenna studies of cases, i.e., printed-dipole and Quasi-Yagi antennas. Additionally, we analyzed the diversity, convergence, and accuracy of the proposed algorithms. We observed the *MOEA/D* has struggled with both *MOOPs*. *OMOPSO* and *SPEA2* presented a superior performance on most of the quality indicators of the printed-dipole problem. In the Quasi-Yagi antenna, visual observation of the Pareto front approximation showed that most of the algorithms have their solutions clustered in specific regions of the objective space. The *OMOPSO* and *SPEA2* presented a more spread set of solutions, with the *OMOPSO* presenting superiority in most of the quality indicators, demonstrating to be the most appropriate *MOEAs* on the *MOOPs* presented in this paper.

For future works, more details about the automatic multi-objective optimization pipeline will be conducted to adapt to different realities. A more refined study on the construction of substitute models can be made, offering greater assertiveness when compared to electromagnetic simulation software.

Acknowledgment. This study was financed in part by the Coordenação de Aperfeiçoamento de Pessoal de Nível Superior - Brasil (CAPES) - Finance Code 001.

References

1. Audet, C., Bigeon, J., Cartier, D., Le Digabel, S., Salomon, L.: Performance indicators in multiobjective optimization. Optim. Online (2018)
2. Carvalho, R.d., Saldanha, R.R., Gomes, B., Lisboa, A.C., Martins, A.: A multiobjective evolutionary algorithm based on decomposition for optimal design of Yagi-Uda antennas. IEEE Trans. Magn. **48**(2), 803–806 (2012)
3. Chand, S., Wagner, M.: Evolutionary many-objective optimization: a quick-start guide. Surv. Oper. Res. Manag. Sci. **20**(2), 35–42 (2015)
4. Coello, C.A.C., Lamont, G.B., Van Veldhuizen, D.A., et al.: Evolutionary Algorithms for Solving Multi-objective Problems, vol. 5. Springer, Boston (2007). https://doi.org/10.1007/978-0-387-36797-2
5. Deb, K., Pratap, A., Agarwal, S., Meyarivan, T.: A fast and elitist multiobjective genetic algorithm: NSGA-ii. IEEE Trans. Evol. Comput. **6**(2), 182–197 (2002)
6. Dong, J., Li, Q., Deng, L.: Fast multi-objective optimization of multi-parameter antenna structures based on improved MOEA/D with surrogate-assisted model. AEU Int. J. Electron. Commun. **72**, 192–199 (2017)
7. Easum, J.A., Nagar, J., Werner, D.H.: Multi-objective surrogate-assisted optimization applied to patch antenna design. In: 2017 IEEE International Symposium on Antennas and Propagation & USNC/URSI National Radio Science Meeting, pp. 339–340. IEEE (2017)
8. Kennedy, J., Eberhart, R.: Particle swarm optimization. In: Proceedings of ICNN 1995-International Conference on Neural Networks, vol. 4, pp. 1942–1948. IEEE (1995)

9. Laszczyk, M., Myszkowski, P.B.: Survey of quality measures for multi-objective optimization. Construction of complementary set of multi-objective quality measures. Swarm Evol. Comput. **48**, 109–133 (2019)
10. Lee, Y.H., Cahill, B.J., Porter, S.J., Marvin, A.C.: A novel evolutionary learning technique for multi-objective array antenna optimization. Progress Electromagn. Res. **48**, 125–144 (2004)
11. Mirjalili, S., Lewis, A.: Novel performance metrics for robust multi-objective optimization algorithms. Swarm Evol. Comput. **21**, 1–23 (2015)
12. Nebro, A.J., Durillo, J.J., Garcia-Nieto, J., Coello, C.C., Luna, F., Alba, E.: SMPSO: a new pso-based metaheuristic for multi-objective optimization. In: 2009 IEEE Symposium on Computational Intelligence in Multi-Criteria Decision-Making (MCDM), pp. 66–73. IEEE (2009)
13. Reyes-Sierra, M., Coello, C.C., et al.: Multi-objective particle swarm optimizers: a survey of the state-of-the-art. Int. J. Comput. Intell. Res. **2**(3), 287–308 (2006)
14. Riquelme, N., Von Lücken, C., Baran, B.: Performance metrics in multi-objective optimization. In: 2015 Latin American Computing Conference (CLEI), pp. 1–11. IEEE (2015)
15. Schott, J.R.: Fault tolerant design using single and multicriteria genetic algorithm optimization. Technical report, Air force inst of tech Wright-Patterson afb OH (1995)
16. Sierra, M.R., Coello Coello, C.A.: Improving PSO-based multi-objective optimization using crowding, mutation and ∈-dominance. In: Coello Coello, C.A., Hernández Aguirre, A., Zitzler, E. (eds.) EMO 2005. LNCS, vol. 3410, pp. 505–519. Springer, Heidelberg (2005). https://doi.org/10.1007/978-3-540-31880-4_35
17. Zhang, Q., Li, H.: MOEA/D: a multiobjective evolutionary algorithm based on decomposition. IEEE Trans. Evol. Comput. **11**(6), 712–731 (2007)
18. Zitzler, E.: Evolutionary algorithms for multiobjective optimization: Methods and applications, vol. 63. Citeseer (1999)
19. Zitzler, E., Künzli, S.: Indicator-based selection in multiobjective search. In: Yao, X., et al. (eds.) PPSN 2004. LNCS, vol. 3242, pp. 832–842. Springer, Heidelberg (2004). https://doi.org/10.1007/978-3-540-30217-9_84
20. Zitzler, E., Laumanns, M., Thiele, L.: SPEA 2: Improving the strength pareto evolutionary algorithm. TIK-report 103 (2001)
21. Zitzler, E., Thiele, L.: Multiobjective evolutionary algorithms: a comparative case study and the strength pareto approach. IEEE Trans. Evol. Comput. **3**(4), 257–271 (1999)

Special Session on Practical Applications of Deep Learning

Cloud Type Identification Using Data Fusion and Ensemble Learning

Javier Huertas-Tato[1]([✉])[ID], Alejandro Martín[2][ID], and David Camacho[2][ID]

[1] Universidad Europea de Madrid, Madrid, Spain
javier.huertas@universidadeuropea.es
[2] Universidad Politécnica de Madrid, Madrid, Spain
{alejandro.martin,david.camacho}@upm.es

Abstract. Cloud type classification is a complex multi-class problem where total sky images are analysed to determine their category such as *Stratus* or *Cirrus*, among others. However, many properties of this domain make high classification accuracy difficult to achieve. In this paper, we design a novel fusion approach, showing that recent image classification architectures based on deep learning, such as Convolutional Neural Networks, can be improved using statistical features directly calculated from images. In this research, three powerful CNNs have been trained on a comprehensive dataset: VGG-19, Inception-ResNet V2 and Inception V3. Simultaneously, a pool of standard machine learning classifiers have been trained on 14 different statistical characteristics on each colour channel. The results evidence that a fusion approach of the predictions of an image-trained CNN and a feature-trained Random Forest classifier improves the classification ability of both methods individually, reaching 95.05% macro average weighted precision over 12 classes in a complex highly imbalanced dataset with noisy examples.

Keywords: Cloud type classification · Data fusion · Ensemble learning · Convolutional Neural Network

1 Introduction

Solar energy production is tightly linked to the current state of the sky, where clouds play a critical role [15]. Therefore, the identification of cloud types has become a relevant field of study within the scientific community. The misrepresentation of cloud types by physical or statistical models may have significant energetic impact [6]. Clouds represent the main source of variability in solar energy [22], by knowing a cloud type, solar energy estimations could be significantly improved. Human classification is possible for this task, however it is slow, inefficient, and inconsistent even between experts [9]. A need to automatically detect cloud types has been identified as crucial to the development of solar energy [23].

Research on this domain frequently uses feature extraction to solve the classification problem. Some frequently referenced works use a set of features extracted

© Springer Nature Switzerland AG 2020
C. Analide et al. (Eds.): IDEAL 2020, LNCS 12490, pp. 137–147, 2020.
https://doi.org/10.1007/978-3-030-62365-4_13

from Total Sky Imagers (TSIs), a technique dating at least a decade back, and frequently used in cloud cover estimation [1]. The amount and quality of features has increased significantly over the years, and these features have been included in several machine learning models. Early approaches, such as that proposed by [14], use a channel from the LANDSAT MSS which is processed with a Multi Layer Perceptron (MLP) to achieve 84% accuracy on 3-class classification. A more recent approach by [19], combines Neural Networks and KNN to classify five classes of clouds. A recent methodology is proposed by [12] that processes the image channels via Fourier transform to train a MLP achieving 90% accuracy on a seven cloud problem. Other approaches use K-Nearest Neighbors (KNN) to classify clouds with feature extraction. Several frequently used works apply KNN as their main inference engine. The proposal by [8] uses a mix of textural and spectral features to achieve a 75% success rate on seven cloud types. These results are improved by [11] where a multi-colour criterion is develops to achieve a 87% using seven cloud types.

Feature extraction and Random Forests [2] have seen some use in this field, with some promise on empirical studies [3]. A recent work by [4] uses RFs have been used along other algorithms to develop a classifier ensemble of cloudy pixels. The proposal by [10] uses a RF to combine features and ceilometer information to improve classification accuracy on an 11-cloud set. Finally, Convolutional Neural Networks (CNN) [13] have been widely used on image recognition problems. These proposals disregard feature extraction to automatically detect relevant image information. Recently, researchers have started adopting CNNs on their cloud classification works [24,25]. Research work developed within this growing trend frequently apply variations of LeNet [13] architectures or direct use of the VGG16 architecture [18]. Recently, more powerful CNN techniques have appeared, which are easily applicable to this domain.

As novel architectures and techniques for CNNs appear, a necessity for newer (and computationally cheaper) approaches to machine learning are needed, different approaches are currently used as those based on evolutionary computing to reduce the complexity of the CNN architectures [16,17]. By fusing cutting-edge CNN architectures, with traditional feature extraction machine learning methods, computer vision classification may be improved with little to none additional computational costs. The main contribution of this paper can be summarized in the design of a novel CNN+RF ensemble for the domain of total sky image classification. The empirical results support the initial hypothesis as the combination of probabilities from both classifiers increased the performance. In addition to our ensemble proposal, the experimental results outperform other mentioned cloud classification algorithms proposed on a harsh 12 sky type labelling, including noisy examples such as Multi-cloud (overlapping cloud types) and Aerosols (Clear sky obscured by aerosol gases, not clouds), achieving a 95.05% macro average weighted precision in this (highly imbalanced) domain.

2   Method

Computer vision has been revolutionized by Convolutional Networks, leaving traditional feature extraction seemingly outdated. However, expert knowledge of image features may still hold some relevant information that CNNs are not yet able to capture. A combined approach is presented here, merging the output probabilities of a Convolutional Neural Network, and the outputs of a Random Forest trained with statistical features extracted from images.

It is hypothesized that CNNs and manual feature extraction approaches classify examples in fundamentally different ways. The correctly classified and misclassified examples differ greatly. Therefore, it may be possible to preserve the true classes, while discarding erroneous classifications, improving accuracy as their mislabeled examples cancel each other. The methods used are described in the next subsections.

2.1   Feature Extraction Method

Feature extraction from images has remained a common approach to perform machine learning on computer vision domains. By extracting a wide collection of relevant image features and by training a machine learning algorithm resistant to irrelevant data, a robust model for classification can be built. This feature extraction process is inspired by similar referential works of the cloud classification domain [8,11] that follow this feature extraction approach. However, the statistical feature set presented here is significantly expanded to let a varied set of machine learning algorithms decide the importance of each feature independently.

We find two feature categories: spectral and textural. Spectral features extract information directly from the colour spectrum of an image, for example via average values, deviation, average differentials or ratios. Textural features rely on Grey Level Co-Variance Matrices (GLCMs) a technique to detect the frequency of contiguous appearances of pixel values. From these, textures such as edges can be detected. These were originally proposed in [7] and have seen frequent use in computer vision.

Specifically, the following characteristics are extracted: Average, Standard deviation, Skewness, Average Colour Difference, Histogram (with 5 bins), Average Colour Ratio, Textural Average, Variance, Homogeneity, Contrast, Dissimilarity, Entropy, Second Movement, and Correlation. These features are calculated on each colour channel, summing up to 60 features. After extraction, they are fed to the standard set of machine learning classifiers, described in the following section.

2.2   Image and Feature Classifiers

There are two distinct approaches to image classification, feature extraction or image processing through CNNs. Two sets of classifiers are used in this problem, CNN classifiers and feature classifiers based on the described approaches.

The CNN classifiers used in this paper are the following: VGG-19 [18], Inception ResNet V2 [20] and Inception V3 [21]. These classifiers learn using randomly augmented batches of sky images until the network automatically stops after the validation error has not improved during a set of cycles. Every network is initialized with transferred weights from the ImageNet domain, a percent of which are frozen and connected to a dense layer with 50% dropout. The optimization algorithm chosen is Adam, while any other remaining hyper-parameter has been tuned empirically.

A set of machine learning classifiers is defined for feature extraction and the following ensemble learning. The following algorithms are used: a regularized logistic regression algorithm, a Linear Discriminant Analysis classifier, a K Neighbors classifier, an optimised version of the CART Decision Tree classifier, a Gaussian Naive Bayes classifier, a Support Vector Machine classifier with linear, polynomial, radial basis function and sigmoid kernels and the Random Forest classifier with 500 internal estimators.

2.3 Ensemble Learning

An overview of the final classifier ensemble pipeline is represented in Fig. 1. From the set of classifiers, the best performing algorithm over the proposed feature set is the Random Forest classifier. This ensemble reports a set of values that represent the voting process made to select the class. This is similar to the categorical outputs of the CNN, where every class has an assigned value from 0 to 1, depending on the previous detection. These probability values can be combined with another algorithm from the set of classifiers to achieve higher performance.

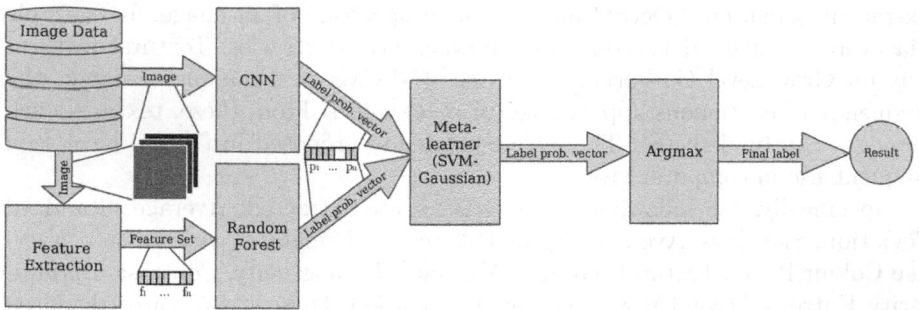

Fig. 1. Visual representation of the pipeline followed, from the image database images to the end classification result.

3 Data

The sky type dataset used contains a set of diverse images and labels. A set of images are blocked by a solar band that blocks radiation, preventing image burns. The rest do not have this band. This first set of images has been gathered from a Yesdas TSI-800 model at a meteorological station of the University of Jaén, Spain, at coordinates 37.7877°N and 3.7782°W and 454 m above mean sea level. When computing features from these images, the sun band is ignored, not being included in any calculation.

The images without solar band have been obtained with an AXIS M30007-PV Network Camera with a fish-eye lens. There are 3 cameras with this model, taking samples at three spatially different locations at coordinates 37.409922°N and 6.272902°W, the second at 37.412107°N, 6.262683°W and the third at 37.4258°N, 6.282963°W, all three at 35m above sea level.

Every image has been curated by meteorology experts from the University of Jaén (MATRAS group) to get their labels and keep relevant images. Images are pre-processed to fit a 256×256 square extracting exclusively useful information of the sky. A total of 4,167 examples of data are used from 12 different classes: Clear sky, Cirrus, Cirrostratus, Cirrocumulus, Altocumulus, Altostratus, Stratus, Stratocumulus, Cumulus, Nimbostratus, Multicloud and Aerosols.

There is a huge imbalance ratio between classes, to prevent overfitting the majority classes we applied a random oversampling method, creating new instances of underrepresented classes. The dataset is split in train validation and test with a 70/10/20 ratio respectively.

4   Experiments

All experimentation presented here has been conducted using Python 3.6.9. Development of CNNs has been made in Keras [5] with the Tensorflow 1.15 backend. A hardware accelerator GPU Nvidia Titan V has been used for training and evaluation.

All models presented here use the default hyper-parameters from their papers without the original dense layer. Every proposal here uses 2048 neurons as the hidden layer with a 50% dropout and ReLU activation. The top 10% of layers from VGG19 and InceptionV3 are frozen with hyper parameters from ImageNet. The top 25% of layers from the InceptionResNet model are also frozen with ImageNet weights.

Three different experiments have been defined in order to compare and select which is the most adequate procedure to address the problem at hand. The goal is to detect which is the proper representation and classification method in order to ensure the best possible performance. **Experiment 1** evaluates the use of a series of Convolutional Neural Network models directly trained on the map of pixels captured by the camera. In contrast, the goal of **Experiment 2** is to assess if the use of a series of statistical characteristics calculated from the image is able to provide a more discriminative representation, thus allowing to better classify the type of cloud. Different standard classifiers, such as Logistic Regression or decision trees, are tested in this second experiment to make a classification based on these statistical features. Finally, **Experiment 3** studies if an ensemble classifier trained on the predictions of the methods deployed in the two previous experiments helps to provide a more accurate classifier.

4.1   Experiment 1: Convolutional Neural Networks

Table 1. Summary table from **Experiment 1**, shows the average results of the executions performed with Inception-ResNet V2, Inception V3 and VGG-10. The results show the average accuracy and macro average precision in the test set.

Model	Accuracy		Macro avg. precision	
	Mean	Std	Mean	Std
Inception-ResNet V2	**0.9229**	0.0123	**0.9274**	0.0106
Inception V3	0.9207	0.0187	0.9223	0.0205
VGG-19	0.9112	0.0104	0.9131	0.0162

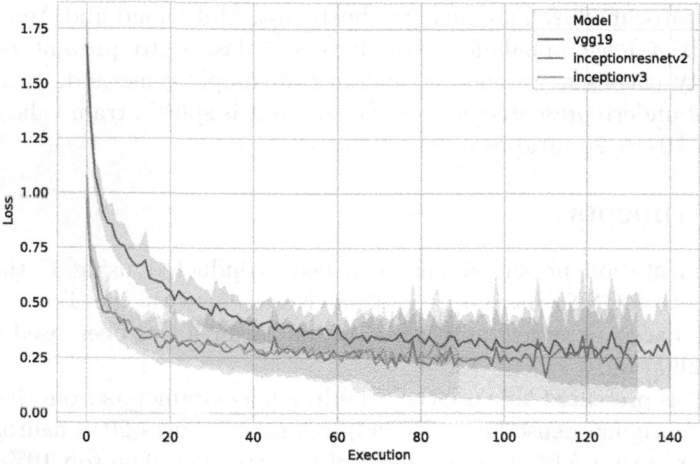

Fig. 2. CNN architectures average learning evolution across epochs.

We have tested three different CNN architectures which have been largely used in image classifications tasks due to their proven performance. Particularly, we run VGG-19, Inception-ResNet V2 and Inception V3. A pre-trained version of all these models was used by employing the Keras library. Table 1 summarises the average results in terms of accuracy and precision. As can be seen, the two Inception based models reach almost equal results, 92.23 ± 2.05 macro average precision with Inception V3 and 92.74 ± 1.06 with the ResNet V2 version.

In order to further evaluate the performance of these three architectures, Fig. 2 shows the loss evolution across epochs. The VGG-19 architecture has a slower convergence if compare to the two Inception based models, with a loss value positioning slightly above these two models throughout the whole training process.

In order to select one model to later combine its predictions with other features, the model maximising the results in the training set in terms of macro average precision was selected. This execution was achieved with the Inception-ResNet V2 architecture, providing 93.44% macro average precision.

4.2 Experiment 2: Standard Classifiers Trained on Statistical Features

In this second experiment we aim to evaluate the use of a series of statistical features or estimators calculated directly from every image, checking its ability to serve as features to classify the cloud images. Given the diverse nature of the features, we run different supervised machine learning classifiers which have provided excellent results in varied domains. The results of running all these classification methods are shown in Table 2. The best result is provided by Random Forests, with 89.81% macro weighted precision. In order to further assess the performance of this classifier in this domain, we run different executions with a different number of internal estimators.

Table 2. Summary table from **Experiment 2**, shows the results of training different standard classifiers with statistical features calculated from every image.

Classifier	Accuracy	Macro avg. precision	Macro avg. weighted precision
LogisticRegression	0.7291	0.6721	0.7505
LinearDiscriminantAnalysis	0.7021	0.6371	0.7387
KNeighborsClassifier	0.7389	0.6805	0.7519
DecisionTreeClassifier	0.8082	0.7823	0.8088
GaussianNB	0.4789	0.4455	0.5900
SVM-linear	0.8093	0.7707	0.8109
SVM-poly	0.6793	0.6381	0.6960
SVM-rbf	0.7703	0.7379	0.7775
SVM-sigmoid	0.3900	0.3771	0.4854
RandomForestClassifier	0.8992	0.8839	**0.8981**

4.3 Experiment 3: Ensemble Approach Based on CNN and Random Forest Predictions Combination

The third experiment tries to improve the classification by combining the predictions of the CNN with the predictions of the Random Forest classifier, which focuses on a series of statistical features. To this purpose, an ensemble learning based approach was followed, aiming to assess if a classifier trained on both predictions can lead to better results in contrast to the use of a single

Table 3. Summary table from **Experiment 3**, shows the results after training different standard classifiers on the predictions delivered by the CNN model and the Random Forest classifier trained on the statistical features. The first row shows the results of making a simple average of the predictions of both prediction sources.

Classifier	Accuracy	Macro avg. precision	Macro avg. weighted precision
Average	0.9404	0.9422	0.9407
LogisticRegression	0.9469	0.9533	0.9472
LinearDiscriminantAnalysis	0.9393	0.9534	0.9408
KNeighborsClassifier	0.9382	0.9508	0.9389
DecisionTreeClassifier	0.8927	0.9146	0.9159
GaussianNB	0.8559	0.8436	0.8637
SVM-linear	0.9502	0.9558	0.9505
SVM-poly	0.9003	0.9560	0.9223
SVM-rbf	0.9415	0.9630	0.9429
SVM-sigmoid	0.9502	0.9530	**0.9505**
RandomForestClassifier	0.9285	0.9333	0.9296

classifier. Table 3 indicates the results obtained by these classifiers, including in the first row the result of performing a simple average between the predictions of both source classifiers. In this experiment, a Support Vector Machine both with a linear or a sigmoid kernel is able to produce the best result so far, reaching a 95.05% macro average precision. This is an increase of 1.5% precision with respect to the best CNN classifier. Thus, the use of the predictions of the Random Forest classifier trained with estimators help the CNN classifier to provide more accurate results.

Comparison with other works in the state of the art reflects the strengths of this fusion method. The referential work presented in [11] shows classification results up to 87% on seven cloud types using camera features. Other modern approaches are more powerful, such as the CloudNet presented by [25], which achieves 87.76% average accuracy on eleven classes. The method presented here reaches 95.05% macro average accuracy on 12 cloud types. Finally, Table 4 summarises the results of the four experiments and show the detailed precision by class. In the last experiment, all classes except one (altostratus) exceed 90% precision.

Table 4. Precision results detailed by class for the three experiments described. The three last rows show the average result by experiment according to different criteria.

Class	Experiment 1 CNN	Experiment 2 RF	Experiment 3 CNN + RF (SVM Sigmoid)
Aerosol	1.0000	0.9524	1.0000
Altocumulus	0.9341	0.8523	0.9451
Altostratus	0.8750	0.9412	0.9355
Clear sky	0.9545	1.0000	0.9545
Cirrocumulus	0.8947	0.6364	0.9474
Cirrus	0.9161	0.9103	0.9290
Cirrostratus	0.9552	0.9688	0.9848
Cumulus	0.9589	0.9429	0.9722
Stratocumulus	0.9630	0.8912	0.9781
Stratus	0.9211	0.9204	0.9375
Multicloud	0.9235	0.9000	0.9355
Nimbostratus	0.9167	0.5556	0.9167
Accuracy	0.9339	0.9014	0.9502
Macro avg. precision	0.9344	0.8726	0.9530
Macro avg. weighted precision	0.9340	0.9012	**0.9505**

5   Conclusions and Future Works

A novel data fusion approach to image classification is applied to the domain of cloud classification. The empirical results obtained outclass similar referential works in the state of the art. Instead of designing a more complex network increasing computational costs, traditional features are used to boost results of classification in this presented domain. By employing cutting-edge CNN architectures and a novel feature extraction methodology, classification accuracy rises significantly in comparison with older works. On top of individual improvements, a macro-average precision of 95.05% is reported for the data fusion method, which outperforms any other methods presented. The data fusion method reports higher classification accuracy than individual models or simple combinations of predictions (such as averaging prediction classes). Future applications may involve extending this work to other common domains. The outputs of CNN and feature extraction can be fused in several other ways, studying further avenues for combination could boost the presented results.

Acknowledgment. This work has been supported by Spanish Ministry of Science and Education under TIN2014-56494-C4-4-P grant (DeepBio), and Comunidad Autónoma de Madrid under S2018/TCS-4566 grant (CYNAMON). We gratefully acknowledge the support of NVIDIA Corporation with the donation of the Titan V GPU used for this research.

References

1. Boers, R., et al.: Optimized fractional cloudiness determination from five ground-based remote sensing techniques. J. Geophys. Res. Atmos. **115**(24) (2010). www.scopus.com
2. Breiman, L.: Random Forests. Mach. Learn. **45**(1), 5–32 (2001). https://link.springer.com/article/10.1023/A:1010933404324
3. Caruana, R., Karampatziakis, N., Yessenalina, A.: An empirical evaluation of supervised learning in high dimensions. In: Proceedings of the 25th International Conference on Machine Learning, pp. 96–103 (2008). www.scopus.com
4. Cheng, H., Lin, C.: Cloud detection in all-sky images via multi-scale neighborhood features and multiple supervised learning techniques. Atmos. Meas. Tech. **10**(1), 199–208 (2017). www.scopus.com
5. Chollet, F., et al.: Keras (2015). https://keras.io
6. Haiden, T., Forbes, R., Ahlgrimm, M., Bozzo, A.: The skill of ECMWF cloudiness forecasts. ECMWF Newsl. **143**, 14–19 (2015). www.scopus.com
7. Haralick, R.M., Dinstein, I., Shanmugam, K.: Textural Features for Image Classification. IEEE Trans. Syst. Man Cybern. SMC **3**(6), 610–621 (1973). www.scopus.com
8. Heinle, A., Macke, A., Srivastav, A.: Automatic cloud classification of whole sky images. Atmos. Meas. Tech. **3**(3), 557–567 (2010). www.scopus.com
9. Hoyt, D.V.: Interannual cloud-cover variations in the contiguous united states. J. Appl. Meteorol. **17**(3), 354–357 (1978)
10. Huertas Tato, J., et al.: Automatic cloud-type classification based on the combined use of a sky camera and a ceilometer. J. Geogr. Res. Atmos. (2017). https://e-archivo.uc3m.es/handle/10016/28557
11. Kazantzidis, A., Tzoumanikas, P., Bais, A.F., Fotopoulos, S., Economou, G.: Cloud detection and classification with the use of whole-sky ground-based images. Atmos. Res. **113**, 80–88 (2012). www.scopus.com
12. Kliangsuwan, T., Heednacram, A.: Feature extraction techniques for ground-based cloud type classification. Expert Syst. Appl. **42**(21), 8294–8303 (2015). www.scopus.com
13. LeCun, Y., Bengio, Y., Hinton, G.: Deep learning. Nature **521**(7553), 436 (2015)
14. Lee, J., Weger, R.C., Sengupta, S.K., Welch, R.M.: A Neural Network Approach to Cloud Classification. IEEE Trans. Geosci. Remote Sens. **28**(5), 846–855 (1990). www.scopus.com
15. Li, Y., Thompson, D.W.J., Stephens, G.L., Bony, S.: A global survey of the instantaneous linkages between cloud vertical structure and large-scale climate. J. Geophys. Res. **119**(7), 3770–3792 (2014). www.scopus.com
16. Martín, A., Lara-Cabrera, R., Fuentes-Hurtado, F., Naranjo, V., Camacho, D.: EvoDeep: a new evolutionary approach for automatic deep neural networks parametrisation. J. Parallel Distrib. Comput. **117**, 180–191 (2018)
17. Martín, A., Vargas, V.M., Gutiérrez, P.A., Camacho, D., Hervás-Martínez, C.: Optimising convolutional neural networks using a hybrid statistically-driven coral reef optimisation algorithm. Appl. Soft Comput. **90**, 106144 (2020)
18. Simonyan, K., Zisserman, A.: Very deep convolutional networks for large-scale image recognition. arXiv preprint arXiv:1409.1556 (2014)
19. Singh, M., Glennen, M.: Automated ground-based cloud recognition. Pattern Anal. Appl. **8**(3), 258–271 (2005). www.scopus.com

20. Szegedy, C., Ioffe, S., Vanhoucke, V., Alemi, A.A.: Inception-v4, inception-resnet and the impact of residual connections on learning. In: Thirty-First AAAI Conference on Artificial Intelligence (2017)
21. Szegedy, C., Vanhoucke, V., Ioffe, S., Shlens, J., Wojna, Z.: Rethinking the inception architecture for computer vision. In: Proceedings of the IEEE Conference on Computer Vision and Pattern Recognition, pp. 2818–2826 (2016)
22. Tzoumanikas, P., Nikitidou, E., Bais, A.F., Kazantzidis, A.: The effect of clouds on surface solar irradiance, based on data from an all-sky imaging system. Renew. Energy **95**, 314–322 (2016). www.scopus.com
23. World Meteorological Organization : World Meteorological Organization/World Weather Research Programme (WMO/WWRP). Recommended methods for evaluating cloud and related parameters World Weather Research Programme (WWRP)/Working Group on Numerical Experimentation (WGNE) Joint Working Group on Forecast Verification Research (JWGFVR) (2012). www.scopus.com
24. Ye, L., Cao, Z., Xiao, Y.: DeepCloud: ground-based cloud image categorization using deep convolutional features. IEEE Trans. Geosci. Remote Sens. **55**(10), 5729–5740 (2017)
25. Zhang, J., Liu, P., Zhang, F., Song, Q.: CloudNet: ground-based cloud classification with deep convolutional neural network. Geophys. Res. Lett. **45**(16), 8665–8672 (2018)

Deep Learning in Aeronautics: Air Traffic Trajectory Classification Based on Weather Reports

Néstor Jiménez-Campfens[1]([✉]), Adrián Colomer[1], Javier Núñez[2],
Juan M. Mogollón[2], Antonio L. Rodríguez[2], and Valery Naranjo[1]

[1] Institute of Research and Innovation in Bioengineering,
Universitat Politècnica de València, Valencia, Spain
jojicam@etsii.upv.es, {adcogra,vnaranjo}@i3b.upv.es
[2] Skylife Engineering S.L., Parque Científico y Tecnológico Cartuja, Sevilla, Spain
{javier.nunez,jmmogollon,antoniorv}@skylife-eng.com

Abstract. New paradigms in aviation, as the expected shortage of qualified pilots and the increasing number of flights worldwide, present big challenges to aeronautic enterprises and regulators. In this sense, a concept known as Single Pilot Operations arises in the task of dealing with these challenges, for which, automation becomes necessary, especially in Air Traffic Management. In this regard, this paper presents a deep learning-based approach to leveraging the job of both ground controllers and pilots. Making use of Meteorological Terminal Air Reports, obtained regularly from every aerodrome worldwide, we created a model based on a multi-layer perceptron capable of determining the approach trajectory of an aircraft thirty minutes prior to the expected landing time. Experiments on aircraft trajectories from Toulouse to Seville, show an accuracy, recall and F1-score higher than 0.9 for the resultant predictive model.

Keywords: Air traffic management · Weather reports · METAR · Trajectory prediction · Deep learning

1 Introduction

Since the 1950s, technological advances in engines, voice communication and navigation equipment have resulted in a decreasing number of cockpit members from 5 to 2 persons. In the last years, further automation and technological developments, added to an expected exacerbation of the existing global shortage of qualified pilots [1] have propitiated an arising interest in a concept known as Single Pilot Operations (SPO), with the aim of reducing the current commercial cockpit crew from 2 to a single pilot, favouring research within this topic [2,3]. In this sense, it is argued that re-conceptualising the flight-deck and the role of the pilot [3] along with an increasing support from a ground-operator [4], are necessary conditions for SPO.

© Springer Nature Switzerland AG 2020
C. Analide et al. (Eds.): IDEAL 2020, LNCS 12490, pp. 148–155, 2020.
https://doi.org/10.1007/978-3-030-62365-4_14

There is great consensus that, in order to implement SPO, several co-pilot functions will have to migrate to ground control stations [2, 5] or be automated in the aircraft. In both cases, it is accepted that the use of automation, through digital assistants, will need to be increased both on-deck and on the ground in order for SPO to be successful [2, 6].

In this regard, decision-support systems have become a necessity with the aim of easing the increasing workload of both pilots and ground operators. In particular, artificial intelligence algorithms could be a great asset in solving difficult, non-trivial tasks such as air traffic conflict resolution, flight delay prediction and trajectory prediction.

With respect to air traffic conflict resolution, Deep Learning (DL) techniques have been proved to be useful to prevent conflicts between two aircraft by using neural networks and genetic algorithms while being time-efficient [7].

With the aim of predicting flight delays, different approaches have been studied based on weather data usage. In this sense, [8] proposed a Machine Learning (ML) based model able to classify airline delays induced by varying weather condition, comparing the efficiency of different algorithms, including Random Forest, AdaBoost, k-Nearest-Neighbours and Decision Trees. From a different perspective, [9] proposed a Long Short-Term Memory (LSTM) architecture to predict flight delays including airport and weather data, among other features.

With regard to trajectory prediction, literature shows that the number of studies applying ML or DL techniques to trajectory prediction and classification, based on weather data, is limited compared to that of delay prediction and conflict resolution. In this regard, in [10] an Encoder-Decoder Recurrent Neural Network (RNN) was used to predict a flight trajectory using the flight plan as input and incorporating weather data of areas close to the planned trajectory. In [11], a LSTM network was used to predict the trajectory of the flight using position data such as altitude, longitude and latitude, as input to the model, stating that not including weather data and its fluctuating features makes the model prone to sharp turns in the predicted trajectory and, in consequence inaccurate predictions. More extensively, [12] presented a wider comparison of multiple ML and DL supervised algorithms in the task of landing runway trajectory classification of arriving aircraft, based on trajectory features (latitude, longitude, speed, altitude and course angle), and specific characteristics of the aircraft and airport.

In this sense, our work proposes a Multi Layer Perceptron (MLP) algorithm trained to predict the approach trajectory of an aircraft thirty minutes prior to its landing, using solely weather reports generated by the destination airport. For our work, the trajectory between the airports of Toulouse (TLS) and Seville (SVQ) was considered.

2   Materials and Methods

For this study, two types of data were necessary: flight trajectory data, including position (latitude and longitude), altitude, speed and direction of each flight at every given moment; and weather data at the destination airport, encoded as Meteorological Terminal Air Reports (METAR). For each flight, two METAR reports were gathered previous to the expected landing time following the criteria in Table 1 and being $X \in \{0 - 24\}$. In total, a set of 237 samples of flights travelling from Toulouse to Seville between the 4th of November, 2017, and the 14th of January, 2020 were included into the experiments.

Table 1. META Reports timestamps

Expected landing	METAR 1 Timestamp	METAR 2 Timestamp
Xh 00 m – Xh 15 m	$(X-1)$h 30 m	$(X-1)$h 00 m
Xh 15 m – Xh 45 m	Xh 00 m	$(X-1)$h 30 m
Xh 45 m – Xh 59 m	Xh 30 m	Xh 00 m

2.1 Data Acquisition

In order to gather the data, a web-scrapping algorithm was used using the *flightradar24* website [13]. Indicating a flight reference, we automatically updated the generated database with any new flights information in *flightradar24*.

2.2 Data Curation

Landing Trajectory for Ground-Truth Generation. Working with a retrospective database, the complete trajectory is known. In order to train a classification model, we labelled each flight according to its Terminal Arrival Point (TAP) using a non-supervised clustering algorithm (*k*-means). The average of the last 20 sampled points for longitude and latitude were fed to the *k*-means method. As initial cluster centroids, we used the coordinates of each point of interest (one for every TAP - in the case of the Seville airport $k = 2$, i.e. Rotex and Santa).

Weather Reports Standarization. METAR reports encode weather information in a semi-structured way. These reports include a combination of multiple fields (separated by blank spaces), some of which are always present and some of which are optional, ones with fixed length and others with variable length depending on the weather condition, and some fields representing quantitative features (e.g. wind speed or temperature) and others qualitative features (e.g. cloud or rain types). A sample of a METAR report from the Airport of Seville can be seen in Table 2:

To structure this data, taking advantage of its particular disposition, a regular-expression search algorithm was used, obtaining an array of 64 features for each report. Quantitative information is combined with one-hot-encoded categorical information for qualitative data (see Table 3).

Table 2. A representative sample of METAR Report

LEZL 132230Z VRB01KT 0400 R27/0450D R09/0325N FG VV003 07/07 Q1023 TEMPO 3000 NSW

Table 3. Information extracted from the META Reports to build the predictive model

Quantitative data	Qualitative data
Expected landing hour	Presence of variable direction wind
Expected landing minute	Ceiling And Visibility OK (CAVOK)
Wind Direction (°)	Weather
Maximum wind direction (°)	Expected change in weather (TEMPO)
Minimum wind direction (°)	Fraction of sky covered by clouds
Wind speed (Knots)	Cloud type
Wind gusts (Knots)	
Visibility (m)	
Expected Visibility (m)	
Temperature (°C)	
Dew point (°C)	
Pressure (hPa)	
Cloud altitude (feet)	

2.3 Data Segregation

With the goal of training and validating our model, we shuffled and split the data into two subsets. The first one for training and validating the model (80% of the complete dataset), to fit the parameters w (weights) and b (bias) of the network and to optimise the MLP hyperparameters, respectively. The remaining 20% of the data comprises the testing subset used to assess the model performance. Our dataset was consequently split into 190 flights to train and validate the model and 47 samples for testing its performance.

Moreover, each field of the input data (structured weather data) was standardised calculating its z-score as $z = \frac{F - \mu}{\sigma}$, where F is the feature sample, and μ and σ are the mean and standard deviation of the feature distribution, respectively. It is important to note that test data was standardised using μ and σ parameters from the training and validation set.

2.4 Input Data Selection

To be useful, a predictive model should be able to classify the approach and landing trajectory with such anticipation that the final trajectory is still unknown. Looking at all 237 trajectories in Fig. 1, we can assume at 30 min before landing the final trajectory is not yet discerned.

Consequently, input data should be taken from the last METAR report available acquired at least 30 min before the expected flight landing time. With this assumption, from the two METAR reports available for each flight, we selected one as input data to the model following the rules presented in Table 4 and being $X \in \{0 - 24\}$:

Table 4. META Report selection

Expected landing	Selected report
Xh 00 m – Xh 15 m	METAR 1
Xh 15 m – Xh 30 m	METAR 2
Xh 30 m – Xh 45 m	METAR 1
Xh 45 m – Xh 59 m	METAR 2

2.5   Classification Model

In this study, we used a DL model based on a MLP with descending number of neurons. Starting with an input layer of 64 neurons (one for each input item from the structured weather report), 4 hidden layers with 32, 16, 8 and 4 neurons, respectively, and an output layer of 2 neurons (one for each possible class/TAP).

We used the ReLU activation function for every fully connected layer except for the last one, for which we used the SoftMax activation function, to calculate the probability of each sample to belong to each class.

After each layer, we applied Batch Normalization in order to prevent Internal Co-variance Shift [14], and a Dropout layer (0.25) with the aim of randomly ignoring 25% of the neurons in each layer during training stage in preventing over-fitting [15]. We initialised the batch size to $\beta = 16$ in order to update the weight parameters of our model after forward propagating 16 samples through

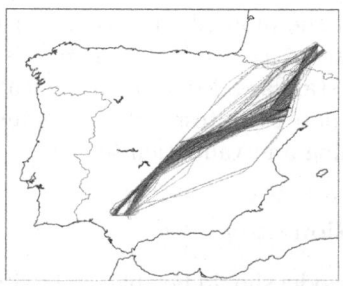

Fig. 1. Visualization of the 237 available trajectories (Toulouse-Seville) in our dataset.

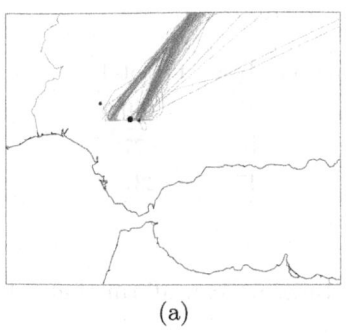 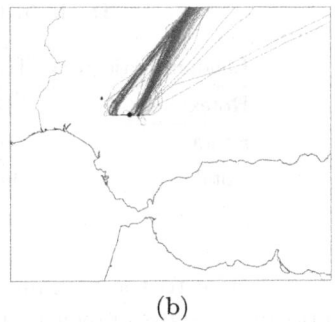

(a)   (b)

Fig. 2. Automatic ground-truth generation (clustering results). (a) Raw trajectories of the training subset and (b) classified trajectories according its terminal arrival point. (Color figure online)

the net and calculating its error using the categorical cross-entropy loss function (1):

$$\mathcal{L}(y, \hat{y}) = -\sum_i y_i \log \hat{y}_i \tag{1}$$

where y is the ground-truth label of a specific sample and $\hat{y}$ is the predicted label of such sample among the i possible classes (i.e. $i = k = 2$). Furthermore, we applied the Adam optimiser [16] to update the network weights after forwarding the inputs of each batch and calculating the error. We trained our model over 200 epochs, using a validation split of 0.2 (20% of the training dataset -38 samples- was used to validate the model).

Note that the experimental part of this work was performed in Python 3.5, using TensorFlow 2.0 and using the high-level Keras API [17] for building and training the MLP.

3 Results

3.1 Ground-Truth Clustering Stage

Regarding the ground-truth generation stage, the k-means algorithm is able to successfully cluster each trajectory to its appropriate class. The results of the classification algorithm are reported in Fig. 2. In green, the aircraft trajectories whose TAP is located at west (Santa) are observed while in red the trajectories entering by the east TAP (Rotex) of the destination aerodrome are represented. This fact allows us to train a predictive model from a gold standard automatically generated using the k-means clustering algorithm taking into advance the longitude and latitude of the flight trajectories.

3.2 MLP Training Stage

This staged is performed after the ground-truth generation for each flight's trajectory and after shuffling and partitioning our METAR dataset (in which each instance learning is composed of 64 features) in two subsets (see Table 5).

Table 5. Data distribution structure

Landing trajectory	Training & Validation set	Test set	Total
Rotex	131	29	160
Santa	59	18	77
Total	190	47	237

With regard to the training and validation stage, we can affirm that, after 200 epochs, no signs of over-fitting are evidenced.

3.3   Inference

When checking the performance of our trained model on new data, belonging to the test set (the 47 data samples separated and not used to train nor validate the model), we obtained the results reported in Table 6.

Table 6. Test results: most suitable terminal arrival point for the 47 samples

Terminal arrival point	Accuracy	Recall	F1-score
Rotex	0.94	0.97	0.96
Santa	0.92	0.85	0.88
Weighted average	0.94	0.94	0.94

Analysing the results, overall values show a promising modelling of the selection of the most suitable TAP according to the weather information of the destination airport. A slightly out-performance of the model when predicting the Rotex TAP is registered. This fact may be due to the data distribution shown in Table 5, where we can see that our dataset is not equally split between the two target classes (160 Rotex samples and 77 Santa learning instances).

4   Conclusion

In this work, we proposed a DL-based classification model able to predict flight approach trajectories based exclusively on meteorological reports. More particularly, this model was trained with historical weather reports of the destination airport and trajectory data of a particular air route, making use of a MLP architecture.

In future works, with the aim of improving the model, a bigger and more balanced dataset should be used in the training and testing stages. In this sense, more trajectories concerning the same destination airport could be added. Using this work as a baseline, future investigations should try to apply the proposed model to different airports with more TAPs. Furthermore, this work could be used as a first step for trajectory regression algorithms making use of RNN architectures to predict the entire sequential approach trajectory.

Acknowledgment. This work has received funding from the Clean Sky 2 Joint Undertaking (JU) under grant agreement No 831884. The Titan V used for this research was donated by the NVIDIA Corporation.

References

1. Pilot and technical outlook: Seattle. Boeing Commercial Airplanes, WA (2015)
2. Wolter, C.A., Gore, B.F.: NASA/TM-2015-218480: A validated task analysis of the Single Pilot Operations concept, no. January 2015 (2015)
3. Harris, D.: A human-centred design agenda for the development of single crew operated commercial aircraft. Aircr. Eng. Aerosp. Technol. **79**(5), 518–526 (2007)
4. Bailey, R.E., Kramer, L.J., Kennedy, K.D., Stephens, C.L., Etherington, T.J.: An assessment of reduced crew and single pilot operations in commercial transport aircraft operations. In: AIAA/IEEE Digital Avionics System Conference - Proceedings, vol. 2017-September, no. February 2018 (2017)
5. Lachter, J., Brandt, S.L., Battiste, V., Ligda, S.V., Matessa, M., Johnson, W.W.: Toward single pilot operations: developing a ground station. In: Proceedings of International Conference on Human-Computer Interactive Aerospace, August (2014)
6. Comerford, D., Brandt, S.L., Mogford, R.: NASA/CP - 2013–216513 NASA's Single -Pilot Operations Technical Interchange Meeting: Proceedings and Findings, April, p. 89 (2013)
7. Durand, N., Alliot, J.M., Médioni, F.: Neural nets trained by genetic algorithms for collision avoidance. Appl. Intell. **13**(3), 205–213 (2000)
8. Choi, S., Kim, Y.J., Briceno, S., Mavris, D.: Prediction of weather-induced airline delays based on machine learning algorithms. In: 2016 IEEE/AIAA 35th Digital Avionics Systems Conference (DASC), pp. 1–6. IEEE, September 2016
9. Gui, G., Liu, F., Sun, J., Yang, J., Zhou, Z., Zhao, D.: Flight delay prediction based on aviation big data and machine learning. IEEE Trans. Veh. Technol. **69**, 140–150 (2019)
10. Liu, Y., Hansen, M.: Predicting aircraft trajectories: a deep generative convolutional recurrent neural networks approach. arXiv preprint arXiv:1812.11670 (2018)
11. Shi, Z., Xu, M., Pan, Q., Yan, B., Zhang, H.: LSTM-based flight trajectory prediction. In: 2018 IEEE International Joint Conference on Neural Networks (IJCNN), pp. 1–8, July 2018
12. Bosson, C.D., Nikoleris, T.: Supervised learning applied to air traffic trajectory classification. In: 2018 AIAA Information Systems-AIAA Infotech@ Aerospace, p. 1637 (2018)
13. FlightRadar24 website. https://www.flightradar24.com/
14. Ioffe, S., Szegedy, C.: Batch normalization: Accelerating deep network training by reducing internal covariate shift. arXiv preprint arXiv:1502.03167 (2015)
15. Srivastava, N., Hinton, G., Krizhevsky, A., Sutskever, I., Salakhutdinov, R.: Dropout: a simple way to prevent neural networks from overfitting. J. Mach. Learn. Res. **15**(1), 1929–1958 (2014)
16. Kingma, D.P., Ba, J.: Adam: a method for stochastic optimization. arXiv preprint arXiv:1412.6980 (2014)
17. Chollet, F., et al.: Keras (2015). https://keras.io

Analysis of Hand-Crafted and Automatic-Learned Features for Glaucoma Detection Through Raw Circumpapillary OCT Images

Gabriel García$^{(\boxtimes)}$, Adrián Colomer, and Valery Naranjo

Instituto de Investigación e Innovación en Bioingeniería (I3B), Universitat Politècnica de València, Camino de Vera s/n, 46022 Valencia, Spain
jogarpa7@i3b.upv.es

Abstract. Taking into account that glaucoma is the leading cause of blindness worldwide, we propose in this paper three different learning methodologies for glaucoma detection in order to elucidate that traditional machine-learning techniques could outperform deep-learning algorithms, especially when the image data set is small. The experiments were performed on a private database composed of 194 glaucomatous and 198 normal B-scans diagnosed by expert ophthalmologists. As a novelty, we only considered raw circumpapillary OCT images to build the predictive models, without using other expensive tests such as visual field and intraocular pressure measures. The results ratify that the proposed hand-driven learning model, based on novel descriptors, outperforms the automatic learning. Additionally, the hybrid approach consisting of a combination of both strategies reports the best performance, with an area under the ROC curve of 0.85 and an accuracy of 0.82 during the prediction stage.

Keywords: Glaucoma · Circumpapillary OCT · Hand-driven learning · Deep learning · Hybrid classification

1 Introduction

Glaucoma is a chronic optic neuropathy characterised by causing several visual field defects and structural changes in the optic nerve, such as a thinning of the retinal nerve fibre layer (RNFL) [1]. Nowadays, this degenerative disease is the leading cause of blindness worldwide and is expected to affect 111.8 million people in 2040 [2]. The glaucoma diagnosis includes different expensive analysis (pachymetry, tonometry and visual field tests, among others) besides a subjective interpretation of expert ophthalmologists who often differ, especially in terms of early identification [3]. Currently, imaging techniques based on fundus image and optical coherence tomography (OCT) have become a powerful tool to address the glaucoma diagnosis.

© Springer Nature Switzerland AG 2020
C. Analide et al. (Eds.): IDEAL 2020, LNCS 12490, pp. 156–164, 2020.
https://doi.org/10.1007/978-3-030-62365-4_15

Related Work. Timely treatment of glaucoma is essential to avoid the irreversible vision loss [2], so several computer-aided diagnosis systems and predictive algorithms focused on OCT and fundus images have been proposed in the literature to achieve early detection. Most of them were performed through traditional machine-learning (ML) techniques based on feature extraction and selection methods [4–6]. All of them had in common the use of additional parameters relevant for glaucoma diagnoses, such as the intraocular pressure (IOP) and visual field (VF) tests, besides the OCT images. Unlike these works, we propose an innovative end-to-end system able to predict the glaucoma disease just from raw circumpapillary OCT images, without taking into account external expensive tests, like VF or IOP. We aim to elucidate the added value that this OCT samples around the retina optic nerve head (ONH) can provide for glaucoma detection. In recent years, the overwhelming irruption of the deep learning (DL) has replaced the traditional hand crafted-based methods, but most of the state-of-the-art studies used fundus images [7,8] or RNFL thickness probability maps extracted by combining fundus images and OCT B-scans [9,10]. However, to the best of the authors' knowledge, we are the first that apply deep learning to evidence glaucoma just from raw circumpapillary OCT images.

Contribution of This Work. In this paper, we propose a comparison between traditional and contemporary machine-learning models to analyse whether hand-driven approaches can outperform deep-learning algorithms for glaucoma detection, especially when addressing small databases. We hypothesise that the CNNs cannot replace the original way in which people can encode the information captured from a subjective point of view; in the same way that CNNs are able to identify hidden patterns that are not within reach of the human eye. For this reason, as the main novelty, we propose a hybrid model by fusing the features extracted from both ML and DL approaches to identify glaucoma.

2   Material

The present work was carried out making use of a private database, coming from the Oftalvist Ophthalmic Clinic, which consists of 392 B-scans around the ONH of the retina. In particular, 194 samples from 97 patients were diagnosed by expert ophthalmologists as glaucomatous, whereas 198 circumpapillary images from 99 patients were associated with normal eyes. Note that each B-scan, of dimensions $M \times N = 496 \times 768$ pixels, was acquired using the *Heidelberg Spectrallis OCT* system, which allows obtaining an axial resolution of 4–5 µm.

3   Methodology

3.1   Data Partitioning

In order to provide reliable results, we carried out a patient-based partitioning of the database to separate the training set from the independent test set.

In particular, $\frac{4}{5}$ of the data (158 healthy and 156 glaucomatous OCT images) were used to train the predictive models, whereas $\frac{1}{5}$ (40 healthy and 38 glaucomatous B-scans) were employed to test them. Additionally, we applied an internal $k = 5$-fold cross-validation technique intending to manage the overfitting and select the best parameters during the validation stage. Finally, we used the entire training set to build the final predictive models, which were assessed on the external set.

3.2 Hand-Driven Learning Methodology

This approach consists of three main phases corresponding to the feature extraction, feature selection and classification through several traditional ML classifiers. Note that we used the RNFL and retina structures segmentation extracted by the Heidelberg Spectrallis system to perform this methodology.

Feature Extraction. For each circumpapillary OCT image I_i, being $i = \{1, 2, 3, ..., P\}$ and $P = 392$ the number of B-scans, we combine, as a novelty, different variables related to four main descriptors: RNFL thickness, texture variables, fractal analysis and demographic data (age and gender). Regarding the RNFL thickness, we propose in this paper an innovative way of codifying the information, unlike [5,6] where the authors employed the measures directly extracted from the hardware system. Let the vector $T_i = \{t_1, t_2, ..., t_j, ..., t_N\}$, where each $t_j \in T$ is the RNFL thickness calculated from the image I_i in the position j, the proposed method is able to group the thickness values in a $h_{i,p}$ histogram vector, where $p = \{1, 2, 3, 4\}$. For each p value, $h_{i,p}$ vector allows quantifying the number of thickness t_j whose distance is ranged between D_p and D_{p+1}, being $D = [0, 15, 30, 45, 100]$ a vector of relevant distances, which were selected after studying the samples of the training set. Worth noting that minimum ($maxT_i$) and maximum ($minT_i$) RNFL thickness values were also calculated for this descriptor category. Concerning the texture variables, grey-level co-occurrence matrix (GLCM) [11] and local binary patterns (LBP) were applied for encoding the textural information contained in the retina region of each I_i. Variables such as contrast, correlation, energy, homogeneity, entropy, mean and standard deviation were calculated from a GLCM of dimensions 8×8, using two offsets of $[-2, 0]$ and $[-2, 2]$ to measure the frequency in which a pixel with an intensity x is adjacent to another with intensity y. Additionally, in order to recognise local texture information, we combined the uniformly invariant to rotation transforms $LBP_{P,R}^{riu2}$ operator, proposed in [12], with the rotational invariant local variance VAR descriptor, to finally compute the LBP variance (LBPV) histogram, proposed in [13] (see Fig. 1). It should be noted that similar texture descriptors have been previously considered for glaucoma detection from fundus images [14,15] but, to the best of the authors' knowledge, this is the first time that GLCM and LBP features are applied to circumpapillary OCT images. Additionally, we analysed the fractal dimension in five directions ($0°$, $30°$, $45°$, $60°$, $90°$) via the Hurst Exponent [16] computation to determine the presence of underlying trends in the complexity of the retinal region of each I_i. After the feature extraction phase, 75 hand-crafted variables per learning instance were taken into account to address the next stage.

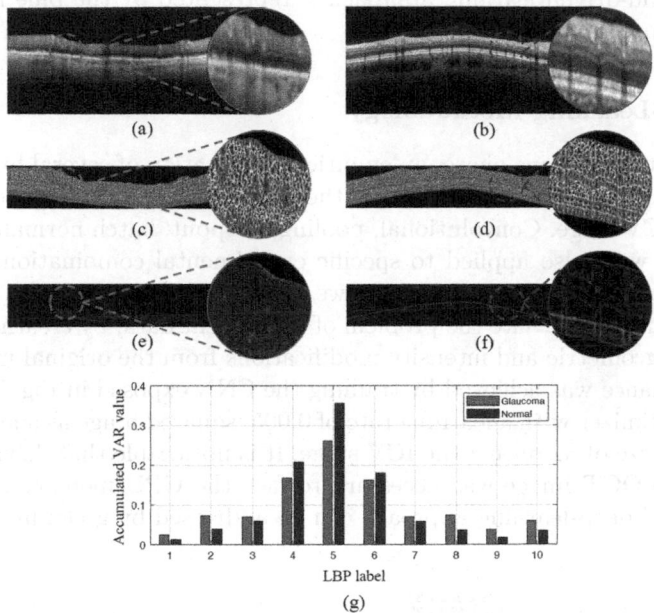

Fig. 1. (a-b) Examples of glaucomatous and normal samples. (c-d) LBP images. (e-f) VAR images. (g) 10-bin histograms of the LBPV operator from LBP and VAR images.

Feature Selection. An in-depth statistical analysis was carried out to select the most relevant variables in order to feed the proposed classifiers. Initially, a *Kolmogorov-Smirnov test* was applied to determine the distribution of the variables. Then, *Student's t-tests* or *Mann-Whitney U tests* were performed to analyse the discriminatory ability of each variable v by comparing means or medians, respectively, depending on whether v followed a normal distribution $N(0,1)$ or not. The correlation coefficient was also calculated to obtain the independence grade between pairs of variables. Note that a level of significance $\alpha = 0.05$ was defined for both hypothesis contrast to discard the non-relevant features, as well as the redundant information when $p\text{-}value < \alpha$.

Model Training. Once the most relevant features were selected, we trained different ML classifiers such as support vector machine (SVM) and multi-layer perceptron (MLP), in line with the state-of-the-art studies [4] and [6]. Several *box constraints* and *kernel scale* parameters for the SVM classifiers, and learning rates, loss functions, optimizers and network structures for the MLP network were considered during the internal cross-validation (ICV) stage. A considerable outperforming of the MLP classifier was achieved, with respect to the SVM, using the gradient descent adaptative optimizer with a learning rate of 0.001 and the binary cross-entropy as a loss function. Concerning the network structure, one hidden layer with 8 neurons reported the best model performance. Note that the proposed hand-driven learning approach is represented by the blue lines in the flowchart exposed in Fig. 3.

3.3   Deep-Learning Methodology

Similarly to the previous phase, an empirical exploration of several hyperparameters was performed in order to build the best predictive deep-learning model during the ICV stage. Convolutional, pooling, dropout, batch normalisation and dense layers were also applied to specific experimental combinations in search of the best network architecture. Also, we considered the use of data augmentation techniques to alleviate the problem of insufficient data, by creating artificial samples via geometric and intensity modifications from the original images. The best performance was achieved by training the CNN exposed in Fig. 2 and using Adadelta optimizer with a learning rate of 0.005, squared hinge as a loss function and a batch size of 32 during the ICV stage. It is noticeable that down-sampling ×0.5 of each OCT image was necessary to face the GPU memory constraints. The proposed deep-learning approach can be addressed by green lines in Fig. 3.

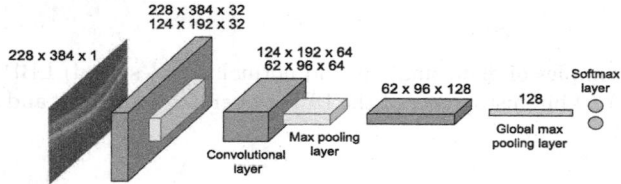

Fig. 2. Illustrative representation of the implemented CNN architecture.

3.4   Hybrid Approach

As the main novelty in this paper, we propose a hybrid model to address the glaucoma detection taking into account both the hand-crafted and automatic-learning features. Our aim is to combine the original human point of view with the hidden potential enclosed in the CNNs. Specifically, we made use of the previously defined deep-learning base model as a feature extractor from each OCT image. Then, we fused the 75 ML and 128 DL extracted features to form the final feature vector from which we performed, in the same conditions, the feature selection and MLP training stages carried out in Sect. 3.2. Finally, the three proposed models were assessed and compared using the test set, according to the flowchart exposed below. Note that the information relative to the hybrid approach can be interpreted by the yellow lines in Fig. 3.

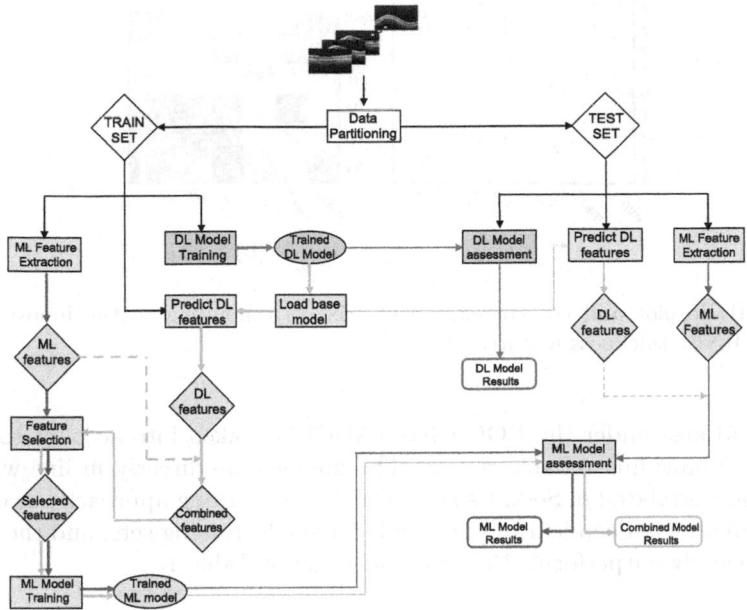

Fig. 3. Flowchart detailing the proposed ML, DL and hybrid approaches, in blue, green and yellow, respectively. (Color figure online)

4   Results and Discussion

4.1   Feature Selection Results

Regarding the hand-driven learning approach, 25 from a total of 75 features that composed a learning instance were selected after the statistical analysis. Otherwise, concerning the hybrid approach, 100 features from a total of 203 were reported as relevant variables to address the MLP training stage. Note that both ML and hybrid-final feature vectors included variables corresponding to the four kinds of descriptors used in this work. In addition, all the proposed features corresponding to the new RNFL thickness histogram-based method resulted statistically significant. A boxplot relative to these features is exposed in Fig. 4 to show the discriminatory ability of the proposed new descriptor. Besides, we also represent the correlation matrix of the same variables to evidence the independence level between them.

4.2   Glaucoma Prediction

Validation Results. Classification results reached during the validation stage are detailed in Table 1 to objectively compare the proposed hand-driven learning (HDL), deep-learning (DL) and hybrid-learning methodologies. Different figures of merit, such as sensitivity (SN), specificity (SPC), F-score (FS), accuracy

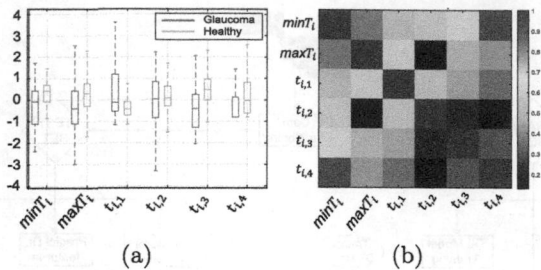

Fig. 4. (a) Boxplot and (b) correlation matrix corresponding to the innovative six proposed RNFL thickness features.

(ACC) and area under the ROC curve (AUC) are taken into account to assess the models providing reliable results. The findings are directly in line with the hypothesis postulated in Sect. 1 since hand-driven learning approach has demonstrated surpassing deep-learning methods for small training sets, and the hybrid strategy clearly outperforms the rest, according to Table 1.

Table 1. Quantitative results reached during the ICV stage from all approaches.

	HDL approach	DL approach	Hybrid approach
SN	**0.802 ± 0.108**	0.747 ± 0.079	0.779 ± 0.110
SPC	0.807 ± 0.070	0.751 ± 0.080	**0.912 ± 0.035**
FS	0.803 ± 0.072	0.748 ± 0.029	**0.830 ± 0.074**
ACC	0.809 ± 0.058	0.748 ± 0.023	**0.847 ± 0.048**
AUC	0.890 + 0.056	0.823 ± 0.046	**0.943 ± 0.018**

Test Results. In this section, we detail an external validation of the three proposed models using the independent test set, as it was previously explained in Fig. 3. The classification results corresponding to the test set are exposed in Table 2. We can observe that, in line with the ICV stage, hand-driven learning provides a slight improvement regarding the deep-learning approach, which reaches values around 0.7 for all measures. Additionally, the hybrid methodology, characterised by the fusion of the features extracted from both ML and DL models, reports the most promising results for almost all figures of merit. It is important to note that an objective comparison with other state-of-the-art studies is not possible because all of them were performed on private databases or using another kind of input data, such as RNFL thickness probability maps or visual field tests.

Table 2. Results comparison between the proposed models during the prediction stage.

	HDL model	DL model	Hybrid model
SN	**0.7632**	0.6750	0.7368
SPC	0.7500	0.6842	**0.9000**
FS	0.7533	0.6835	**0.8000**
ACC	0.7564	0.6795	**0.8205**
AUC	0.8138	0.7480	**0.8467**

5   Conclusion

In this work, three different learning methodologies have been proposed with the aim of elucidating that, under specific circumstances, hand-driven learning approaches can outperform deep-learning algorithms. The reported results evidenced that a combination of hand-crafted and data-learning strategies can improve the models' performance, especially for small databases.

References

1. Weinreb, R.N., Khaw, P.T.: Primary open-angle glaucoma. Lancet **363**(9422), 1711–1720 (2004)
2. Jonas, J.B., Aung, T., Bourne, R.R., Bron, A.M., Ritch, R., Panda-Jonas, S.: Glaucoma-authors' reply. Lancet **391**(10122), 740 (2018)
3. National GAU. Glaucoma: diagnosis and management (2017)
4. Bizios, D., Heijl, A., Hougaard, J.L., Bengtsson, B.: Machine learning classifiers for glaucoma diagnosis based on classification of retinal nerve fibre layer thickness parameters measured by stratus oct. Acta Ophthalmol. **88**(1), 44–52 (2010)
5. Asaoka, R., Hirasawa, K., Iwase, A.E.A.: Validating the usefulness of the "random forests" classifier to diagnose early glaucoma with optical coherence tomography. Am. J. Ophthalmol. **174**, 95–103 (2017)
6. Kim, S.J., Cho, K.J., Oh, S.: Development of machine learning models for diagnosis of glaucoma. PLoS ONE **12**(5), e0177726 (2017)
7. Diaz-Pinto, A., Colomer, A., Naranjo, V., Morales, S., Xu, Y., Frangi, A.F.: Retinal image synthesis and semi-supervised learning for glaucoma assessment. IEEE Trans. Med. Imaging **38**(9), 2211–2218 (2019)
8. Medeiros, F.A., Jammal, A.A., Thompson, A.C.: From machine to machine: an oct-trained deep learning algorithm for objective quantification of glaucomatous damage in fundus photographs. Ophthalmology **126**(4), 513–521 (2019)
9. Muhammad, H., Fuchs, T.J., De Cuir, N., De Moraes, C.G.E.A.: Hybrid deep learning on single wide-field optical coherence tomography scans accurately classifies glaucoma suspects. J. Glaucoma **26**(12), 1086 (2017)
10. Wang, P., Shen, J., Chang, R., Moloney, M., Torres, M., Burkemper, B.E.A.: Machine learning models for diagnosing glaucoma from retinal nerve fiber layer thickness maps. Ophthalmol. Glaucoma **2**(6), 422–428 (2019)

11. Haralick, R.M., Shanmugam, K., Dinstein, I.H.: Textural features for image classification. IEEE Trans. Syst. Man Cybern. (6), 610–621 (1973)
12. Ojala, T., Pietikainen, M., Maenpaa, T.: Multiresolution gray-scale and rotation invariant texture classification with local binary patterns. IEEE Trans. Pattern Anal. Mach. Intell. **24**(7), 971–987 (2002)
13. Guo, Z., Zhang, L., Zhang, D.: A completed modeling of local binary pattern operator for texture classification. IEEE Trans. Image Process. **19**(6), 1657–1663 (2010)
14. Ali, M.A., Hurtut, T., Faucon, T., Cheriet, F.: Glaucoma detection based on local binary patterns in fundus photographs. In: Medical Imaging 2014: Computer-Aided Diagnosis, vol. 9035, p. 903531. International Society for Optics and Photonics (2014)
15. Kavya, N., Padmaja, K.: Glaucoma detection using texture features extraction. In: 2017 51st Asilomar Conference on Signals, Systems, and Computers. IEEE (2017)
16. Hurst, H.E.: Long term storage. An experimental study (1965)

Video Semantics Quality Assessment Using Deep Learning

Rui Jesus[1,2]([⊠]), Bárbara Silveira[1], and Nuno Correia[1]

[1] NOVA LINCS, FCT, Universidade NOVA de Lisboa, Caparica, Portugal
[2] ADEETC - Instituto Superior de Engenharia de Lisboa, IPL, Lisbon, Portugal
rjesus@deetc.isel.ipl.pt

Abstract. This work proposes a method to assess the quality of user-generated videos (UGVs) of specific social events. The method is based on matching the semantic information extracted from videos and the information obtained from text news of the same event. Deep learning techniques are used to detect objects in the video scenes. News articles are represented by a set of relevant terms automatically extracted from the news. This paper describes our method and an evaluation of it.

Keywords: User generated content · Video quality assessment · Semantic deep learning

1 Introduction

Internationally famous events such as the soccer World Cup final, the Eurovision Song Contest or the Prince of England's royal wedding are events that lead people to produce and share vast amounts of content. The message contained in videos produced by professionals is generally richer, but the videos generated by normal users contain details that are interesting to other users. Thus, UGVs are one of the best ways of sharing relevant episodes of these events. However, each video has a different quality, either by the way it was filmed, by the brand of the devices or by the captured message, greatly influenced by the subjectivity of the user who produced it. UGVs on social networks easily earn great popularity among users. Thus, videos with incomplete or subjective event messages can quickly proliferate a little-right message of a great international event. Therefore, it is also essential to evaluate the quality of the message [9] that the video conveys of the event.

The human vision [2] can be one of the best ways to assess video quality. However, this assessment may be subjective and does not provide a standard measure. One way of objectively measure the meaning of a video can be by the automatic extraction of semantic information from the video content.

The use of semantics for assessing video quality is the main focus of this paper. Semantic concepts (objects) are detected in UGVs of a specific event, and this detection is compared to the information found in the news articles made by professionals, related to the same event. The result of this comparison is used to provide an assessment of the video quality.

C. Analide et al. (Eds.): IDEAL 2020, LNCS 12490, pp. 165–172, 2020.
https://doi.org/10.1007/978-3-030-62365-4_16

The following sections present related approaches, our proposal and its evaluation.

2   Related Work

Although the human visual system is the one that performs best, when it comes to assessing the quality of a video, it is not the most reliable since it is subjective to each person [8]. This fact created the need to develop and improve an objective assessment, which is accomplished by the use of measurements, where the goal is to give results that correlate closely with results obtained through human perception [12]. Objective evaluation of video quality has been carried out based on three different types of characteristics: (1) aesthetic [5,13], (2) video production [10] and (3) semantics [9].

To assess the quality of the video through the use of semantics, a combination between the subjective and objective evaluation techniques are used so that it is possible to compare the performance of both methods. In some cases, the importance of the content can be compared with the importance of the perceptual quality for the user satisfaction [2].

Semantics proposals for video quality assessment [9] are based on video semantics extraction methods [1,3,6,10]. The concept-extraction schemes were primarily carried out on the news video since those have content structures. Most of these studies are operated in a procedure composed of two stages. The first consists of segmenting the video clip into certain analysis units and also extracting their representative features (usually keyframes). The following stage is the decision-making process where the extraction of the semantic index from the feature descriptors occurs concerning the improvement of the framework robustness [10].

More recent works seek to solve the problem of extracting video events using deep learning techniques [3,4,6]. Our method to evaluate the quality of the video also uses deep learning algorithms. It is based on the Tapaswi [11] approach to align the text obtained in synopses and books with movies and TV series.

3   Video Quality Assessment

To assess the quality of a video given its semantics, it is proposed the system presented in Fig. 1. Given a set of videos (UGVs) of an event and a collection of text news produced by journalists from the same event, the system evaluates the quality of each video through a semantic measure that compares text and video semantics.

The **Video Collection** in Fig. 1, $V_{event} = \{V_1, V_2, ..., V_M\}$ is composed of a set of N videos (UGVs) related to an event. Where V_n denotes a video. The **News Collection**, $N_{event} = \{A_1, A_2, ..., A_K\}$, related to the same event (K denotes the number of news articles) are produced by experts in the field. It is assumed the collection of news represents the best way to report the semantics of the event. The **Video Semantics Detection** block represents the detection

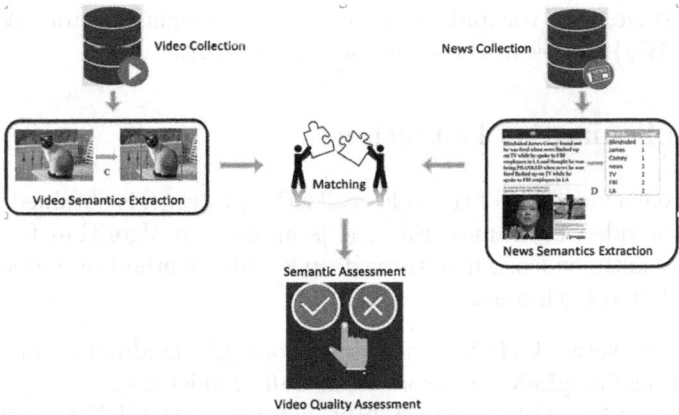

Fig. 1. Video quality assessment system.

algorithms used to extract objects from the video collection. The **News Semantics Detection** is the component related to the extraction of the most relevant terms of the news. The **Matching** block uses the information from the **Video Semantics Detection** and the **News Semantics Detection** components to measure the distance between the video and the text news, in terms of semantics. The **Video Quality Assessment** element in Fig.1 is used to define the video quality based on the result of the semantic measure obtained in the **Matching** block.

4 News Semantics Detection

The text of the news collection, N_{event}, related to a specific event, is analyzed to build a vocabulary with the semantic meaning of the event. This vocabulary $Vcb = \{wd_1, wd_2, ..., wd_N\}$ is composed of the N most relevant words extracted from these articles, captured from online news websites such as CNN, BBC or Independent. Each word, wd_n, has a weight, W_n. The following steps compute them:

1. Extraction of the words, wd_n, and the number of the terms occurrence from each website, $W_{n,k} = n(wd_n, A_k)$;
2. Removal of meaningless words (words that have less than two letters and more than fifteen letters). Then, some characters of the words are eliminated (for example, the word "hello€" would be replaced after the "hello" substitution);
3. Computation of the weights, W_n. For repeated words in different news, the number of occurrences is summed $W_{nT} = \sum_{j=1}^{K} W_{n,j}$ and normalized, based on the TF-IDF scoring measure,

$$W_n = \frac{W_{nT}}{\sum_{j=1}^{N} W_{jT}}. \tag{1}$$

At this stage, the vocabulary, Vcb, and the weights vector, $W_{NEWS} = \{W_1, W_2, ..., W_N\}$ represent the semantics of the event.

5   Video Semantics Detection

Semantics from a video, V, of the video collection, $Vevent$, is obtained by detecting objects in video keyframes. First, it is applied an algorithm for detecting scene changes in the video. Then, three deep learning methods are used to detect objects in selected keyframes:

- Deep Neural Network (**DNN**) algorithm (openCV module) - This module is trained using GoogLeNet network from Caffe Model Zoo;
- Convolutional Neural Networks (**CNN**) algorithm (MobileNet module) - This module is trained using the framework TensorFlow;
- **Yolo** system [7] - Fast and region-based algorithm for object detection, which is based on a single neural network with several convolutional layers.

To compare the semantics of the video with the semantics of the news, these algorithms are trained to detect a set of concepts $C = \{C_1, C_2, ..., C_N\}$, which are associated with the terms of the news vocabulary Vcb. That is, for each entry in Vcb, a correspondent concept is detected in video keyframes. Thus, each video, V, is also represented by a set of concepts, C, and weights of each concept, $W_V = \{W_{c1}, W_{c2}, ..., W_{cN}\}$. The following steps compute these weights:

1. The weight of a concept is equal to the confidence level of the object detection algorithm (each detection algorithm provides a confidence level between 0 and 1);
2. For multiple detections of an object in a video, the confidence level is summed up each time the object is detected;
3. Normalization of the weights is similar to the normalization used in the weights of the terms of the news. It is provided by dividing them by the sum of the weights of all concepts detected in a video.

6   Matching Between News and Videos

The assessment of the semantics quality of specific videos of an event, produced by users, is performed by matching the semantics extracted in the videos, with the semantics obtained from the news of the event, written by specialized journalists. From the previous section, news of an event are represented by a vocabulary of isolated words and their weights. Videos are represented by correspondent concepts (objects detected) and their weights (detection level of confidence).

The semantic correlation between a video, V, represented by C and W_V, and a collection of news, N_{event}, of the same event, described by Vcb, and W_{NEWS} is computed by the cosine similarity of the weights vectors,

$$Sim(W_{NEWS}, W_V) = \frac{\sum_{j=1}^{N} W_j W_{cj}}{\sqrt{\sum_{j=1}^{N} W_j^2} \sqrt{\sum_{j=1}^{N} W_{cj}^2}}. \tag{2}$$

Equation 2 is computed for all the videos that report one specific event. Videos with a higher value of $Sim(W_{NEWS}, W_V)$ are selected as the videos with more semantic quality.

7 Evaluation

The data used for the tests are videos and articles news. The accuracy of the object detection algorithms are an important part of the method proposed to evaluate the video quality. Therefore, an evaluation of the performance of the three object detection methods is performed. To evaluate the object detection accuracy of the three algorithms, 17 videos are selected from the database of the Cognitus project. These videos are about street concerts and football matches. For the video quality assessment, three videos related to the 2018 Eurovision contest event are used

The news articles chosen are related to the event, 2018 Eurovision contest. The main sources of the news are online news websites (e.g., CNN, BBC or Independent) The vocabulary and the weights are obtained from 10 news.

From all the experiments related to object detection, the following conclusions were obtained:

- **DNN** classifier can identify a grand diversity of objects, but is not very accurate and has a low confidence level in the detection;
- **CNN** classifier can identify with a high confidence level and mostly the right objects in the scene, but it has some inaccuracies;
- **Yolo** is the best in identifying the "person" object in the scene, but it also has inaccuracies and lack in the detection of other objects.

Figure 2 shows the sum of the confidence level obtained by all the object detected correctly by each of the methods used to extract semantics from a video. It shows the detection results for each video of the Eurovision contest. Best results are obtained when all the methods are combined, and the confidence level is set to 1 (100%).

The algorithm for video quality assessing is evaluated with videos and news of the 2018 Eurovision contest event. From the 3 videos used, first, 1 and 3 keyframes per scene is analysed. DNN, CNN and Yolo are evaluated individually as the methods of semantics detection. Then, the information extracted by each is joint to evaluate a version combining the different semantics obtained by each method. The confidence level returned by each method, regarding the objects detected, is not always as expected. Therefore, a version with all detectors combined and the confidence level equal to 1 (100% of confidence), for all objects detected, is also evaluated. Table 1 summarizes the results obtained.

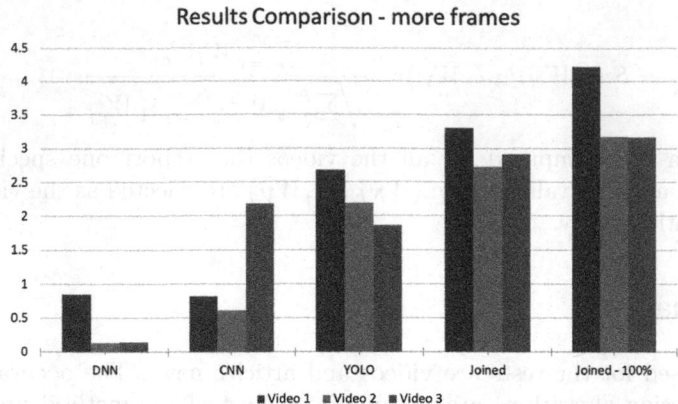

Fig. 2. Object detection results with 3 keyframes per scene.

Table 1. Video quality assessing: results obtained using the object detection methods individually and combined (all detectors).

	DNN	CNN	All detectors	All detectors (100%)
Video 1 (1 keyframe)	0,445	0,476	0,504	0,521
Video 2 (1 keyframe)	0,444	0,473	0,470	0,500
Video 3 (1 keyframe)	0,190	0,467	0,506	0,480
Video 1 (3 keyframes)	0,444	0,473	0,492	0,518
Video 2 (3 keyframes)	0,443	0,478	0,426	0,460
Video 3 (3 keyframes)	0,190	0,444	0,546	0,542

From the Table 1, when using 1 keyframe per scene and consider the DNN or the CNN detectors individually, the video with more quality is the first. The Euclidean distance metric was also tested, instead of the cosine similarity. In this case, the selected one was the second video. When combining all the detectors (for repeated concepts, the concept is only referenced one time, but its value is the sum of the normalized value of the object detected by each detector), the best video was the third video. When doing the same but using the Euclidean distance metric, the best video was the second one.

Different detectors give different levels of confidence, to analyze the influence of the confidence level, the next experience use for all detector a level of confidence equal to 1 (100%). The first video is the best (last column of Table 1) like when using only DNN or CNN. While for the Euclidean metric, the best video was always the second.

In some videos, one keyframe per scene/shot does not represent all the semantics of the scene. Thus, new tests are performed with 3 keyframes per scene. More objects are detected (see Fig. 2) but also more errors. Nevertheless, more different concepts are obtained, which translates into different results, in terms of video quality (see Table 1). The best video for the CNN detector change to the second video. DNN and YOLO detectors have almost the same results.

Video 3 is the one that earns the most by increasing the number of keyframes, in particular with the CNN detector. This increases the correlation with the news when the 3 detectors are joined.

8   Conclusions and Future Work

This paper describes a method to assess the quality of a video of a specific event, in terms of the meaning of the message of the video. The experiments conducted to evaluate the method proposed result in the following conclusions:

- The evaluation of the method proposed shows the accuracy of the object detection algorithms is important to the performance of the method;
- The methodology used to assess the quality of the videos, using the information of the algorithms and the content of the news, provided some interesting results when the structure of the event is not relevant to understand the meaning of the event;
- The combination of different object detection algorithms improve the value of the semantic detected on a video.

For future work, it is important to improve the model that represents the news and the way it is aligned with the video. It is also relevant to include the opinion of the users.

Acknowledgments. This work is mainly supported by NOVA LINCS (UIDB/0 4516/2020) with the financial support of FCT - Fundação para a Ciência e a Tecnologia, through national funds. It was also partially supported by Cognitus project.

References

1. Chen, M., Chen, S., Shyu, M., Wickramaratna, K.: Semantic event detection via multimodal data mining. IEEE Sig. Process. Mag. **23**(2), 38–46 (2006). https://doi.org/10.1109/MSP.2006.1621447
2. Evans, M., Kerlin, L., Larner, O., Campbell, R.: Feels like being there: viewers describe the quality of experience of festival video using their own words. In: Extended Abstracts of the 2018 CHI Conference on Human Factors in Computing Systems, pp. LBW029:1-LBW029:6. CHI EA 2018. ACM, New York (2018). https://doi.org/10.1145/3170427.3188507
3. Jiang, H., Lu, Y., Xue, J.: Automatic soccer video event detection based on a deep neural network combined CNN and RNN. In: 2016 IEEE 28th International Conference on Tools with Artificial Intelligence (ICTAI), pp. 490–494, November 2016. https://doi.org/10.1109/ICTAI.2016.0081
4. Liu, L., et al.: Deep learning for generic object detection: a survey. arXiv abs/1809.02165 (2018)
5. Sun, L., Yamasaki, T., Aizawa, K.: Photo aesthetic quality estimation using visual complexity features. Multimedia Tools Appl. **77**(5), 5189–5213 (2017). https://doi.org/10.1007/s11042-017-4424-4

6. Pouyanfar, S., Chen, S.: Semantic event detection using ensemble deep learning. In: 2016 IEEE International Symposium on Multimedia (ISM), pp. 203–208, December 2016. https://doi.org/10.1109/ISM.2016.0048
7. Redmon, J., Farhadi, A.: Yolo9000: better, faster, stronger. In: 2017 IEEE Conference on Computer Vision and Pattern Recognition (CVPR), pp. 6517–6525 (2017)
8. Seshadrinathan, K., Soundararajan, R., Bovik, A.C., Cormack, L.K.: Study of subjective and objective quality assessment of video. IEEE Trans. Image Process. **19**(6), 1427–1441 (2010). https://doi.org/10.1109/TIP.2010.2042111
9. Shahid, M., Khatibi, S., Tuemay, Y.: Popularity index through video semantic quality assessment. In: 2014 IEEE China Summit International Conference on Signal and Information Processing (ChinaSIP), pp. 344–348, July 2014. https://doi.org/10.1109/ChinaSIP.2014.6889261
10. Shyu, M., Xie, Z., Chen, M., Chen, S.: Video semantic event/concept detection using a subspace-based multimedia data mining framework. IEEE Trans. Multimedia **10**(2), 252–259 (2008). https://doi.org/10.1109/TMM.2007.911830
11. Tapaswi, M.: Story Understanding through semantic analysis and automatic alignment of text and video. Ph.D. thesis, Karlsruhe Institute of Technology (2016)
12. Webster, A., Jones, C., Pinson, M., Voran, S., Wolf, S.: Objective video quality assessment system based on human perception. In: SPIE's Symposium on Electronic Imaging: Science and Technology, vol. 1913 (1993). https://doi.org/10.1117/12.152700
13. Yeh, H., Yang, C., Lee, M., Chen, C.: Video aesthetic quality assessment by temporal integration of photo- and motion-based features. IEEE Trans. Multimedia **15**(8), 1944–1957 (2013). https://doi.org/10.1109/TMM.2013.2280250

A Deep Learning Approach for Intelligent Cockpits: Learning Drivers Routines

Carlos Fernandes[1]([✉]), Flora Ferreira[2], Wolfram Erlhagen[2], Sérgio Monteiro[1], and Estela Bicho[1]

[1] Center Algoritmi, University of Minho, Guimarães, Portugal
`a74892@alunos.uminho.pt, estela.bicho@dei.uminho.pt`
[2] Center of Mathematics, University of Minho, Guimarães, Portugal
`wolfram.erlhagen@math.uminho.pt`

Abstract. Nowadays an increasing number of vehicles are being equipped with powerful cockpit systems capable of collecting drivers' footprints over time. The collection of this valuable data opens effective opportunities for routine prediction. With the growing ability of vehicles to collect spatial and temporal information solving the routine prediction problem becomes crucial and feasible. It is then extremely important to advance and take advantage of the capabilities of these cockpit systems. A vehicle that is capable of predicting the next destination of the driver and when the driver intends to leave to that destination can prepare the journey in advance. Previous studies tackling the next location prediction problem have made use of Traditional Markov models, Neural Networks, Dynamic models, among others. In this work, a framework based on the hierarchical density-based clustering algorithm followed by a Long Short-Term Memory (LSTM) recurrent neural network is proposed for spatial-temporal prediction of drivers' routines. Based on real-life driving scenarios of three different users, the proposed approach achieved a test set accuracy of 96.20%, 90.23%, and 86.40% when predicting the next destination and a R^2 Score of 93.69, 79.21, and 28.81 when predicting the departure time, respectively. The results indicate that the proposed architecture can be implemented on the vehicle cockpit for the assistance of the management of future trips.

Keywords: Human mobility patterns · Next destination prediction · Departure time prediction · Deep learning · Intelligent vehicles

1 Introduction

The popularity of human routine prediction is rising due to its significant application value and the enhanced ability of systems to collect and process information in various spatial and temporal contexts. Understanding human mobility patterns and routines help to capture human necessities, this information can be used by intelligent systems to support individual and social events. Spatial and

© Springer Nature Switzerland AG 2020
C. Analide et al. (Eds.): IDEAL 2020, LNCS 12490, pp. 173–183, 2020.
https://doi.org/10.1007/978-3-030-62365-4_17

temporal context (where and when) are key factors for describing events. Collecting these factors is essential for analyzing and predicting human routines in practical applications. Many studies have reported that individuals' daily routines exhibit a high level of spatial-temporal regularity, a high probability of returning to a few highly frequented locations and a tendency to visit specific locations at specific hours [4,8,15,20,24]. For example, in the scope of this work, drivers may have very similar weekdays routines consisting of driving their children to school, driving to work, driving back to pick their children at school, driving to the gymnasium to exercise and drive back home at the end of the evening. A daily routine is also typically dependent on temporal constraints such as going to the church every Sunday. Recently, human routines and mobility patterns have been investigated by experts in various fields such as economics [11], automotive [6], urban computing [7], criminology [16], among others. There are several possible uses for systems able to predict human routines. In terms of automotive systems, the predictions might be used to prepare the car before the driver leaves to the next destination, the predictions could also be used alongside traffic estimation systems to suggest the best driving routes to the next destination, all of this without requiring input from the driver.

In this work, we propose an efficient driving routine prediction framework based on a hierarchical density-based clustering algorithm followed by a deep neural network (DNN). Our approach focuses on building a deep learning framework through the observation of past driving scenarios using real-life data based on GPS coordinates and temporal information from three different drivers. The developed framework can be used to predict the driver's next destination and departure time to that destination.

2   Background

2.1   Clustering Approaches for Points of Interest Extraction

Some points of interest (POI) might be quite predictable and easy to extract due to the high level of repetition occurring in a driver's daily routine, for instance, places like home or work. Other POIs might be more difficult to extract due to the low level of repetition, these might be places like the museum or the cinema. Clustering techniques including K-Means clustering, hierarchical clustering, and density-based clustering are commonly applied for POI extraction [13,25,26]. The most popular clustering technique due to its capability of detecting clusters of varying shapes and sizes is density-based spatial clustering (DBSCAN) [5,22]. However, the performance of DBSCAN is strongly correlated with two parameters - radius and minimum neighbors - both are always set with empirical values [13,27]. To obtain more meaningful clusters the hierarchical clustering algorithm provides a more intuitive way for clustering exploration, additionally, various studies have shown that urban areas and human movements present a hierarchical nature meaning that hierarchical clustering approaches have a strong theoretical foundation in human routine analysis [1,19,21]. Hierarchical density-based clustering (HDBSCAN) is a recent clustering method that seeks to integrate both hierarchical clustering and density-based spatial clustering, this algorithm has been emerging with great success [3,14,17]. A driver's routine presence in some locations may be dense but dispersive in others so, using DBSCAN with empirical and fixed values might lead to poor POIs extraction since this method shows weakness when identifying clusters of different density. Contrarily, hierarchical density-based clustering can achieve better performance in POIs extraction when handling with varying density data. Therefore, HDBSCAN will be used to identify POIs in this work.

The performance of the HDBSCAN algorithm relies on the few implicit assumptions it makes about the clusters. This algorithm looks for regions of the data that are denser than the surrounding space and assumes that noise is present in the data. This clustering algorithm relies on one main parameter called m_{pts} which indicates the minimum number of points that a region must have to be considered a cluster [3].

2.2 Deep Neural Networks

Recently, the deep learning field has received a lot of attention and various DNNs have been applied successfully in multiple industries [9]. Among the several DNNs architectures recurrent neural networks (RNNs) are the most widely used when tackling temporal dynamics and time-series data. However, vanilla RNN architecture suffers from learning problems when tackling long sequences of data due to exploding gradients. This limitation was solved by developing several RNN architectures. The most successful architecture is known as long short-term memory (LSTM) [12]. The latter has been applied to analyze temporal dynamics and sequential structured data for various applications and industries such as speech recognition, natural language processing, finance, medical, among others [10,18,23]. The LSTM architecture is based on a set of recurrently connected cells, each cell contains one self-connected memory state, known as the internal cell state C^t, and three multiplicative units, the input, output and forget gates. These provide write, read and reset operations, and also allow the internal cell state to store and access information over long periods of time mitigating the vanishing gradient problem of vanilla RNN [12].

3   Methodology

3.1   Problem Definition

The fundamental assumption behind this work is that driving is a routine and drivers' routines show a high degree of temporal and spatial regularity. So, if a DNN can recall and learn from past routines the DNN might be able to predict the drivers' future intent.

It is important to note that perfect predictions are not realistic due to the nature of life, even if a driver has a very strict set of daily routines it is still possible for the driver to occasionally make deviations. For instance, a driver might always go to work during the weekdays at 8 a.m., however, a day will eventually come when the driver might not be able to go to work due to an illness, a car malfunction, among other factors. This rationale indicates that the prediction of drivers' routines must be a probabilistic approach. For that reason, the developed framework will analyze past routines using a hierarchical density-based clustering algorithm to identify points of interest (POIs) in the driver routine, afterwards, the DNN will be trained based on the past data and identified POIs and lastly, the outputs of the DNN will be the probability for each of the identified POIs as being the next likely destination. The DNN will also output the departure time to the next destination. When evaluating the proposed approach due to the probabilistic nature of human routines the Top-n accuracy will be considered when evaluating the next destination classification task.

This work aims to predict the next destination that a driver will visit at a specific future time based on the driver's past routine. To introduce the proposed framework some basic definitions are clarified as follows:

- **Definition 1:** A driver's routine is composed of the trip information, a single record R can be expressed as:

$$R = (c, e, t, d) \tag{1}$$

where c denotes the car coordinates (longitude and latitude), e denotes if the car engine was turned On or Off, t denotes the time in minutes and d denotes the day of the week.

- **Definition 2:** An unlabelled routine sequence (URS) is a driver's time-ordered sequence of records which can be defined as:

$$URS = < R_0, R_1, ..., R_k > \tag{2}$$

where k denotes the length of the unlabelled routine sequence.

- **Definition 3:** A POI is a destination that has been visited by a driver with significant frequency. A driver can have multiple POIs, these create a collection which is defined as:

$$Collection = \{POI_0, POI_1, ..., POI_n\} \tag{3}$$

where n denotes the number of POIs.

- **Definition 4:** A driver's labelled routine item (LRI) is the combination of a record and its corresponding POI, which can be defined as:

$$LRI = (R, POI) \tag{4}$$

- **Definition 5:** A series of labelled routine items make up the driver's labelled routine sequence (LRS), which can be defined as:

$$LRS = < LRI_0, LRI_1, ..., LRI_m > \tag{5}$$

where m denotes the length of the labelled routine sequence.

3.2 Framework of the Proposed Approach

Figure 1 illustrates the working steps of the proposed driver routine prediction framework. In this work, the car trip information consists of the GPS coordinates, the current day of the week, the current time in minutes (where the starting time t = 0 min is at 0:00 a.m. of each day) and information about the current car engine state (On/Off). The framework outputs the most likely next destination and the departure time of the driver to the next destination. Learning occurs implicitly, the driver does not need to be asked for his/her regular destinations or departure schedule. The framework works as follows, firstly the data is cleaned and divided into training data, validation data, and test data. Afterward, the training data is feed into the clustering algorithm. The clustering algorithm will identify the POIs visited by the user in the past and creates the training driver routine sequence. The validation and test data are then fed into the clustering algorithm creating the respective routine sequences. The training and validation data are then used to train and fine-tune the DNN. The DNN is composed by an LSTM layer and two dense layers, the role of the LSTM layer is to extract a meaningful representation of the input routine sequence, this layer extracts higher-level features that are then fed into the dense layers to produce the two outputs: next destination and departure time to that destination. The loss functions defined for the next destination and leave time outputs are cross-entropy and mean squared error, respectively. Since the network is trained using gradient descent which minimizes a scalar, the loss functions are combined and the gradients are calculated based on the sum of the loss functions.

The training of the deep learning model was performed on GPU NVidia 1060GTX with TensorFlow, to discover the optimal hyperparameters that lead to the best performance the deep learning model was trained using a randomized search method. This method consists in training the model using a defined number of combinations of hyperparameters, the hyperparameters are chosen randomly from specified distributions [2]. The hyperparameters combination that achieves the best validation performance is considered the optimal combination.

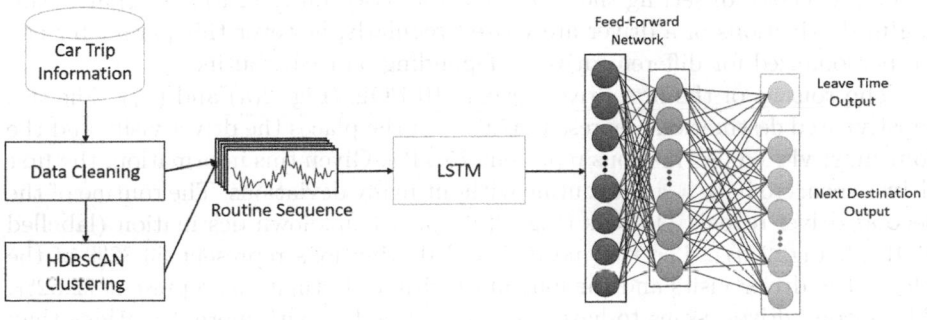

Fig. 1. Proposed Deep Learning framework.

To evaluate the framework Top-n accuracy and F1 score were the defined metrics for evaluating the performance of the framework when predicting the next destination (classification task). The metrics used to evaluate the performance of the framework when predicting the departure time (regression task) were mean squared error (MSE) and coefficient of determination R^2. The MSE metric is given by $\frac{1}{n} \sum_{i=1}^{n} (y_i - \hat{y}_i)^2$, where y_i are the true values and $\hat{y}_i$ are the predicted values. For the MSE metric lower scores are preferred. The R^2 metric measures the percentage of variance in the dependent variable that the model explains and is given by $1 - \frac{\sum_{i=1}^{n}(y_i-\hat{y}_i)^2}{\sum_{i=1}^{n}(y_i-\bar{y}_i)^2}$, where y_i are the true values, $\hat{y}_i$ are the predicted values, and $\bar{y}_i$ is the mean of the sample. The best possible R^2 score is 1.00.

4   Experiments and Results

To validate the proposed framework, datasets from three different drivers were considered. A simple mobile application was developed and installed on the driver's Android phones. The volunteers were asked to press a button in the mobile application when they turn the car on or off. When the button is pressed information about GPS coordinates, date, and time were recorded. The experiment lasted for 10 weeks. Because the car is turned off and on at the same place each dataset must be a sequence of multiple two very similar records in terms of GPS coordinates (very similar because of GPS noise), this must hold true except for the first and last dataset records. Taking this reasoning into consideration, a record can be considered missing if it reflects an isolated GPS record different from the previous and the next records (this might happen due to a read-out error or a driver forgetting to use the application). In the data cleaning process, these isolated records are removed.

The data clustering was performed with the m_{pts} parameter set to 4, meaning that a destination is only considered a POI if the driver has been at that destination at least 4 times, destinations with less than 4 visits are considered as unknown. We hypothesize that clustering less-visited destinations as unknown locations will help the deep learning model to learn the underlying routine structure. The choice of setting the m_{pts} to 4 was based on the rationale that meaningful destinations of a driver are visited regularly, however this parameter can be personalized for different driver's depending on their routine.

The routine of the first driver reveals 10 POIs (Fig. 2(a) and (b)). The two most visited destinations represent 47.67% of the places the driver visits and the four most visited destinations represent 75.11%. Given this information, the first driver seems to have a strict routine without many deviations. The routine of the second driver reveals 11 POIs (Fig. 2(c)) plus 1 unknown destination (labelled POI 12, Fig. 2(d)). The two most visited destinations represent 53.82% of the places the driver visits and the four most visited destinations represent 72.52%. The second driver seems to have a strict routine but with more deviations than the first driver given the higher number of destinations. Lastly, the routine of the third driver reveals 11 POIs (Fig. 2(e)) plus 1 unknown destination (labelled POI 12, Fig. 2(f)). The two most visited destinations represent 51.58% of the places the driver visits and the four most visited destinations represent 81.37%. The third driver also seems to have a strict base routine of 4 destinations and deviations between the other less visited 8 destinations, contrarily to the second driver the unknown destinations of this driver are not close to each other.

To evaluate the performance of the framework the datasets were divided into training, validation, and test sets. The first six weeks of data were used for training purposes, the 7th and 8th weeks were used for validation and the 9th and 10th week were used for the testing phase.

Using a hyperparameter optimization technique known as randomized search, around 1500 combinations of different hyperparameters were considered to search for the best DNN performance. The considered hyperparameters were: number of hidden LSTM nodes (ranging from 32 to 100), number of hidden FFN nodes (ranging from 32 to 100), learning rate (ranging from 0.001 to 0.1), batch size (ranging from 8 to 32), FFN hidden nodes activation functions (ReLU, SeLU, tanh, sigmoid) and hidden nodes dropout rates (ranging from 0.0 to 0.5). The best combination of hyperparameters was selected based on the validation set performance, the best DNN was then used to make predictions on the test set. The performance of the best DNN for each driver is summarized in Table 1. From the analysis of Table 1, it can be concluded that the DNN is able to capture the underlying routines of each driver. The classification performance is generally very good, for the first and second drivers the training performance is slightly better than the validation and test performances, this might be a signal of over-fitting however, it might also indicate that these drivers had a slight change on their routine on the last month of the collected data. The classification performance was the lowest when tackling the third driver, this driver might have the routine with most deviations since it also was the only driver with multiple unknown destinations far from each other, nevertheless, the DNN was able to achieve a very good top-1 accuracy of 86.40% on the test set. Overall, the DNN shows very good performance when predicting the next destination of the drivers. If we consider the 3 most probable predictions of the model (Top-3 accuracy) the accuracy is 100%, 98.50%, and 94.40% on the test set of the first, second, and third driver, respectively. In terms of the departure time prediction the DNN achieved a wide range of performances, when tackling the routine of the first driver the DNN was able to understand very well the complex dynamics of departure time, it achieved a very good R^2 Score of 93.69 and 0.0059 MSE on the test set. The DNN struggled the most when tackling the routine of the third driver, the DNN was able to achieve a reasonably R^2 Score of 28.81 and 0.0486 MSE on the test set.

The approach proposed in this work can achieve very good performance when predicting future destinations and departure time of drivers. However, this work has some limitations that should be addressed in future experiments. One limitation of this work is the reduced number of datasets and the duration of data collection (10 weeks). We have shown that the developed framework is able to capture weekly routines but we cannot state that monthly routines are well captured, in the future it would be beneficial to evaluate the framework with larger

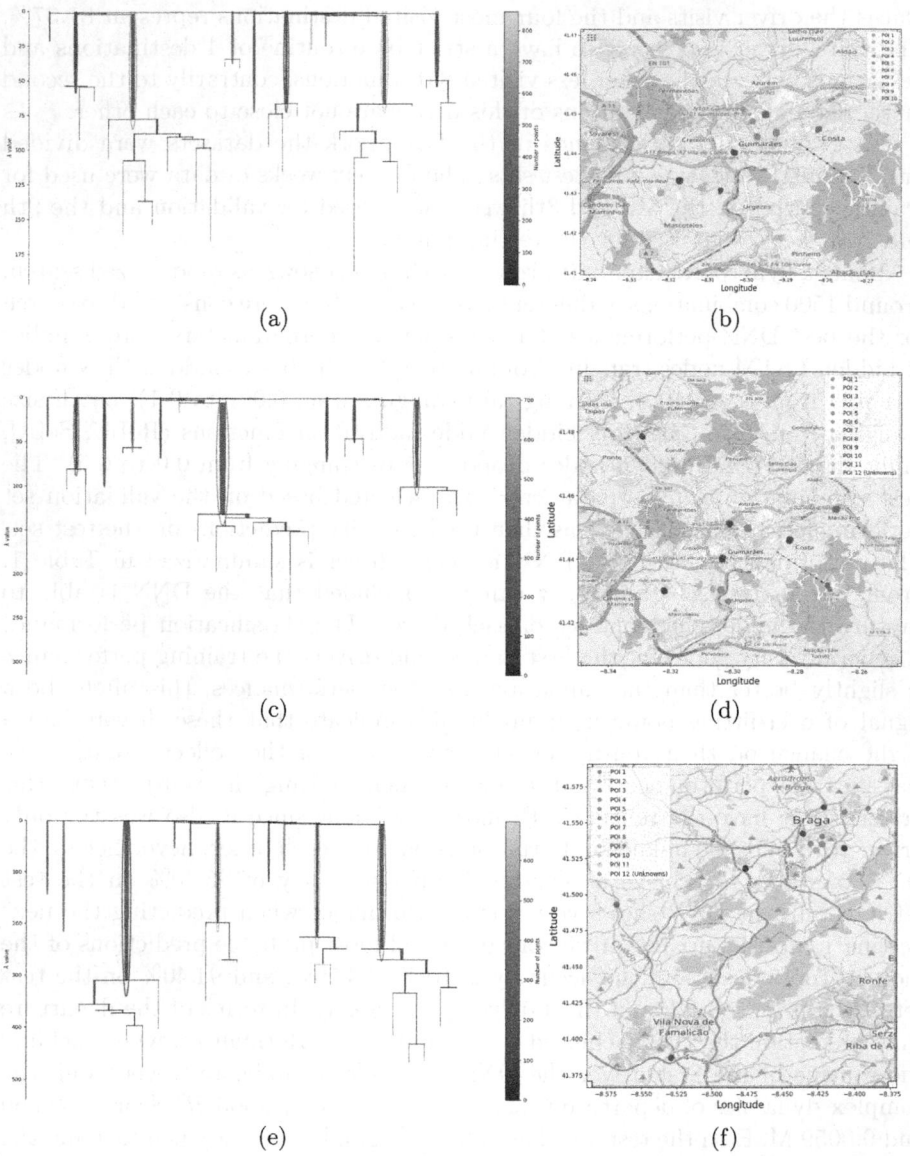

Fig. 2. Clustering results obtained by HDBSCAN algorithm with the number of minimum points for cluster equal to 4. Condensed tree for Driver 1 (a), Driver 2 (c), and Driver 3 (e) and respectively representation of the points of interest for Driver 1 (b), Driver 2 (d), and Driver 3 (f).

datasets. Another limitation of this work is that the dataset has the underlying assumption that the driver is always the same, which might not be realistic since a car can be shared, for example, between husband and wife. To mitigate this

Table 1. Performance metrics for the proposed deep learning approach on the training, validation and test sets.

	Training set			Validation set			Test set		
	Driver 1	Driver 2	Driver 3	Driver 1	Driver 2	Driver 3	Driver 1	Driver 2	Driver 3
Top-1 Acc.	98.43%	95.37%	86.57%	93.84%	85.95%	85.95%	96.20%	90.23%	86.40%
Top-2 Acc.	100%	99.54%	93.98%	98.63%	92.56%	96.69%	99.46%	97.74%	92.00%
Top-3 Acc.	100%	100%	96.06%	100%	97.52%	99.17%	100%	98.50%	94.40%
Top-4 Acc.	100%	100%	99.07%	100%	100%	100%	100%	100%	97.60%
Top-5 Acc.	100%	100%	99.07%	100%	100%	100%	100%	100%	98.40%
F1 score	95.76%	91.25%	62.65%	83.09%	73.27%	67.58%	92.41%	72.91%	71.90%
R^2 score	92.41	73.55	53.74	91.27	54.31	38.37	93.69	79.21	28.81
MSE	0.0063	0.0241	0.0173	0.0075	0.0483	0.0439	0.0059	0.0195	0.0486

limitation, future work might focus on collecting data from the vehicle that identifies the driver and feeds that information into the model, this way the model can also understand that different drivers have different routines and habits.

5 Conclusion

In this paper, we proposed a deep learning framework based on hierarchical density-based clustering and LSTMs to predict the future destinations and departure time of drivers'. The proposed approach was able to implicitly learn the complex dynamics of daily routines and achieved very good performance both when predicting the next destination and the departure time. These findings constitute a very promising basis to further develop and experiment with driving data acquired from the vehicle's cockpit. Provided that the stated limitations can be overcome, our work opens many perspectives for future development, first by being applied to data acquired by the vehicle's cockpit and then to be implemented in the vehicle's cockpit to assist drivers in their daily routines. Future work will focus on starting the implementation and testing of the proposed approach on vehicles cockpits in the scope of the joint project UMinho & Bosch – "Easy Ride:Experience is everything" (ref POCI-01-0247-FEDER-03933).

Acknowledgments. This work received financial support from European Structural and Investment Funds in the FEDER component, through the Operational Competitiveness and Internationalization Programme (COMPETE 2020) and national funds, through the ADI Project Bosch & UMinho **"Easy Ride: Experience is everything"** , ref POCI-01-0247 FEDER-039334, and by FCT – Fundação para a Ciência e Tecnologia within the R&D Units Project Scope: UIDB/00319/2020 and UIDB/00013/2020.

References

1. Bao, J., Zheng, Yu., Wilkie, D., Mokbel, M.: Recommendations in location-based social networks: a survey. GeoInformatica **19**(3), 525–565 (2015). https://doi.org/10.1007/s10707-014-0220-8

2. Bergstra, J., Bengio, Y.: Random search for hyper-parameter optimization. J. Mach. Learn. Res. **13**(Feb), 281–305 (2012)
3. Campello, R.J.G.B., Moulavi, D., Sander, J.: Density-based clustering based on hierarchical density estimates. In: Pei, J., Tseng, V.S., Cao, L., Motoda, H., Xu, G. (eds.) PAKDD 2013. LNCS (LNAI), vol. 7819, pp. 160–172. Springer, Heidelberg (2013). https://doi.org/10.1007/978-3-642-37456-2_14
4. Eagle, N., Pentland, A.S.: Eigenbehaviors: identifying structure in routine. Behav. Ecol. Sociobiol. **63**(7), 1057–1066 (2009)
5. Ester, M., Kriegel, H.P., Sander, J., Xu, X., et al.: A density-based algorithm for discovering clusters in large spatial databases with noise. KDD **96**, 226–231 (1996)
6. Ferreira, F., et al.: A dynamic neural model for endowing intelligent cars with the ability to learn driver routines: where to go, when to arrive and how long to stay there? In: Towards Cognitive Vehicles Workshop (TCV2019), IROS2019, pp. 15–18 (2019)
7. Gao, S.: Spatio-temporal analytics for exploring human mobility patterns and urban dynamics in the mobile age. Spatial Cogn. Comput. **15**(2), 86–114 (2015)
8. Gonzalez, M.C., Hidalgo, C.A., Barabasi, A.L.: Understanding individual human mobility patterns. Nature **453**(7196), 779–782 (2008)
9. Goodfellow, I., Bengio, Y., Courville, A.: Deep Learning. MIT press, Cambridge (2016)
10. Graves, A., Jaitly, N., Mohamed, A.r.: Hybrid speech recognition with deep bidirectional LSTM. In: IEEE Workshop on Automatic Speech Recognition and Understanding, pp. 273–278. IEEE (2013)
11. Heckman, J.J., Mosso, S.: The economics of human development and social mobility. Annu. Rev. Econ. **6**(1), 689–733 (2014)
12. Hochreiter, S., Schmidhuber, J.: Long short-term memory. Neural Comput. **9**(8), 1735–1780 (1997)
13. Huang, Q.: Mining online footprints to predict user's next location. Int. J. Geogr. Inf. Sci. **31**(3), 523–541 (2017)
14. Järv, P., Tammet, T., Tall, M.: Hierarchical regions of interest. In: 19th IEEE International Conference on Mobile Data Management (MDM), pp. 86–95 (2018)
15. Jiang, S., Ferreira, J., González, M.C.: Clustering daily patterns of human activities in the city. Data Min. Knowl. Discov. **25**(3), 478–510 (2012)
16. Kadar, C., Pletikosa, I.: Mining large-scale human mobility data for long-term crime prediction. EPJ Data Sci. **7**(1), 1–27 (2018). https://doi.org/10.1140/epjds/s13688-018-0150-z
17. Korakakis, M., Spyrou, E., Mylonas, P., Perantonis, S.J.: Exploiting social media information toward a context-aware recommendation system. Soc. Netw. Analy. Min. **7**(1), 1–20 (2017). https://doi.org/10.1007/s13278-017-0459-9
18. Lipton, Z.C., Kale, D.C., Elkan, C., Wetzel, R.: Learning to diagnose with LSTM recurrent neural networks. arXiv preprint arXiv:1511.03677 (2015)
19. Louail, T., et al.: Uncovering the spatial structure of mobility networks. Nat. Commun. **6**(1), 1–8 (2015)
20. Rinzivillo, S., Gabrielli, L., Nanni, M., Pappalardo, L., Pedreschi, D., Giannotti, F.: The purpose of motion: Learning activities from individual mobility networks. In: International Conference on Data Science and Advanced Analytics (DSAA), pp. 312–318. IEEE (2014)
21. Roth, C., Kang, S.M., Batty, M., Barthélemy, M.: Structure of urban movements: polycentric activity and entangled hierarchical flows. PloS ONE **6**(1), e15923 (2011)

22. Schubert, E., Sander, J., Ester, M., Kriegel, H.P., Xu, X.: Dbscan revisited, revisited: why and how you should (still) use DBscan. ACM Trans. Database Syst.(TODS) **42**(3), 1–21 (2017)
23. Siami-Namini, S., Namin, A.S.: Forecasting economics and financial time series: Arima vs. LSTM. arXiv preprint arXiv:1803.06386 (2018)
24. Song, C., Qu, Z., Blumm, N., Barabási, A.L.: Limits of predictability in human mobility. Science **327**(5968), 1018–1021 (2010)
25. Xu, Y., Shaw, S.-L., Zhao, Z., Yin, L., Fang, Z., Li, Q.: Understanding aggregate human mobility patterns using passive mobile phone location data: a home-based approach. Transportation **42**(4), 625–646 (2015). https://doi.org/10.1007/s11116-015-9597-y
26. Yuan, J., Zheng, Y., Xie, X.: Discovering regions of different functions in a city using human mobility and POIS. In: 18th ACM SIGKDD International Conference on Knowledge Discovery and Data Mining, pp. 186–194 (2012)
27. Zhou, C., Frankowski, D., Ludford, P., Shekhar, S., Terveen, L.: Discovering personally meaningful places: an interactive clustering approach. ACM Trans. Inf. Syst. (TOIS) **25**(3), 12-es (2007)

Predicting Recurring Telecommunications Customer Support Problems Using Deep Learning

Vitor Castro[1]($\boxtimes$) (ID), Carlos Pereira[2] (ID), and Victor Alves[3] (ID)

[1] University of Minho, Campus Gualtar, 4710 Braga, Portugal
vitorcastro.it@gmail.com
[2] NOS Comunicações, Senhora da Hora, Portugal
carlos.migpereira@nos.pt
[3] Algoritmi Center, University of Minho, Campus Gualtar, 4710 Braga, Portugal
valves@di.uminho.pt

Abstract. In search of a better quality of experience and more revenue, telecommunication companies are searching for proactive ways of dealing with unsatisfactory user experiences and predicting customer's behavior. Customer Support (CS) is one of the key areas of customer satisfaction. A good CS enables customers to have a smooth interaction with the company and the services provided when there are doubts or malfunction. Frequently, the problems reported by customers are not resolved in the first interaction, which leads to greater dissatisfaction with the service provider and possibly to future churn. If the company knows in advance of a possible recurrence, it can respond and try to fix the problem without customers noticing or being affected. In this article, a data set of customer data, CS data, and historical service are used to create a deep learning-based model for predicting customer recurrence. Deep neural networks are well-known for their capability to model complex problems when compared to classical machine learning algorithms. The obtained model, with a decision threshold most appropriated for the business needs, presented an F1-score of 60% and AUC-ROC of 61%, with a Recall and Precision of the recurrent class of 29% and 21%, respectively.

Keywords: Deep neural networks · Machine learning · Customer support · Telecommunications · Quality of Experience

1 Introduction

In search of a better Quality of Experience (QoE) and more sales, Telecommunications Companies (TC) are looking for proactive ways to deal with unsatisfactory user experiences and predict customer behavior [3,11].

The use of digital content has increased rapidly in the last decade. The amount of data that is transferred daily between devices is enormous and leaves an interesting gap for companies [1]. In the digital age, companies that collect data about their business activities in a well-structured manner gain a strong

© Springer Nature Switzerland AG 2020
C. Analide et al. (Eds.): IDEAL 2020, LNCS 12490, pp. 184–193, 2020.
https://doi.org/10.1007/978-3-030-62365-4_18

advantage over their competitors because knowledge of how to deal with competition has always been a key factor.

TC has made a huge contribution to the increasing use of digital devices by providing the infrastructure for ordinary users. They end up with large data pool from which information can be extracted and converted into knowledge.

Customer Support (CS) services have evolved significantly in recent years, and most simple problems can usually be solved with specific software without direct human interaction. Nevertheless, customers sometimes still have to contact a specialized operator directly.

Nowadays, companies store a lot of information about their interactions with customers, including a detailed history of previous calls. This information can be used to determine how often customers return and why the complaints were motivated. The perfect scenario is that no customers complain because the contracted service is fully functional. Therefore, companies are working to get closer to this goal. One way to address this problem is to make the recurring complaints less common. This could be done by trying to predict the repetition in advance. With this information, companies can use automated processes that are typically used by a technical support agent and can sometimes solve the problems described.

Deep Neural Networks (DNN) have been used in several applications in which there are complex relationships between features, in contrast to the most common classic models of machine learning. The main goal of this work is to create a predictive model that uses customer and service data to predict whether further interaction with the same problem will be required after a customer's interaction with CS within a certain period of time. This information enables the company to solve the problem in advance and thus contribute to customer satisfaction.

2   Background

In order to provide the most cost-effective support service, customers usually first contact the general support operator, who is not specialized in any type of problem. As a rule, generalist CS operators are unable to solve technical problems and to schedule a call between technical support and the customer. Accordingly to the protocol, operators typically apply a number of automated processes that can solve the reported problem. Sometimes the problem seems to be solved, but the solution is temporary and the devices show the same symptoms later on. According to the data in our study, customers call the CS again in about 10% of cases within a month because they repeatedly occur in the previously claimed problem. Operator time was then wasted and the solution was unsatisfactory, which added to the customer's annoyance. It is therefore important to avoid these situations, but also to predict whether customers will recur. In this way, technical teams could try to solve the problem before the customer feels the problem again or has to file a new complaint.

TC have many use cases and challenges where data mining and machine learning have already been used to solve problems such as fraud and churn. Fraud is one of the main problems that TC faces today and they lose millions of dollars every year. A. H. Elmi et al. designed a neural network that could predict

SIM box fraud with an accuracy of 98.71% [5]. The proposed result of the work is that multi-layer perceptrons (MLP) can typically perform well in fraud detection scenarios. Churn is also one of the most important problems that companies have to deal with. J.W. Denton et al. [4] used two approaches to predict churn, which consisted of a decision tree and an artificial neural network. To better classify the likelihood of churn, a K-means clustering algorithm was used to divide customers into different churn sections that replicate loyalty segments. The division considered the billing amount as the customer's current raw value, the tenure mounts to emphasize the importance of loyalty in the model, and payment behavior to replicate credit risks. This division made it possible to assess whether the behavior of the customer is different in the different clusters, which would lead to the creation of different models for each individual. Although many efforts have been made in recent years to predict customer churn, this only allows reactive measures when a customer is in danger of leaving.

The importance of using more or less historical data is addressed in [13], where the use of more data in the same time window and the expansion of this time window are discussed. In this study, it was concluded that using only 50% of the data available in the defined time window did not contribute to a loss of performance. When trying to get more historical data by increasing the time window, a new problem arises: the customer may have a tenure less than the time window. The solution found was to create several aggregations taking into account several time windows and shifted moments in relation to the test data. With extractions from different moments and with different time windows, the memory requirement and the training difficulties increase. The conclusion was that the ensemble method contributed to better classification and should be taken into account in future or similar approaches like the one being studied.

3   Case Study

This section describes the entire process of creating the prediction model. The CRISP-DM methodology was followed, which describes approaches commonly used by specialists to address such problems [12]. The main advantage of this method is that it is independent of the industry, tools, and data. Accordingly, the following section describes the business understanding, data understanding, data preparation, modeling, and evaluation using the CRISP-DM methodology. At the end of this section, the various models developed are described, which provide an effective overview of the different results and their meaning.

3.1   Business Understanding

Improving CS efficiency and customer relationships is very important as it reduces churn and increases company profits. The aim of this work is to improve customer satisfaction by reducing the number of recurring problems, i.e., problems that occur for the same reasons within a period of time. A period of one month was considered, accordingly to company standards. A monthly timeframe is useful in TC because all processes and payments are in the same timeframe. This is accomplished by predicting when a problem will recur after each interaction with the CS. With this information, the CS can take action and try to solve the problem before the customer notices or has to call again.

In the historical service data of the data set used, all interactions between customers and support service operators as well as the history of service use of each individual customer were recorded. The goal is to use this data to train a deep learning model to predict when the customer problem reported in a call will recur. The entire process is shown in Fig. 1.

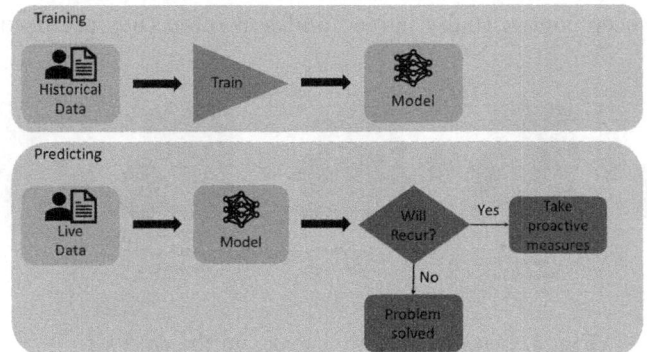

Fig. 1. Predictive model usage in business

3.2   Data Understanding

The data was retrieved from various databases that contained information about CS interactions, personal customer data, service usage, contracted services, and installed devices. Merging all of this retrieved data into one data set resembled 5.9 million rows, which after cleanup processes became 3.4 million rows with 10% positives in the binary target class. The positive class is the one that indicates that a recurrence will occur because the question to be answered is whether customers will recur in the problem or not. When merging the data, the different moments in which data is updated in the company's systems were taken into account, as some are updated in real-time and others only monthly. Main data characteristics are the following:

- Over 1000 distinct types of problems reported by customers in the cleaned data set;
- The 100 most common problems cover over 91% of the reports;
- There is no seasonality;
- About 43% of the customers who have recurred only had a single recurrence, with about 65% recurring three or fewer times;
- Most recurrences usually occur in the first three days after a customer service call.

3.3   Data Preparation

Since the data comes from different sources, the main data set must be selected, in which all other information is added. In this data set, for each entry that corresponds to an interaction between customer service and the customer, there is information about: the contracted service and technology, the data for the start and end of the interaction and the corresponding follow-up, the problem described and which of the support areas took care of the situation, as well as information about the start and the end of the expected term of the contract. Finally, to learn more about the customer, information about service usage of the last few months was collected. This data source contains information about which services have been contractually agreed and how often they are used (Fig. 2).

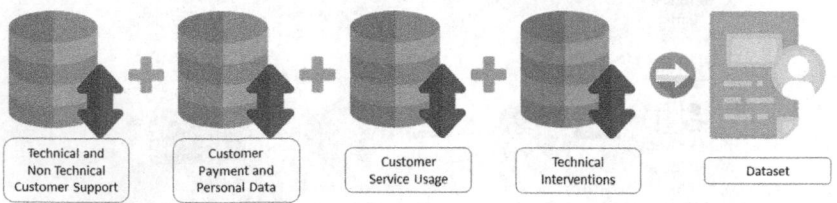

Fig. 2. Data sources that make up the data set

Features were extracted after collecting the data from the data sources. This allowed to engineer the following other important features:

Customer Support
- Time between the last call and the corresponding previous one with the same problem;
- Frequency with which the same problem has been reported in the last 30, 60, 90 and 180 days; The same number, but in the intervals 30–60, 60–90 and 90–180 days;
- Deviation of service consumption in the percentage of the current month compared to the average usage of the previous months;
- Number of days that the customer account has associated with the service provider; Number of days until the end of customer loyalty obligations;
- Date-related attributes such as public holidays and time of day.

Customer's Service Usage
- Number of SMS, cell phone data, cell phone call time, fixed telephone call time, television media time, downloading Internet, data and uploading the Internet data used.

Technical Interventions
- Number of previous required technical maintenance, service installations, service changes, and re-establishments.

3.4 Modeling

To get the best model, several experimental configurations of different fully connected MLPs were tested. When designing the models, two main architectures were selected: the first with 3 wide hidden layers and the second with 15 narrow hidden layers. According to [7], architectures with one, two, or three layers with about 2/3 of the neurons of the input size in each layer can achieve similar results compared to deeper architectures, which is why the first architecture was created. Fewer layers also contribute to a less computing-hungry model with shorter training times [10]. However, deeper neural networks can sometimes find complex relationships within the data [2], which led to the second architecture being created for our tests.

Batch normalization layers, which are not shown in Table 1, were used between all the MLP layers in order to improve the speed, performance, and stability of the artificial neural network [6].

Table 1. Description of the different MLP structures used

Structure ID	#Hidden layers	Description	#Connections
1	6	4673-3100-3100-3100-3100-1000	4.32E20
2	3	4673-3100-3100	4.49E10
3	15	4673-3100-2000-(2x)1000-(5x)500-(5x)100	9.05E39
4	2	3100-3100	9.61E6

The learning process is determined by the loss function, which in turn is affected by class imbalance. Various class weights have been tested to address this problem. After pre-evaluating the results obtained using equal class weights in comparison with balanced weights [8], the latter was chosen due to the better performance obtained. For this reason, all experiments were carried out using balanced class weights. Binary cross-entropy loss function was considered suitable for this problem. The ADAM optimizer was also chosen since it has been showing the best performances in similar problems [9].

Since the architecture of the DNN is not the only factor that needs to be considered when creating a model, several hyper-parameter configurations were tested for each of the structures specified (Table 2). Hold-out was the method chosen for testing. The data set was divided in a 65%–15%–20% way, with the first set being used to train the model, the second for validation purposes, and the latter for the final test. It is important to notice that no cross-validation was applied because that would not allow keeping the temporal sequence of data in the training phase.

The usual method for evaluating several hyperparameters in a neural network is to do some kind of grid-search. However, this method has a high computational cost and was not feasible due to time and computational limitations. Therefore, after analyzing some previous tests and leveraging the existing team knowledge, a set of hyperparameter configurations were chosen (Table 2).

Table 2. Hyperparameter configurations used for each different structure

ID	Structure ID	Batch size	Learning rate (LR)	Reduce LR	Train size
1	1	64	0.001	50%	320k
2	2	64	0.001	50%	320k
3	2	1024	0.001	50%	320k
4	2	10000	0.001	50%	320k
5	3	10000	0.001	50%	320k
6	3	64	0.001	50%	320k
7	2	64	0.0001	No	320k
8	2	64	0.01	No	320k
9	2	64	0.001	50%	640k
10	2	1024	0.1	No	320k
11	4	1024	0.001	50%	320k
12	4	64	0.001	50%	320k
13	4	64	0.001	50%	640k

Because the problem to be addressed is an unbalanced problem with only 9.1% of cases recurring, accuracy is not the best metric to evaluate model performance. Instead, AUC-ROC and F1 scores were used to evaluate the models. Since these models are to be deployed by the company, the computing resources must also be taken into account. For this reason, the memory and the time that each model needs to train were taken into account.

3.5 Evaluation

The best results achieved are highlighted in Table 3, with an AUC-ROC of 73%, and an F1-score of 35% with a Precision of 23% and a Recall of 72%. When comparing with the other experimental setups it is clear that the main factor is the amount of data the model was trained with. When comparing it with the same model structure and same parameters but using only 320k (ID 12), it is possible to see an increment of 5% in the AUC-ROC, as well as an increase in both Precision and Recall for the positive class.

The majority of the models were evaluated using 500k entries. From those, 320k were used to train, 80k to validate, and 100k to test. In a deployment state, the models would likely be trained with the full 3.9 million entries. Increasing the training size would quickly increase the training time, as seen in this paper. If the network had to be re-trained due to a sudden event, it can take weeks, which is a lot of downtime. For this reason, it can be useful to have a model

Table 3. Results obtained with the different configurations for the validation set

ID	Best epoch	Memory	Time	Prec. Pos	Rec. Pos	F1 Pos	AUC-ROC
1	5	18 GB	16 H	21%	60%	32%	69%
2	7	13 GB	16 H	23%	47%	31%	65%
3	7	13 GB	2 H	19%	55%	28%	65%
4	7	15 GB	2:30 H	20%	56%	29%	66%
5	11	16 GB	1:30 H	24%	45%	31%	65%
6	6	13 GB	25 H	21%	64%	31%	59%
7	5	13 GB	10 H	20%	58%	30%	57%
8	3	13 GB	10 H	20%	64%	30%	69%
9	5	22 GB	32 H	21%	68%	32%	70%
10	No convergence	13 GB	0:30 H	–	–	–	–
11	3	13 GB	1 H	19%	67%	29%	68%
12	4	13 GB	6 H	21%	60%	32%	68%
13	**5**	**20 GB**	**12 H**	**23%**	**72%**	**35%**	**73%**

with the same characteristics as the highlighted model, but with a larger batch size to allow a faster train.

Regarding batch size, it can be seen that increasing from 64 to 1024 shortens training time because the calculation used to measure loss is less frequent. However, an increase from 1024 to 10000 shows an extension of the training time. This is due to the fact that extensive calculations are carried out for a much larger number of instances that have to be taken into account.

Increasing or decreasing the learning rate from the standard 0.001 to 0.01 or 0.0001 does not have a major impact on the training time. However, if you increase this value to 0.1, it is possible that the model train is faster, but loses all of its predictive power.

Our previous tests showed that a decrease in the learning rate is essential to prevent the model from being set to a local minimum. This can easily happen because it is an unbalanced problem.

The best model, with respect to the validation set, resulted in Fig. 3, which relates precision and recall as a function of the decision threshold. The default threshold is 0.5 but to better apply the model to the case study a threshold of 0.75 was chosen to be optimal. This trade-off allows to identify fewer customers who would probably recur, but with more confidence in the prediction, making the process more efficient for the company. By doing so, a Precision of 58% and a Recall of 61% are achieved, with respect to the final test set. The final F1-score is 60% and the AUC-ROC is 61%.

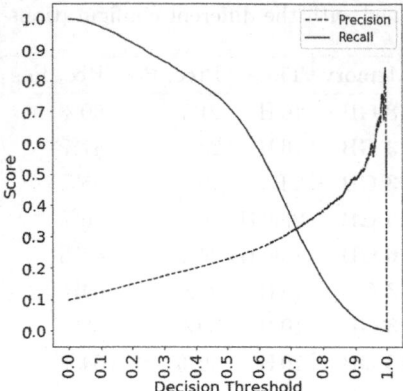

Fig. 3. Precision and recall Scores as a function of the decision threshold for the validation set

4   Conclusions

TC providers have experienced an increase in their customer base, facing more complex challenges. In a highly competitive market, customers can easily switch between service providers, reducing potential profits. For this reason, satisfaction is very important for these companies. In recent years, some efforts have been made to predict customer churn. However, a better approach would be to not motivate the customer to churn.

In this article, an attempt to predict the recurrence of problems was made, in order to increase their satisfaction. The results achieved are promising and could be used in daily operations to improve customer experience by taking proactive measures in solving predicted recurring problems. In Table 3 it is possible to observe that using 3 layers (ID 2) instead of 2 layers (ID 12) resulted in worse performance, suggesting that deeper architectures do not always improve performance. Deeper and narrower neural networks (ID 6) showed a large degradation in performance, which could be due to the excessively narrow end. However, deeper but not narrow neural networks (ID 1) showed comparable performance to the 2-layer model (ID 12), the one with the best performance from above. This suggests that deeper neural networks can result in a better performance model (comparing ID 12 to ID 1), but the number of nodes in the layers is very important to it. It is also understandable that, when using layers of the same width, the deeper the neural network the more computationally hungry it gets. For certain applications where training is often needed, this could be an aspect to be taken into consideration. All in all, less deep neural networks with a large enough number of nodes are preferred since it allows for a lower computational-hungry model that is able to train faster and generalize better with this type of data.

Better results could be achieved if access to device information was granted in real-time. The complexity of the problem, with hundreds of different reported problems, each one with its particular characteristics and distribution, makes predictions more difficult. If a different model was trained for each of the most common recurrent problems, those models would be specialized in finding recurrences in the problem that they were trained on, allowing for a better separability of the overall recurrences.

Acknowledgments. This work has been supported by FCT – Fundação para a Ciência e Tecnologia within the R&D Units Project Scope: UIDB/00319/2020.

References

1. Banerjee, A.: Big data & advanced analytics in telecom: a multi-billion-dollar revenue opportunity prepared by. Technical report December, Heavy Reading (2013)
2. Bianchini, M., Scarselli, F.: On the complexity of neural network classifiers: a comparison between shallow and deep architectures. IEEE Trans. Neural Netw. Learn. Syst. **25**(8), 1553–1565 (2014)
3. Chen, C.M.: Use Cases and Challenges in Telecom Big Data Analytics (2016)
4. Denton, J.W., Hung, M.S., Osyk, B.A.: A neural network approach to the classification problem. Expert Syst. Appl. **1**(4), 417–424 (1990). https://doi.org/10.1016/0957-4174(90)90050-5
5. Elmi, A.H., Ibrahim, S., Sallehuddin, R.: Detecting SIM box fraud using neural network. In: Kim, K.J., Chung, K.-Y. (eds.) IT Convergence and Security 2012. LNEE, vol. 215, pp. 575–582. Springer, Dordrecht (2013). https://doi.org/10.1007/978-94-007-5860-5_69
6. Ioffe, S., Szegedy, C.: Batch normalization: accelerating deep network training by reducing internal covariate shift. In: 32nd International Conference on Machine Learning, ICML 2015, vol. 1, pp. 448–456 (2015)
7. Karsoliya, S.: Approximating number of hidden layer neurons in multiple hidden layer BPNN architecture. Int. J. Eng. Trends Technol. **3**(6), 714–717 (2012)
8. King, G., Zeng, L.: Logistic regression in rare events data. Pol. Anal. **9**, September 2002. https://doi.org/10.1093/oxfordjournals.pan.a004868
9. Kingma, D., Ba, J.: Adam: a method for stochastic optimization. In: International Conference on Learning Representations, December 2014
10. Stathakis, D.: How many hidden layers and nodes? Int. J. Remote Sens. **30**, 2133–2147 (2009). https://doi.org/10.1080/01431160802549278
11. Wang, J., et al.: Efficient alarm behavior analytics for telecom networks. Inf. Sci. **402**, 1–14 (2017). https://doi.org/10.1016/j.ins.2017.03.020
12. Wirth, R., Hipp, J.: CRISP-DM: towards a standard process model for data mining. In: Proceedings of the 4th International Conference on the Practical Applications of Knowledge Discovery and Data Mining, January 2000
13. Yan, L., Miller, D.J., Mozer, M.C., Wolniewicz, R.: Improving prediction of customer behavior in nonstationary environments. In: Proceedings of the International Joint Conference on Neural Networks, vol. 3, pp. 2258–2263 (2001). https://doi.org/10.1109/ijcnn.2001.938518

Pre- and Post-processing on Generative Adversarial Networks for Old Photos Restoration: A Case Study

Robinson Paspuel[1]($\boxtimes$) (iD), Marcelo Barba[1] (iD), Bryan Jami[1] (iD),
and Lorena Guachi-Guachi[1,2,3] (iD)

[1] Yachay Tech University, Hacienda San José, Urcuquí 100119, Ecuador
robinson.paspuel@yachaytech.edu.ec
[2] SDAS Research Group, Ibarra, Ecuador
[3] Department of Mechatronics, Universidad Internacional del Ecuador,
Av. Simon Bolivar, Quito 170411, Ecuador
http://www.sdas-group.com/

Abstract. Old historical images are an invaluable source of knowledge that allows people to learn about past events and, in general, the form of the world in the past. In the case of townscapes, the photos may depict specific details as building appearance prior to their reconstruction, enlargement or demolition, or even former appearance of cities (buildings, inhabitants, transportation, among others). In this sense, more and better details of the image lead to an exact representation of a city in a given time. Generative Adversarial Networks (GANs) are a category of deep artificial neural networks (DANNs) that show great success in generating realistic characteristics into image, video and voice data. This work explores how the pre- and post-processing techniques influence the overall effectiveness of GANs-based techniques for restoring and coloring old photos. Pre- and post-processing based on traditional image processing methods preserve and enhance the information contained in old photographs; however, their effectiveness is limited by the amount of information retained in the original photograph. On the other hand, GANs-based techniques offer the ability to increase the amount of information and thus boost the effectiveness of traditional methods. Experiments are performed referring to the old photos of Quito's city. The preliminary results show that pre- and post-processing algorithms are essential even in artificial intelligence approaches, eliminating undesirables artifacts and increasing visual quality.

Keywords: GANs · Image processing · Old image restoration

1 Introduction

Most of our knowledge as society comes from past events, nonetheless this knowledge comes mostly from recovered written registers. However, nowadays, many historians have found in photography another way to understand the closer past, discovering in this way, the testimonial and documentary value of images.

© Springer Nature Switzerland AG 2020
C. Analide et al. (Eds.): IDEAL 2020, LNCS 12490, pp. 194–201, 2020.
https://doi.org/10.1007/978-3-030-62365-4_19

New technologies provide a wide range of image processing approaches to offer a real solution for restoring, normalizing, or stylising photographs, particularly those old ones that have suffered wear, discoloration, or other types of damage. In the last decades, several works have introduced traditional approaches based mainly on sharpening, contrast enhancement, image denoising, contrast stretching, histogram processing, among others [1–3]. These methods are useful for image restoration. However, they are mostly combined individually with other methods to obtain an appropriate photo quality. On the other hand, some Deep Artificial Neural Networks (DANNs) approaches have been introduced to support techniques for automatic restoration of damaged old photos such as Convolutional Neural Networks (CNNs) [4], Multi-scale Iterative Network [5], Generative adversarial networks (GANs) [6], among others.

Recently, some innovative methods based on GANs, that belong to deep neural network techniques, have been introduced to automatically create realistic characteristics into real image, video and voice data, as well as for restoration purposes. For instance, a GAN-based restoration scheme (GAN-RS) introduced in [7] aims to cope with the difficulty of adaptive and real-time performance and to preserve the image content removing the underwater noise. A high-resolution medical image synthesis using progressively grown GANs stated in [8] produces realistic medical images such as eye photographs or magnetic resonance images. In addition, some optimization studies have been introduced aiming to improve low visual quality and achieve superior restoration performance. For instance, Super-Resolution Generative Adversarial Network (SRGAN) [9] was introduced to recover photo-realistic natural images from single image for 4× upscaling factors and handle limitations of the Peak Signal-to-Noise Ratio (PSNR) on image super-resolution. Another SRGAN approach, called Enhanced SRGAN (ESRGAN), is stated in [10] to remove unpleasant artifacts added in the super-resolution generation.

In order to determine how the pre- and post-processing techniques influence the overall effectiveness of GANs-based techniques for restoring and coloring old photos. This work presents a hybrid methodology to evaluate the effectiveness of combining traditional image processing techniques (histogram equalization, median filter, power law transformation, and Contrast Limited Adaptive Histogram Equalization) with GANs-based techniques (ESRGAN [11], and DeOldify [12]) to improve visual quality of old photos. As a case study, this work uses old photos of Quito's city, since the old townscapes will allow to know the appearance of the city for urban history purposes. Preliminary results show that the post-processing stage significantly reduces the artifacts introduced by the GANs-based methods used to increase photo resolution, as well as notoriously enhances the color contrast after the coloring process.

2   Proposed Methodology

This work presents a hybrid methodology with three main phases as shown in Fig. 1. The photos collected are pre-processed with traditional image

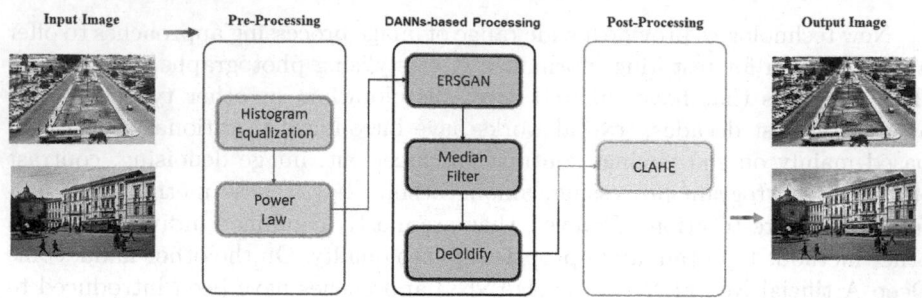

Fig. 1. Main computational flow of the proposed methodology.

enhancement techniques before being processed using DANNs-based techniques to generate a high-resolution and color image. Finally, the photos are post-processed using conventional methods to enhance the overall quality of the GANs results. In addition, a comparison of the results of pre-processed, post-processed and unprocessed photos is conducted to determine the effect of pre- and post-processing stages on the overall effectiveness achieved by GANs techniques for restoration purposes.

1. **Pre-Processing**
   Two mathematical algorithms are applied to the photos to highlight attributes and adjust the contrast. Firstly, histogram equalization is performed to improve the contrast of the given photo. It extends out the intensity range of the photo. Then, a power law transformation (gamma) is used to expand the range of the output values for a particular input, and vice versa. In particular, this work makes use of a transformation of dynamic power law that consists of taking the average intensity value of the image (AIV) and a fixed threshold (THR). The factor (γ) is given by $\gamma = AIV/THR$. It increases the intensity value of dark images, and decreases the intensity of light images.

2. **DANNs-based Processing and Median Filter**
   This stage fed an ESRGAN with the photo obtained in the pre-processing stage. ESRGAN produces a high resolution photo. Nevertheless, it introduces some artifacts in the photo as it augments its resolution, such as folds; therefore, this work uses a median filter of 5×5 to obtain a photo free of noise and folds. Finally, this stage carries out DeOldify architecture [12] to obtain a colored photo.

   - **Enhanced Super-Resolution Generative Adversarial Networks (ESRGAN)**
     ESRGAN is a GAN-based architecture introduced in Wang et al. [10]. ESRGAN consists of Residual-in-Residual Dense Blocks (RRDB) and a relativistic discriminator. This architecture allows the network to process and convert low-resolution photos into high resolution ones. Figure 2 shows a view of the ESRGAN architecture.

– **DeOldify.** DeOldify [12] is a network architecture distributed free under MIT license. This network is based on a pre-trained U-net, and is able to color black and white images successfully. It was trained using a novel method called NoGAN which focuses on training the generator and the discriminator separately using traditional training methods and the training model with GANs techniques.

3. **Post-processing**

This stage applies the Contrast Limited Adaptive Histogram Equalization (CLAHE) algorithm [13]. CLAHE consists of dividing the image into small regions called tiles, and process each tile separately instead of processing the whole image. Then, the tiles are joined them together again using bilinear interpolation. Before applying the CLAHE algorithm, DeOldify output is transformed into CIELAB colorspace which its L channel is known to represent better the human perception of lightness. CLAHE is applied to this channel, and finally, the image is converted back to RGB.

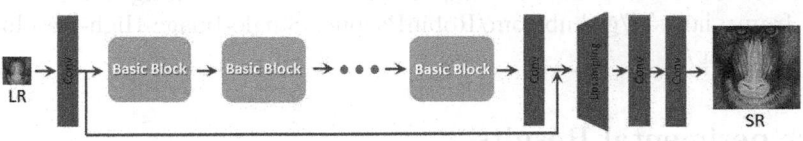

Fig. 2. Architecture of ESRGANs. Image from [10]

3 Experimental Setup

3.1 Dataset

The old photos of Quito's city were collected from google images to reconstruct and enhance them with the aim of creating a urban history of the city. The collected photos have low resolution in gray scale with varied resolution from 275×450 to 550×810 pixels. Old photos represent the culture and history of this Ecuadorian city, such as the center of the city, its churches, a panoramic view of the town, or an important monument in construction. Figure 3 shows some samples of the old photos to be reconstructed.

All experiments were done using Google Colab service. This service includes a graphic card Nvidia Tesla P-100 with 16 Gb of capacity and a RAM with 30 Gb of capacity. To test the proposed methodology, Python software routines with Opencv image processing library have been implemented. In addition, Python interfaces of Pytorch deep learning framework, and Keras were also used to face with research code and rapid prototyping based on deep learning architectures. In order to determine the impact of pre- and post-processing techniques

| (a) | (b) | (c) | (d) |

Fig. 3. Old photo samples related to: (a) City Hall's street on a cloudy day; (b) Frontside of a Church from Quito colonial, (c) view of the city and Panecillo hill; (d) construction of the monument to the virgin of the Panecillo.

on GAN-based techniques, this work explores old photos reconstruction and coloring using: a) ESRGAN + DeOldity (DANNs); b) DANNs + median Filter; c) DANNs + median Filter with pre-processing; d) DANNs + median Filter with post-processing; and e) DANNs + median Filter with pre- and post-processing.

All of the developed code and dataset used for testing can be download from https://github.com/RobinPaspuel/Single-Image-High-Resolution-and-Colorizing/.

4   Experimental Results

To determine the ability of the proposed methodology to reconstruct old photos and provide a colored and high resolution photo (4× original), 20 colored high resolution photos were degraded to a quarter of its original resolution and converted to grayscale. Additionally, a power law transformation with $\gamma = 0.7$ was applied to increase their illumination. The degraded photos underwent the experiments described in the previous section, the resulting images were compared with respect to their corresponding colored high resolution photos in terms of Root Mean Square Error (RMSE) and Structural Similarity Index (SSIM) metrics given by Eqs. (1) and (2), respectively. RMSE refers to the error or difference between the colored high resolution photo and the reconstructed one. A small value implies a minimal difference between both images. SSIM evaluates the similarity between two images. Both metrics are given in average percentage values among 20 evaluated photos.

$$RMSE = \sqrt{\frac{\sum_{t=1}^{T}(x_{1,t} - x_{2,t})^2}{T}} \tag{1}$$

where, x_1 and x_2 correspond to the pixels in the same position of the reference image and the reconstructed image. T is the total of pixels.

$$SSIM(x,y) = \frac{(2\mu_x\mu_y + c_1)(2\sigma_{xy} + c_2)}{(\mu_x^2 + \mu_y^2 + c_1)(\sigma_x^2 + \sigma_y^2 + c_2)} \tag{2}$$

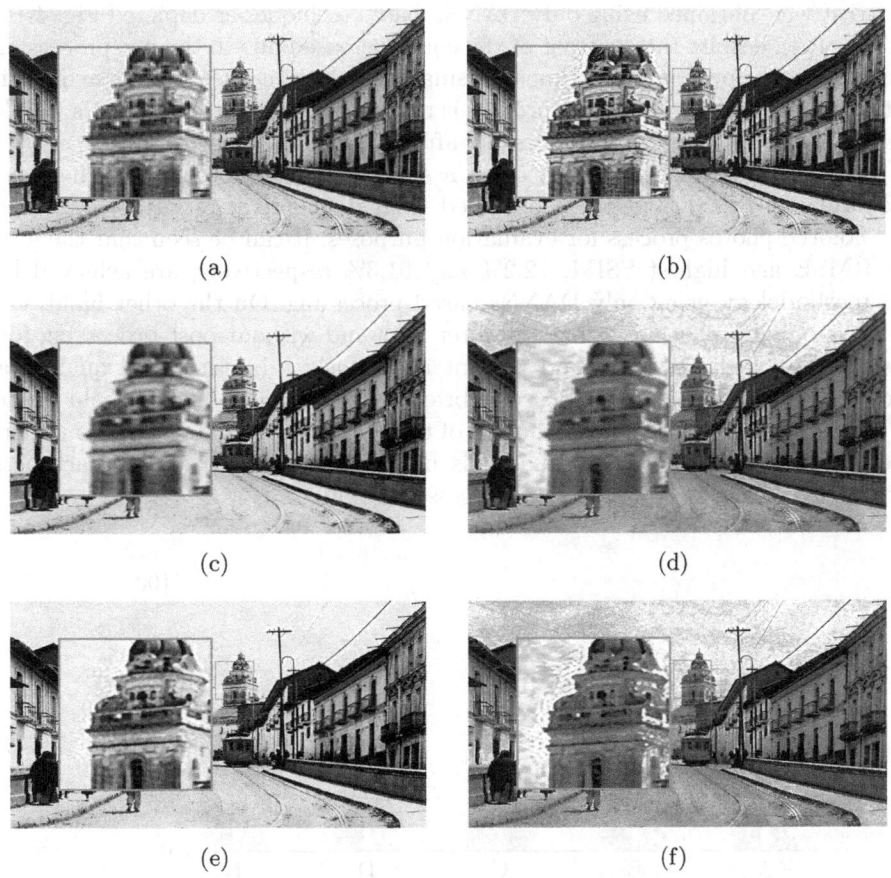

Fig. 4. Results related to: (a) Old photos of Quito City; (b) results achieved by DANNs without median filter (c) results achieved by DANNs + median filter; (d) results obtained by DANNs + median filter with pre-processing; (e) results obtained by DANNs + median filter with post-processing; and (f) results achieved by DANNs + median filter with pre- and post-processing.

where, μ is the average, σ is the variance and c_1 and c_2 are variables used to avoid division by zero, they are given by; $c_1 = (k_1 L)^2$, $c_2 = (k_2 L)^2$. L is the dynamic range of the pixel values, $k_1 = 0.01$ and $k_2 = 0.03$ are constant variables.

Some results obtained from applying the proposed methodology are depicted in Fig. 4. It can be observed that, the proposed hybrid methodology with pre- and post processing visually leads to less noisy results and better contrast as shown Fig. 4(e). The experimental tests demonstrated that the photos tend to decrease their average intensity after applying histogram equalization, but this value increases after applying the Dynamic Power Law transformation with $THR = 150$. Moreover, this increasing does not considerably affect the achieved contrast, as shown Fig. 4(d)–(f). The photos with the lowest contrast were obtained using only DANNs-based techiques as depicted Fig. 4(b). In Fig. 4(d), despite the contrast of the color increased due to the pre-processing techniques, the presence of artifacts as small flat blocks caused by coarse quantization through super resolution process is notorious. The contrary effect is visible in Fig. 4(e) applying a post-processing after DANNs-based processing.

The results furnished in Fig. 5, on a scale from 0 to 100%, show the average values of RMSE and SSIM obtained from the comparison of the degraded and colored photos process for evaluation purposes. It can be seen that the lowest RMSE and highest SSIM, 12.2% and 61.3% respectively, are achieved by the methodology using only DANNs-based processing. On the other hand, the DANNs-based processing + median filter with and without post-processing follow the best metrics values and present notoriously a better visual quality as shown Fig. 4(c), (d). The proposed hybrid methodology modifies the old photo to a greater extent due to high number of transformations that the photo undergoes. Despite this, the visual quality is noticeably improved and the artifacts introduced in the super-resolution process are reduced.

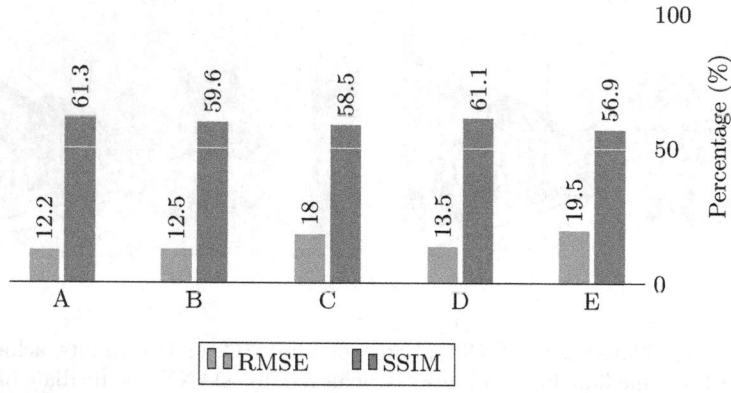

Fig. 5. Performance reached by the proposed methodology: A) DANNs without median filter; B) DANNs + median filter; C) DANNs + median filter with pre-processing; D) DANNs + median filter with post-processing; and E) DANNs + median filter with pre- and post-processing.

5   Conclusion

In this paper, we show the advantages of using traditional image enhancement methods along with IA-based methods. The application of Histogram Equalization and a Dynamic Power Law transformation improve the quality of coloring in comparison with photos with a lack of previous treatment. The median filter successfully eliminates the artifacts generated by the process of increasing resolution with no significant changes in the overall quality of the photo. CLAHE increases the contrast and accuracy of the colors after the process of coloring.

The results obtained show that despite the increase of the RMSE and the decrease of the SSIM, the visual details and the quality perception of the treated photos are better in comparison with the no treated photos.

The proposed hybrid methodology based on ESRGAN + median filter + DeOldity with post-processing was used to reconstruct old photos of Quito 's city with a good visual quality to create urban history that focuses on providing visual information on buildings, streets and transportation that are part of early years' cultural heritage.

In future works, we hope to explore the benefits of pre-processing techniques aimed to reduce the computational requirements to process images with IA based models.

References

1. Wang, F.: A study of digital image enhancement for cultural relic restoration. Int. J. Eng. Tech. Res. **7**(11), 41–44 (2017)
2. Patel, P., Bhandari, A.: A review on image contrast enhancement techniques. Int. J. Online Sci. **5**(5), 14–18 (2019)
3. Raj, S., Kumar, S., Raj, S.: An Improved Histogram Equalization Technique for Image Contrast Enhancement. ResearchGate, January 2015
4. Kuo, T.Y., Wei, Y.J., Lee, M.J., Lin, T.H.: Automatic damage recovery of old photos based on convolutional neural network. In: 2019 International Symposium on Intelligent Signal Processing and Communication Systems (ISPACS), pp. 1–2. IEEE (2019)
5. Hong, M., Qu, Y., Li, C., Chen, S.: Multi-scale iterative network for underwater image restoration. In: 2019 2nd China Symposium on Cognitive Computing and Hybrid Intelligence (CCHI), pp. 201–206. IEEE (2019)
6. Jolicoeur-Martineau, A.: GANs beyond divergence minimization, no. 1, pp. 1–14 (2018). http://arxiv.org/abs/1809.02145
7. Chen, X., Yu, J., Kong, S., Wu, Z., Fang, X., Wen, L.: Towards real-time advancement of underwater visual quality with GAN. IEEE Trans. Industr. Electron. **66**(12), 9350–9359 (2019)
8. Beers, A., et al.: High-resolution medical image synthesis using progressively grown generative adversarial networks. arXiv abs/1805.03144 (2018)
9. Ledig, C., et al.: Photo-realistic single image super-resolution using a generative adversarial network. In: 2017 IEEE Conference on Computer Vision and Pattern Recognition (CVPR), pp. 105–114 (2017)
10. Wang, X., et al.: ESR-GAN: enhanced super-resolution generative adversarial networks. In: The European Conference on Computer Vision Workshops (ECCVW), September 2018
11. Jolicoeur-Martineau, A.: The relativistic discriminator: a key element missing from standard GAN. CoRR abs/1807.00734 (2018). http://arxiv.org/abs/1807.00734
12. Antic, J.: Deoldify (2018). https://github.com/jantic/DeOldify
13. Zuiderveld, K.: Contrast Limited Adaptive Histograph Equalization. Graphic Gems IV, pp. 474–485 (1994)

On Analysing Similarity Knowledge Transfer by Ensembles

Danilo Pereira[1]([✉]), Flávio Santos[1], Leonardo N. Matos[2], Paulo Novais[3],
Cleber Zanchettin[1], and Teresa B. Ludermir[1]

[1] Federal University of Pernambuco, Recife, PE, Brazil
{dcp2,faos,cz,tbl}@cin.ufpe.com
[2] Federal University of Sergipe, São Cristóvão, SE, Brazil
leonardo@dcomp.ufs.br
[3] University of Minho, Braga, Portugal
pjon@di.uminho.pt

Abstract. Knowledge transfer is the task of transferring the knowledge learned by a model A to a new model B. This task is essential in Deep Learning, since there are complex models with excellent results, but computationally costly to be executed. The Similarity Knowledge Transfer (SKT) method proposes an approach to transfer the knowledge layer-by-layer between a donor model and a receiver model. This transfer is carried out through the representations learned by the layers from the teacher model. Despite presenting good results, the SKT method proposes just a way to transfer knowledge between two models. Therefore, this work presents the Similarity Knowledge Transfer Ensemble (SKTE) method, a generic form of SKT that allows the transfer from several teachers to a single student model. We carried out experiments with the CIFAR10 benchmark, where the results obtained showed promising results in this activity.

Keywords: Knowledge transfer · Ensembles · Deep learning

1 Introduction

Deep learning is a sub-field of Machine learning. It consists of computational models composed of multiple layers responsible for learning representations of data with a high level of abstraction. Deep Learning has achieved state of the art (SOTA) results in different domains such as Natural Language Processing and Computer vision. Although DL presents SOTA results on various tasks, it needs a high computational power to process all layers, as opposed to simple approaches to analyze stress [2] and mental fatigue [13], for example. Some approaches have been proposed to transfer the knowledge from pre-trained models to new models small to mitigate the computational power needed.

Supported by organization CAPES (Brazilian research agency).
D. Pereira and F. Santos—Equal contribution.

The notion of transfer the knowledge from a bigger and trained network (teacher) to a smaller one (student) is not a recent topic. Indeed, it was first described in a seminal paper on model compression by Buciluǎ et al. [1]. The key idea behind model compression is to train a small and fast model in a way that it could mimic an ensemble without or with a minimal drop in its performance. Another well-known model compression technique was introduced by Hinton et al. [6]. Passalis et al. [12] proposed the method SKT for unsupervised knowledge transfer between two deep learning models. It uses similarity-induced embeddings to transfer the knowledge between any two layers of neural networks even when both have a different number of neurons.

A widespread approach in machine learning is using ensemble methods to produce a better model. Ensemble methods are a class of strategies to combine several models in a single one, similar to a group decision making process [8]. Since ensemble methods are also conventional in deep learning, it is interesting to produce methods to transfer knowledge from the ensemble to a unique model. Although the SKT method presents promising results, it did not offer an approach using the ensemble scenario. To overcome this challenge, we propose an SKT extension using an ensemble scenario, named SKT Ensemble.

To evaluate our proposed method, we perform the experiments using the CIFAR-10 dataset. We chose this dataset because it is a universal benchmark in the DL literature, and the baseline also uses this dataset.

This work is organized in the following way: Sect. 2 presents the relevant related work. Next, the proposed method is present in Sect. 3, and Sect. 4 discusses the experiments and results. Section 5 presents the final remarks about the work.

2   Related Work

Hinton et al. [6] showed that a trained network keeps more information in its soft probabilities than just in the class label. Then, transfer knowledge is achieved when these smoothed probabilities (also known as *dark knowledge*) are used as targets to train a small network. In this method, to raise a temperature T plays a vital role in *softer* the label probabilities.

Tang et al. [15] showed it is possible to train a student model from a weaker teacher model without applying any trick in the learning scheme and with limited train data. In Romero et al. [14], beyond the soft probabilities, the student model is trained with *hints* from intermediate representations that are used to guide its feature activations. Another technique, called Attention Transfer [18], aims to force a student network to mimic the pattern responses (here denoted as "attention") from a teacher network instead of using the soft targets. In Net2Net [3], the weights of an already trained model are used to initialize a higher student model.

In [11], the authors showed that the performance of the student model degrades when the gap between the teacher is large. Then, they introduced the concept of Teacher Assistant Knowledge Distillation (TAKD), which employs an intermediate-sized assistant model that reduces the gap between student and teacher. In fact, another different multi-step step knowledge distillation were proposed as in [5] and [17]. Furlanello et al. [5] trains an ensemble of networks using a sequence of knowledge distillation steps in a way that the previous student model is the teacher of the subsequent one. Yang et al. [17] modify the teacher's loss function sequentially to make it more "tolerant" by merely adding loss terms. That facilitates a few secondary classes to emerge and complement to the primary class.

In this direction, other works proposed improvement in the knowledge distillation employing modifications in the loss function. In [16], the authors demonstrated that the standard knowledge distillation loss ignores important structural knowledge (complex interdependencies) of the teacher. They suggest an alternative contrastive-based loss for transferring knowledge between models. The method called Contrastive Representation Distillation (CRD) can capture more information about the teacher's representation of the data. Malinin et al. [10] proposed a new task named Ensemble Distribution Distillation (EnD) that enables a single model to mimic a distribution of an ensemble of teacher models. This allows that total uncertainty in predictions to be decomposed into *knowledge uncertainty* and *data uncertainty* and retains information about its *diversity* beyond the classification performance. It is important to notice that the EnD's authors based this work in a previous work of the Prior Networks [9].

Passalis and Tefas [12] recently introduced a novel unsupervised knowledge transfer method based on similarity-induced embeddings, named SKT. This method can transfer knowledge between any two layers of a neural network regardless of the number of neurons. In other words, the knowledge can be transferred without any lossy dimensionality reduction operation.

Finally, in [4], a study about the efficacy of the knowledge distillation is presented. The authors concluded that a higher accuracy in larger teaching models does not mean that the student model will achieve high performance as well.

3   Proposed Method

3.1   SKT

The SKT method's goal is to transfer the knowledge from a trained donor model D to a receiver model R (initialized with random weights). It performs this task in a non-supervised way using the similarity from the intermediate layer representations (embeddings).

Let D the donor (also refered as teacher) model and R the receiver (also refered as student) model. Given a input vector x, we can obtain the representation (embedding) from the i-th layer through the functions $D(x, i)$ and $R(x, i)$. The input vector x from both models (D and R) must be the same, but there is no restriction regarding the D and R's output dimensions.

The SKT main step is to transfer the knowledge from a specific layer M of the D model to a specific layer L of the R model. Thus, the layer L is trained to replicate the behavior of layer M. To learn the M behavior, the layer L from R must learn the representation similarity from a transfer set $X = x_1, x_2, x_3, ..., X_n$. The transfer set X is a set of samples, for example, if D is a model to perform object detection then each x_i in the transfer set X is a input image sample. Given the transfer set X with N elements, firstly a similarity matrix T of dimension $N\dot{N}$ is computed, where T_{ij} means the similarity between the representations of x_i and x_j obtained from layer M of the donor D. The similarity is obtained according to Eq. 1:

$$T_{ij} = |\langle D(x_i, M), D(x_j, M)\rangle| \qquad (1)$$

After the similarity matrix T be calculated, it is computed a similarity matrix P from layer L of the receiver R using the same transfer set X. The Eq. 2 illustrates how we can compute the similarity matrix P.

$$P_{ij} = |\langle R(x_i, L), R(x_j, L)\rangle| \qquad (2)$$

After computing the T and P similarity matrix, the loss function 3 is used to calculate the loss from the receiver model R. This loss function forces the model R to learn representations alike to model D, thus generating very closed similarity matrices, therefore, transferring knowledge from D to R. After computing the loss, the Gradient Descent optimization method is used to calculate the gradient of J_{TP} with respect to the layer L of R ($\frac{\partial J_{TP}}{\partial \theta_R}$) and update them.

$$J_{TP} = \frac{1}{N^2} \sum_{i=1}^{N} \sum_{j=1}^{N} (T_{ij} - P_{ij})^2 \qquad (3)$$

3.2   SKT Ensemble

The SKT [12] method presents a layer-to-layer knowledge transfer approach from a donor model T to a receiver model R. However, in some problems it is advantageous to use an ensemble of models, and the SKT method does not deal with this scenario. Therefore, we propose an SKT Ensemble (SKTE) whose goal is to transfer the geometry properties from an ensemble of donors D to a receiver model R.

The SKTE method is structured in the following way: Given a transfer set $X = x_1, x_2, x_3, ..., x_n$ and a donors set $D = D_1, D_2, ..., D_z$, firstly we calculate the similarity matrix T^y of each donor in D and the similarity matrix from the receiver model R.

$$JE_{TP} = \frac{1}{N^2} \sum_{i=1}^{N} \sum_{j=1}^{N} [(T_{ij}^1 + T_{ij}^2 + ... + T_{ij}^z)/z - P_{ij}]^2 \qquad (4)$$

From the Eq. 4, we compute the loss to approximate the similarity matrix P (receiver) to the average similarity matrix of all donors. So, we can force the receiver model to learn the most frequent information contained in the layers of all donors.

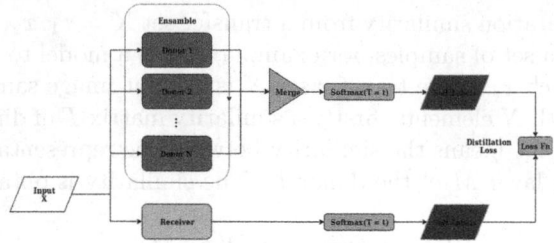

Fig. 1. Visual representation of the model distillation process for unsupervised knowledge transfer from a ensembles to a single student model.

3.3   Model Distillation

The concept of the distillation was introduced by [6]. It consists of an approach for knowledge transfer in which a small model learns to imitate a large trained model. This training configuration is often refered as a teacher-student approach, where the big model is the teacher, and the smaller model is the student.

The main idea of the authors in [6] is that the relative probabilities of a model's incorrect answers tell us a lot about its generalizability. In other words, consider a model trained to identifying animals and it confuses an image of a wolf with a dog at a given probability, however that probability is higher than the likelihood of the model mistaking the labeling of a dog as a plane.

In the model distillation process, transferring the knowledge is done by minimizing a cost function in order to reproduce the distribution function of the classes that were predicted by the teacher model. In neural networks, this class distribution is usually achieved from a layer using the function *softmax*, which is given by

$$q_i = \frac{exp(\frac{z_i}{T})}{\sum exp(\frac{z_j}{T})} \tag{5}$$

in which z_i is the logit calculated for each class i, that are converted into a probability q_i.

When calculating the cost function from the smoothed outputs of the teacher model, the same temperature value is used for both teacher and student models, and this cost is called distillation loss. Then, the student minimizes the cross-entropy between its output probabilities and the ones of the teacher.

However, in our work, we only considered the unsupervised version, i.e., only the *distillation* cost function, since it was compared with the SKT (Sect. 3.1) which it is a purely unsupervised approach.

3.4   Distillation Ensemble

The distillation [6] can be easily extended to transfer knowledge not only from a model but also from an ensemble of teachers to an apprentice model. This idea is not new and was brought up with strong arguments by Caruana and collaborates [1].

Specifically for *distillation*, the apprentice model regulate its learning process by minimizing a cost function between its outputs and the smoothed output of the teacher's models from the ensemble. These outputs are combined using an arithmetic or geometric average. In this work, we used arithmetic average and trained the model in an unsupervised fashion, i.e., only minimizing the distillation cost function, as shown in Fig. 1.

4 Experiments and Results

4.1 CIFAR-10

The CIFAR-10 database consists of 60,000 color images of 32×32 pixels arranged in 10 mutually exclusive classes [7]. This base has 6,000 images for each class. In total, there are 50,000 images in the training set and 10,000 images in the test set. By default, the CIFAR-10 database has to be split into five training batches and one test batch containing 10,000 images each. The test batch contains exactly 1,000 images per class, while the remaining images were randomly separated among the training batches. A training batch may contain an imbalance between classes. However, in total, there are 5,000 images per class.

Experiments. To carry out the knowledge transfer experiment, we use two architectures, which are: (i) donor (D) and (ii) receiver (R). The architecture of D, present in the Fig. 2a, is composed of the following layers sequence: two convolutional (conv) layers with 32 filters of size 3×3, a 2×2 max-pooling layer, two conv layer with 64 filters of size 3×3, a 2×2 max-pooling layer, and two fully-connected layers (512×10 units). The R network, Figure 2b, is similar to D, but there is only one conv layer before the max-pooling, and the two fully-connected layers have less units (128×10).

In the experiments, we trained the models in 50 epochs, set the batch size to 64, the learning rate to 10^{-4}, and used the Adam algorithm with batch normalization as optimizer. It is important to note that we used three neural networks as donors and a single neural network as receiver with the above mentioned architecture.

4.2 Results and Discussion

This section presents all results obtained during the experiments and discuss them. Table 1 shows the results achieved using the CIFAR-10 dataset. The first two rows are the results from the baseline paper.

Our replication of the basiline SKT and Distillation presented similar performance to the reported in the baseline paper. The principal observation is regarding to the *distillation* case, whose difference was a little higher compared to the baseline paper. We also have conducted experiments with SKT Ensemble using EnDLoss [10] and Dirichlet Loss [9] using different temperatures to find the best setting.

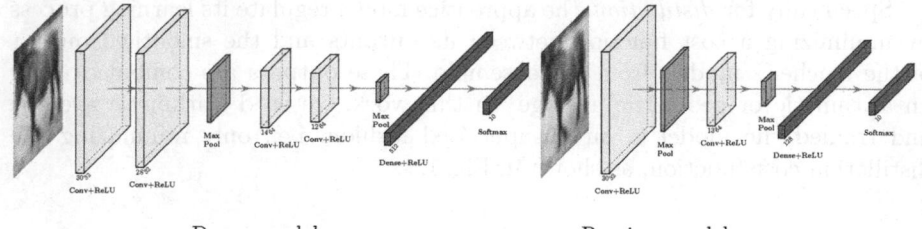

Donor model   Receiver model

Fig. 2. Donor and receiver models architectures.

Table 1. Accuracy error (%) to the experiments using CIFAR-10 dataset.

Model	Accuracy error (%)
Baseline Distill [12]	31.38
Baseline SKT [12]	27.50
Distill Reproduced ($T = 2$)	34.97
Distill Ensamble ($T = 2$)	**27.99**
SKT Reproduced	28.73
SKT Ensamble	**25.80**
SKT Ensamble EnDLoss $T = 2$	**25.54**
SKT Ensamble EnDLoss $T = 5$	**23.30**
SKT Ensamble EnDLoss $T = 10$	**22.68**
SKT Ensamble Dirichlet $T = 2$	**25.72**
SKT Ensamble Dirichlet $T = 5$	**23.87**
SKT Ensamble Dirichlet $T = 10$	**26.28**

From Table 1, we can observe that all SKT Ensemble configurations achieved better results than SKT and Distill baseline. The global best result was an accuracy error of 22.68% from SKT Ensemble EnDLoss using a temperature equal to 10. Since the SKT forces the receiver model to learn the geometry of intermediate layers of donor models and the EnDLoss helps to emulate the complete ensemble model, the achieved results indicate that both information results in a better model.

The results presented in Table 1 shows that the temperature parameter has a high impact on the results. Using a high value to temperature produces a more uniform output distribution, thus, turning out all output predictions very close. According to [6], when the receiver uses a lower temperature, it pays less attention to negative logits than the average. Therefore, as the impact of temperature choice depends on the donor models predictions, it is necessary to conduct an ablation study with this parameter to find the best configuration. Besides, we observed that higher temperatures had produced worse results for distillation, probably because when using only this method, there is a more robust smoothing in the donors' outputs, causing the loss of crucial information.

5   Conclusion

This paper presents a method to transfer the knowledge from an ensemble of neural networks to a single receiver model. We evaluate the SKTE method using the well-know dataset CIFAR-10. The obtained results show that the combination of the SKTE method using EnDLoss with temperature equal to 10 achieved a better performance than the base SKT and all others reference models.

This work can be extended in a variety of ways. For example, we can apply the SKTE method in other challenge datasets and neural networks such as ImageNet and ResNet. Besides, it is necessary to perform experiments in different domains such as Natural Language Processing.

Acknowledgments. This work has been supported by FCT - Fundação para a Ciência e a Tecnologia within the R&D Units project scope UIDB/00319/2020. The authors from affiliations 1 and 2 thank CAPES and CNPq for the financial support.

References

1. Buciluǎ, C., Caruana, R., Niculescu-Mizil, A.: Model compression. In: Proceedings of the 12th ACM SIGKDD International Conference on Knowledge Discovery and Data Mining, pp. 535–541. ACM (2006)
2. Carneiro, D., Novais, P., Pêgo, J.M., Sousa, N., Neves, J.: Using mouse dynamics to assess stress during online exams. In: Onieva, E., Santos, I., Osaba, E., Quintián, H., Corchado, E. (eds.) HAIS 2015. LNCS (LNAI), vol. 9121, pp. 345–356. Springer, Cham (2015). https://doi.org/10.1007/978-3-319-19644-2_29
3. Chen, T., Goodfellow, I., Shlens, J.: Net2net: accelerating learning via knowledge transfer. arXiv preprint arXiv:1511.05641 (2015)
4. Cho, J.H., Hariharan, B.: On the Efficacy of Knowledge Distillation (2019)
5. Furlanello, T., Lipton, Z.C., Tschannen, M., Itti, L., Anandkumar, A.: Born Again Neural Networks (2018)
6. Hinton, G., Vinyals, O., Dean, J.: Distilling the knowledge in a neural network. arXiv preprint arXiv:1503.02531 (2015)
7. Krizhevsky, A.: Learning Multiple Layers of Features From Tiny Images. University of Toronto (05 2009)
8. Lima, L., Novais, P., Costa, R., Bulas-Cruz, J., Neves, J.: Group decision making and quality-of-information in e-health systems. Log. J. IGPL **19**, 315–332 (2011). https://doi.org/10.1093/jigpal/jzq029
9. Malinin, A., Gales, M.: Predictive uncertainty estimation via prior networks. In: Advances in Neural Information Processing Systems, pp. 7047–7058 (2018)
10. Malinin, A., Mlodozeniec, B., Gales, M.: Ensemble Distribution Distillation (2019)
11. Mirzadeh, S.I., Farajtabar, M., Li, A., Levine, N., Matsukawa, A., Ghasemzadeh, H.: Improved Knowledge Distillation Via Teacher Assistant (2019)
12. Passalis, N., Tefas, A.: Unsupervised knowledge transfer using similarity embeddings. IEEE Trans. Neural Netw. Learn. Syst. **30**(3), 946–950 (2019). https://doi.org/10.1109/TNNLS.2018.2851924
13. Pimenta, A., Carneiro, D., Novais, P., Neves, J.: Monitoring mental fatigue through the analysis of keyboard and mouse interaction patterns. In: Pan, J.-S., Polycarpou, M.M., Woźniak, M., de Carvalho, A.C.P.L.F., Quintián, H., Corchado, E. (eds.) HAIS 2013. LNCS (LNAI), vol. 8073, pp. 222–231. Springer, Heidelberg (2013). https://doi.org/10.1007/978-3-642-40846-5_23

14. Romero, A., Ballas, N., Kahou, S.E., Chassang, A., Gatta, C., Bengio, Y.: Fitnets: hints for thin deep nets. arXiv preprint arXiv:1412.6550 (2014)
15. Tang, Z., Wang, D., Zhang, Z.: Recurrent neural network training with dark knowledge transfer. In: 2016 IEEE International Conference on Acoustics, Speech and Signal Processing (ICASSP), pp. 5900–5904. IEEE (2016)
16. Tian, Y., Krishnan, D., Isola, P.: Contrastive Representation Distillation (2019)
17. Yang, C., Xie, L., Qiao, S., Yuille, A.: Knowledge Distillation in Generations: More Tolerant Teachers Educate Better Students (2018)
18. Zagoruyko, S., Komodakis, N.: Paying More Attention to Attention: Improving the Performance of Convolutional Neural Networks Via Attention Transfer (2016)

Special Session on New Trends and Challenges on Social Networks Analysis

Social Network Recommender System, A Neural Network Approach

Alberto Rivas[1,2], Pablo Chamoso[1,2], Alfonso González-Briones[1,2,3(✉)],
Juan Pavón[3], and Juan M. Corchado[1,2,4,5]

[1] BISITE Research Group, University of Salamanca, Edificio I+D+i, Calle Espejo 2,
37007 Salamanca, Spain
rivis@usal.es
[2] Air Institute, IoT Digital Innovation Hub (Spain),
Carbajosa de la Sagrada, 37188 Salamanca, Spain
[3] Research Group on Agent-Based, Social and Interdisciplinary Applications
(GRASIA), Complutense University of Madrid, Madrid, Spain
alfonsogb@ucm.es
[4] Osaka Institute of Technology, Osaka 535-8585, Japan
[5] Universiti Malaysia Kelantan, Kota Bharu, Kelantan, Malaysia

Abstract. Social networks have increased considerably due to the development of networks with specific purposes and represent a high percentage of daily communications between people. Due to the large amount of content in any type of social network, it is necessary to guide users to find the content that best suits their needs. The inclusion of artificial intelligence techniques greatly facilitates the task of finding relevant content. This document presents a recommendation system (RS) for a business and employment-oriented social network. Therefore, job offers are recommended to users, but other users are also encouraged to follow them. The system presented is based on virtual agent organizations, and uses artificial neural networks to determine whether job offers and users should be recommended or not. The system has been evaluated on a real social network and has provided a high acceptance rate of both job offers and user recommendations.

Keywords: Artificial neural network · Recommender system · Social networks · Virtual organizations

1 Introduction

The Internet has changed the way people communicate and get to know each other even in the workplace. Social networks such as Monster, XING or LinkedIn have a large volume of the offers available today around the world. However, at the local level, there are many labour social networks in the most developed countries. For example, in Spain, where this work has been done, there is a social network called beBee [2], which although it operates internationally, much of its traffic comes from Spanish users.

© Springer Nature Switzerland AG 2020
C. Analide et al. (Eds.): IDEAL 2020, LNCS 12490, pp. 213–222, 2020.
https://doi.org/10.1007/978-3-030-62365-4_21

Users can create a profile in beBee with a candidate or company role. Therefore, users can register and publish job offers. In addition, beBee allows users to share interesting or related content related to the user's work sector, thus also promoting the creation of labour synergies between people in the same sector. In this way, in addition to allowing registration in job offers, it allows the creation of relationships between users who work with related topics and who may be interested in following themselves. These relationships are very important to keep abreast of the latest trends or to create new working relationships.

The large volume of offers and users is a major drawback when it comes to searching and finding the most relevant content on a social network. For example, in the case of beBee, there are more than twelve million registered users. Therefore, it is practically a necessity that this type of portals have RSs. The inclusion of a referral system is vital in providing users with a positive experience, offering what they need without having to search for it. This increases the retention rate of users and makes them talk well about the social network, making new users decide to use it as a social network. To achieve this, it is necessary that the RS provides very high results in terms of: i) offers in which the user registers in the case of job offers recommendations; and ii) users to whom after having been recommended, have actually been followed by the user who received the recommendation. In order to evaluate the work developed, the system has been integrated into the social network beBee and the evaluation has been separated for each of the two types of recommendations, user-supplied on the one hand and user-user on the other hand.

The following section presents the background analyzing previously proposed techniques in RSs and their application in other environments. Section 3 presents the system designed, the proposed system is described and in Sect. 4 the results of its evaluation are presented. Finally, conclusions are drawn on the performance of the proposed system and future work is discussed in Sect. 5.

2 Background

In recent years, the Deep Learning paradigm [9] has been an authentic revolution in online recommending [5], the fields of pattern recognition, artificial vision [8] and natural language processing [10]. In fact, Deep Learning can be a solution for many learning problems. In essence, Deep Learning field models, and more particularly deep artificial neural networks, are new computational models that have a number of layers or processing stages never before used. This allows deep neural networks to achieve unprecedented success rates in different tasks in the field of pattern recognition, as the use of an increasing number of processing layers allows a progressive refinement of representations and input characteristics. In this way, neurons in the deeper layers of Deep Learning models are able to recognize concepts or patterns of high complexity and abstraction, allowing them to detect the presence of different objects in images or classify text according to their emotional load or highly recommend. In this way, in multiple areas, RSs are being applied using Deep Neural Networks so that the system is able to learn multiple input parameters, such as for example in YouTube [4].

3   Proposed System

3.1   Information Extraction

This section presents two distinct parts: the extraction of information from users and the extraction of information from job vacancies. Although they have certain similar parameters, they are two well-differentiated cases and the extraction of information is carried out by two agents of different types, *User_information_agent* for users and *Job_information_agent* for job offers.

Users' Information. When users register on the social network, they provide their information from a structured form. The more information they provide, the more adjusted to the profile will be the recommendations made by the system. Once a user indicates an update of their information, the *User_information_agent* starts an extraction of information.

The information extracted is associated with different parameters that the system will use later to evaluate whether a relationship (user-user or user-employment offer) should be recommended or not. These parameters are listed below:

- Education ($\bar{e}$): each of its studies including the year and place.
- Languages ($\bar{l}$): it is necessary to extract and unify each of the languages indicated by the user who is able to use.
- Skills ($\overline{sk}$): list of skills indicated by the user.
- Years of experience (ye): years of experience in the sector.
- Salary range (sr): information from the salary range of the current.
- Geographic area (ga): is a determining factor, both when it comes to interacting with other users and when recommending offers.

In addition to this information extracted by the *User_information_agent*, there is more information about the user associated with its behavior and the content it shares on the social network. For this reason, each *User_agent_n* extracts information from each user, completing the information extracted with:

- Strong ties ($\bar{s}$): the information on each user's strong links is kept up to date (other users that already follows).
- Strong ties profile ($\overline{sp}$): in addition, the system obtains a list of keywords from the profiles of each user followed by a user.
- Groups ($\bar{g}$): beBee offers its users the possibility of joining "hives", which are groups of users who share the same interests or professional sector.
- Published offers ($\bar{o}$): job offers can also be posted by users.
- Job applications ($\bar{a}$): two users that apply to the same job offers, could be interested in establishing professional links.
- Publication of contents ($\overline{co}$): the content published can comprise of a text, an image or a link.
- Interests ($\bar{i}$): when a user recommends (or likes) the content published by another user and they do not share a strong tie, the level of affinity increases.

All of this information is taken into account in the later stages of the system when determining whether or not a user should be referred to another user or a job offer.

Jobs' Information. One of the main goals of employment-oriented social networks is that users apply to job offers. Therefore, the proposal of suitable job offers to users is an essential role of the social network.

One of the main problems when it comes to extracting information from job offers is that whoever publishes them usually uses the same text for a set of job portals, so they copy and paste the information from the offer in an unstructured way. In beBee, offers can be sent in a structured way, but not mandatory, allowing the insertion of an offer in free format. This is so as not to limit the insertion of offers, because from a strategic point of view, it is important that there are as many offers as possible so that users do not have to use other portals. Despite this disadvantage, text mining tools are used, such as the use of dictionaries, the elimination of stop words, stemming techniques, etc.

So, in the same way that user profile factors are analyzed, job offer factors also have to be analyzed and evaluated so that the system is able to provide suitable recommendations. The following job offer factors are identified:

- Required main education ($\overline{me_o}$): the level of education required by an offer.
- Required skills ($\overline{rs_o}$): abilities or skills that make them suitable for the job.
- Desirable skills ($\overline{ds_o}$): this factor does not have the weight of the previous factor, but it does have the same reasoning.
- Previous experience (years) (ye_o): level of experience that guarantees the user's ability to work in the required field.
- Required languages ($\overline{ln_o}$): the number of languages a user knows is also a key factor for offers that involve the knowledge of languages.
- Salary range (sr_o): it can be inferred that the higher the salary offered, the greater the salary of the user should be.
- Geographic area (ga_o): geographical location and whether the candidates are required to reside in that location in order to determine which users should be recommended.

With all this information extracted, along with the user information described above, the system is able to determine whether or not an offer should be recommended to a user.

3.2   Recommendation

One of the advantages of agent based development is that different agents can take the same role concurrently and independently from other agents. In this case, the recommendations are made by a group of agents of the following type *ANN_agent*. These agents have been implemented in such a way that they act as TensorFlow wrappers, which provides a flexible framework for experimenting with various deep neural networks architectures using large-scale distributed training [1].

In this case, the recommendation must be explained separately depending on whether it is a user-user recommendation or a user-user recommendation, since they are not carried out in exactly the same way.

User-Job Offer Recommendation. The system starts when a user is identified in the social network in order to present him/her the offers best adapted to his/her profile. Cache mechanisms are implemented that keep the recommendations for 4 h, so it only runs if the user logs on for the first time in at least 4 h. Once started, this RS is based on two neural networks, following the architecture presented in Fig. 1.

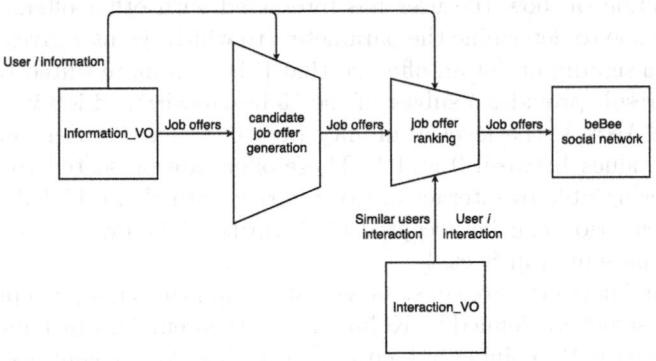

Fig. 1. User-job offer recommender schema.

Firstly, the system retrieves information about the user who has just logged on (user i) and the job vacancies published or renewed in the last 15 days and which are classified in the sector to which they belong (tens of thousands of offers). The dataframe o_j defined in (1) is build and associated to every job offer (j), and it is used as input for the ANN. Every user (i) has an associated information dataframe (u_i), defined in (2).

$$o_j = \{\overline{me_o}, \overline{rs_o}, \overline{ds_o}, ye_o, \overline{ln_o}, sr_o, ga_o\} \tag{1}$$

$$u_i = \{\overline{e}, \overline{l}, \overline{sk}, ye, sr, ga\} \tag{2}$$

These dataframes are part of the entry of the first neural network that as an exit provides a set of job offers candidates to be recommended to the user. Hundreds of job offers are provided as an output. These job vacancies are best suited to the user's profile based on their profile information. However, not so many offers can be recommended to users, as the acceptance rate of the recommendation would not be good. From these offers, therefore, another call is made to an ANN that takes as input the user's information (u_i), the set of job vacancies candidates, and the information provided by Interaction_VO on how that user relates (including history of offers to which he has registered, relations with other users, content that have interested in the social network, etc.) and other users with similar information. This type of information is structured in a array $ou_{i,j}$ that is defined as shown in (3).

$$ou_{i,j} = \{me_{i,j}, ln_{i,j}, sk_{i,j}, ye_{i,j}, sr_{i,j}, ga_{i,j}\} \tag{3}$$

Each of the values contained in the ou_{ij} array is the result that the neural network assigns to the parameters in terms of the adaptability of the offers to the profile, so that the candidate offers are limited to those that best suit them, since they will have a higher probability of acceptance.

Another ANN compiles a ranking on the basis of this information, along with information on how the user has interacted with other offers. This ANN makes it possible to determine the parameters to which the user gives the highest priority when signing up for an offer, so that it is even more suited to the user's profile. The result provides a subset of the 15 best-positioned job vacancies. The result of the ANN is the level of affinity (af) between the user and the offer, which takes values between 0 and 1. These offers are presented to the user in the beBee, being able to interact or not interact with them. If it does interact, the recommendation can be accepted or discarded. The evaluation carried out to the RS is presented in Sect. 4.

Both neural networks are constructed using a multilayered perception (MLP) following the structure defined by Kolmogorov's theorem [7], which indicates that it can be solved with a single hidden layer that has $2n + 1$ neurons, where n is the number of neurons in the neurons in the output layer. The first ANN has 13 neurons in the input layer containing information on the offer j and the user i, thus the hidden layer has been configured with 27 neurons and the output layer has 6 neurons, one for each of the parameters of the array $ou_{i,j}$. The second ANN has 12 neurons in the input layer, the 6 that relate their profile to the offer are the output of the previous neural network and another 6 that define the weight that historically has been assigned to each of the parameters in the offers to which it has been noted in advance. There are 25 neurons in the hidden layer and one, the value of af in the output layer. In both cases, softmax [6] is the selected activation function, which is defined in Eq. (4). y_i represents the result of the ANN for neuron i of the output layer.

$$P(i) \equiv \sigma(y, i) = \frac{e^{y_i}}{\sum_{n=0}^{N} e^{y_n}} \tag{4}$$

To evaluate the configuration of neural networks, the information available in the databases of the social network has been used. More specifically, the information available from 1st March 2016 to 28th February 2017 has been selected. This information includes the profiles of the offers, users and offers to which each user has registered. It should be noted that the selection process is generally not monitored by the social network, so the information on whether the user was finally hired for an offer is unknown.

The information used for the training and validation of the system includes 133,217 users (although the social network has more than 12 million users, most of them are not active) who made 687,632 job applications to 57,324 published offers. This provides an average of almost 12 job applications per published offer and each user signed up for an average of 5.16 job offers.

To test and validate the system, 712,368 non-existent relationships were generated (which should result in not recommending the offer to the user), completing a data set of 133,217 users, 57,324 job offers and 1,400,000 relationships. The total of these relationships was estimated to be balanced between actual entries (which should be suggested by the system) and non-existent relationships (which should not be suggested by the system). It should be noted that non-existent relationships were always created with offers and users from the same sector, for example, IT offers were only provided to users specialized in IT. It should also be considered that the non-existent relationships were related to the offers that users chose not to respond to although they had seen the offer on the day it had been sent. For the evaluation, the 70% of the data was used for training and the 30% of the data was used for the validation, providing the results presented in Table 1.

Table 1. Job offer recommendation - validation accuracy

Validation accuracy	Recommended	Not recommended
Real applications	94.84%	5.16%
Fake applications	9.32%	90.68%

On the last day of each month, the neural networks are retrained in order to adjust them to new variations such as new labour market trends.

User-User Recommendation. In the same way that it happens with the job posting recommendation to users, the user referral to other users runs every time a user logs on to the social network, with a 4-hour cache system. Therefore, when connected and the system runs, the flow described in Fig. 2 occurs.

In this case, the neural network inputs are the dataframes formed by user information, defined in (2) and by user interaction information, defined in (5). From the user's interaction dataframe, existing weak ties are evaluated and recommended. The neural network is implemented by one *ANN_agent* and it provides a level of affinity for each of them. The social network presents the recommendations of 15 in 15 sorted by affinity, so that as it interacts with the recommendations, the social network always shows 15 recommended users.

$$u_{i,j} = \{\overline{s}, \overline{sp}, \overline{g}, \overline{o}, \overline{a}, \overline{co}, \overline{i}, sr\} \tag{5}$$

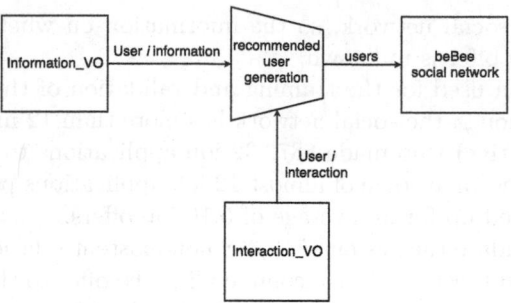

Fig. 2. User-user recommender schema.

The structure of the ANN, as well as that of the job advertiser, is based on Kolmogorov's theorem. In this case, the information of the two users whose possible relationship is evaluated, together with the history of interactions with users in the social network, is used as inputs. Thus, there are 24 neurons in the input layer and 49 neurons in the hidden layer. There is a neuron in the output layer and its result is a numerical value between 0 and 1 that determines the affinity between the two evaluated users. When the 15 recommendations are presented to the user on the social network, the most valuable ones are shown.

To evaluate the configuration of neural networks, information available in the social network databases has been used. More specifically, the information available between 1 March 2016 and 28 February 2017 has been selected. This information includes the profiles of the users and the users they follow on the social network.

The information used for the training and validation of the system includes 87,171 users (although the social network has more than 12 million users, most of them are not active) who made 270,230 follow ups on other users. This provides an average of 3.1 user relationships.

To test and validate the system, 229,770 non-existent follow up relationships were generated (which should result in not recommending the offer to the user), completing a data set with 500,000 follow up relationships. The total of these relationships was estimated to be balanced between real follows (which should be suggested by the system) and non-existent follow up relationships (which should not be suggested by the system). For the evaluation, the 70% of the data was used for training and the 30% of the data was used for the validation.

4   Experiment and Results

This section describes the dataset used to evaluate this work. The results obtained for the proposed system are presented below, which in turn are structured according to the type of recommendation: user-user and user-job offer.

4.1   Experiment Data Set

The dataset used in the system evaluation was obtained from the social network beBee. As far as the study between users and job offers is concerned, the 250 users who have had the most activity in the six most outstanding professional areas have been selected, in addition to having a relevant activity on the social network during the last month. In this way, a total of 1,500 users have been selected. In addition, all the offers published during the last 15 days have been selected, filtered by the six most relevant and most active professional areas on the social network. Specifically, 114,872 offers have been selected divided by professional area. On the other hand, in order to analyze relations between users, 2,000 users whose activity is high enough to be the object of this study have been randomly selected.

4.2   Results

In this section the results obtained from this exemplary dataset are presented and the two proposed RSs are evaluated separately. First, the user-job offer and then the user-user relationship RSs.

User-Job Offer Recommendations. The 250 most active users in each of the different areas have been selected. The flow to be followed by the user when receiving an offer is as simple as accepting or rejecting it. In this case, it is noted that the number of recommendations made for each group of 250 users does not vary much, regardless of the number of offers available on the social network. This is because the system filters all offers and finally establishes a ranking with the 15 most relevant ones for the user. As the user interacts with them, new offers appear, hence the small oscillation between different categories. On the other hand, it can be seen that for all categories the system is quite precise, as the acceptance rate is about 90% on average across all areas.

User-User Recommendations. For the study of relations between users, it is based on a set of 2,000 users with an activity that could be used in this evaluation.

A total of 43,873 recommendations have been made to the full set of users, with an acceptance rate of 88.27%. In addition, it should be noted that for each user the 15 most relevant recommendations are offered at all times, adding new recommendations as the user interacts with the system. This data is reflected in the fact that for each user during this evaluation period almost 22 recommendations have been made on average, of which approximately 19 have been accepted.

5   Conclusion

When determining how efficient each RS (user-user and user-job offer) is, it is necessary to compare it with other RSs. However, since the RS presented is very context-dependent, it compares with the previous system that was developed a year ago for the same social network, whose results were presented in [3]. In this sense, both of these recommenders provided an improvement in the percentage of user acceptance of the suggestion. The RS user-user presented in this paper, obtained 88.27% acceptance of the recommendation, compared to the 84.11% acceptance that initially obtained the previous system. With respect to the user-job offer RS presented in this paper, 90% acceptance of the recommendation was obtained, in the same way as slightly higher than the system proposed in [3]. The main difference is that the previous work is based on a Case Based Reasoning System (CBR), in which in a first approximation of the system the recommendation is not completely precise, but is learning as the user interacts. However, in this paper the recommendation is as precise as possible in a first recommendation.

The results obtained with users and offers in a completely new environment have been slightly less accurate than the value obtained in the validation phase of the neural network, where 70% of the information was available for training and 30% for validation. This is probably because the values used within 30% of the validation may be related to the information used for training.

Acknowledgments. This work has been supported by project "BeEMP". ID: RTC-2016-5642-6. Project co-financed with Min. of Economy, Industry and Competitiveness and ERDF funds.

References

1. Abadi, M., et al.: Tensorflow: large-scale machine learning on heterogeneous distributed systems. arXiv preprint arXiv:1603.04467 (2016)
2. beBee: bebee, affinity networking (2020). https://www.bebee.com/. Accessed 16 Feb 2020
3. Chamoso, P., Rivas, A., Rodríguez, S., Bajo, J.: Relationship recommender system in a business and employment-oriented social network. Inf. Sci. **433–434**, 204–220 (2018)
4. Covington, P., Adams, J., Sargin, E.: Deep neural networks for youtube recommendations. In: Proceedings of the 10th ACM Conference on Recommender Systems, pp. 191–198. ACM (2016)
5. Goodfellow, I., Bengio, Y., Courville, A., Bengio, Y.: Deep Learning, vol. 1. MIT press, Cambridge (2016)
6. Jung, H., Lee, S., Yim, J., Park, S., Kim, J.: Joint fine-tuning in deep neural networks for facial expression recognition. In: 2015 IEEE International Conference on Computer Vision (ICCV), pp. 2983–2991. IEEE (2015)
7. Kolmogorov, A.N.: Foundations of the Theory of Probability: Second English Edition. Courier Dover Publications, Mineola (2018)
8. Russakovsky, O., et al.: Imagenet large scale visual recognition challenge. Int. J. Comput. Vis. **115**(3), 211–252 (2015)
9. Schmidhuber, J.: Deep learning in neural networks: an overview. Neural Networks **61**, 85–117 (2015)
10. Zhang, X., Zhao, J., LeCun, Y.: Character-level convolutional networks for text classification. In: Advances in Neural Information Processing Systems, pp. 649–657 (2015)

Exploring Multi-objective Cellular Genetic Algorithms in Community Detection Problems

Martín Pedemonte[1]([⊠]), Ángel Panizo-LLedot[2], Gema Bello-Orgaz[2], and David Camacho[2]

[1] InCo, Universidad de la República, Montevideo, Uruguay
mpedemon@fing.edu.uy
[2] Departmento de Sistemas Informáticos, Universidad Politécnica de Madrid, Madrid, Spain
{angel.panizo,gema.borgaz,david.camacho}@upm.es

Abstract. Interest in network analysis has not stopped increasing over the last decade. The Community Detection Problem (CDP) has been a hot topic in network analysis, so many different approaches have been proposed. Among them, optimization methods have proven to be highly effective for this task. Traditionally, the CDP has been tackled as a single-objective optimization problem. Nevertheless, this trend has started to change, and new methods have appeared following multi-objective approaches. Genetic Algorithms have been applied to the CDP with relative success, especially NSGA-II. However, cellular Genetic Algorithms (cGAs) have yet received little attention. In cGAs, the population is structured in small overlapping neighborhoods producing a slow spread of high-quality solutions. The main contribution of this paper is understanding if the smooth diffusion scheme of MoCell (a multi-objective cGA) can provide any benefit over current multi-objective GAs for the CDP. To verify the effectiveness of MoCell, an evaluation was conducted on 21 synthetically generated networks and two real-world ones. The experiments show that MoCell is able to outperform NSGA-II, especially in large networks scenarios.

Keywords: Static community detection problem · Network analysis · Multi-objective Evolutionary Algorithms · Cellular genetic algorithms

1 Introduction

Research in network analysis has increased dramatically in recent years caused mainly by the growing interest in social network analysis and the massive public availability of data. Among the different problems that have been addressed in this topic, the detection of communities has attracted great attention [4,8].

The Community Detection Problem (CDP) consists of finding a partition in groups of the nodes of a network satisfying both intra and inter-group connectivity requirements. These groups of nodes are commonly known as communities.

© Springer Nature Switzerland AG 2020
C. Analide et al. (Eds.): IDEAL 2020, LNCS 12490, pp. 223–235, 2020.
https://doi.org/10.1007/978-3-030-62365-4_22

There is no consensus regarding a single definition of community within the scientific community. However, many of them typically imply that nodes of a community have to be strongly connected to each other, while the nodes that belong to different communities have to be weakly connected [13]. In the classical CDP formulation, nodes can only belong to a single community, nevertheless, other formulations involve different criteria. For example, overlapping community detection [3,5] or dynamic community detection [23,24], among others.

The CDP has been tackled in many formulations using specific heuristics [9,21] and metaheuristics [22]. The literature contains both algorithms that use a single optimization criterion, such as modularity [20], and multiple conflicting objectives [28]. In the latter case, a set of non-dominated solutions (Pareto Front, PF) is pursued [15]. The PF solutions correspond to different trade-offs between the goals, and hence to network partitions with different numbers of communities. This provides a great opportunity to analyze networks at various hierarchical levels. When analyzing the performance of a multi-objective optimization method, the distribution of the PF is considered [15]. A multi-objective algorithm has a better performance when the PF found is more evenly scattered among the objective space and reaches its extreme values.

Bio-inspired methods, and especially Genetic Algorithms (GAs), have proven successful to solve multi-objective optimization problems, and have already been applied to the CDP with relative success [22]. There are several Multi-Objective Genetic Algorithms (MOGAs) in the literature but undoubtedly the best known and used in community detection is NSGA-II [10].

Cellular GAs (cGAs) have received little attention for the CDP despite showing promising results in several problems [2]. The population in cGAs is structured in many small overlapping neighborhoods that produces a slow spread of high-quality solutions along the grid. Although cGAs were originally conceived for parallel implementation in massively parallel devices [29], they have gained interest in non-parallel environments. Like other parallel evolutionary algorithms, cGAs can improve the quality of results obtained by traditional sequential algorithms due to their enhanced search engine [1]. Multi-Objective Cellular Genetic Algorithm (MoCell) [18,19] adapts cGA to multi-objective problems incorporating an external archive that is used for storing the non-dominated solutions found during the execution, which are also used as a part of the mating process.

Our goal is to understand if the smooth diffusion scheme of MoCell is able to provide any benefit over current NSGA-II approaches for the CDP. Up to our knowledge and the exhaustive literature review on [22], there are no previous efforts tackling community detection problems in any of its formulations with MoCell nor cGAs. For this reason, in this work, we address the classical static community detection problem using MoCell to asses the effectiveness of this algorithm on this type of problems. The experimental evaluation conducted on 21 synthetically generated and two real-world networks shows that MoCell can outperform the NSGA-II approach, especially when considering large networks.

The rest of the article is structured as follows. The next section presents the static community detection problem. Section 3 describes the MOCell algorithm and how it is instantiated for tackling the static CDP. Then, Sect. 4 presents the numerical results and the analysis of these results. Finally, in Sect. 5, we outline the conclusions of this work and suggest future research directions.

2   Static Community Detection Problem

Let N be a network modeled as a graph $G = (V, E)$, where V is a set of vertices, called nodes, and $E = \{\{u, v\} : u, v \in V\}$ is a set of links, called edges, that connect two elements of V. The goal of the CDP is grouping the vertices of G in such a way that vertices in the same group have more edges among them than with the rest of vertices of G. In this work, we will focus on *non-overlapping* communities, so each vertex or node in G can only belong to a single group.

This article studies the CDP as a multi-objective optimization problem. The partition of the network is evaluated using two competing objectives. On one hand, the number of connections between members of the same community needs to be maximized. On the other hand, the number of connections between different communities needs to be minimized. Thus, in this formulation, there is not a single best solution, instead a Pareto front $F = \{C_1..C_m\}$ is found, i.e., a set of m non-dominated solutions, where each $C_i = \{S_1..S_n\}$ corresponds to a different partition of the network into n groups or communities S_i.

In order to select the two objective functions, the study presented in [28] was followed. The study shows that using two negatively correlated fitness functions as objectives gives the best results for the CDP. Following the guidelines of that work, the Community Score (CS) [26] and Average-ODF (ODF) [12] were selected as the objective functions. Next both functions are presented.

- **Community Score**: is the function selected as the intra-community objective. It measures the density of connections in relation to the number of nodes inside a community. The higher the CS of a partition, the higher the number of connections between members of the same community. Thus, this objective has to be maximized. Equation 1 shows how CS is calculated.
- **Average-ODF**: is the function selected as inter-community objective. It measures the average fraction nodes' edges pointing outside their communities. The lower the ODF, the lower the number of connections between communities. Therefore, this objective has to be minimized. Equation 2 shows how ODF is calculated, where $d(u) = |\{\{u, v\} \in E\}|$ is the degree of a node, i.e., the number of edges incident with the node u.

$$CS(C_i) = \sum_{S_i \in C_i} (\frac{2 * |\{\{u, v\}|u, v \in S_i\}|}{|S_i|})^2 \tag{1}$$

$$ODF(C_i) = \sum_{S_i \in C_i} (\frac{1}{|S_i|} \sum_{u \in S_i} (\frac{|\{\{u, v\}|v \notin S_i\}|}{d(u)})) \tag{2}$$

3   Multi-objective Cellular Genetic Algorithm

In cGA, the population is structured in many small overlapped neighborhoods. Each individual is placed in a cell on a toroidal n-dimensional grid and belongs to several neighborhoods. The interaction between individuals is limited since selection for reproduction is local to each neighborhood. Therefore, mating is restricted to each of the neighborhoods. The effect of finding high-quality solutions gradually spreads to other neighborhoods along the grid due to the use of a diffusion model that is a consequence of the neighborhoods overlapping [2].

Two classical and popular neighborhood structures are L5 and C9 [25]. L5 has five cells, including the central cell and the adjacent ones located at N, E, W and S. C9 includes the cells of L5 and the four adjacent ones along the diagonals (NW, NE, SW and SE). The cGA population on a 2D grid is shown in Fig. 1.

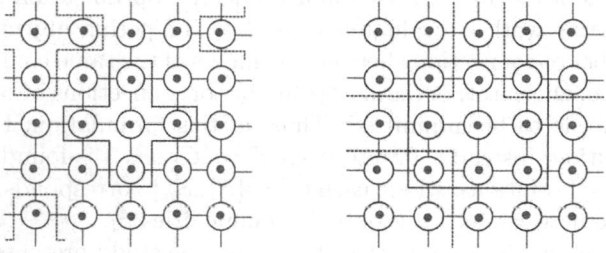

Fig. 1. cGA population and neighborhood on a 2D grid. Left: L5. Right: C9.

The Multi-Objective Cellular Genetic Algorithm (MoCell) is a cGA specially tailored for multi-objective optimization. It uses an external archive for storing the set of non-dominated solutions found by the algorithm, as in SPEA2 [30]. The number of solutions of the archive is bounded and it uses the crowding distance of NSGA-II to decide which solution remove when the archive is full.

The asynchronous MoCell is presented in Algorithm 1 [19]. The selection mechanism takes an individual from the neighborhood of the solution being processed and one from the external archive. Then, the crossover and mutation operators are applied to generate a new offspring. The current solution is replaced by the new solution only if it is dominated by the offspring. In such case, the new solution is inserted into the external archive. In case the current solution and the new offspring are non-dominated, the worst solution in the neighborhood is replaced by the new one. Also in this case, the new individual is inserted into the archive.

Since our approach is based on [24], it uses the "locus-based adjacency" encoding, in which genes are associated to nodes of the network. Thus, the length of the solutions is equal to the number of nodes. The allele values also correspond to nodes of the graph. The encoding requires a decoding stage, in which nodes belonging to the same connected component are assigned to the same community, but it simplifies the design of evolutionary operators as it is compatible with classical crossover methods like two-point crossover. The population is initialized randomly selecting a node from the set of adjacent nodes for each gene of each individual. The evolutionary operators are the two-point crossover and a mutation operator specially tailored for the CDP. In the mutation operator, some genes of the individual are chosen randomly, and their value is selected randomly from the set of adjacent nodes of the gene. Finally, the fitness functions are the ODF and the inverse CS (the goal is transformed in a minimization).

```
1  paretoFront = emptySet()
2  popGrid = generateInitialPopulation()
3  evaluateFitness(popGrid)
4  generation = 0
5  insertInFront(paretoFront,popGrid)
6  while not stopCondition() do
7  │   for individual = 1 to popSize do
8  │   │   currentPosition = getPosition(individual)
9  │   │   nList = getNeighborhood(popGrid,currentPosition)
10 │   │   parent1 = selection(nList)
11 │   │   parent2 = selection(paretoFront)
12 │   │   offspring = crossover(cp,parent1,parent2)
13 │   │   offspring = mutation(mp,offspring)
14 │   │   evaluateFitness(offspring)
15 │   │   replacement(popGrid,currentPosition,offspring,nList)
16 │   │   if individual not dominates offspring then
17 │   │   │   insertInFront(paretoFront,offspring)
18 │   │   end
19 │   end
20 │   generation = generation + 1
21 end
```

Algorithm 1: Pseudocode of MoCell.

The MoCell algorithm for the CDP was implemented using *jMetalPy* framework [7] since this tool already has an implementation of MoCell. *jMetalPy* is a recent Python implementation of the well-known *jMetal* framework [11].

4   Experimental Results

This section describes the instances used for our experiments and the parameters setting of the algorithms. Then, the results obtained are presented and analyzed.

4.1   Community Detection Problem Instances

The instances used in this work are taken from [24] and are publicly available[1]. The repository contains 66 synthetic time-evolving networks with ten snapshots

---

[1] https://www.kaggle.com/apanizolledot/dancer-synthetic-dynamic-networks.

each that has been generated with the DANCER benchmark [6]. In this work, we are not interested in the dynamics of the networks as we are considering the static formulation of CDP. Hence, only the first snapshot of the time-evolving networks is used, which are generated following the guidelines proposed in [16].

While all the dataset is interesting for evaluating dynamic CDPs, this is not true for the static case. Some instances are meant for testing events that do not occur if time is not considered. In the static CDP, we are interested in ascertaining the behavior when the communities in a network become fuzzier. Hence, we have selected 21 instances with different levels of fuzziness divided into three groups. In order to measure the fuzziness of a network, we have calculated the Average-ODF of the ground truth of each network. The lower the value of ODF, the less fuzzy the community structure of the network, representing an easier scenario for the algorithms. Table 1 presents the instances grouped by their difficulty level, including the mean and standard deviation of the Average-ODF.

Table 1. Synthetic datasets selected for the evaluation.

Difficulty	Dataset	Avg-ODF$_{std.dev}$
Easy	nf_sp_03, pd_03, nf_mx_md_03, nf_pa_eq_02, nf_mx_eq_03, nf_ga_eq_01, nf_mx_mc_01	$0.851_{0.037}$
Medium	mx_02, nf_ga_eq_03, mx_03, sm_mm_02, nf_pa_md_01, nf_sm_mm_01, nf_mx_mc_02	$1.616_{0.039}$
Hard	nf_mg_03, nf_sm_eq_03, nf_mg_01, sm_eq_02, nf_sm_eq_01, mg_03, nf_mg_02	$2.345_{0.123}$

These networks are relatively small and have between 100 and 200 nodes; 208 and 633 edges; and 5 and 10 communities. Hence, we have also included two real-world networks from Travian [14], a popular real-time strategy online game. In these networks, nodes are players and edges represent two players that have successfully done a trade transaction (market) or communicated via messages, depending on the network. The alliance membership of the users is used as the ground truth. Table 2 presents the features of the real-world networks.

Table 2. Real-world datasets selected for the evaluation.

Network	Nodes	Edges	Communities	Avg-ODF
Messages	1789	4697	93	49.62
Market	1091	2478	107	91.13

4.2 Algorithms, Parameter and Experimental Settings

In our experimental evaluation, we have included two different MoCell algorithms, MoCell-L5 (using the L5 neighbourhood) and MoCell-C9 (with the C9 neighbourhood). In order to set an actual comparison basis, we also include the numerical results of the NSGA-II from our previous work [24]. Although these results proceed from an algorithm that was designed for a *dynamic* CDP (and that many of its features are not used in an static scenario), they have been included because both algorithms, share the same evolutionary operators (crossover and mutation), but they use different underlying search mechanisms.

The parameter settings of MoCell algorithms, which correspond to standard parametrization values, are included in Table 3. In order to provide a fair comparison between the results of the algorithms, the stopping criterion used for MoCell algorithms is to generate the same number of solutions than in NSGA-II (corresponding to 90,000 function evaluations).

Table 3. Parametrization of the MoCell algorithms.

Parameter	Value
Population size	100 individuals (10×10)
Archive size	100 individuals
Neighborhoods	L5 (4 Surrounding cells) or C9 (8 Surr. cells)
Selection of parents	Binary tournament and binary tournament
Crossover probability (cp)	0.9
Mutation probability (mp)	$\frac{1.0}{l}$ (where l is the length of the individual)

Since the algorithms are stochastic, statistical tests are used to assess the significance of the results. Thirty independent runs for each algorithm and each instance have been performed. Then, the normalized hypervolume (HV) is computed, which considers both the convergence and spread of the obtained non-dominated sets. Two different statistical procedures are used, one for analyzing the results across multiple instances and another one for each instance independently. All the statistical tests are performed with a confidence level of 95%.

In the former approach, the Friedman's test ranks the algorithms according to the normalized HV [27] in the 21 synthetic instances. This test is used to check if the differences in the metric are statistically significant among the algorithms for the whole set of instances. Since more than two algorithms are involved in the study, a pairwise comparison using the Holm's post-hoc procedure is performed.

For the latter approach, which is used for the two real-world instances, the following statistical procedure [27] determines if the distribution of the HV for each algorithm and each instance independently is statistically different. First a Kruskal-Wallis test is performed. A post-hoc testing phase consisting in a pairwise comparison of all the cases compared using the Holm's correction method on the Wilcoxon-Mann-Whitney test is also performed.

4.3 Experimental Analysis

The analysis begins with the comparison of the numerical performance of the algorithms on the synthetic instances. Table 4 shows the median and interquartile range of the HV values of the algorithms for each instance. The HV values are rounded to three decimal places in the table. A dark gray and a light gray background indicate the best and the second-best performing algorithm considering the unrounded number (these values are also used in the tests).

The results show that MoCell algorithms systematically outperform NSGA-II in all the instances considered. MoCell-C9 is the best performing algorithm, obtaining the higher HV in 15 out of 21 instances (5 out of 7 for each difficulty group). MoCell-L5 is also competitive and outperform both MoCell-C9 and NSGA-II in the 6 remaining instances (2 out of 7 for each difficulty group).

If the results for each group are globally analyzed, the mean HV of NSGA-II degrades more than a 10% between easy and medium instances (0.875 to 0.783), while the deterioration is slighter in MoCell (MoCell-L5 from 0.931 to 0.906 and MoCell-C9 from 0.930 to 0.908). Finally, the mean HV of NSGA-II is stable between medium and hard instances (0.783 to 0.769), and it improves a bit in MoCell (MoCell-L5 from 0.906 to 0.909 and MoCell-C9 from 0.908 to 0.910).

Since in each group, the number of instances in which MoCell-C9 outperforms MoCell-L5 is the same, the mean Friedman's ranking for the whole set of synthetic instances and for each group is exactly the same. For the same reason, p-values are exactly the same for each of the groups.

Table 5 presents the mean Friedman's ranking according to the HV values, while Table 6 presents the p-values adjusted with Holm's procedure. In Table 6, '$\lhd$' indicates that the HV value of the first algorithm is statistically better than the second algorithm, and '$\rhd$' states that the opposite is true. When no statistically significant differences are found, the '$-$' symbol is used. The tests show that both MoCell algorithms are significantly better than the NSGA-II, however there is not enough statistical evidence that MoCell-C9 outperforms MoCell-L5.

Next, we continue with the comparison of the numerical performance for the real-world instances. Table 7 shows the median and interquartile range of the HV values of the algorithms for the two instances. Table 8 shows the p-values adjusted with Holm's procedure.

The results on the real-world instances allow to confirm the trend showed in the synthetic networks. MoCell algorithms systematically outperform NSGA-II, obtaining larger HV values. It should be noted that in these large scenarios, the difference in the numerical performance is larger than in the small instances. Also, the results show that MoCell-C9 is the best performing algorithm, being superior to MoCell-L5 and NSGA-II for both instances. The tests show that

Table 4. Median and Interquartile Range of the HV quality indicator. Dark grey color is used to indicate the best result obtained.

Difficulty	Instance	NSGA-II	MoCell-L5	MoCell-C9
Easy	nf_sp_03	$0.910_{0.029}$	$0.951_{0.008}$	$0.954_{0.006}$
	pd_03	$0.724_{0.051}$	$0.889_{0.020}$	$0.875_{0.025}$
	nf_mx_md_03	$0.875_{0.026}$	$0.915_{0.011}$	$0.917_{0.013}$
	nf_pa_eq_02	$0.916_{0.020}$	$0.943_{0.008}$	$0.944_{0.009}$
	nf_mx_eq_03	$0.898_{0.030}$	$0.935_{0.009}$	$0.933_{0.010}$
	nf_ga_eq_01	$0.893_{0.062}$	$0.945_{0.015}$	$0.949_{0.015}$
	nf_mx_mc_01	$0.906_{0.021}$	$0.937_{0.011}$	$0.939_{0.009}$
Medium	mx_02	$0.666_{0.051}$	$0.865_{0.021}$	$0.867_{0.016}$
	nf_ga_eq_03	$0.805_{0.039}$	$0.915_{0.016}$	$0.912_{0.022}$
	mx_03	$0.745_{0.047}$	$0.883_{0.025}$	$0.889_{0.022}$
	sm_mm_02	$0.858_{0.050}$	$0.943_{0.007}$	$0.944_{0.014}$
	nf_pa_md_01	$0.860_{0.044}$	$0.936_{0.012}$	$0.934_{0.015}$
	nf_sm_mm_01	$0.748_{0.062}$	$0.921_{0.013}$	$0.926_{0.012}$
	nf_mx_mc_02	$0.798_{0.050}$	$0.877_{0.017}$	$0.882_{0.018}$
Hard	nf_mg_03	$0.762_{0.032}$	$0.920_{0.021}$	$0.920_{0.021}$
	nf_sm_eq_03	$0.784_{0.037}$	$0.923_{0.019}$	$0.921_{0.014}$
	nf_mg_01	$0.741_{0.049}$	$0.902_{0.019}$	$0.902_{0.018}$
	sm_eq_02	$0.836_{0.036}$	$0.922_{0.014}$	$0.922_{0.015}$
	nf_sm_eq_01	$0.739_{0.057}$	$0.942_{0.014}$	$0.939_{0.014}$
	mg_03	$0.815_{0.050}$	$0.889_{0.018}$	$0.891_{0.015}$
	nf_mg_02	$0.702_{0.059}$	$0.868_{0.028}$	$0.873_{0.026}$

Table 5. Average Rankings of Friedman test of the algorithms.

NSGA-II	MoCell-L5	MoCell-C9
3.000	1.714	1.286

Table 6. Statistical assessment of the algorithms.

	Easy-Medium-Hard	Overall
NSGA-II vs. MoCell-L5	3.23e–02 ▷	6.20e–05 ▷
NSGA-II vs. MoCell-C9	4.02e–03 ▷	<1.00e–16 ▷
MoCell-L5 vs. MoCell-C9	0.42e01 −	0.16e01 −

Table 7. Median and Interquartile Range of the HV quality indicator.

Instance	NSGA-II	MoCell-L5	MoCell-C9
messages	$0.379_{0.078}$	$0.831_{0.031}$	$0.850_{0.045}$
market	$0.581_{0.039}$	$0.765_{0.029}$	$0.782_{0.023}$

Table 8. Statistical assessment of the algorithms.

	messages	market
NSGA-II vs. MoCell-L5	9.00e–09 ▷	9.00e–09 ▷
NSGA-II vs. MoCell-C9	9.00e–09 ▷	9.00e–09 ▷
MoCell-L5 vs. MoCell-C9	5.40e–02 –	4.60e–03 ▷

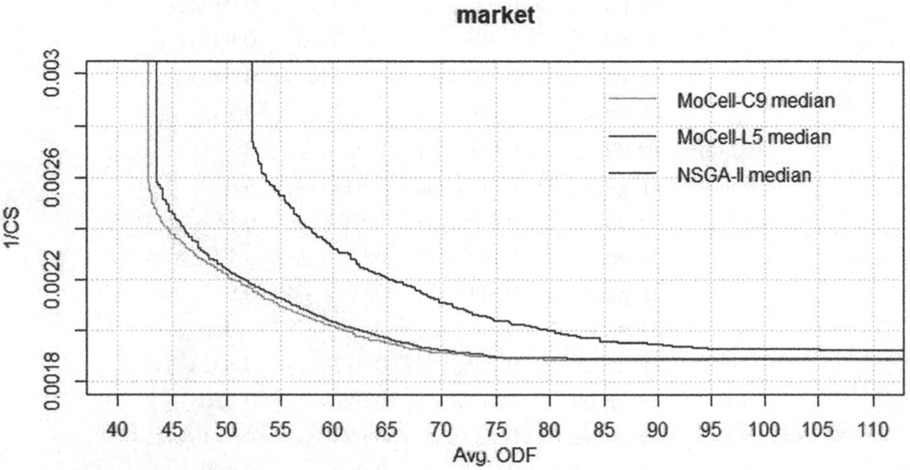

Fig. 2. 50%-attainment surface for market instance.

MoCell algorithms are significantly better than the NSGA-II and that MoCell-C9 outperforms MoCell-L5 with statistical significance in the market instance.

Since HV does not provide information about the shape of the front, we analyze the Empirical Attainment Function (EAF) [17]. EAF graphically shows a boundary between points that are dominated by at least a fraction of the runs from those that are not. In particular, the median EAF represents the curve where the set of attained points is 50% of the total [17] (equivalent to the median). Figure 2 presents the 50%-attainment surface for market instance (for messages is similar). It can be seen how the average surface covered by the MoCell algorithms dominates that of NSGA-II along the entire curve.

Finally, Fig. 3 shows the differences between EAF [17] of NSGA-II and MoCell-C9 for market instance (for messages is similar). The right side plot shows the differences in favor of MoCell-C9, giving insight of the objective space regions in which MoCell-C9 is able to obtain better solutions than NSGA-II.

Within the context of this experimental evaluation, it is noticeable that MoCell algorithms outperform NSGA-II, specially in the largest scenarios considered. In particular, MoCell-C9 has shown promising results, being the best performing algorithm. Also, the results obtained show that the use of a

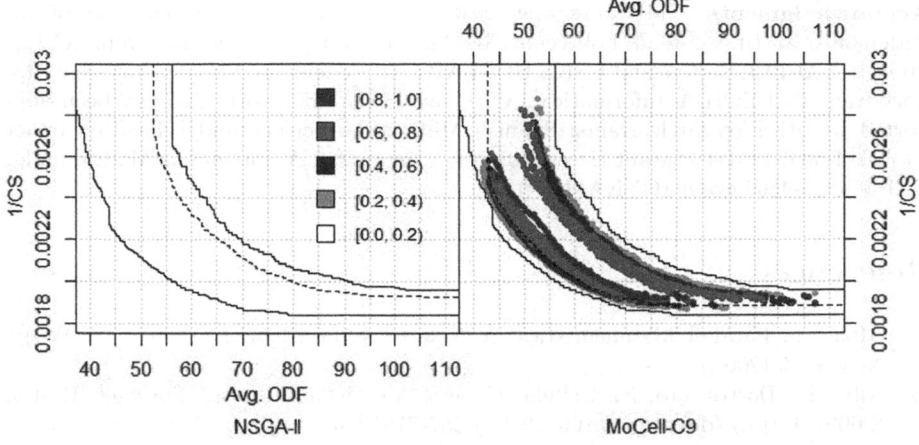

Fig. 3. Difference between EAF of MoCell-C9 and NSGA-II on `market` instance.

structured population, like in MoCell, can be advantageous for solving the static CDP.

5   Conclusions and Future Work

In this work, we have studied the MoCell algorithm for solving the static community detection problem. Up to our knowledge, it represents the first approach to this type of problems with MoCell or cellular GAs. The experimental evaluation conducted over 21 synthetically generated and two real-world networks, with up to 1789 nodes and 4692 edges, showed that MoCell algorithms obtained promising results for the static CDP, being able to outperform the NSGA-II approach in all the instances considered. In particular, MoCell using the C9 neighborhood has been the best performing algorithm within the context of this experiments.

Three main areas that deserve further study were identified. First, we aim to extend our experimental evaluation of MoCell studying the scalability of the numerical performance of the algorithm. For this, new set of synthetic networks should be addressed, especially considering large and complex networks. Based on the high-quality results obtained by MoCell, a second line of interest consists in tackling other formulations of the CDP with MoCell algorithms. In particular, it is quite interesting to evaluate if the diffusion scheme of MoCell could provide advantages for more difficult variants, like dynamic or overlapping community detection. Finally, and since the approach proposed for solving the CDP does not incorporate any knowledge of the problem in its operation, it is interesting to study the effect of incorporating a local search mechanism to the algorithm in order to provide better solutions and to speed up the search.

Acknowledgments. This work was written as part of a research stay of M. Pedemonte at Universidad Politécnica de Madrid (funded by grants from ANII - MOV_CA_2019_1_156657 and CSIC, UDELAR). M. Pedemonte also acknowledge support from PEDECIBA Informática, ANII, and SNI. This work has also been supported by other research grants: Spanish Ministry of Science and Education under TIN2014-56494-C4-4-P grant (DeepBio), and Comunidad Autónoma de Madrid under P2018/TCS-4566 grant (CYNAMON).

References

1. Alba, E.: Parallel Metaheuristics: A New Class of Algorithms, vol. 47. Wiley, New York (2005)
2. Alba, E., Dorronsoro, B.: Cellular Genetic Algorithms, vol. 42. Springer, Boston (2009). https://doi.org/10.1007/978-0-387-77610-1
3. Bello-Orgaz, G., Camacho, D.: Evolutionary clustering algorithm for community detection using graph-based information. In: 2014 IEEE Congress on Evolutionary Computation (CEC), pp. 930–937. IEEE (2014)
4. Bello-Orgaz, G., Jung, J.J., Camacho, D.: Social big data: recent achievements and new challenges. Inf. Fus. **28**, 45–59 (2016)
5. Bello-Orgaz, G., Menéndez, H.D., Camacho, D.: Adaptive k-means algorithm for overlapped graph clustering. Int. J. Neural Syst. **22**(05), 1250018 (2012)
6. Benyahia, O., Largeron, C., Jeudy, B., Zaïane, O.R.: DANCer: dynamic attributed network with community structure generator. In: Berendt, B., Bringmann, B., Fromont, É., Garriga, G., Miettinen, P., Tatti, N., Tresp, V. (eds.) ECML PKDD 2016. LNCS (LNAI), vol. 9853, pp. 41–44. Springer, Cham (2016). https://doi.org/10.1007/978-3-319-46131-1_9
7. Benítez-Hidalgo, A., Nebro, A.J., García-Nieto, J., Oregi, I., Del Ser, J.: jMetalPy: a Python framework for multi-objective optimization with metaheuristics. Swarm Evol. Comput. **51**, 100598 (2019)
8. Camacho, D., Panizo-LLedot, A., Bello-Orgaz, G., Gonzalez-Pardo, A., Cambria, E.: The four dimensions of social network analysis: an overview of research methods, applications, and software tools. arXiv preprint arXiv:2002.09485 (2020)
9. Cordasco, G., Gargano, L.: Community detection via semi-synchronous label propagation algorithms. In: 2010 IEEE International Workshop on: Business Applications of Social Network Analysis (BASNA), pp. 1–8. IEEE (2010)
10. Deb, K., Pratap, A., Agarwal, S., Meyarivan, T.: A fast and elitist multiobjective genetic algorithm: NSGA-II. IEEE Trans. Evol. Comp. **6**(2), 182–197 (2002)
11. Durillo, J.J., Nebro, A.J.: jMetal: a Java framework for multi-objective optimization. Adv. Eng. Softw. **42**, 760–771 (2011)
12. Flake, G.W., Lawrence, S., Giles, C.L.: Efficient identification of web communities. In: Proceedings of the Sixth ACM SIGKDD International Conference on Knowledge Discovery and Data Mining, pp. 150–160 (2000)
13. Fortunato, S.: Community detection in graphs. Phys. Rep. **486**(3–5), 75–174 (2010)
14. Hajibagheri, A., Sukthankar, G., Lakkaraju, K., Alvari, H., Wigand, R.T., Agarwal, N.: Using massively multiplayer online game data to analyze the dynamics of social interactions. Soc. Interact. Virtual Worlds Interdisc. Perspect. (2018)
15. Jiang, S., Ong, Y.S., Zhang, J., Feng, L.: Consistencies and contradictions of performance metrics in multiobjective optimization. IEEE Trans. Cybern. **44**(12), 2391–2404 (2014)

16. Largeron, C., Mougel, P.N., Rabbany, R., Zaïane, O.R.: Generating attributed networks with communities. PloS one **10**(4) e0122777 (2015)
17. López-Ibáñez, M., Paquete, L., Stützle, T.: Exploratory analysis of stochastic local search algorithms in biobjective optimization. In: Bartz-Beielstein, T., Chiarandini, M., Paquete, L., Preuss, M. (eds.) Experimental Methods for the Analysis of Optimization Algorithms, pp. 209–222. Springer, Heidelberg (2010). https://doi.org/10.1007/978-3-642-02538-9_9
18. Nebro, A., Durillo, J., Luna, F., Dorronsoro, B., Alba, E.: MOCell: aellular genetic algorithm for multiobjective optimization. Int. J. Intell. Syst. **24**(7), 726–746 (2009)
19. Nebro, A.J., Durillo, J.J., Luna, F., Dorronsoro, B., Alba, E.: A cellular genetic algorithm for multiobjective optimization. NICSO **2006**, 25–36 (2006)
20. Newman, M.E.: Modularity and community structure in networks. Proc. Natl. Acad. Sci. **103**(23), 8577–8582 (2006)
21. Newman, M.E., Girvan, M.: Finding and evaluating community structure in networks. Phys. Rev. E **69**(2), 026113 (2004)
22. Osaba, E., Del Ser, J., Camacho, D., Bilbao, M.N., Yang, X.S.: Community detection in networks using bio-inspired optimization: latest developments, new results and perspectives with a selection of recent meta-heuristics. Appl. Soft Comput. **87**, 106010 (2020)
23. Osaba, E., Ser, J.D., Panizo, A., Camacho, D., Galvez, A., Iglesias, A.: Combining bio-inspired meta-heuristics and novelty search for community detection over evolving graph streams. In: Proceedings of the Genetic and Evolutionary Computation Conference Companion, pp. 1329–1335 (2019)
24. Panizo-LLedot, A., Bello-Orgaz, G., Camacho, D.: A Multi-Objective Genetic Algorithm for detecting dynamic communities using a local search driven immigrant's scheme. Future Gener. Comput. Syst. **110**, 960–975 (2019)
25. Pedemonte, M., Cancela, H.: A cellular ant colony optimisation for the generalised Steiner problem. Int. J. Innov. Comput. Appl. **2**(3), 188–201 (2010)
26. Pizzuti, C.: GA-Net: a genetic algorithm for community detection in social networks. In: Rudolph, G., Jansen, T., Beume, N., Lucas, S., Poloni, C. (eds.) PPSN 2008. LNCS, vol. 5199, pp. 1081–1090. Springer, Heidelberg (2008). https://doi.org/10.1007/978-3-540-87700-4_107
27. Sheskin, D.J.: Handbook of Parametric and Nonparametric Statistical Procedures. Chapman and Hall/CRC, fifth edition edn. (2011)
28. Shi, C., Yu, P.S., Yan, Z., Huang, Y., Wang, B.: Comparison and selection of objective functions in multiobjective community detection. Comput. Intell. **30**(3), 562–582 (2014)
29. Soca, N., Blengio, J.L., Pedemonte, M., Ezzatti, P.: PUGACE, a cellular evolutionary algorithm framework on GPUs. In: IEEE Congress on Evolutionary Computation, CEC 2010, pp. 1–8 (2010)
30. Zitzler, E., Laumanns, M., Thiele, L.: SPEA2: improving the strength pareto evolutionary algorithm. Technical report 103, TIK, ETH, Switzerland (2001)

Special Session on Machine Learning in Automatic Control

Under-Actuation Modelling in Robotic Hands via Neural Networks for Sign Language Representation with End-User Validation

Jennifer J. Gago$^{(\boxtimes)}$ ⓘ, Bartek Łukawski ⓘ, Juan G. Victores ⓘ, and Carlos Balaguer ⓘ

RoboticsLab, University Carlos III of Madrid, 28911 Leganés, Spain
jgago@ing.uc3m.es

Abstract. This paper presents a study on under-actuation modelling applied to robotic hands aimed at sign language representation. Prior studies using a simulated TEO humanoid robot for representing sign language have shown positive comprehension and satisfaction responses among the deaf and hearing impaired community. The under-actuated mechanics of the robotic fingers were not contemplated in the simulated model, thus the correspondence problem arises as the previous joint space positions cannot be directly sent to the physical system. In addition to the 3:1 and 2:1 ratio of the under-actuation of the finger mechanisms, tendons and springs involve stiffness and elasticity that are difficult or unfeasible to model, and justify the need for a data-driven approach. Three motor command generators using three different neural network models are analysed and evaluated. Two of the generators are trained in a supervised fashion, and the third involves variational self-supervision and a transformation upon the latent space. The simulated joint space positions are translated into motor commands for the physical embodied robot to represent a sign language dactylology, which is in turn evaluated by deaf and hearing impaired end-users.

Keywords: Humanoid robot · Robotic hands · Sign language · Under-actuation

1 Introduction

TEO is an assistive household-companion full-sized humanoid robot which can develop tasks such as ironing, folding, and painting, among others [1]. Designed to work for and with humans, human-robot interaction requires the use of oral and visual-gestural languages. For the latter, the use of hands is fundamental. The ultimate purpose of this research is to develop a fully functional signing humanoid robot capable of interacting with users with all kind of abilities, including deaf and hearing impaired end-users. Hence, it is crucial to find a way to reproduce sign language as effectively as possible.

Several studies using robots and robotic limbs have attempted to enable communication via sign language, as it is the case of the assistant android developed by Toshiba in 2014 [2], and the robots Robovie R3 and NAO trained for educational purposes by the Istanbul Technical University [3], which were able to reproduce some basic word signs; or the Aslan robotic arm [4], which is capable of representing dactylology using

© Springer Nature Switzerland AG 2020
C. Analide et al. (Eds.): IDEAL 2020, LNCS 12490, pp. 239–251, 2020.
https://doi.org/10.1007/978-3-030-62365-4_23

a fully-actuated 16 servomotor mechanism which cannot be easily attached to a full-sized humanoid robot. These attempts have shown that reproducing sign language is a complex task for a robot, especially due to the need to reproduce hand movements as similar as possible to a human hand.

The humanoid robot TEO has also been used in simulation with communicative purposes. Previous studies showed positive outcomes, where dactylology and word vocabulary tests resulted in 82% and 83% correct answer rate respectively [5], neglecting outliers. Simulation allowed researchers to control the hand directly in the joint position space (considering only joint limit constraints, while obviating under-actuation as well as other mechanical phenomena) to generate a dactylology and vocabulary dataset where the position of each joint is defined. In order to control the real hand, a model that maps from the joint position space to the motor space is required. This model will allow the hand to perform the configurations contained in the original dataset.

Hands play a fundamental role in sign language [6]. Under-actuated hands are widespread in multiple robotic dexterous manipulation applications. Many popular hands use this underlying mechanism, such as the Robonaut Hand [7], with under-actuation 20:12; or the Cyberhand [8], with a steeper under-actuation 16:6. TEO possesses two 15:6 under-actuated robotic hands (Dextra TPMG90-2). The under-actuation is 3:1 for all of the fingers except for the thumb, which is 2:1. The hand does not move exactly as a human hand, so finding an equivalence between joint-by-joint movement and the actual kinematics of the hand is fundamental to achieve a human-like behaviour.

Under-actuation is a classical field in robotics which itself has been covered from different approaches [9]. A field intimately bound up with under-actuation is the study of mapping motor commands to world poses in mobile robotics [10], or to the end-effector pose in articulated mechanisms [11]. Another field affected by the issues of modelling under-actuation is balancing, or mechanical stability, as shown in the control studies developed for biped robots in simulation [12,13], where dynamic modelling and

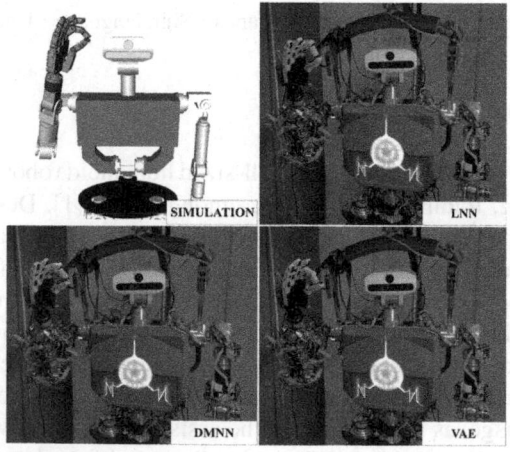

Fig. 1. TEO humanoid robot signing letter "O" in Spanish Sign Language. Comparison between simulation and real robot representations generated by 3 neural networks.

Lyapunov stability theory are used, respectively; or with real robots like MARLO [14], which uses a control design method based on virtual constraints.

The task of solving under-actuation issues in robotic hands has been usually linked to hand design for grasping, and trajectory tracking related to path-planning. Some examples of mechanism design are the study on determining optimal stiffness for grasping [15], or the synthesis of robot hand models and grasp strategy evaluation based on neural networks [16]. Trajectory tracking literature is similar to that of articulated robots, but typically includes tracking of intermediate joints and links [17]. In all these cases, the solution to the under-actuation issue can be derived from the initial design parameters and the final objective is defined beforehand. Nevertheless, there is an uncertainty regarding how to effectively define the final goal in relation to efficient communication. The final goal could consist in maximisation of various parameters, such as the precision of the end-effector versus the under-actuated joints, which are mutually exclusive. This uncertainty, linked to the intrinsic complexity of modelling the mechanical constraints such as tendons and springs that involve stiffness and elasticity, leads this study to opt for a data-driven approach and end-user validation as a way to reach effective communication while minimising the amount of hand-crafted approximations.

From these works, the main techniques used for under-actuation modelling consist in exact dynamic modelling, control theory and, particularly within the last few years, neural networks. Based on the covered state of the art, the preferable method to achieve this goals without leading into exact modelling is the neural networks alternative. Consequently, the contribution of this work can be summed up in the three following points:

- Use of neural networks to model the correspondence between a fully-actuated and an under-actuated hand.
- Research on the correlation between the latent space of a variational autoencoder and the motor space.
- Study on the reliability of the research towards effective communication via end-user validation through a sign language comprehension questionnaire.

2 Proposed Neural Network Models

Three neural network models have been used to map input features (joint positions) to motor commands, to which we refer to as "generators". These alternatives are proposed in order to detect which method generates a more recognisable pose of the sign. The first two generators are implemented as regression models and are trained in a supervised fashion with joint space positions as inputs and motor commands as outputs. The third generator is implemented as a variational self-supervised model, where the correlation between the latent space and corresponding motor command is studied.

2.1 Linear Neural Network (LNN)

The general working setup of a neural network is comprised of (one or more) layers of neurons with some activation such that $a^{[l]} = g^{[l]}(z) = g^{[l]}(W \cdot x + b)$, where $g^{[l]}(\cdot)$ denotes the activation function of layer l, W is the weight matrix, x represents the vector of input features, b is referred to as the bias, and $a^{[l]}$ is the output of the layer.

Updating the weights, while training the neural network model, is accomplished by an optimiser algorithm, such as stochastic gradient descent (SGD), RMSprop, or Adam. The optimisation process is also known as *loss minimisation* since an objective *loss function* is called in order to navigate the space of weights and attain the least *loss*. Typical loss functions include mean squared error (MSE), binary cross-entropy, and categorical cross-entropy. The mean squared error, typically used in regression problems, is expressed by Eq. 1:

$$MSE = \frac{1}{n_y} \sum_{i=1}^{n_y} (y_i - \hat{y}_i)^2 \tag{1}$$

where y represents the vector of n_y labelled values and $\hat{y}$ is the vector of n_y predictions. In our Linear Neural Network (LNN) model, we define the following one-to-one linear mapping for the activation function at layer $l = 1$, thus $g^{[1]}(z) = z$. In this simple neural network model, no hidden layers are involved. The $n_x = 3$ input features (joint positions) are directly mapped to $n_y = 1$ output (motor command). The model corresponds to a linear regression and, as such, to a *hyperplane* given by $\hat{y} = a = W \cdot x + b = g^{[1]}(W \cdot x + b)$, where $\hat{y}$ is the predicted output. This correlates n_x relative angles between consecutive phalanges to n_y motor commands, as detailed in following sections. A representation of these inputs and the output layer, the latter comprised by one neuron, was generated[1] and depicted in Fig. 2.

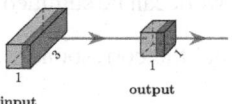

input   output

Fig. 2. Linear Neural Network (LNN).

This network is trained in a supervised fashion as we want to force the inputs (joint positions) to a known space (motor commands). Prior to training the model, the input features are normalised in the range $[0, 1]$. In the same manner, given the regression setup, output features are denormalised to fit the original range. We adopted the Min-Max normalisation given by:

$$X_{norm} = \frac{X - X_{min}}{X_{max} - X_{min}} \tag{2}$$

$$X_{scaled} = X_{norm} \cdot (max - min) + min$$

where X_{min} and X_{max} are the lowest and the highest input, respectively, according to which X is conveniently rescaled to X_{norm} and later shifted to fit the $[min, max]$ feature range, typically $[0, 1]$.

---

[1] Using the tool https://github.com/HarisIqbal88/PlotNeuralNet.

2.2 Dense Multilayer Neural Network (DMNN)

Non-linearities may be addressed by increasing the number of hidden layers in the neural network. In this vein, the DMNN adds three intermediate layers to the previous linear model, comprised by 24, 12 and 6 neurons, respectively, as shown in Fig. 3.

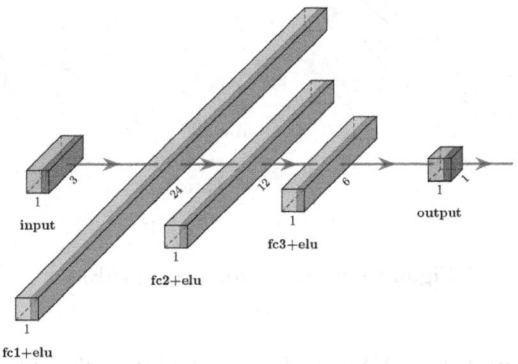

Fig. 3. Dense Multilayer Neural Network (DMNN).

These intermediate layers, as well as the output layer, are *fully connected* (the term *dense layer* is used interchangeably). $a^{[l]} = g^{[l]}(z) = g^{[l]}(W \cdot x + b)$ applies to each neuron of this model. Accounting for the non-linearities, a different activation function was chosen for the neurons of the intermediate layers of this model. Typical activation functions are the sigmoid function, the Rectified Linear Units (ReLU), and the Exponential Linear Units (ELU). We chose the latter (ELU) since it allows negative values to push mean unit activations closer to zero as in batch normalisation, but with lower computational complexity [18].

2.3 Variational Autoencoder (VAE)

The naïve autoencoder neural network approach relies on a composition of two parts: the encoder and the decoder. The full autoencoder model is trained to reconstruct the input through one or more hidden layers, trying to achieve the least reconstruction error. The encoder is trained to learn a compressed representation of the input, whereas the decoder will reconstruct it back to resemble the original state as closely as possible; therefore, the outputs are the same as the inputs. These models are said to be data-specific, lossy and learned automatically from examples. Common applications include denoising and data compression. Typically, the decoder part is discarded, and the encoder model is used to generate a compact, low dimensional representation of the input that correlates parameters of the *latent space* to the output features of the network. Since this model aims to reconstruct its input, instead of predicting target values given some input features, it is said to be a *self-supervised* learning model. In our model, a variational autoencoder (VAE), a symmetric neuron layout was adopted, that is, the encoder is modelled by a dense multilayer network as described in the previous section, and the same fully connected structure is arranged in the opposite order to build the decoder. This configuration is shown in Fig. 4.

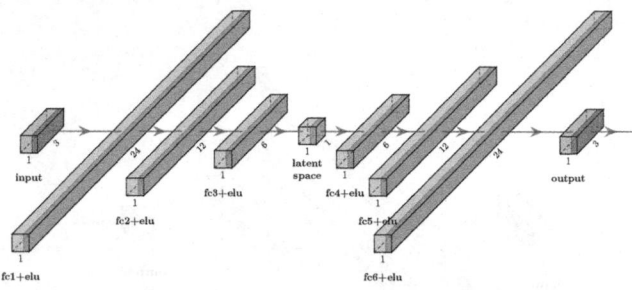

Fig. 4. Variational Autoencoder (VAE).

The general theory behind VAE models evolved from the autoencoder by adding constraints on the parameters being learned [19]. These models aim to learn a lower dimensional latent Gaussian representation of the input data. Since new data is generated based on previously seen data, it is said that the VAE is a *generative* network. In addition to being trained so that the decoded samples match the input features, the Kullback-Leibler (KL) divergence between the learned latent distribution and the prior distribution is regarded as a loss function to train this model with, acting as a regularisation term. In this regard, a VAE model is trained by optimising the sum of the reconstruction loss ($\mathscr{L}_{rec}$, also referred to as *expected log-likelihood*) and KL divergence loss ($\mathscr{L}_{KL}$) with regard to variational parameters ϕ and generative parameters θ, with the Stochastic Gradient Variational Bayes (SGVB) algorithm. $\mathscr{L}_{rec}$ is used to update parameters of both the encoder and the decoder, whereas $\mathscr{L}_{KL}$ is responsible for updating the encoder parameters, only. The resulting (variational) lower bound on the marginal likelihood ($\mathscr{L}_{VAE}$) is given by:

$$\mathscr{L}_{rec} = -\mathbb{E}_{q_\phi(\mathbf{z}|\mathbf{x})}\left[\log p_\theta(\mathbf{x}|\mathbf{z})\right]$$
$$\mathscr{L}_{KL} = D_{KL}(q_\phi(\mathbf{z}|\mathbf{x})\|p_\theta(\mathbf{z})) \tag{3}$$
$$\mathscr{L}_{VAE} = \mathscr{L}_{rec} + \mathscr{L}_{KL}$$

where z is interpreted as the latent representation or *code*, over which a distribution (e.g. Gaussian) is produced by the probabilistic distribution of the encoder $q_\phi(z|x)$ given the datapoint x, and given which other distribution is produced by the probabilistic decoder $p_\theta(x|z)$ over the corresponding values of x. Lastly, D_{KL} stands for the KL divergence, calculated with a closed-form expression since both the prior z and the approximate posterior $q_\phi(z|x)$ are Gaussian:

$$D_{KL}(q_\phi(z|x)\|p_\theta(z)) = -\frac{1}{2}\sum_{k=1}^{K}\left[1 + log\sigma_k^2 - \mu_k^2 - \sigma_k^2\right] \tag{4}$$

where μ_k and σ_k are the k-th components of vectors $\mu_\phi(x)$ and $\sigma_\phi(x)$, respectively, originating from the variational distribution $q_\phi(z_n|x_n)$ with local variational parameters $\phi_n = \{\mu_n, \sigma_n\}$ representing the mean and standard deviation, which leads to:

$$q_{\phi_n}(z_n|x_n) \sim \mathcal{N}(z_n|x_n, diag(\sigma_n^2)) \tag{5}$$

An inference network is introduced to reduce the number of local variational parameters that would have to be optimised otherwise. Two global variational parameters for each datapoint, $\mu_\phi(x_n)$ and $\sigma_\phi(x_n)$, are learned in order to approximate the local parameters μ_n and σ_n, respectively. These global parameters constitute the weights of the inference network. Equation 5 derives into $q_\phi(z_n|x_n) \sim \mathcal{N}(z_n|\mu_\phi(x_n), diag(\sigma_\phi^2(x_n)))$. In our VAE model, the output layer is not used for prediction of the required motor commands as in the previous models. The joint position data is instead compressed into a latent space at the network innermost layer $l = 4$ of dimension $n_l = 1$, corresponding to that of the motor commands. We then study the correlation between the emerged latent space features (ℓ) and those of the motor commands $\hat{y} = f(\ell) = f(a^{[4]})$.

3 Laboratory Experiments

The first experiment consists in training the proposed neural network models and comparing their precision in terms of the error defined by their loss function. The training data is obtained by tracking and analysing the under-actuated movement of the fingers.

3.1 Experimental Setup

The tracked fingers are the pinky (most visible) and the thumb (required as it is different from the other four). Computer vision tracks the position of the strategically placed coloured fiducials. The experimental setup for the pinky finger is shown in Fig. 5.

Fig. 5. Experimental setup. Bottom left: side view of the hand showing each moving element of the pinky finger (links) plus five red markers (joints). (Color figure online)

3.2   Experimental Results

Figure 6 depicts the resulting scatter plots of angle measurements on the pinky finger and the thumb, respectively, after 10 demonstrations. Each demonstration consists in an initially extended finger, made to gradually retract until a "closed" pose is reached. Equal increments of $0.1°$ are applied, totalling in 200 measured poses in the case of the pinky finger (range from 0 to $20°$) and 100 in regard to the thumb (range from 0 to $10°$). Angle values between consecutive phalanges are estimated by obtaining the pixel coordinates of the markers.

Tendons in Dextra TMPG90-2 follow a meandering path through the phalanges, which explains the sequential movement of proximal, middle and distal phalange (which is the last joint to rotate).

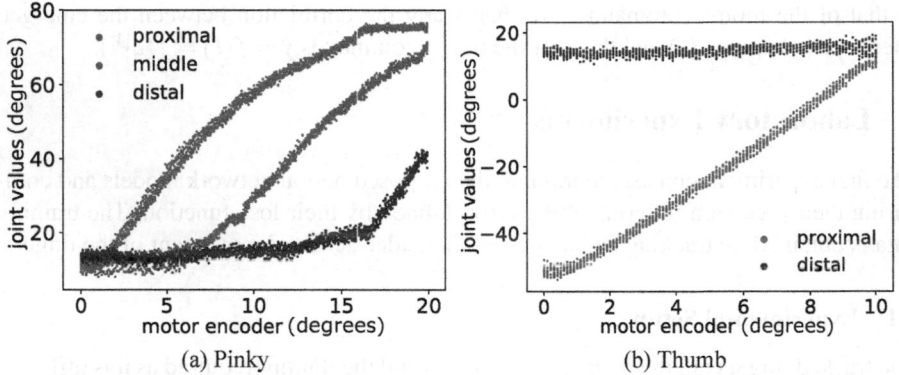

(a) Pinky   (b) Thumb

Fig. 6. Joint position measurements, motor values versus phalange angles.

The LNN and DMNN generators are trained using the observed joint values as input features x, and the corresponding sent motor commands as the output feature y. The missing joint of the thumb model is fed with the servo-controlled abduction joint position.

The VAE generator is trained using solely the observed joint positions x on both of its ends, in a self-supervised fashion, and unaware of motor commands. Once the VAE is trained, the encoder part of the VAE computes the latent space feature $\ell^{(i)}$ for each training sample $x^{(i)}$. Figure 7 depicts each $\ell^{(i)}$ versus the corresponding unused $y^{(i)}$ motor command, which can therefore be understood as the correlation between the latent space ℓ and the motor space y.

The insight from the plot leads us to model the correlation function f of the end of Sect. 2.3 as a simple linear transformation: $\hat{y} = m \cdot \ell + b = f(\ell)$.

The physical meaning of this relation is that the VAE is capable of capturing and modelling the non-linearities of the mechanism without knowledge of the motor commands, and only a scaling factor and offset are required to match the actual motor command. As the process involves stochastic steps, repeatability of the transformation may be achieved from training the model only once, or by fixing the initial random seeds.

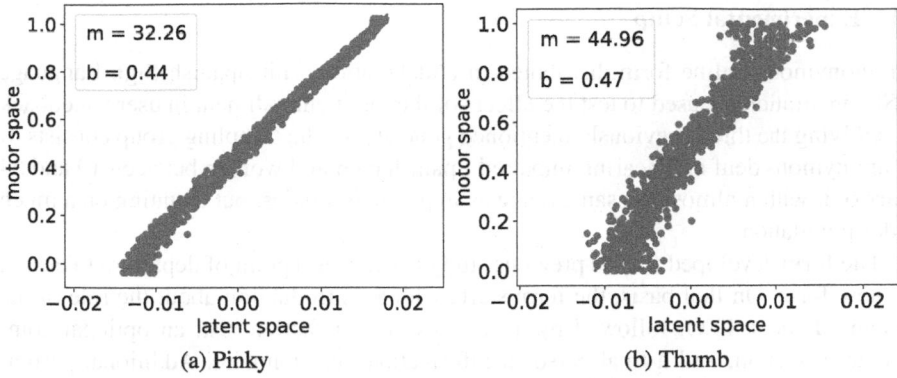

Fig. 7. Latent versus motor space for VAE generator.

The neural network models are implemented and trained using the Keras infrastructure [20]. Each generator is trained for 50 epochs, with fixed initial random seeds, unshuffled data, Glorot uniform kernel initialisers, and a 20% cross-validation split. The $\hat{y}$ predictions of these generators are applied to the original dactylology dataset, used as x, for the experiments of the next section. Upper and lower limits are established for finger joint positions due to mechanical constraints and to avoid breakages and overstrains. Small, but insightful variations, are encountered among letters represented by different generators, as shown in Fig. 1.

4 End-User Experiments

The second experiment consists in end-user validation via questionnaire. User validation is a fundamental step in order to test the reliability of the results obtained in the laboratory experiments.

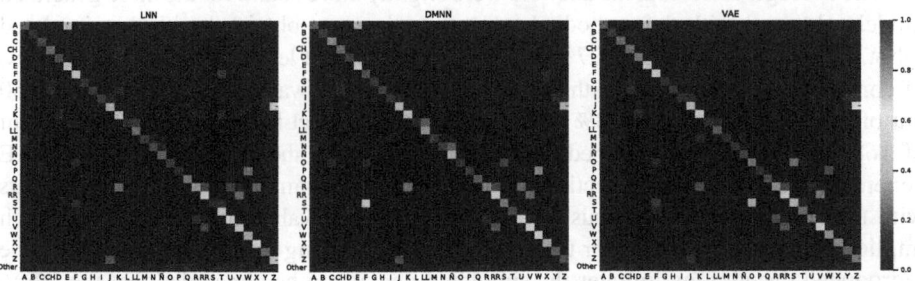

Fig. 8. Confusion matrix heatmaps for LNN, DMNN and VAE (from left to right). The columns are named according to the expected result, and the rows are named according to the experimental results.

4.1 Experimental Setup

An anonymous online form distributed in collaboration with Spanish Sign Language (LSE) institutions is used to test the effects of the robot embodiment in user's feedback by applying the three previously mentioned generators. The sampling group consists of 56 anonymous deaf and hearing impaired Spanish men and women between 19 and 54 years old; within almost the same range as in previous studies, but counting on a much wider population.

The form developed for the previous study is taken as a point of departure to design the new form. On that basis, the form starts with an introduction about the role of the humanoid robot TEO, followed by a LSE test, a satisfaction test, an optional commentaries section, and a final personal information questionnaire. Additional participant characteristics collected within the questionnaire include: Age, which is a relevant factor regarding generational attachment to technology. Educational level, which may influence user's predisposition towards robotics. And gender, in case it could be useful to track some significant tendencies.

The dactylology test is designed to test the performance of the under-actuation modelling. This decision is taken considering the intrinsic importance of finger joint positions in dactylology, where slight variations may result in transcendent letter misunderstandings. Since the Spanish alphabet consists of 30 letters, counting on three generators, 90 letters are tested per user. The distribution of letters in the form is random for each alphabet and among generators, to avoid sampling bias. There are some letters in LSE which share relatively similar hand configurations. Therefore, transitions and arm orientation are treated carefully to obtain reliable results.

4.2 Dactylology Results

The results obtained from the dactylology comprehension experiment can be observed in Fig. 8, where the warmest colours represent a high correct answer rate and the coldest ones represent a low answer rate. As can be observed at first sight, the warmest colours are concentrated on the diagonal of the matrix, which is a positive outcome.

The averages show that the answers were slightly more sound for the three generated models obtained with the embodied robot than the ones obtained with the simulated robot. In quantitative terms, a 77% (369 out of 480 correct letters, counting on outliers) of correct answers obtained with the simulated robot [5] was beaten by an 80% (1350 out of 1680 correct letters), 78% (1315 out of 1680 correct letters) and 79% (1330 out of 1680 correct letters) obtained with the real robot for the LNN, DMNN and VAE generators, respectively. Neglecting the difference in the sampling size (16 participants with the simulated robot, versus 56 with the real robot validation experiment, which implies a 350% of increase) for the comparison, and taking into consideration that the reproduced dactylology imitates the simulated one, it can be concluded that the 1% to 4% improvement range is due to the effect of the robot embodiment.

The difference between the maximum and minimum number of correct letters across generators is 35 (2% over total asked letters), so this outcome is not considered to be significant enough to compare the proficiency among generators. Despite the distribution of correct answers, it is considered enlightening to quantify the numbers of misunderstood letters, since it is simpler to correct sign reproduction abnormalities in letters that show errors correlated to other determined letters, than in letters that present a random recognition pattern. The LNN and DMNN generators produced confusions with 76 incorrect letters, while the VAE generator produced confusions with only 67 incorrect letters, which implies a 12% divergence reduction and tips the balance towards the VAE generator. The number of outliers, and the number of letters that produced outliers, is also important in terms of evaluating how the generators produced unclassifiable letters. The LNN, DMNN and VAE generators produced 19, 14 and 16 outliers, in 11, 8 and 10 letters, respectively, which represents around 1% of the outcome sampling, so it is not significant for this study. Another important consideration is the lowest percentages of correct answers shown in each generator, which are 46%, 20% and 27% for the LNN, DMNN and VAE generators, respectively, which means that letter classification have been considerably more critical for DMNN than for LNN generations.

5 Conclusions

In the search to develop a model which enables sign language representation from a fully-actuated simulated model to the under-actuated real robot, three models are proposed based on three different neural networks: LNN, DMNN and VAE. The first two generators are trained using the known joint positions and motor commands, and the last one is trained using only the known joint positions and identifying a correlation with the motor commands as a separate process. The dactylology multimedia set generated with the embodied robot is validated by deaf and hearing impaired users.

The purpose of this work was to reuse previous signing poses and, therefore, comprehension results were expected to be at most as good as the previous ones due to the correspondence problem. Surprisingly, the outcome has shown general improvements in user comprehension (78–80% correct answers with respect to previous 77%), which indicates the success of the under-actuation modelling. This amelioration may be attributed to the positive influence of the robot embodiment. The main difference among generators relies on how confusing the generated dactylology results, which ends up being clearer for the VAE generator (12% clearer), the overall most competitive and salient generator. While results of the validation experiments are positive for the presented objectives, certain advances are required to achieve effective sign language communication. With a view to this objective, satisfaction metrics and further feedback collected in the delivered form will be analysed and applied in further studies.

Acknowledgments. The research leading to these results has received funding from RoboCity2030-DIH-CM Madrid Robotics Digital Innovation Hub ("Robótica aplicada a la mejora de la calidad de vida de los ciudadanos. Fase IV"; S2018/NMT-4331), funded by "Programas de Actividades I+D en la Comunidad de Madrid" and cofunded by Structural Funds of the EU. The authors thank CNSE (The Spanish Confederation of the Deaf) and LSE organizations such as Signapuntes Lengua de Signos and Mediación Comunicativa Cádiz for their kind collaboration with this project.

References

1. Estevez, D., Victores, J.G., Fernandez-Fernandez, R., Balaguer, C.: Ironing, robotic, with 3D perception and force, torque feedback in household environments. In: 2017 IEEE/RSJ IROS International Conference on Intelligent Robots and Systems, Vancouver, BC, Canada, 24–28 September, pp. 6484–6489 (2017)
2. Kelion, L.: Toshiba's robot is designed to be more human-like, BBC News, 9 March 2016. https://www.bbc.com/news/technology-35763917. Accessed 24 Feb 2019
3. Kose, H., Yorganci, R.: Tale of a robot: humanoid robot assisted sign language tutoring. In: 2011 11th IEEE-RAS International Conference on Humanoid Robots, Bled, Slovenia, 26–28 October (2011)
4. Goossens, M.: Optimisation of a humanoid sign language robot. B.S. thesis, Universiteit Antwerpen, Antwerpen, Belgium (2016)
5. Gago, J.J., Victores, J.G., Balaguer, C.: Sign language representation by TEO humanoid robot: end-user interest, comprehension and satisfaction. Electronics **8**, 57 (2019)
6. Kyle, J.G., Woll, B.: Sign Language: The Study of Deaf People and Their Language. Cambridge University Press, Cambridge (1985)
7. Bridgwater, L.B., et al.: The robonaut 2 hand - designed to do work with tools. In: 2012 IEEE ICRA International Conference on Robotics and Automation (2012)
8. Carrozza, M.C., Dario, P., Vecchi, F., Roccella, S., Zecca, M., Sebastiani, F.: The cyber-hand: on the design of a cybernetic prosthetic hand intended to be interfaced to the peripheral nervous system. In: 2003 IEEE/RSJ IROS International Conference on Intelligent Robots and Systems (2003)
9. Birglen, L., Lalibert, T., Gosselin, C.M.: Underactuated Robotic Hands, vol. 40. Springer, Heidelberg (2008). https://doi.org/10.1007/978-3-540-77459-4
10. Casarez, C.S., Fearing, R.S.: Steering of an underactuated legged robot through terrain contact with an active tail. In: 2018 IEEE/RSJ IROS International Conference on Intelligent Robots and Systems (2018)
11. Jalani, J., Herrmann, G., Melhuish, C.: Robust trajectory following for underactuated robot fingers, pp. 495–500 (2010)
12. Chevallereau, C., Grizzle, J.W., Shih, C.: Asymptotically stable walking of a five-link underactuated 3D bipedal robot. IEEE Trans. Robot. **25**(1), 37–50 (2009)
13. Kolathaya, S., Reher, J., Hereid, A., Ames, A.D.: Input to state stabilizing control Lyapunov functions for robust bipedal robotic locomotion. In: IEEE Annual American Control Conference (2018)
14. Buss, B.G., Ramezani, A., Hamed, K.A., Griffin, B.A., Galloway, K.S., Grizzle, J.W.: Experiments, preliminary walking, with underactuated 3D bipedal robot MARLO. In: 2014 IEEE/RSJ IROS International Conference on Intelligent Robots and Systems (2014)
15. Malvezzi, Monica., Salvietti, Gionata, Prattichizzo, Domenico: Evaluation of grasp stiffness in underactuated compliant hands exploiting environment constraints. In: Arakelian, Vigen, Wenger, Philippe (eds.) ROMANSY 22 – Robot Design, Dynamics and Control. CICMS, vol. 584, pp. 409–416. Springer, Cham (2019). https://doi.org/10.1007/978-3-319-78963-7_51
16. Yao, S., Zhan, Q., Ceccarelli, M., Carbone, G., Lu, Z.: Analysis and grasp strategy modeling for underactuated multi-fingered robot hand. In: 2009 IEEE International Conference on Mechatronics and Automation (2009)
17. Li, Z., Du, R., Yu, H., Ren, H.: Statics modeling of an underactuated wire-driven flexible robotic arm. In: 5th IEEE RAS/EMBS International Conference on Biomedical Robotics and Biomechatronics (2014)

18. Clevert, D., Unterthiner, T., Hochreiter, S.: Fast and accurate deep network learning by exponential linear units (ELUs). In: 2016 ICLR International Conference Learning Representations (2016)
19. Kingma, D.P., Welling, M.: Auto-encoding variational bayes. In: 2014 ICLR International Conference on Learning Representations (2014)
20. Chollet, F.: Keras: the python deep learning library. Astrophysics Source Code Library, June 2018

Exploratory Data Analysis of Wind and Waves for Floating Wind Turbines in Santa María, California

Montserrat Sacie, Rafael López[ID], and Matilde Santos[(✉)][ID]

Complutense University of Madrid, 28040 Madrid, Spain
{msacie,rlopez,msantos}@ucm.es

Abstract. Offshore wind turbines, and particularly floating wind turbines (FOWT) are subjected to strong wind and wave loads that affect the structural stability and energy efficiency of these renewable energy devices. Although wind -and less often waves- forecasting models have been developed, a deep analysis of the relationship between both external disturbances is necessary to consider the combined effect on the fatigue of the offshore WT. This work presents a study of the most relevant features of wind and waves using distribution analysis and ML techniques on wind and waves real data from an offshore buoy. Linear regression and SVM have been applied to the modelling of the data. These models may be very useful for the design of these floating structures and to study the impact of these external loads on the fatigue. The results lead us to consider the necessity of generating short-term models in specific geographical locations.

Keywords: Data analysis · Correlation · SVM · Machine learning · Wind energy · Floating offshore wind turbines

1 Introduction

Nowadays, one of the greatest global challenges that society is facing is climate change. Governments worldwide are fostering the exploitation and use of renewable energies, reducing the emissions of greenhouse gases and mitigating global warming [1, 2].

In contrast to the onshore wind energy, an already mature technology [3], offshore wind energy has appeared in the last two decades as a promising alternative. Marine wind devices can benefit from stronger and more stable winds and thus produce greater energy. But the cost of installation and maintenance of the most common offshore bottom-fixed turbines is still very high. That is why recently floating offshore wind turbines (FOWT) have been proposed. These floating devices can be installed in deeper waters, avoiding the visual and noise impact, harnessing more constant wind [4].

However, FOWTs pose new challenges [5]. Due to the strong loads they are subjected to, mainly wind and waves, induced vibration may damage the structure and jeopardize the stability of the turbine. This requires a deep analysis of the wind and wave data available. Realistic dynamics models must include metocean conditions. Moreover, how

© Springer Nature Switzerland AG 2020
C. Analide et al. (Eds.): IDEAL 2020, LNCS 12490, pp. 252–259, 2020.
https://doi.org/10.1007/978-3-030-62365-4_24

the combination of both external loads affects the performance and stability of the floating turbines must be studied.

In this work, we have carried out an exploratory analysis of wind and waves at an offshore location, Santa María (CA, USA), to obtain the relationships among these loads and to model their combined effect.

The structure of the paper is as follows. Section 2 presents a brief state of the art on wind and waves on FOWT. Section 3 describes the real metocean data used and the main features. The Exploratory Analysis of the Data is presented in Sect. 4. Linear regression and SVM models are obtained in Sect. 5. Conclusions end the work.

2 Winds and Waves on Floating Offshore Wind Turbines

Metocean conditions affect the stability of the FOWT to a greater degree than if they were anchored to the seabed. The wind is normally analyzed previously to the installation of the turbine [6]. However, the effect of waves is not usually considered, though they are the main source of disturbances for floating structures. In the best of cases, the wave model is simplified just defining sea states (SSN) based on wind speed. Moreover, the directionality of the waves is typically represented by a misalignment relative to the wind that, until recently, has been considered null in most designs of FOWTs [7]. But in high seas, this value is rarely null due to different physical and orographic conditions. This misalignment may cause large loads on the tower.

Few works do consider wind and waves misalignment, as in [8]; though it is focused on structural control, it highlights the importance of considering wind–wave misalignment when analyzing loads of offshore wind turbines. In [9], the effect of directional spreading of waves on an offshore wind farm is studied. In [10] and [11], some floating wind turbines dynamics in selected misaligned wind and wave conditions are investigated, showing that aligned wind and waves cause the largest short-term tower base fatigue damage. More recently, a statistical analysis for different properties of wind and wave loading including wind-wave misalignment has been carried out on monopile WTs [12]. Results presented in [13] show that measured motions in the surge, heave, and pitch are similar for the aligned and misaligned cases on water tank tests.

As it is possible to see, most of these works study wind and waves on bottom-fixed WTs or are focused on vibration reduction more than on predicting these loads.

3 Materials

The information used in this project corresponds to the standard meteorological and descriptive data obtained from the National Oceanic and Atmospheric Administration (NOAA) in the USA [14]. In this work, the offshore data measurements are taken by floating buoys distributed through the U.S and international waters maintained by the National Data Buoy Center (NDBC). Although they are not in the open sea, NDBC buoys are installed about 20–40 km off the coast.

Out of the more than 300 National Data Buoys Stations, we have selected one that has real-time data available. The chosen NDBC station with a moored offshore buoy for data gathering is Santa Maria station (34.956 N 121.019 W), located at the Pacific Ocean, in the Northwest of California.

We have downloaded the meteorological files with historical data for the last 9 years as well as real-time files from January to April 2020. In addition, we need wind and waves features obtained at the same geographic location to develop a forecasting model of the misalignment between wind and waves.

3.1 Features Description

Standard Meteorological Data files contain one row for hourly atmospheric, wind and waves averaged values, generally measured every 8 min. Historical data files are classified by year while real-data files contain the last 45 days measures. The same meteorological features are available in both types of files. Real-time measures are given every 10 min instead of each hour. As an example, Fig. 1 shows some of the historical data for the year 2019.

dataRead2019 = 16214x18 table

	YY	MM	DD	hh	mm	WDIR	WSPD	GST	WVHT	DPD	APD
1	2018	12	31	23	50	321	5.8000	7.0000	2.6700	10.8100	7.8000
2	2019	1	1	0	50	332	7.1000	8.4000	2.6400	11.4300	7.6100
3	2019	1	1	1	50	329	7.2000	9.1000	2.7400	10.0000	7.6600
4	2019	1	1	2	50	3	6.2000	8.4000	2.9400	10.0000	7.7500
5	2019	1	1	3	50	36	7.4000	8.4000	2.9100	10.8100	7.9700
6	2019	1	1	4	50	52	7.3000	8.7000	2.8900	10.8100	7.7200
7	2019	1	1	5	50	35	8.0000	9.7000	2.7800	10.8100	7.5300
8	2019	1	1	6	50	36	6.7000	8.5000	2.5000	11.4300	7.2800
9	2019	1	1	7	50	58	6.6000	8.2000	2.4200	10.0000	7.3500

Fig. 1. Samples of the data.

The variables and their corresponding meaning are as follows [15]:

- YY: Year, MM: Month, DD: Day, hh: Hour and mm: Minute, when measurements were taken.
- WDIR: Wind direction (° clockwise from North)
- WSPD: Wind speed (m/s)
- GST: Gust speed (m/s)
- WVHT: Significant wave height $H_{1/3}$ (m)
- DPD: Dominant wave period or Wave peak-spectral period (s)
- APD: Average wave period (s)
- MWD: Wave direction in DPD (°)
- PRES: Sea level pressure (hPa)
- ATMP: Air temperature (°C)
- WTMP: Sea surface temperature (°C)
- DEWPT: Dewpoint temperature (°C)
- TIDE: Water Level (ft) above or below Mean Lower Low Water (MLW) (feets)

The variables we are going to work with are the following: wind speed (WSPD); significant wave height (WVHT, $H_{1/3}$); wave peak-spectral period (DPD); wind direction (WDIR), and wave direction (MWD).

4 Exploratory Analysis of the Data

Exploratory Analysis of the Data (EDA) consists of the visualization of the row data, the calculation of metrics that summarize the distribution of data, such as the median or standard deviation, and scaling, cleaning missing values or normalizing variables values [16]. The interest in applying EDA is the flexibility to explore the data without having a hypothesis or prior knowledge of the domain, which will allow us to make some decisions regarding the data to be included, feature selection, etc. [17].

Taking a first look at data and features (Fig. 1) has helped us to establish the ranges of the variables. Some variables do not have any information (999), particularly dewpoint temperature (DEWP), station visibility limited for buoys from 0 to 1.6 nautical miles (VIS), and water level above or below Mean Lower Low Water (TIDE). Those variables are not closely related to wind and waves forecasting; indeed, they are not included in the generation of metocean data set for FOWT simulations [15]. So they have been directly discarded. The five columns that represent time information have been coded as one variable, Date, taking into account that minutes variable (mm) is always equal to 50 in historical files and thus it is not relevant.

In Fig. 2 the distribution of the variables at Santa María station is represented by boxplots. The two boxes in the upper left corner show the wind speed without preprocessing, boxplot(1,1), and after eliminating outliers and erroneous data (boxplot(2,2)). Note the variation in the average wind speed. The misalignment (°) has been also calculated, boxplot(2,3).

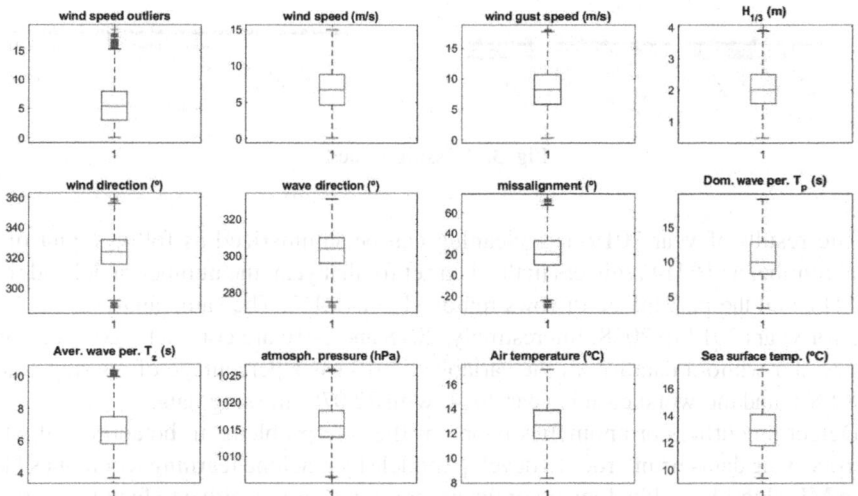

Fig. 2. Santa María station features distribution.

4.1 Missing Values and Outliers Detection

The missing values in metocean data set can be caused by a specific fault in the sensors or by the lack of measuring devices in the selected station buoy.

Among the three types of techniques for handling missing data: Imputation, Interpolation and Deletion, we have applied deletion. This decision is taken because of the huge amount of data available on the website (annual data sets since 1980). It would have been necessary to be an expert on the domain to interpolate or estimate the missing values. Therefore, rows containing missing values in at least one of the selected columns have been deleted. According to the percentage of data discarded, loading more data from previous years has been considered.

For example, data of year 2019 are represented in Fig. 3 (left). Features are represented in the x-axis and two categories, missing value (rejected) or valid (useful), in the y-axis. As it is possible to see, the variables that contain more missing values are Wind direction, Wind speed and Peak of gust speed. As it is possible to see, in the month of October there are no measurements on any variable so it is necessary to get data from other years for the month of October. The same happens in other years, as it is possible to see in Fig. 3, right, for year 2015.

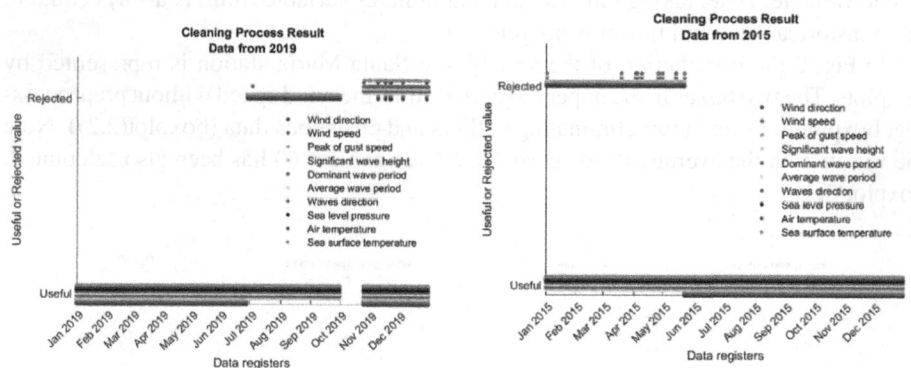

Fig. 3. Missing values.

The results of year 2019 data cleaning can be summarized as follows. Out of an initial number of 16214 registers in the data set for that year, the number of deleted rows is 10411 and the percentage of rows removed is 64,21%. The same analysis has been done for years 2014 to 2018. Interestingly, 2018 and 2016 are complete, 2017 has only one register without data for all the variables, 2015 has a percentage of missing values of 38.68% and the worst case is year 2014, with 72.07% missing data.

Detecting outliers or anomalies is one of the core problems to be addressed when preprocessing datasets in order to develop models by machine learning techniques [18]. Some ML algorithms, like logistic or linear regression, are sensitive to features distributions of the input examples and pattern detection depends on them. We have discarded outliers that were over the 90th percentile or below the 10th percentile. After data cleaning, we have plotted each feature time-series to ensure that there have been no mistakes during the cleaning process in the data distribution (Fig. 4, data distribution of waves and wind features for year 2018).

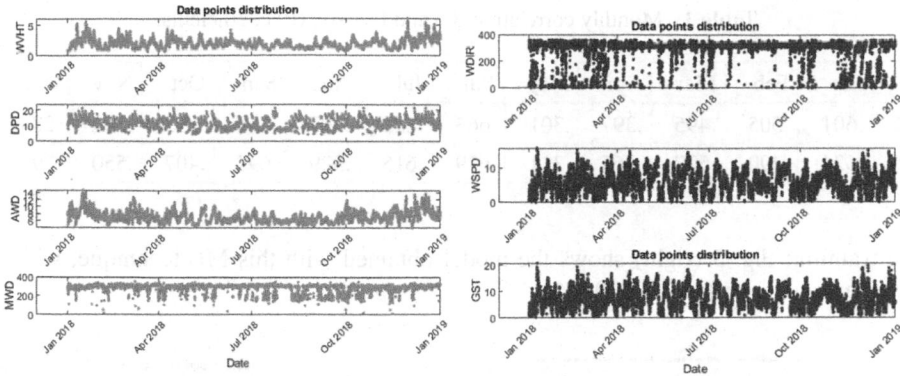

Fig. 4. Waves (left) and Wind (right) features time series.

For the most relevant features, some descriptive basic statistics have been also obtained. The corresponding histograms have been adjusted by probability distributions. Rayleigh and Weibull distributions seem to be the ones that best fit most of the variables.

5   Correlation Analysis of Wind and Waves and Models

Pearson's linear correlation coefficients among all the features allow us to conclude that some of the wind and wave variables of interest are correlated, particularly WSPD with WVHT (0.47). Other variables are also related, as it can be expected (for instance, WSPD and GST, etc.).

We have calculated the annual correlation between WSPD and WVHT for three different years, 2016, 2017 and 2018. The correlation is very similar for these three years (≈ 0.4387). This parametric correlation is equivalent to the Spearman correlation analysis when the amount of data is big enough, as it is the case, with Spearmen coefficients 0.4596 (2018), 0.4458 (2017) and 0.3683 (2016). As the ratio is not so high, this means that there is some direct correlation between these two variables but other factors may be also influencing this correlation between them.

Based on the hypothesis that correlation may vary with the seasons, the monthly correlation has been obtained for each month of year 2018 independently (that year did not have any missing value), Table 1. According to the results obtained for the annual correlation, conclusions drawn for 2018 can be extrapolated for other years. Correlation varies significantly from one season of the year to another (Table 1), mainly due to the strength of the wind on the sea that also varies; in summer months it reaches the highest value, 0.7154 in September, while in spring is around 0.3 in the three months or in winter it gets the lowest value of .2844. The fact that these correlation values are not too high leads us to consider that it would be necessary to obtained short-term models for each season of the year [19].

Once the linear model has been obtained, regression with SVM [20] has been applied to the significant wave height. Figure 5 (left) shows the configuration of the SVM during

Table 1. Monthly correlation (C) and Pearson (P) coefficients.

	Jan	Feb	Mar	Apr	May	Jun	Jul	Aug	Sep	Oct	Nov	Dec
C	.601	.505	.495	.397	.301	.663	.622	.567	.715	.392	.640	.284
P	.624	.490	.527	.365	.311	.679	.615	.529	.697	.407	.550	.299

the training; Fig. 5 (right) shows the model obtained with this ML technique, which gives a small error (MSE $\approx$ 0. 34).

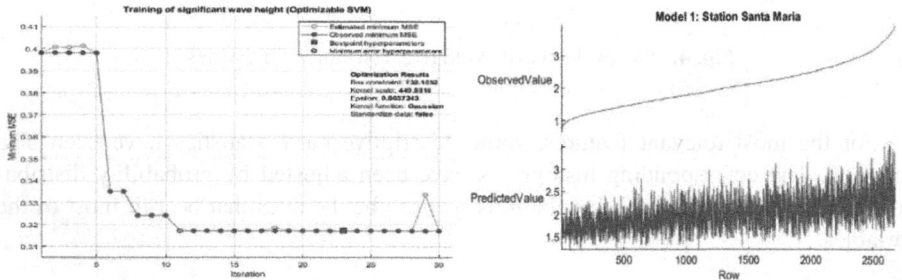

Fig. 5. SVM configuration (left) and wave height regression model.

6   Conclusions and Future Works

The application of Machine Learning techniques has been proved useful to find patterns in historical metocean data. Nevertheless, due to the stochastic nature of wind and waves, the results obtained with the correlation analysis carried out in this work suggest that short-term models (for each season) should be obtained to get good accuracy. Besides, pre-processing of the data has been shown to be essential to improve the accuracy of the models, reducing errors, such as in the case of the SVM model, here presented, that predicts the significant wave height with a small error.

As future works, the training of the models should be carried out for a larger number of stations to avoid drawing conclusions on the relationships between wind and waves closely related to the geographic location. Other ML techniques could be applied, and seasonal models should be developed.

Acknowledgments. This work was partially supported by the Spanish Ministry of Science, Innovation and Universities MCI/AEI/FEDER Project RTI2018-094902-B-C21.

References

1. Mikati, M., Santos, M., Armenta, C.: Modelado y simulación de un sistema conjunto de energía solar y eólica para analizar su dependencia de la red eléctrica. Revista Iberoamericana de Automática e Informática Industrial **9**(3), 267–281 (2012)

2. Ackermann, T.: Wind Power in Power Systems. Wiley, Hoboken (2005)
3. Mikati, M., Santos, M., Armenta, C.: Electric grid dependence on the configuration of a small-scale wind and solar power hybrid system. Renewable Energy **57**, 587–593 (2013)
4. Tomás-Rodríguez, M., Santos, M.: Modelado y control de turbinas eólicas marinas flotantes. Revista Iberoamericana de Automática e Informática Industrial **16**(4), 381–390 (2019)
5. Rubio, P.M., Quijano, J.F., López, P.Z., et al.: Intelligent control for improving the efficiency of a hybrid semi- submersible platform with wind turbine and wave energy converters. Revista Iberoamericana de Automática e Informática Industrial **16**(4), 480–491 (2019)
6. Gomes, I.L.R., Melicio, R., Mendes, V.M.F.: Wind power with energy storage arbitrage in day-ahead market by a stochastic MILP approach. Logic J. IGPL **208**, 570–582 (2019)
7. Li, Z., Adeli, H.: Control methodologies for vibration control of smart civil and mechanical structures. Exp. Syst. **35**(6), e12354 (2018)
8. Stewart, G.M., Lackner, M.A.: The impact of passive tuned mass dampers and wind–wave misalignment on offshore wind turbine loads. Eng. Struct. **73**, 54–61 (2014)
9. Trumars, J., Jonsson, J.O., Bergdahl, L.: The effect of wind and wave misalignment on the response of a wind turbine at Bockstigen. In: 25th International Conference on Offshore Mechanics and Arctic Engineering, pp. 635–641. American Society of Mechanical Engineers Digital Collection (2006)
10. Bachynski, E.E., Kvittem, M.I., Luan, C., Moan, T.: Wind-wave misalignment effects on floating wind turbines: motions and tower load effects. J. Offshore Mech. Arct. Eng. **136**(4), 041902 (2014)
11. Bachynski, E.E., Moan, T.: Second order wave force effects on tension leg platform wind turbines in misaligned wind and waves. In: ASME 33rd International Conference on Ocean, Offshore and Arctic Engineering. American Society of Mechanical Engineers Digital Collection (2014)
12. Sun, C., Jahangiri, V.: Fatigue damage mitigation of offshore wind turbines under real wind and wave conditions. Eng. Struct. **178**, 472–483 (2019)
13. Oh, S., Iwashita, T., Suzuki, H.: Numerical modelling and validation of a semisubmersible floating offshore wind turbine under wind and wave misalignment. J. Phys. Conf. Ser. **1104**(1), 012010 (2018). IOP Publishing
14. NOAA: National oceanic and atmospheric administration. https://www.noaa.gov/. Accessed 19 Jun 2020
15. Stewart, G.M., Robertson, A., Jonkman, J., Lackner, M.A.: The creation of a comprehensive metocean data set for offshore wind turbine simulations. Wind Energy **19**(6), 1151–1159 (2016)
16. Tukey, J.W.: Exploratory data analysis, vol. 2. Reading, Mass (1977)
17. Aguilar, R.M., Torres, J.M., Martin, C.A.: Automatic learning for the system identification. A case study in the prediction of power generation in a wind farm. Revista Iberoamericana de Automática e Informática Industrial **16**(1), 114–127 (2019)
18. Guevara, C.B., Santos, M., Lopez, V.: Negative selection and Knuth Morris Pratt algorithm for anomaly detection. IEEE Latin Am. Trans. **14**(3), 1473–1479 (2016)
19. Zhou, J., Shi, J., Li, G.: Fine tuning support vector machines for short-term wind speed forecasting. Energy Convers. Manag. **52**(4), 1990–1998 (2011)
20. Dormido-Canto, S., et al.: TJ-II wave forms analysis with wavelets and support vector machines. Rev. Sci. Instrum. **75**(10), 4254–4257 (2004)

Wind Turbine Pitch Control First Approach Based on Reinforcement Learning

J. Enrique Sierra-García[1](✉) and Matilde Santos[2]

[1] Department of Electromechanical Engineering, University of Burgos, 09006 Burgos, Spain
jesierra@ubu.es
[2] Institute of Knowledge Technology, Complutense University of Madrid, 28040 Madrid, Spain
msantos@ucm.es

Abstract. The control strategy defined for a wind turbine (WT) aims to achieve the highest energy efficiency and at the same time to ensure safe operation under all wind conditions. The goal of the pitch control of a WT is to stabilize the output power around its nominal (rated) value by means of the position of the rotor blades with respect to the wind. In this work, a pitch control strategy based on reinforcement learning (RL) is proposed. The control system consists of a state estimator, a reward mechanism, a policy table and policy update algorithm. Different reward strategies and policy update algorithms for the RL controller have been tested and compared with a PID regulator. The proposed controller stabilizes the output power of the wind turbine around the rated power more accurately and with smaller overshoot than the traditional one.

Keywords: Control · Pitch angle · Reinforcement learning · Wind turbine

1 Introduction

Wind turbines (WT) are one of the most widely renewable energy systems used. This clean energy has been proved essential to the sustainability of the current and future worldwide energy demand [1]. Although a mature technology, there are still many engineering challenges related to wind energy that must be addressed [2].

For each turbine, certain operational objectives are defined in order to obtain the best possible performance. These energy goals are achieved thanks to the control strategy defined to generate energy from the wind. The WT control system is designed to get the highest efficiency and at the same time to ensure safe operation under all wind conditions [3]. This may be critical for floating offshore wind turbines (FOWT) as it has been proved that the control can affect the stability of the floating device [4, 5]. There are different actuating mechanisms, namely pitch (pitch angle), yaw (yaw angle) and generator torque that must be controlled.

This work is focused on pitch control, that is, the angle of the blades [6]. The goal of the pitch control system is to stabilize the output power around its nominal (rated) value by means of the position of the rotor blades with respect to the wind. In this way, the efficiency of the installation is optimized in relation to the prevailing wind force.

© Springer Nature Switzerland AG 2020
C. Analide et al. (Eds.): IDEAL 2020, LNCS 12490, pp. 260–268, 2020.
https://doi.org/10.1007/978-3-030-62365-4_25

But this is not an easy task as WTs are highly non-linear systems with coupled internal variables and unknown parameters and, on top of that, they may be subjected to external disturbances and strong loads [7, 8].

The WT control problem has been addressed in the literature using different classical and intelligent control techniques, mainly fuzzy logic and neuro-fuzzy, such as in [9–16]. Among them, some researches have applied reinforcement learning (RL) to WT control. For instance, Sedighizadeh proposes an adaptive PID controller tuned by RL [17]. In [18] an artificial neural network based on RL for WT yaw control is presented. The paper by Kuznetsova proposes a RL algorithm to plan the battery scheduling in order to optimize the use of the electric grid [19]. Tomin et al. propose adaptive control techniques to extract the stochastic behaviour of wind speed using a trained RL agent, and then apply their obtained optimal policy to the wind turbine adaptive control design [20]. In Hosseini et al., passive RL solved by particle swarm optimization policy (PSO–P) is used to handle an adaptive type-2 neuro-fuzzy inference system with unsupervised clustering for controlling the pitch angle of a real wind turbine [21]. Chen et al. also propose a robust wind turbine controller that adopts adaptive dynamic programming based on RL and system state data [22].

In this work we propose a controller inspired by the RL approach to generate the pitch control signal. Different reward strategies and policy update algorithms have been tested. The main contribution of our work is that we generate the control targets directly from the controller without the use of any other regulator. Simulation results show how the proposed controller stabilizes the power of the installation to the rated value and it performs better than a conventional PID controller.

The rest of the paper is organized as follows. Section 2 describes the wind turbine model. The RL controller, the reward strategies and the policy update algorithms are presented in Sect. 3. Results are discussed in Sect. 4. The work ends with the conclusions and future works.

2 Wind Turbine Model

The model of a small turbine of 7 kW is used. The equations of the model are summarized in (1–3) [11].

$$\dot{I}_a = \frac{1}{L_a}\big(K_g \cdot K_\phi \cdot w - (R_a + R_L)I_a\big), \tag{1}$$

$$\dot{w} = \frac{c_1}{2 \cdot J \cdot w}\left(\left(\frac{v \cdot c_2}{w \cdot R} - c_3\theta - c_4\theta^{c_5} - c_6\right)e^{-\frac{v \cdot c_7}{w \cdot R}} \cdot \rho \cdot \pi R^2 \cdot v^3\right)$$
$$- \frac{1}{J}\big(K_g \cdot K_\phi \cdot I_a + K_f w\big), \tag{2}$$

$$P_{out} = R_L \cdot I_a^2. \tag{3}$$

Where L_a is the armature inductance (H), K_g is a dimensionless constant of the generator, K_ϕ is the magnetic flow coupling constant (V · s/rad), R_a is the armature resistance (Ω), R_L is resistance of the load (Ω), considered in the study as purely resistive, w is the angular rotor speed (rad/s), I_a is the armature current (A). The values of the coefficients c_1 to c_7 depend on the characteristics of the wind turbine, J is the rotational inertia (Kg.m^2), R is the radius or blade length (m), ρ is the air density (Kg/m^3), v is wind speed (m/s), K_f is friction coefficient (N m/rad/s), and θ is the pitch (rad).

The wind turbine parameters used during the simulations are shown in Table 1 [23].

Table 1. Parameters of the wind turbine model

Parameter	Description	Value/Units
L_a	Inductance of the armature	13.5 mH
K_g	Constant of the generator	23.31
K_ϕ	Magnetic flow coupling constant	0.264 V/rad/s
R_a	Resistance of the armature	0.275 Ω
R_L	Resistance of the load	8 Ω
J	Inertia	6.53 kg m^2
R	Radio of the blade	3.2 m
ρ	Density of the air	1.223 kg/m^3
K_f	Friction coefficient	0.025 N m/rad/s
$[c_1, c_2, c_3]$	C_p constants	[0.73, 151, 0.58]
$[c_4, c_5, c_6, c_7]$	C_p constants	[0.002, 2.14, 13.2, 18.4]

3 Description of the RL Based Controller

The structure of the controller inspired by the RL strategy [24] is shown in Fig. 1. The state variables are I_a and w, the control input is the pitch angle θ, and the controlled variable is the power P_{out}. It is composed of a state estimator, a reward calculator, a policy table, an actuator, and a method to update the policy.

The state estimator receives the power error (P_{err},), defined as the difference between the nominal power P_{ref} and the current power P_{out}, its derivative ($\dot{P}_{err}$), and the wind speed (V_{wind}). These signals are discretized and they define the state $s_t \in S$, where t is the current time.

The policy $\pi : S \rightarrow A$ is a function which assigns to each state in S an action a_t that belongs to a finite set of actions A. This action a_t is selected to maximize the long term expected reward. The actuator transforms the discrete action a_t to a control signal for the pitch θ_{ref} in the range $[0, \pi/2)$ (rad). Each time an action is executed, at the next iteration the reward calculator observes the new P_{err} and $\dot{P}_{err}$ and calculates a reward/penalty r_t for the action a_{t-1}. This reward is used by the policy update algorithm to modify the policy for the state s_{t-1}.

The policy has been modelled as a table $T^\pi_{(s,a)} : S \times A \rightarrow \mathbb{R}$, together with a function $f_P : S \rightarrow A$. The table relates each pair $(s, a) \in S \times A$ with a real number that represents

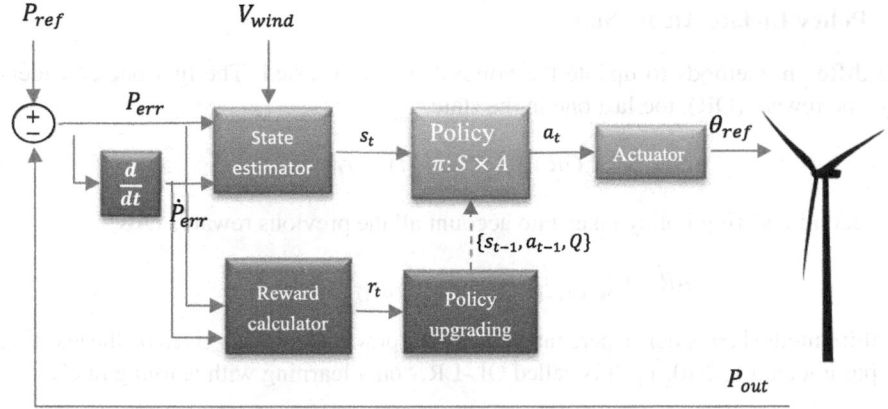

Fig. 1. Architecture of the RL based controller

an estimation of the long term expected reward to be received when action a is executed in the state s, also known as Q. The type of estimation depends on the policy update algorithm. The table has s rows (states) and a columns (actions). Given a state s, the function f_P searches the action with the maximum Q value.

The performance of the controller varies with the reward mechanism and the policy update algorithm. These components are individually explained below.

3.1 Reward Strategies

Different rewarding strategies have been defined. The first one considers only the error, P_{err}. The reward is bigger as the output power is closer to the expected value, in a proportional way. It is called "Position reward strategy" (PRS), given by (4):

$$r_t = r_{PRS} = K_{r1} \cdot \left(P_{err_{MAX}} - |P_{err}(t)|\right) \tag{4}$$

Where the factor K_{r1} is used to adjust the weight of the reward and $P_{err_{MAX}}$ is the maximum expected power error.

The second proposed alternative increases the reward when the controller action makes the output closer to the expected value (positive reward), and punishes it when it is moving away (negative reward). The reward is related to the error derivative, and it is called "Velocity reward strategy" (VRS), calculated as (5):

$$r_t = r_{VRS} = \begin{cases} -K_{r2} \cdot \dot{P}_{err}(t) & P_{err}(t) > 0 \\ K_{r2} \cdot \dot{P}_{err}(t) & P_{err}(t) \leq 0 \end{cases} \tag{5}$$

Where the gain K_{r2} weights the reward.

Both strategies can be combined by applying a linear operator (multiplication or addition). We have defined the third reward strategy PVRS as (6):

$$r_t = r_{PVRS} = r_{PRS} * r_{VRS} \tag{6}$$

3.2 Policy Update Algorithms

Five different methods to update the policy have been tested. The first one considers only one reward (OR), the last one in the state,

$$OR : T^{\pi}_{(s_{t-1}, a_{t-1})}(t) = r_t \tag{7}$$

The second updating policy takes into account all the previous rewards (AR):

$$AR : T^{\pi}_{(s_{t-1}, a_{t-1})}(t) = T^{\pi}_{(s_{t-1}, a_{t-1})}(t-1) + r_t \tag{8}$$

The third method considers a percentage of all the previous rewards given by the learning rate parameter $\alpha \in \mathbb{R}[0, 1]$. It is called OL-LR, "only learning with learning rate",

$$OL - LR : T^{\pi}_{(s_{t-1}, a_{t-1})}(t) = T^{\pi}_{(s_{t-1}, a_{t-1})}(t-1) + \alpha \cdot r_t \tag{9}$$

The fourth updating policy, called LF-LR, uses the learning rate α and the forgetting factor $(1 - \alpha)$.

$$LF - LR : T^{\pi}_{(s_{t-1}, a_{t-1})}(t) = (1 - \alpha) \cdot T^{\pi}_{(s_{t-1}, a_{t-1})}(t-1) + \alpha \cdot r_t \tag{10}$$

The last update method is the Q-learning algorithm, where γ is the discount factor:

$$a_{max} = argMAX_a \left(T^{\pi}_{(s_t, a)}(t-1) \right) \tag{11}$$

$$QL : T^{\pi}_{(s_{t-1}, a_{t-1})}(t) = (1 - \alpha) \cdot T^{\pi}_{(s_{t-1}, a_{t-1})}(t-1) + \alpha \left(r_t - \gamma \cdot T^{\pi}_{(s_{t-1}, a_{max})}(t-1) \right) \tag{12}$$

4 Discussion of the Simulation Results

Simulation results of the RL based controller and a PID regulator are shown and compared. The simulation time is 100 s. The wind speed is randomly generated between 11.5 and 14.8 m/s. The power error signal P_{err} is discretized in a range of 50 values, the error derivative $\dot{P}_{err}$ and wind speed V_{wind} are discretized in 20 values. Learning rate α is set to 0.5 and the discount factor γ is set to 0.1. The parameters of the PID, Kp, Ki, and Kd, have been tuned by trial and error to [1, 0.2, 0.9], respectively.

Figure 2 (left) shows the power output with different reward strategies and AR policy update algorithm; output power with the PRS reward mechanism (blue line), and with VRS reward policy (red line). Figure 2 (right) shows the responses with the VRS reward and different policy update algorithms, OR (blue), AR (red) and Q-learning (green). The PID output is the yellow line and the reference rated power is purple.

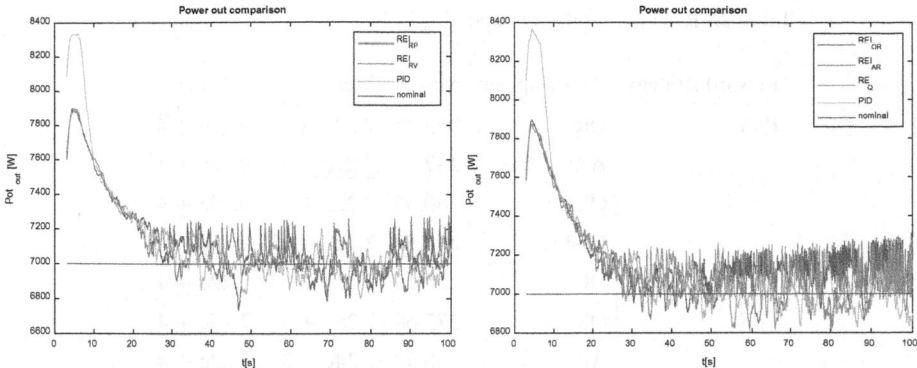

Fig. 2. Power output comparison for different rewards (left) and policy updates (right). (Color figure online)

In both figures it is possible to observe how the proposed RL controller stabilizes the output power around its rated value. Simulations are shown from 2 s on due to the fact that the estimator has some delay to compute the output (because of the policy update algorithm), so the response is delayed. Although both controller have the same starting point, the PID controller reacts faster than the RL one. That is why the output power value obtained by the PID at 3 s is higher than the other output values obtained with the RL controller and different configurations.

The PID gives larger overshoot although its output is less noisy. In general VRS produces bigger but fewer peaks than PRS, maybe because VRS takes positive and negative values (rewards and penalties), and PRS only uses positive values. Another expected result is that OR policy update produces noisier results than AR; the OR algorithm only considers the last value updated in the policy table, while AR takes into account all previous values, so the Q value is more likely to change with OR, and thus the actions vary more often. Q-learning performs between forgetting and remembering all previous values, somehow in the middle and therefore its results are less noisy than OR but noisier than AR.

Numerical results have been also obtained for different combinations of rewards strategies and policy update algorithms (Table 2). The best results have been boldfaced.

All the combinations give values close to the 7 kW rated power. As it may be seen the proposed RL based controller provides better performance than the PID in every case. This may be due to the larger overshoot given by the PID (Fig. 2). Regarding the MSE error, the best performance is obtained by the combination (PRS, LF-LR). In general PRS strategy provides smaller error than VRS but larger variance. This can be explained as with VRS the reward changes faster than with PRS, producing actions that change more frequently.

When the reward mechanisms are combined, with the PVRS proposal the results are in the middle of the obtained by each individual reward strategy.

Table 2. Results for different updating policies and reward strategies

Reward strategy	Policy update	MSE	Mean	Variance
PRS	OR	255.75	7.22e + 3	9.79e + 4
	AR	257.69	7.21e + 3	9.86e + 4
	OL-LR	260.35	7.22e + 03	9.99e + 4
	LF-LR	**251.53**	**7.21e + 03**	9.75e + 4
	QL	256.19	7.22e + 03	9.22e + 4
VRS	OR	277.68	7.28e + 3	**7.45e + 4**
	AR	266.98	7.24e + 3	9.23e + 4
	OL-LR	261.96	7.25e + 3	8.12e + 4
	LF-LR	269.83	7.27e + 3	7.83e + 4
	QL	270.92	7.27e + 3	7.523 + 4
PVRS	OR	268.25	7.26e + 3	8.31e + 4
	AR	256.82	7.24e + 3	8.16e + 4
	OL-LR	267.81	7.26e + 3	8.26e + 4
	LF-LR	261.55	7.26e + 3	7.57e + 4
	QL	261.81	7.24e + 3	8.95e + 4
PID		346.11	7.30e + 3	2.13e + 05

5   Conclusions and Future Works

In this work a new control strategy inspired by RL has been designed and applied to the pitch control of a wind turbine. The RL based controller includes a reward mechanism and a policy updating algorithm. The policy acts as the memory of the controller and the policy update algorithm allows the system to learn how to make the right decisions.

Different reward strategies have been evaluated as well as several policy update algorithms. Simulation results of the proposed controller have been compared with the performance of a PID regulator obtaining smaller error and lower overshoot when stabilizing the output power of the WT to its rated value.

Among other possible future works, we may highlight the study of other reward strategies, testing this control system in a real wind turbine, and the generalization of the approach to implement general purpose tracking controllers.

Acknowledgments. This work was partially supported by the Spanish Ministry of Science, Innovation and Universities MCI/AEI/FEDER Project RTI2018-094902-B-C21.

References

1. Gomes, I.L.R., Melicio, R., Mendes, V.M.F.: Wind power with energy storage arbitrage in day-ahead market by a stochastic MILP approach. Logic J. IGPL **28**, 570–582 (2019)

2. Mikati, M., Santos, M., Armenta, C.: Electric grid dependence on the configuration of a small-scale wind and solar power hybrid system. Renew. Energy **57**, 587–593 (2013)
3. Menezes, E.J.N., Araújo, A.M., da Silva, N.S.B.: A review on wind turbine control and its associated methods. J. Clean. Prod. **174**, 945–953 (2018)
4. Tomás-Rodríguez, M., Santos, M.: Modelado y control de turbinas eólicas marinas flotantes. Revista Iberoamericana de Automática e Informática Industrial **16**(4), 381–390 (2019)
5. Kim, C., Muljadi, E., Chung, C.C.: Coordinated control of wind turbine and energy storage system for reducing wind power fluctuation. Energies **11**(1), 52 (2018)
6. Acho, L.: A proportional plus a hysteretic term control design: a throttle experimental emulation to wind turbines pitch control. Energies **12**(10), 1961 (2019)
7. Li, Z., Adeli, H.: Control methodologies for vibration control of smart civil and mechanical structures. Exp. Syst. **35**(6), e12354 (2018)
8. Aguilar, R.M., Torres, J.M., Martin, C.A.: Automatic learning for the system identification. A case study in the prediction of power generation in a wind farm. Revista Iberoamericana de Automática e Informática Industrial **16**(1), 114–127 (2019)
9. Santos, M.: Un enfoque aplicado del control inteligente. Revista Iberoamericana de Automática e Informática Industrial RIAI **8**(4), 283–296 (2011)
10. Navarrete, E.C., Perea, M.T., Correa, J.J., Serrano, R.C., Moreno, G.R.: Expert control systems implemented in a pitch control of wind turbine: a review. IEEE Access **7**, 13241–13259 (2019)
11. Sierra-García, J.E., Santos, M.: Wind turbine pitch control with an RBF neural network. In: Herrero, Á., Cambra, C., Urda, D., Sedano, J., Quintián, H., Corchado, E. (eds.) SOCO 2020. AISC, vol. 1268, pp. 397–406. Springer, Cham (2021). https://doi.org/10.1007/978-3-030-57802-2_38
12. Rubio, P.M., Quijano, J.F., López, P.Z., et al.: Intelligent control for improving the efficiency of a hybrid semi- submersible platform with wind turbine and wave energy converters. Revista Iberoamericana de Automática e Informática Industrial **16**(4), 480–491 (2019)
13. Moodi, H., Bustan, D.: Wind turbine control using TS systems with nonlinear consequent parts. Energy **172**, 922–931 (2019)
14. Rocha, M.M., da Silva, J.P., De Sena, F.D.C.B.: Simulation of a fuzzy control applied to a variable speed wind system connected to the electrical network. IEEE Latin Am. Trans. **16**(2), 521–526 (2018)
15. Sierra, J.E., Santos, M.: Modelling engineering systems using analytical and neural techniques: Hybridization. Neurocomputing **271**, 70–83 (2018)
16. Asghar, A.B., Liu, X.: Adaptive neuro-fuzzy algorithm to estimate effective wind speed and optimal rotor speed for variable-speed wind turbine. Neurocomputing **272**, 495–504 (2018)
17. Sedighizadeh, M., Rezazadeh, A.: Adaptive PID controller based on reinforcement learning for wind turbine control. In: Proceedings World Academy of Science, Engineering and Technology, vol. 27, pp. 257–262 (2008)
18. Saénz-Aguirre, A., Zulueta, E., Fernández-Gamiz, U., Lozano, J., Lopez-Guede, J.M.: Artificial neural network based reinforcement learning for wind turbine yaw control. Energies **12**(3), 436 (2019)
19. Kuznetsova, E., Li, Y.F., Ruiz, C., Zio, E., Ault, G., Bell, K.: Reinforcement learning for microgrid energy management. Energy **59**, 133–146 (2013)
20. Tomin, N., Kurbatsky, V., Guliyev, H.: Intelligent control of a wind turbine based on reinforcement learning. In: 2019 16th Conference on Electrical Machines, Drives and Power Systems ELMA, pp. 1–6. IEEE (2019)
21. Hosseini, E., Aghadavoodi, E., Ramírez, L.M.F.: Improving response of wind turbines by pitch angle controller based on gain-scheduled recurrent ANFIS type 2 with passive reinforcement learning. Renew. Energy **157**, 897–910 (2020)
22. Chen, P., Han, D., Tan, F., Wang, J.: Reinforcement-based robust variable pitch control of wind turbines. IEEE Access **8**, 20493–20502 (2020)

23. Mikati, M., Santos, M., Armenta, C.: Modelado y simulación de un sistema conjunto de energía solar y eólica para analizar su dependencia de la red eléctrica. Revista Iberoamericana de Automática e Informática Industrial 9(3), 267–281 (2012)
24. Santos, M., López, V., Botella, G.: Dyna-H: A heuristic planning reinforcement learning algorithm applied to role-playing game strategy decision systems. Knowl. Based Syst. 32, 28–36 (2012)

Intelligent Fuzzy Optimized Control for Energy Extraction in Large Wind Turbines

Carlos Serrano-Barreto[1] and Matilde Santos[2]([⊠]) [iD]

[1] Complutense University of Madrid, 28040 Madrid, Spain
`serranobarretocarlosluis@gmail.com`
[2] Institute of Technical Knowledge, Complutense University of Madrid, 28040 Madrid, Spain
`msantos@ucm.es`

Abstract. In this paper an intelligent controller is designed to obtain the maximum power of a large floating offshore wind turbine. The control of these turbines is more complex due to the strong loads they are subjected to and the uncertainty that comes from the environment, mainly wind and waves, and from its non-linear dynamics. In this case, the control goal is to maximize the output power of the wind turbine by controlling the rotor speed. An incremental PD-type fuzzy controller has been implemented; it generates the pitch angle reference. The performance of this control scheme on the NREL 5 MW floating offshore wind turbine has been compared with the internal control that is provided within the FAST software. Results are encouraging, showing that the intelligent control strategy is able to produce more energy.

Keywords: Intelligent control · Fuzzy logic · Pitch angle · Floating offshore wind turbine · Renewable energy

1 Introduction

The demand for energy continues to grow and many countries have opted to promote renewable energies to eliminate pollution and carbon residues [1, 2]. Wind energy has proven to be a very efficient clean energy [3, 4]. But most common and widely installed onshore wind turbines (WT) have some limitations that have made it jump to offshore wind energy [5].

Within offshore wind turbines, floating offshore wind turbines (FOWT) have a series of advantages, such as eliminating the visual and acoustic impact, that they can be installed in deep waters, they take advantage of a stronger and more constant wind, etc. However, the fact that these turbines maybe much bigger due to the unlimited space in the high sea poses new control challenges. They are highly non-linear systems, with time changing parameters, and complex dynamics [6]. Besides, they are subjected to strong loads that produce undesirable vibrations [7–9].

This has led to investigate knowledge-based techniques to address the control of these floating turbines and to deal with the uncertainty that comes from both their dynamics and the environment (mainly wind and waves) [10].

© Springer Nature Switzerland AG 2020
C. Analide et al. (Eds.): IDEAL 2020, LNCS 12490, pp. 269–276, 2020.
https://doi.org/10.1007/978-3-030-62365-4_26

This paper addresses the problem of large turbines control. Specifically, it seeks to obtain maximum power by controlling the speed of the rotor, which in turn depends on the pitch control of the blades. An incremental fuzzy-PD controller has been designed for this purpose. It calculates the pitch reference that will feed the NREL 5 MW wind turbine model in order to control the rotor speed and maximize the energy. The simulation results with the fuzzy control strategy have been compared with the obtained for the same turbine defined within the FAST software (Fatigue, Aerodynamics, Structures, and Turbulence), with encouraging results.

The main contribution of this paper is the indirect rotor speed control by generating the pitch reference signal, similarly to [11], but using a fuzzy logic controller. Other papers are mainly focused on the pitch angle control, with conventional [12–14] or intelligent techniques [5, 15, 16].

The structure of the paper is as follows. Section 2 describes how the wind turbine performs. Section 3 presents de design of the fuzzy logic control scheme. In Sect. 4 results are presented and discussed. The paper ends with the conclusions and future works.

2 Wind Turbine Power Equations

We work with the offshore wind turbine NREL 5-MW, whose parameters are listed in Table 1.

Table 1. Wind turbine parameters [17].

Nominal output power	5 MW
Nominal wind speed	12.5 m/s
Blade number	3
Initial rotor speed	12.1 rpm
Generator nominal output	5 MW
Blade length	61.609 m
Cut in wind speed	3.5 m/s
Generator number	1
Tip Rad	63 m

The power that can be obtained from the wind is given by [4]:

$$P = 0.5\rho A v^3 \tag{1}$$

Where P is the power (W), ρ is the air density (Kg/m^3), A is the area of the blades (m^2) and v is the wind speed (m/s).

There is a limit of the theoretical maximum efficiency of a wind turbine (Betz limit), meaning that at most only 59.3% of the kinetic energy from wind can be used to spin the turbine and generate electricity. The amount of power that can be taken from the turbine is determined by the Cp coefficient. The Cp coefficient is a function of blade angle β and the blade tip ratio speed (TRS), λ [18]. Thus, the mechanical power that can be obtained is given by:

$$P_{\omega t} = 0.5\rho A v^3 Cp(\beta, \lambda) \tag{2}$$

There are different approximation of the power coefficient Cp of the turbine; in this work we have used the following [18].

$$Cp = 0.5176\left(\frac{116}{\lambda i} - 0.4\beta - 5\right)e^{-\frac{21}{\lambda i}} + 0.0068\lambda \tag{3}$$

The variable λi, that has no physical meaning, is determined by,

$$\frac{1}{\lambda i} = \frac{1}{\lambda + 0.08\beta} - \frac{0.035}{3\beta + 1} \tag{4}$$

The tip ratio speed λ is the relationship between the angular velocity of the blades and the wind speed, given by [2]:

$$\lambda = \frac{w.R}{v} \tag{5}$$

Where w (rpm) is the rotation speed of the rotor, and R the radius of the rotor (length of the blades).

Finally, the angle of attack of the blades that is going to be determined by the control law is:

$$\beta = \mathrm{atan}\left(\frac{v}{w.R}\right) \tag{6}$$

3 Fuzzy Control Design

The control structure is shown in Fig. 1. An intelligent control system based on fuzzy logic has been implemented. It controls the rotor speed by means of the blade pitch angle trying to obtain the maximum output power.

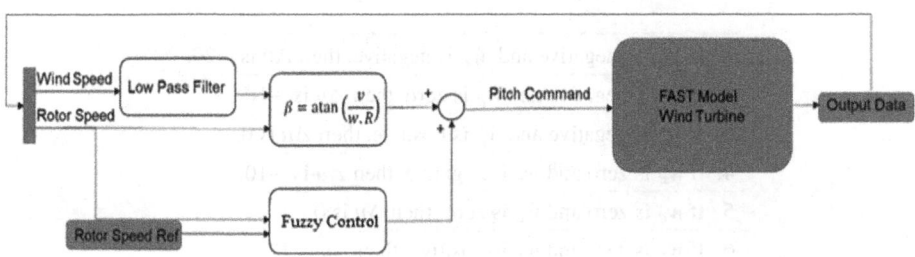

Fig. 1. Control structure

The fuzzy controller calculates the increase or decrease of the pitch angle of the blades, that will be the reference for the rotor speed control. We have implemented a

Takagi-Sugeno incremental Fuzzy-PD regulator that is equivalent to a PI controller [19]. The inputs of the controller are the error of the rotor speed, w_e (rpm), and the derivate, $\dot{w}_e$ (rpm/s), defined as,

$$w_e = w_{ref} - w \tag{7}$$

Where the reference w_{ref} has been set to 12.5 rpm, according to [11].

Three triangular fuzzy sets have been assigned to each input: Positive (green), Negative (blue) and Zero (red), as shown in Fig. 2.

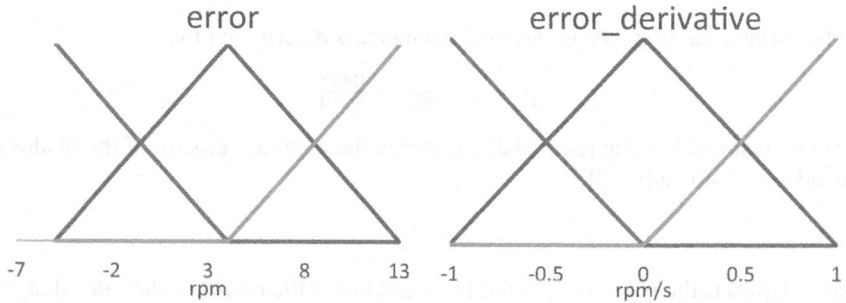

Fig. 2. Error (left) and error derivate (right) fuzzy sets. (Color figure online)

The error range is in the interval $[-5, 13]$ rpm, reaching its maximum at 12.5 rpm (constant generator torque) and the error derivate has been normalized between $[-1, 1]$ rpm/s to produce a smooth response.

The rules of the Takagi-Sugeno fuzzy controller are expressed as,

$$Rule(i) : if \; x_i \, is \, A_{i1}, \ldots, x_n \, is \, A_{in} \; the \; y_i = c_{i0} + c_{i1}x_i + \ldots + c_{in}x_n \tag{8}$$

Where the output is the pitch angle increment, Δu (°). They are listed in Table 2.

Table 2. Fuzzy rules of the fuzzy controller.

1. If w_e is negative and $\dot{w}_e$ is negative, then Δu is -20.
2. If w_e is negative and $\dot{w}_e$ is zero, then Δu is -10.
3. If w_e is negative and $\dot{w}_e$ is positive, then Δu is 0.
4. If w_e is zero and $\dot{w}_e$ is negative, then Δu is -10.
5. If w_e is zero and $\dot{w}_e$ is zero, then Δu is 0.
6. If w_e is zero and $\dot{w}_e$ is positive, then Δu is 10.
7. If w_e is positive and $\dot{w}_e$ is negative, then Δu is 0.
8. If w_e is positive and $\dot{w}_e$ is zero, then Δu is 10.
9. If w_e is positive and $\dot{w}_e$ is positive, then Δu is 20.

The rules for pitch angle were designed taking into account that absolute value of the maximum pitch angle for this turbine in FAST is 25° (typically it is between 0° and 10°).

4 Simulation Results

NREL FAST v8 [20] and Matlab software packages are used for the simulation. The proposed fuzzy controller is going to be compared with the internal controller of the Bladed-style DLL library embedded in the FAST simulator, test 24.

The wind profile is shown in Fig. 3. The average wind speed is 12.5 m/s. As it can be expected, it is a noisy signal. For that reason, a low-pass filter is applied (Fig. 1).

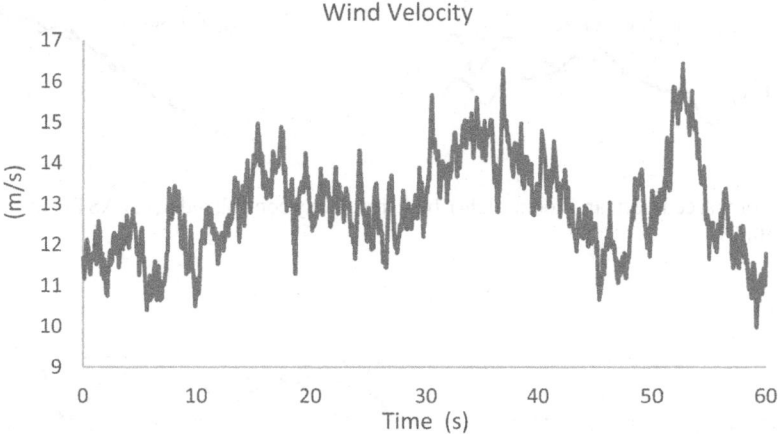

Fig. 3. Wind speed profile

In order to compare the response of the wind turbine with and without the fuzzy control strategy, the blade pitch angle with both approaches is shown in Fig. 4. It can be observed how from 22.5 s on the FAST regulator (red line) starts sending control commands to the blades to change the pitch angle, and the fuzzy controller starts at around 35 s (blue line).

According to the characteristics of the wind turbine (Table 1), the rotor starts at an initial speed of 12.1 rpm. The corresponding responses of both control schemes are shown in Fig. 5 (left) and a detailed view of it from 21.5 s on is presented in Fig. 5 (right), where the red line is the FAST response and the blue line the fuzzy proposed one. It is possible to see how the responses are quite similar until 30 s. At that time, the fuzzy controller starts to perform and then the difference between both controllers are noticeable.

As a result, the fuzzy controller extracts more power in comparison to the controller included in the FAST simulator (Fig. 6). Indeed, by analyzing Fig. 6 we can see that the rotor is able to extract more power from the wind than its maximum output power. However, operating above the rated power would induce excessive loads on the structure,

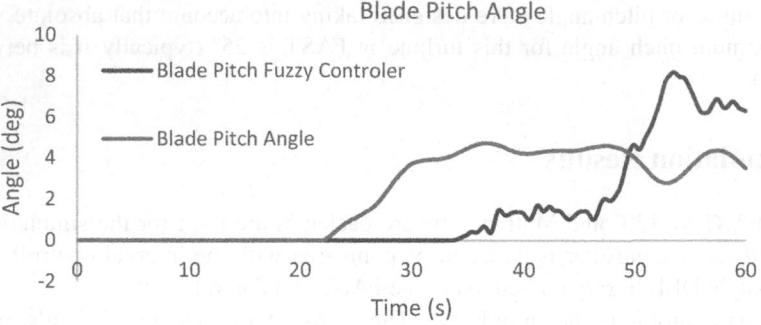

Fig. 4. Pitch angle. (Color figure online)

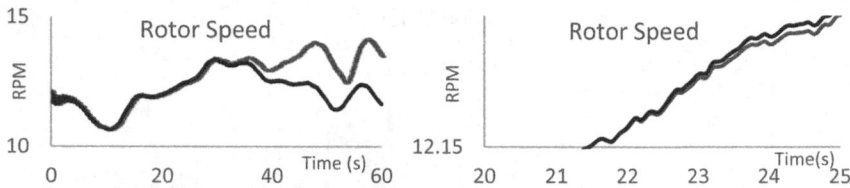

Fig. 5. Rotor speed (left) and zoom (right), blue line, fuzzy control; red line, FAST control. (Color figure online)

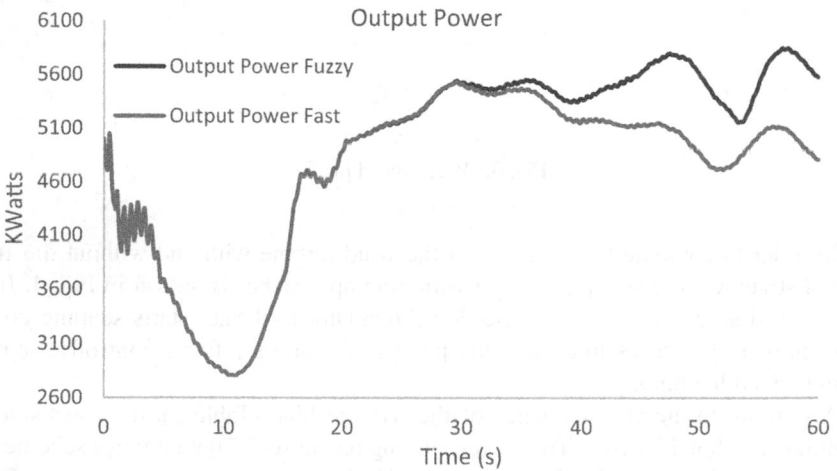

Fig. 6. Output power with FAST (red) and fuzzy (blue) controllers. (Color figure online)

compromising the life cycle of the turbine. In that case, the turbine should be shut down to avoid structural damage though.

In this simulation we have not done it in order to show how the fuzzy control is able to obtain more energy but this saturation of the actuators is included in the WT module.

5   Conclusions and Future Works

Floating turbines are oversized to better harvest the wind. This makes its control more complex. In this work a fuzzy system has been designed that obtains the reference of the pitch angle of a wind turbine to control the rotor speed. The goal is to obtain the output power that maximizes the energy efficiency of the wind turbine.

The fuzzy system implemented is an incremental PD fuzzy control. Its output feeds the NREL 5 MW turbine model.

The fuzzy controller shows a good performance, being able to deal with the non-linearity and uncertainty of the wind turbine. It response has been compared with a conventional controller that is embedded in the system under the same conditions, giving a smoother response and greater efficiency.

This work is a first approach to the application of intelligent control to floating turbines. As future works, different fuzzy type controllers [21] and adaptive fuzzy controllers could be tried to enhance the wind turbine response.

Acknowledgments. This work was partially supported by the Spanish Ministry of Science, Innovation and Universities MCI/AEI/FEDER Project number RTI2018-094902-B-C21.

References

1. Aguilar, R.M., Torres, J.M., Martin, C.A.: Automatic learning for the system identification. A case study in the prediction of power generation in a wind farm. Revista Iberoamericana de Automática e Informática Industrial **16**(1), 114–127 (2019)
2. Gomes, I.L.R., Melício, R., Mendes, V.M.F., Pousinho, H.M.I.: Wind power with energy storage arbitrage in day-ahead market by a stochastic MILP approach. Log. J. IGPL **28**(4), 570–582 (2020)
3. Mikati, M., Santos, M., Armenta, C.: Electric grid dependence on the configuration of a small-scale wind and solar power hybrid system. Renew. Energy **57**, 587–593 (2013)
4. Mikati, M., Santos, M., Armenta, C.: Modelado y simulación de un sistema conjunto de energía solar y eólica para analizar su dependencia de la red eléctrica. Revista Iberoamericana de Automática e Informática Industrial **9**(3), 267–281 (2012)
5. Rubio, P.M., Quijano, J.F., López, P.Z., et al.: Intelligent control for improving the efficiency of a hybrid semi- submersible platform with wind turbine and wave energy converters. Revista Iberoamericana de Automática e Informática Industrial **16**(4), 480–491 (2019)
6. Tomás-Rodríguez, M., Santos, M.: Modelado y control de turbinas eólicas marinas flotantes. Revista Iberoamericana de Automática e Informática Industrial **16**(4), 381–390 (2019)
7. Li, Z., Adeli, H.: Control methodologies for vibration control of smart civil and mechanical structures. Expert Syst. **35**(6), e12354 (2018)
8. Kim, C., Muljadi, E., Chung, C.C.: Coordinated control of wind turbine and energy storage system for reducing wind power fluctuation. Energies **11**(1), 52 (2018)
9. Quiles, E., Garciia, E., Cervera, J., Vives, J.: Development of a test bench for wind turbine condition monitoring and fault diagnosis. IEEE Lat. Am. Trans. **17**(06), 907–913 (2019)
10. Santos, M.: Un enfoque aplicado del control inteligente. Revista Iberoamericana de Automática e Informática Industrial RIAI **8**(4), 283–296 (2011)
11. Galvani, P.A., Sun, F., Turkoglu, K.: Aerodynamic modeling of NREL 5-MW wind Turbine for nonlinear control system design: a case study based on real-time nonlinear receding horizon control. Aerospace **3**(3), 27 (2016)

12. Acho, L.: A proportional plus a hysteretic term control design: a throttle experimental emulation to wind turbines pitch control. Energies **12**(10), 1961 (2019)
13. Nasiri, M., Mobayen, S., Zhu, Q.M.: Super-twisting sliding mode control for gearless PMSG-based wind turbine. Complexity **2019** (2019). Article ID 6141607
14. Kim, D., Lee, D.: Hierarchical fault-tolerant control using model predictive control for wind turbine pitch actuator faults. Energies **12**(16), 3097 (2019)
15. Civelek, Z.: Optimization of fuzzy logic (Takagi-Sugeno) blade pitch angle controller in wind turbines by genetic algorithm. Eng. Sci. Technol. Int. J. **23**(1), 1–9 (2020)
16. Rocha, M.M., da Silva, J.P., De Sena, F.D.C.B.: Simulation of a fuzzy control applied to a variable speed wind system connected to the electrical network. IEEE Lat. Am. Trans. **16**(2), 521–526 (2018)
17. Jonkman, J., Butterfield, S., Musial, W., Scott, G.: Definition of a 5-MW reference wind turbine for offshore system development (No. NREL/TP-500-38060). National Renewable Energy Lab. (NREL), Golden, CO (United States) (2009)
18. Civelek, Z., Lüy, M., Çam, E., Mamur, H.: A new fuzzy logic proportional controller approach applied to individual pitch angle for wind turbine load mitigation. Renew. Energy **111**, 708–717 (2017)
19. Santos, M., De la Cruz, J.M., Dormido, S., De Madrid, A.P.: Between fuzzy-PID and PID-conventional controllers: a good choice. In: Proceedings of North American Fuzzy Information Processing, pp. 123–127. IEEE (1996)
20. NREL FAST. https://www.nrel.gov/wind/nwtc.html. Accessed 20 Aug 2020
21. Santos, M., Dormido, S., De La Cruz, J.M.: Fuzzy-PID controllers vs. fuzzy-PI controllers. In: Proceedings of 5th International Fuzzy Systems, vol. 3, pp. 1598–1604. IEEE (1996)

Special Session on Emerging Trends in Machine Learning

Autoencoder Latent Space Influence on IoT MQTT Attack Classification

María Teresa García-Ordás[2,3], Jose Aveleira-Mata[2(✉)],
José-Luis Casteleiro-Roca[1], José Luis Calvo-Rolle[1],
Carmen Benavides-Cuellar[1,3], and Héctor Alaiz-Moretón[2,3]

[1] CTC, Department of Industrial Engineering, CITIC, University of A Coruña,
Avda. 19 de febrero s/n, 15405 Ferrol, A Coruña, Spain
{jose.luis.casteleiro,jlcalvo}@udc.es
[2] Research Institute of Applied Sciences in Cybersecurity (RIASC) MIC,
Universidad de León, 24071 León, Spain
[3] Department of Electrical and Systems Engineering, Escuela de Ingenierías,
Campus de Vegazana, University of León, 24071 León, Spain
{mgaro,jose.aveleira,carmen.benavides,hector.moreton}@unileon.es

Abstract. IoT (Internet of Things) alludes to many different devices and systems connected to Internet, being 5 billion the number of these devices working around the world actually. The security policies applied to this kind of systems can be improve due to their behaviour, usually associated to their low price and low computing capacity.

This work addresses the behaviour and impact of latent space of an auto-encoder for creating a classification model based on decision trees, in order to include it in a IDS (Intrusion Detection System) specialized in IoT environments. A validate IoT dataset, based on MQTT (Message Queue Telemetry Transport), has been used for applied the techniques implemented for extracting an optimal model oriented to detect the attacks over this protocol with a suitable results.

Keywords: IoT · MQTT · Cybersecurity · Classification · Auto-encoders

1 Introduction

"Internet of Things" (IoT) devices have a special behavior due to their low price and low computing capacity, reducing their encryption functionality. For these reasons, the security policies included in this kind of devices are not the best. It is certainly true that standard IoT devices are smaller and efficient, however, they can be a security breach in our network environment. Moreover, use different lightweight communication protocols, which consume less resources and bandwidth, in the application layer with protocols such as MQTT, CoAP, and other network level protocols such as zigbee or Bluethooth [1].

IDS (Intrusion Detection Systems) are a common solution for detecting treats in a network. This tool is based on matching patterns of well known attacks.

© Springer Nature Switzerland AG 2020
C. Analide et al. (Eds.): IDEAL 2020, LNCS 12490, pp. 279–286, 2020.
https://doi.org/10.1007/978-3-030-62365-4_27

Therefore an IDS is able to detect intrusions and stranger behaviours in a network [5].

Usually, the IDS are used in two ways:

– Signature-based detection: effective when the attacks are well-known finding specific patterns in the network traffic.
– Anomaly-based detection systems: based on the identification of treats thanks to continue monitoring of whole system behaviour, comparing it with a standard status previously known [13].

Anomaly-based IDS can achieve satisfactory results with attacks detection purposes including zero-day attacks [16]. The machine learning approaches are utilized for developing the models that can be included in the IDS, in order to improve the common detection methods [7]. Moreover, there are several deep learning architectures for developing new models to be included in the IDS, using Auto-encoders and Deep Belief Networks (DBN) for reducing dimensions for classifying purposes [15] Long Short Term Memory (LSTM) networks is other deep learning technique applied to detect treats in the network [10].

It is necessary to use datasets for implementing mentioned intelligent techniques [6] for IoT protocols, that contain traffic data [11] and not only, information from sensors and actuators.

With purpose of applying the set of technique addresses in this work, an own dataset that has been validated in previous research has been used [2].

The main focus of this paper is to understand the latent space defined by an auto-encoder for classifying three different attacks and normal frames at the same time, using an IoT environment dataset.

This work is structured as follows: the case study is described in the next section. The methods used are introduced in Sect. 3. Experiments and their results are described in Sect. 4, and the conclusions and some future works are showed in Sect. 5.

2   Case Study

One of the most commonly used protocols in the IoT application layer is the Message Queue Telemetry Protocol (MQTT) [8]. It is a publication/subscription messaging protocol designed for lightweight machine-to-machine (M2M) communications and ideal for connecting small devices to networks with minimal bandwidth. The MQTT architecture follows a star topology with a central node that works as a server or broker. The MQTT protocol can be encrypted [3], which requires more processing power to the devices.

With the aim of generating a dataset with attacks to the MQTT protocol vulnerabilities, a test environment is developed simulating a real situation with a IoT system using this protocol. The IoT environment is composed of a broker server, a distance sensor, an actuator consisting of a relay and a smartphone and a PC used to interact with this environment. All traffic generated is captured from the router with the OpenWRT in pcap files. Different attacks on the

MQTT protocol are related to this system: Denial of Service Attack (DoS): The Broker is in control of managing all the messages that are generated, therefore, a denial of service attack on it will cause a bad service in the IoT system. A DoS attack is performed using the MQTT-malaria program [12] simulating that 1000 devices are connected simultaneously sending 1000 messages for each one. Intrusion attack by other clients: In MQTT systems that do not have authentication or is in plain text in LAN an attacker can pass for another client of the system, with the special character '#' see all the messages of the broker [4]. The attacker can send false information and see the content of the messages managed by the broker. This attack is made using a program that works as a MQTT client named "mosquito" from it the messages are captured and it sends false information both the relay and the distance sensor.

Man-in-the-middle attack (MitM): the attacker interposes between the broker and the distance sensor, to know and modify the messages that he has intercepted, a different PC using the Kali-linux Distribution. Using the Ettercap tool, he can intercept the communication and the Nfqsed program to detect the MQTT packets and modify the values of the distance sensor.

The attacks are carried out as all traffic is captured. With the obtained pcap files it dissects the frames of all the generated traffic, taking common fields to all the frames, include the system times, the relative time of the collection and all the fields from the MQTT protocol. The frames are tagged considering the timestamp of when the attacks were made, indicating if they are under attack or not.

After the previous process, a dataset composed by 86,186 frames of MQTT is generated. It is composed of frames from the three different types of attacks and normal frames: 45.514 DoS frames, 3.855 MitM frames 1.898 Intrusion frames and 234.919 normal frames.

3   Machine Learning Methods

3.1   Feature Reduction: Autoencoder

There is a group of deep learning techniques widely used known as autoencoders [14], which tries to learn a deep representation of the data by compressing the features. Autoencoder architecture is symmetric. The number of input neurons is the same as the number of output neurons and in the middle, one of the hidden layers is composed by less neurons than the input or the output layer. This hidden layer is the latent space. The part of the autoencoder that goes from the input layer to the latent space constitutes the encoder, while the rest constitutes the decoder.

The goal of an autoencoder is to reconstruct the input data. For this reason, if after encoding the data using the encoder and decoding it using the decoder, that data is correctly reconstructed, the autoencoder is able to reduce the input data to the number of neurons in the latent space.

3.2 Classification: Decision Tree

Decision trees are a popular tool in machine learning [9]. The tree structure represents a workflow or process so it can be considered as a step-by-step approach to solving a task. Each internal node represents a test on an attribute, each branch represents the result of the test, and each leaf node represents a class label. The path from the root node to leaf, represents a classification rule. One of the reasons why trees have been chosen is their simplicity of understanding and the necessity of a white box model.

4   Experiments

In this section the experiment setup is explained in deep, as well as the final results and a briefly discussion of these.

4.1   Methodology

The dataset used is quite unbalanced and contains missing values and multiple categorical variables that a machine learning model cannot interpret. Therefore, the following preprocessing steps have been carried out.

First of all, IP and MAC addresses not containing data, were filled with value 0. After that, in order to provide more information than that provided by an IP converted into a numerical value, the IP is divided into 4 blocks WWW.XXX. YYY.ZZZ converting each IP into 4 numerical values.

MAC addresses were modified removing the ":" character and converting them into a numerical value.

The following categorical variables have been converted into a numerical value using label encoding: mqtt-clientid, mqtt.conack.flags, mqtt.conflags, mqtt.hdrflags, mqtt.protoname, mqtt.topic, and mqtt.msg.

In addition, a study of the 66 characteristics has been carried out to eliminate those that do not have a sufficiently representative variety (for example, constant columns).

After this process, a dataset with 48 numeric features per data is obtained.

The next step is to normalize the data so that training based on machine learning can be carried out. In this case, the MinMax normalization has been used. MinMax normalization limits each of the characteristics between 0 and 1. Once the data was normalized, the autoencoder feature reduction study was carried out.

A study of the importance of the AE latent space has been carried out for the correct classification of cyber attacks.

All autoencoder layers except the latent space, have a relu activation function while the latent space has a linear activation function. Each autoencoder has been trained for 50 epochs with a batch size of 50, using the mean square error as loss function and adam as optimizer.

Our autoencoder is composed of 7 layers of 48-35-28-N-28-35-48 neurons being $N \in [1, 27]$

Once the network is trained, the dataset has been encoded to reduce its features from 48 to n (n goes from 1 to 27) and this resulting reduced dataset will be used in the classification step.

In the classification step, decision trees have been used performing a hyperparameter search through GridSearchCV using crossvalidation and varying the maximum number of leafs and the criteria (gini or entropy).

To solve que problem of imbalanced classes, a class weighting has been carried out using the "calculate_weights" function of sklearn.

4.2   Results and Discussion

For each classification, Precision, Recall and F-Score metrics have been calculated. The experiments were evaluated for each attack class and also for the average by weighting each value by the proportion of data in each class. In Table 1, the average results for the different sizes of the autoencoder latent space (LS Size) can be shown.

Table 1. Results over all different sizes of the latent space of the autoencoder

LS Size	Precision Av	Recall Av	F-Score Av	LS Size	Precision Av	Recall Av	F-Score Av
1	0.9868	0.9887	0.9867	15	0.9944	0.9943	0.9944
2	0.9918	0.9919	0.9918	16	0.9941	0.9940	0.9940
3	0.9924	0.9923	0.9924	17	0.9941	0.9942	0.9942
4	0.9939	0.9939	0.9939	18	0.9936	0.9935	0.9935
5	0.9936	0.9934	0.9935	19	0.9948	0.9947	0.9948
6	0.9938	0.9939	0.9938	20	0.9943	0.9943	0.9943
7	0.9943	0.9943	0.9943	21	0.9942	0.9942	0.9942
8	0.9941	0.9939	0.9939	22	0.9941	0.9940	0.9940
9	0.9938	0.9936	0.9937	23	0.9941	0.9939	0.9940
10	0.9945	0.9943	0.9944	24	0.9949	0.9948	0.9949
11	0.9938	0.9939	0.9938	25	0.9943	0.9943	0.9943
12	0.9938	0.9938	0.9938	26	0.9943	0.9944	0.9943
13	0.9940	0.9941	0.9941	27	0.9938	0.9940	0.9939
14	0.9942	0.9943	0.9942				

The evolution of the average values for each of the latent spaces can be seen more clear in Fig. 1.

It can be observed that all the metrics are practically constant from 27 neurons to a value of 7 neurons in the latent space, from which the performance drops.

In addition, a study of each attack has been carried out to see how affects separately the encoding made by the autoencoder. See Fig. 2.

As it can be seen in Fig. 2, the DoS attack is perfectly identified in all cases, even when there is a single neuron in the latent space. However, MitM attacks

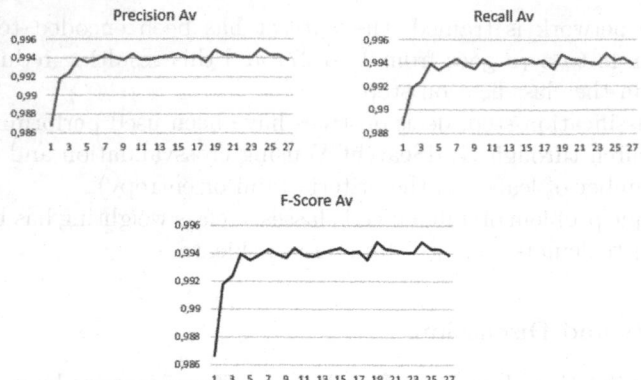

Fig. 1. Precision, recall and F-Score weighted average results. On the X-axis, the number of neurons in the latent space and on the Y-axis the corresponding metric value.

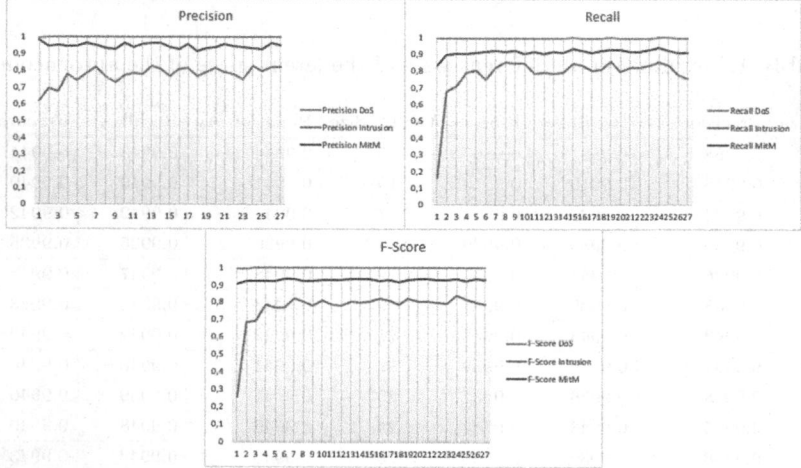

Fig. 2. Precision, recall and F-Score for each attack class. On the X-axis, the number of neurons in the latent space and on the Y-axis the corresponding metric value.

detection decrease when reducing the data to few features, although it maintains very similar performance anyway.

The clearest case can be seen in the case of intrusion attacks, the least representative class, which, in the case of latent spaces with less than 7 neurons, the metrics sink to values below 20% in the Recall, causing values very low on the F-Score.

This behavior means that the autoencoder is capable of encoding the dataset with little loss of performance, with up to 7 characteristics for all attacks.

This is very important because the smaller the number of characteristics in a data, the faster the inference of new data is carried out. In this case, being able to predict packets in real time is of vital importance in order to detect in time the attacks that are carried out and being able to block them.

The study confirms that the number of features could be further reduced in the case of wanting to detect only DoS or MitM attacks.

5   Conclusions and Future Works

In this paper, an exhaustive analysis of the influence of the feature reduction via autoencoders in IoT MQTT attacks has been carried out. The main goal is to determine the minimum number of features needed to describe MQTT packages in order to be able to deploy a real time detection system for IoT attacks.

Several sizes of the latent space of an autoencoder have been used to reduce the original dataset. Results have demonstrated than when less than 7 features are used, the performance of the system decreases, but from 7, the performance is very constant and close to a 100% of accuracy.

Observing the results for each type of attack, it can see how DoS and MitM attacks are less sensitive to the change of the number of features whereas Intrusion attacks accuracy decreases fast as the number of features is reduced.

For that reason, using an autoencoder with 7 neurons in the latent space can lead us to reduce the total amount o features in the dataset and consequently the time on the training and prediction steps. This allows us to deploy a real time system to defend our network from IoT attacks based on MQTT protocol successfully.

Acknowledgements. This work is partially supported by:
– Spanish National Cybersecurity Institute (INCIBE) and developed Research Institute of Applied Sciences in Cybersecurity (RIASC).
– Junta de Castilla y León - Consejería de Educación. Project: LE078G18. UXXI2018/000149. U-220.

References

1. Al-sarawi, S., Anbar, M., Alieyan, K., Alzubaidi, M.: Internet of Things (IoT) Communication Protocols : Review, pp. 685–690 (2017)
2. Alaiz-Moreton, H., Aveleira-Mata, J., Ondicol-Garcia, J., Muñoz-Castañeda, A.L., García, I., Benavides, C.: Multiclass classification procedure for detecting attacks on MQTT-IoT protocol. Complexity **2019**, 1–11 (2019). https://doi.org/10.1155/2019/6516253
3. Alqazzaz, A., Aloufi, E., Alharthi, R., Zohdy, M.A., Alrashdi, I., Ming, H.: A practical evaluation of a secure and energy-efficient smart parking system using the MQTT protocol. ACM Int. Conf. Proc. Ser. 165–170 (2019). https://doi.org/10.1145/3325917.3325937
4. Andy, S., Rahardjo, B., Hanindhito, B.: Attack scenarios and security analysis of MQTT communication protocol in IoT system, 19–21 September 2017
5. Ben-Asher, N., Gonzalez, C.: Effects of cyber security knowledge on attack detection. Comput. Hum. Behav. **48**, 51–61 (2015). https://doi.org/10.1016/j.chb.2015.01.039

6. Bhuyan, M.H., Bhattacharyya, D.K., Kalita, J.K.: Towards generating real-life datasets for network intrusion detection. Int. J. Network Secur. **17**(6), 683–701 (2015)
7. Chakrabarty, B., Chanda, O., Saiful, M.: Anomaly based intrusion detection system using genetic algorithm and K-centroid clustering. Int. J. Comput. Appl. **163**(11), 13–17 (2017). https://doi.org/10.5120/ijca2017913762, http://www.ijcaonline.org/archives/volume163/number11/chakrabarty-2017-ijca-913762.pdf
8. Hamdani, S., Sbeyti, H.: A comparative study of COAP and MQTT communication protocols. In: 7th International Symposium on Digital Forensics and Security, ISDFS 2019, pp. 1–5 (2019). https://doi.org/10.1109/ISDFS.2019.8757486
9. Han, L., Li, W., Su, Z.: An assertive reasoning method for emergency response management based on knowledge elements c4.5 decision tree. Expert Syst. Appl. **122**, 65–74 (2019). https://doi.org/10.1016/j.eswa.2018.12.042, http://www.sciencedirect.com/science/article/pii/S0957417418308108
10. Kim, J., Kim, J., Thu, H.L.T., Kim, H.: Long short term memory recurrent neural network classifier for intrusion detection. In: 2016 International Conference on Platform Technology and Service (PlatCon), pp. 1–5 (2016). https://doi.org/10.1109/PlatCon.2016.7456805, http://ieeexplore.ieee.org/document/7456805/
11. Koroniotis, N., Moustafa, N., Sitnikova, E., Turnbull, B.: Towards the development of realistic botnet dataset in the Internet of Things for network forensic analytics: Bot-IoT dataset. Future Generation Comput. Syst. **100**, 779–796 (2019). https://doi.org/10.1016/j.future.2019.05.041
12. Palsson, K.: mqtt-malaria @ github.com (2018). https://github.com/remakeelectric/mqtt-malaria
13. Prabha, K., Sudha, S.: A Survey on IPS methods and techniques. Int. J. Comput. Sci. Issues, **13**(2), 38–43 (2016). https://doi.org/10.20943/01201602.3843, http://ijcsi.org/contents.php?volume=13&&issue=2
14. Pumsirirat, A., Yan, L.: Credit card fraud detection using deep learning based on auto-encoder and restricted boltzmann machine. Technical Report 1 (2018). www.ijacsa.thesai.org
15. Tao, X., Kong, D., Wei, Y., Wang, Y.: A big network traffic data fusion approach based on fisher and deep auto-encoder. Information **7**(2), 20 (2016). https://doi.org/10.3390/info7020020, http://www.mdpi.com/2078-2489/7/2/20
16. Zhou, Q., Pezaros, D.: Evaluation of machine learning classifiers for zero-day intrusion detection - an analysis on CIC-AWS-2018 dataset (2019). http://arxiv.org/abs/1905.03685

A Recommendation System of Nutrition and Physical Activity for Patients with Type 2 Diabetes Mellitus

João Godinho[1], Sara Batista[1], Diogo Martinho[1], and Luís Conceição[1,2]([✉])

[1] GECAD – Research Group on Intelligent Engineering and Computing for Advanced Innovation and Development, Institute of Engineering, Polytechnic of Porto, 4200-072 Porto, Portugal
{1141150,114119,diepm,msc}@isep.ipp.pt
[2] ALGORITMI Centre, University of Minho, 4800-058 Guimarães, Portugal

Abstract. The diseases related to how we eat are a major public health concern and continue to endanger the health of the population, as well as the sustainability of health systems due to the significant increase in the burden of anti-diabetic drugs as well as new therapeutic classes that have a high cost. An unbalanced diet can result in metabolic disorders, malnutrition, overweight, mental problems and other medical risk factors, such as cardiovascular disease, type 2 diabetes mellitus or, in the worst case, cancer. In 2019, according to the International Diabetes Federation statistic, the estimated prevalence of diabetes in the world population over the age of 20 was 10.4%, covering about 463 million people. Therefore, there is a need for a greater care and attention for this disease, both in terms of the respective treatment and prevention. This kind of need is dependent of the decisions that either the patient or the health professional make daily. With the rise of big data and data analysis technologies, recommendation systems have become essential and necessary for a custom data management, according to the user. These systems play a key role because of its ability to increase the amount of information available. In this paper, we propose a conceptual definition of a recommendation system that aims to assist the monitoring of Type 2 Diabetes Mellitus disease thus enabling a more effective management of the disease.

Keywords: Recommendation systems · Diabetes mellitus · Healthy lifestyle · Physical activity

1 Introduction

Diabetes is a disease that is characterized by the inability of the body to produce and use. As a result, whenever there is a lack of insulin, glucose stays in the blood instead of providing energy to cells necessary for all the actions we do in our daily lives. The Diabetes can be classified in two different ways, namely, Diabetes Type 1 and Type 2 [1]. The focus of the work here proposed is related to the Type 2 Diabetes Mellitus condition. This condition is characterized by a relative and progressively increasing deficiency of

C. Analide et al. (Eds.): IDEAL 2020, LNCS 12490, pp. 287–297, 2020.
https://doi.org/10.1007/978-3-030-62365-4_28

insulin, in which pancreas becomes unable to produce enough insulin to control blood sugar levels. It may also happen that the body become resistant to insulin as there is a deregulation of metabolism of carbohydrates, lipids and proteins which can result in impaired insulin secretion, insulin resistance or a combination of both [2, 3].

The risk factors for having this type of disease are varied and among them are age, obesity, sedentary lifestyle and family history related to this disease. It is known that about 80% of patients with Diabetes Mellitus Type 2 are overweight or are obese. The correct practice of healthcare in this context is relevant to control the disease since patients suffering from diabetes that are regularly monitored and evaluated were capable of decreasing further complications associated to the disease [4]. In this sense, changing food habits and increasing the amount of physical exercise done every day is essential to avoid a more sedentary lifestyle and to maintain a good health status. However, a restrictive diet may have a negative effect on the health condition of the person in case individual needs are forgotten. Therefore, an adequate individual nutritional assessment is necessary to avoid unbalanced diets that can lead to specific nutritional deficiencies, overweight or obesity [5].

Proper diet and physical exercise are the foundation for the right control of Diabetes. These two factors may be enough to control the levels of glucose in the blood at the initial stage of diabetes. It is necessary to re-educate patients that carry this disease so that they can be independently achieve a balanced diet. Thus, it is important to pay attention on the following aspects [5]:

- Favour the consumption of carbohydrates from sources such as whole or slightly refined cereals, fruits, milk and low-fat yogurt.
- Limit the consumption of sugary products and carbohydrates;
- Ensure a high fiber intake;
- Limit the consumption of foods rich in salt (sodium);
- Avoid excessive consumption of alcohol;
- Energy balance is essential to weight control.

Interventions and changes to the daily routine of a patient must be supported by both structured programs, self-management and with psychological support if necessary. Structured programs improve biomedical and psychosocial outcomes, but it depends on reinforcement to assure a more continued benefit [6].

Education can be obtained using technology that allows patients to acquire new practices and routines. In this context, it makes sense to consider the use of recommendation system that supports patients in their daily activities and manage diabetes including the register of the glucose, daily exercises, meals recommendations, notifications to remind to take the medication and register of the different emotional states.

2 Recommender Systems

With the rise of big data and data analysis technologies, recommendation systems have become essential and necessary for a custom data management, according to the user preferences and needs. They can play an essential role in various contexts due to increased information available. This kind of systems can help a user in daily decision-making, presenting a more effective way to manage his preferences. We can conclude that recommendation systems consist in special software programs designed to recommend items to users according to their interests based on an analysis of their daily routines and behaviours [7].

In this context, context-ware recommendation systems (CARS – Context-Aware Recommender Systems) provide more advanced features than traditional ones as they include contextual information such as time, location and user routines in order to better understand their habits and the influence of the preferences of the same. The incorporation of context information in recommendation systems leverages can improve the relevance of the recommendations with regard to the needs of users [8].

There are different known approaches to the development of recommendation systems, including collaborative filtering, content-based filtering or a combination of both.

2.1 Collaborative Filtering

Collaborative filtering is independent of the domain, and is based on opinions of other users, which cannot simply be described by metadata. It identifies people with similar preferences to assign similar ratings to an element. This process uses profiles to specify the relationship or similarity between users and begins to create an array of user elements and the respective preferences. Then similarity between the profiles of users is then measured to detect and get users with common interests which will serve as recommendations. This group of users is also defined as "neighborhood" [9].

The algorithms used in the development of collaborative filtering are algorithms based on memory and recommendations based on model.

Memory-Based Algorithms
Memory-based algorithms analyse all information on certain items, users and respective classifications are stored in memory. It tries to find users that are similar to the active user and uses their preferences to predict items or classifications for the active user [9].

Model-Based Recommendations
The model-based algorithms involve building a model based on the dataset of ratings. In other words, it extracts some information from the dataset, and use that as a "model" to make recommendations without having to use the complete dataset every time. This approach potentially offers the benefits of both speed and scalability.

Two different techniques can be used to build a model: Machine Learning and Data Mining. These Algorithms analyze the array of user items to identify the respective relationships between them. Then use these detected relationships to create a list of the N key recommendations found [9].

2.2 Content-Based Filtering

The content-based technique is an algorithm which analyses attributes of the items to correctly make predictions. The content of each item is represented as a set of descriptors or terms, typically the words that occur in a document. When documents such as web page, publications and news are recommended, the content-based filtering technique is more effective [9].

In content-based filtering technique, the recommendation is based on the user's profile, through a historic, using resources to extract the contents of the items that the user had evaluated in the past. If those item's evaluation has a positive rating, they will be recommended to the user. Different types of models are used are used to find similarities between different documents in order to generate significant recommendations, including the Model of Vector Space [9], which models the relationship between different documents. These techniques give rise to recommendations according to the underlying model or statistical analysis machine learning. No profiles of other users are required as they have no influence on the recommendation. If a user profile changes, the technique of content-based filtering can easily adjust to the recommendations accordingly [9].

2.3 Hybrid Filtering

The hybrid Filtering technique combines different recommendation techniques to get a better system optimization, in order to avoid some limitations and problems of recommendation systems [9].

For the content-based filtering, three problems can be distinguished [10]:

- Content description - In some domains generating a useful description of the content can be challenging. In domains where the items consist of music or video, for example, a representation of the content is not always possible with today's technology;
- Over-specialization - A content-based filtering system will no select items if the previous user behavior does not provide evidence for this. Additional techniques would have to be added in order to give the system the capability to make suggestion outside the scope of what the user has already shown interest in.
- Subjective domain problem - Content-based filtering techniques have difficulty in distinguishing between subjective information such as points of views and humor.

A collaborative filtering system does not have these shortcomings. That is because there is no need for a description of the items being recommended, so the system can deal with any kind of information. Nevertheless, the collaborative system is not perfect either and it does introduce problems of its own such as [10]:

- Early rate problem - Collaborative filtering systems cannot provide recommendations for new items since there are no user ratings on which to base a prediction. Even if users star rating the item it could take time before the item has received enough ratings in order to make accurate recommendations. Similarly, recommendations will also be inaccurate for new users who have rated only few items;
- Sparsity problem - The existing number of items exceeds the amount that a person is able and willing to explore by far. This makes it hard to find items that are rated by enough people on which to base predictions.

• Gray sheep - Groups of users are needed with overlapping characteristics. Even if such groups exist, users who do not consistently agree or disagree with any group of people will receive inaccurate recommendations.

The use of various recommendation techniques can suppress the weakness of individual technique in a combine model. A combination of approaches may be made in the following ways: separate implementation of algorithms and combination of income, using content-based filtering in collaborative approach or using collaborative filtering in content-based approach [9].

3 Diabetes Recommender System

Recommendation systems with the application of the techniques described above can become an added value in various areas, in order to assist the user on topics that are of interest. One area where this could be applied is healthcare, where it is possible that patients with certain diseases can benefit from it, causing a more precise and accurate monitoring. Patients can benefit from the use of recommendation systems, for instance, to receive recommendations as they perform activities in their daily lives. The recommendations are receive automatically with suggestions that allow them to improve their health status through the adoption of healthier habits and behaviors.

Thus, it is suggested through data management techniques, the development of an information system, applied to patients with type 2 diabetes, which behaves as a recommendation system and indicates or suggests types and amounts of physical exercises and food diets to help the patient adopt healthier behaviors and be able to manage his/her disease more effectively.

The advance of technology has proven to have numerous advantages to the daily life of people around the globe. Thanks to this, we can create ubiquitous health systems in order to simplify the life of patients of have chronic diseases. As explained before, the recommendation systems for diabetes, aims to record and support a healthy diet, suggesting the best recommendations according to the specific condition of a patient. A dietary management is the key factor to prevent and treat chronic diseases such as obesity and diabetes. As a complement, the system should suggest and encourage the practice of physical activity according to the records previously acquired about the user. Thus, these systems will consist of a software that helps the patient through a continuous record of their routine, to make decisions about the food they must eat and the other above-mentioned goals [11].

Recently, the use of smartphones with enhanced features, coupled with advances in computer technology has enabled the development of new dietary recommendation systems for diabetes in the individual mobile phones. For example, these systems now offer features that can capture and receive meal's images and classifies them as a whole or segments and recognizes the food separately. The nutritional content of the meal can then be estimated through a nutritional database, and this information is finally returned to the patient through different mechanisms such as through reports or statistics. Other alternative could be using gamification strategies to increase the user's medication adherence and promote records practices of their routines [12].

In fact, one of the biggest challenges in new technologies for the consumer health is the low knowledge, resulting in a low adhesion problem. However, there are techniques that can help overcome this challenge and associated difficulties which have proven to be successful and gamification is one of them. The use of game-based technology can reinforce behavior changes and encourage the user to use the application itself. Gamification is a method that can be used to add interest to the users, especially those suffering from chronic diseases [13].

That method promises a great improvement that help making the routines and it's recorded a lot easier, still guaranteeing a certain commitment for the long run, with tasks previously considered demotivating. The use of gamification has also proven to offer a wide number of emotional, cognitive and social benefits to the user. Gamification can increase learner engagement for the user as they transform dull content into engaging and interesting experiences. It can encourage friendly competition among colleagues and can lead learners to feel pride in completing a goal after a series of gamified challenges and tasks [14].

Another advantage of gamification is that provides instant feedback, whether positive or negative. It allows learners to progress, not by chance, but by having the right knowledge or correct response to a question or scenario. Similarly, the lack of knowledge or an incorrect response does not allow learners to move forward [15].

Badges can be a form of gamification as it can boost user's motivation to continue moving forward and achieving some kind of "virtual" goal. It gives learners a sense of completion as well as a sense of authority, as they are a tangible symbol of the learner's accomplishments [15].

4 Proposed Solution

Based on the theory and context discussed in previous chapters, we proposed, in this article, the development of an application to support patients with type 2 diabetes. This application aims to assist and allow continuous recording of patient's routine, through the development of a recommendation system to assist the patient when making decisions throughout his/her daily life, such as recommending a certain food diets or recommend the practice of different physical exercises.

The application will initially consider a user profile, and for that, the following data will be included to describe the user: Gender (F/M), Age/Birth Date, Weight (kg), Height (cm).

In addition to register the patient's profile, there is a need to create a medical record associated with it, where it will be possible to register your daily routine using the data: Glucose, Diabetes, Tension, Food intake, Physical Exercise, Emotional State.

The registry of these data, which will constitute the medical record for each patient who uses the application, will allow him to manage his disease better, creating his "diary". It also allows the patient's companions, if it happens, to follow the same way, and understand the patient's routines.

These values will be equally important for the recommender system that will be implemented in this application, to recommend according to the needs and patient profile.

4.1 Mobile Application Architecture

In order to achieve the main goal of this article it is necessary to design an architecture of the software and hardware components that will be used and their respective relationship. Therefore, we propose an API service-oriented architecture (SOA). Examples of features which will be supported by the proposed system are to obtain and subsequent suggest certain foods for a meal to the user according to their nutritional value, or simply recommend practice of physical exercise.

For this, we consider the existence of two independent databases, where one contains the information of the patient register values (mentioned in the segment above) and the different kind of physical activities, when the other one will consist in the storage of the different kind of foods.

The connection between the mobile client and the databases/recommendation system will be performed by REST API micro services, which, in turn, using the same type of connection, would allow a user to call the respective application in a mobile device interface as shown down below (Fig. 1).

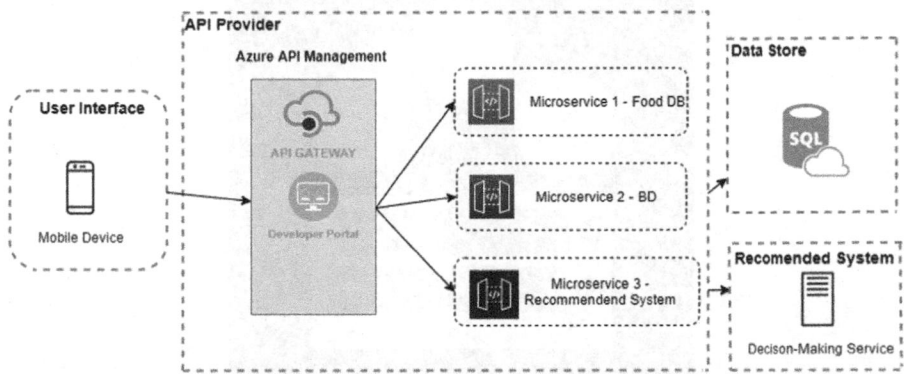

Fig. 1. Mobile application architecture

The connection between the mobile client and the different databases will be managed by an API gateway deployed in the Azure DevOps and the access to this gateway will be granted with a simple *API Key*. The junction of a API gateway, which allows to authenticate traffic as well as control and analyze how the API are being used and the API Key that is needed to grant access to the gateway itself, provides a better security to the communication made by the application. This is beneficial in a way that the mobile client does not have the need to communicate directly with the different databases, causing a better performance of the application itself and at the same time, granting a better security to the application. The API Management can then redirect the API Request to the different micro services. The *API Management* connection with the FoodDB will be made by a REST API micro service that uses an *API* service provided by *EDAMAM*[1], which has a food database of its own. Using *SQL Server Management Studio,* we can

---

[1] EDAMAM - https://developer.edamam.com/.

create a BD that is responsible for storing all the information relative to the user and his physical activities. This way, there is no need to store information on the mobile client internal memory, which allows the application to run more smoothly.

After obtaining the user data and the respective recommendations, these would be manipulated by the user on a mobile device on which he may give a feedback about the recommendations made by the system. In turn, the system will register and learn from the provided feedback to give more accurate and efficient recommendations. The following picture represents an example of a mobile interface where the user can obtain the registered values and its following recommendations (Fig. 2).

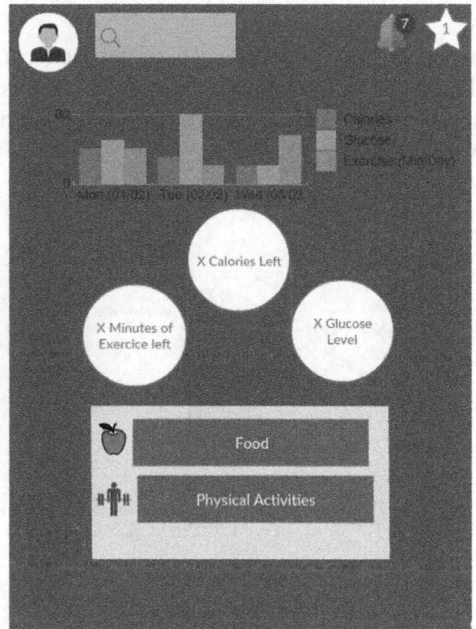

Fig. 2. Mobile interface mock-up

In this representation, a mobile interface is suggested for the user, where a daily graphic representation of the calorie intake, glucose level value and the amount of physical exercise appears, indicating the value that may be missing to reach an ideal daily record and provide a healthier and stable lifestyle. It also shows a section that the user can use to find suggestions of the best food to eat in his daily meals, as well as physical exercises that they can practice according to their characteristics and profile.

Additionally, the application will also provide a feature to measure the amount of calories of a single dish through a "Text Recognition" feature that can be used to detect the nutritional composition of certain food products, meals, or even food menus. With the help of *Mobile Vision Api*, which provides a framework for finding text in photos, the number of calories and the respective macronutrients of a single dish can be identified by matching the objects' description with the information retrieved from the EDAMAM database. This allows the user to have a better knowledge on his diet and, at the same time, provides a better management of his daily calorie intake.

The recommendations can be viewed though notifications that appear according to a continuous record of the user's daily routine so that the daily goals are not exceeded, and a healthy lifestyle is established. By following the suggested recommendations, the user is rewarded, and points are awarded for each step well exceeded. This is possible due to a gamification system implemented and represented, on the interface, by the "star" that contains the number of points established.

4.2 Recommendation System Architecture

The objective of building a recommender system for diabetes disease is to be able to suggest the best foods to eat in a meal and physical exercises to practice, according to the respective profile. At the moment of having a meal, there are several precautions that need to be taken, and several decisions that need to be made to choose the right foods to maintain a healthy diet. On the other hand, and at the same time, it is important to practice physical exercise, at least once a day, with food intake before and after doing so, to provide a healthier lifestyle and avoid sedentarism. The recommender system then includes all these factors, which include hydration, food nutrition, the level of calories consumed as well as those spent by the user through physical exercise.

The diets will be initially defined, and will indicate which foods should be consumed, the number of calories that the user should eat according to their lifestyle, and the type of physical exercise they should perform. Thus, a diet plan will address these issues and the recommendations will be suggested according to that plan and the values defined. Therefore, the recommender system will obtain a diet that corresponds to the nutritional requirements of the user's profile and recommend based on the established values.

Regarding the strategy followed to implement the recommendations, a hybrid system is considered since it contemplates more than one recommender technique, such as collaborative filtering and content-based filtering. For this recommender system, the items will correspond to the food to eat and the possible physical exercises for each user to practice. The architecture of this recommendation system is represented and described below (Fig. 3).

- **Content-based filtering**: through this technique, the algorithm recommends a food or physical exercise that the user has already eaten or practiced, respectively. These algorithms are based on behaviours and classifications that the user has already assigned to the items and recommends based on those registered and stored preferences.
- **Collaborative filtering**: this type of filtering is applied to provide food or physical exercises for the user based on the classifications of those previously made by other users. One of the advantages of this algorithm is that it does not limit the user to just his preferences and manages to obtain items different from those that have already been recommended to him.

The recommender system for this application will be implemented based on an existing library called librec, with a hybrid recommender system method implemented. This library will constitute the service developed in Java, placed on another DevOps server.

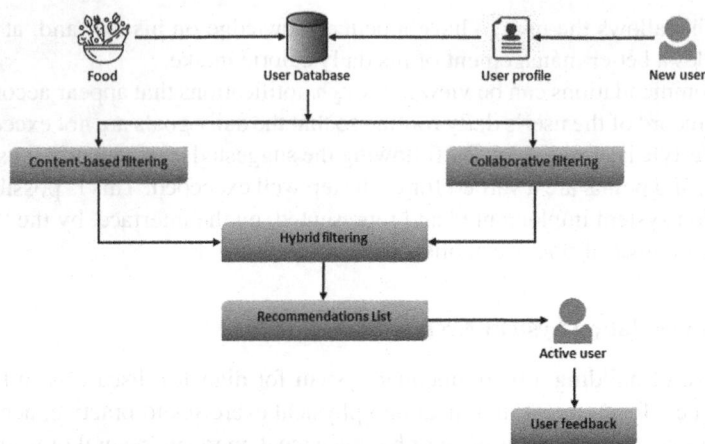

Fig. 3. Recommendation system architecture

Subsequently, the application will communicate with this server by REST API. Communications from the application to the DevOps server will result in the recommendations that will appear whenever the user resort this functionality and need some suggestion at the time of a decision, whether for a meal, or for the practice of physical exercise [16]. The list of recommendations is finally proposed to the user, who accepts or rejects them. When assigning ratings to recommendations, these will be stored in a repository, and used by the system to update and shape its profile.

5   Conclusions and Future Work

Recommender Systems in health has been subject of study over the past years in order to serve as a solution to help both healthcare professionals and patients.

In this paper, we propose the design of a recommendation system in the area of healthcare, specifically for the management of diabetes mellitus type 2 disease. The proposed recommendation system will aim to facilitate and assist the user in the management of diabetes disease by supporting the user and provide suggestions to improve daily behaviours both related to eating habits and practice of physical exercise. Regarding eating habits, the proposed system will provide recommendation of food intake or recommendation of new recipes based on the existing level of glucose in the user's blood as well as the number of calories needed daily user always considering the healthy lifestyle. Even if the appropriate diet for the user to is one of the key points for the management of this type of disease, physical exercise also remains crucial since following the practice of exercise, can prevent the occurrence of Hypo/Hyperglycemia. The amount of the aerobic exercise and/or force varies from user to user and should be adapted to physiological characteristics of the user.

Furthermore, the proposed recommendation system follows a novel micro services architecture which combines simultaneously both tracking and user diet and daily routine monitoring features to make recommendations. There are not many approaches in the literature that follow this kind of approach based on micro services and by doing so it becomes possible for the proposed solution to be easily integrated with other existing approaches.

As future work, we suggest the implementation of a recommendation system in a mobile application, where the patient can monitor his/her diet and daily routine. Besides that, the patient will also be to access data such as physical activity performed, the existing level of glucose in the blood while still receiving recommendations to support decision-making in order to manager his/her disease more effectively and easy.

Acknowledgments. The work presented in this paper has been developed by National Funds through FCT (Fundação para a Ciência e a Tecnologia) within the Projects UIDB/00319/2020, UIDB/00760/2020 and the Luís Conceição Ph.D. Grant with the reference SFRH/BD/137150/2018.

References

1. Roche, L.: Farmacêutica Química. https://www.roche.pt/corporate/index.cfm/infosaude/patologias/diabetes-tipo-2/. Accessed 13 June 2019
2. DeFronzo, R.A., et al.: Type 2 diabetes, 9 Fevereiro 2017
3. Neuenschwander, M., et al.: Role of diet in type 2 diabetes incidence: umbrella review of meta-analyses of prospective observational studies (2019)
4. Lisa, M., Leontis, R.N.: ANP-C, Amy Hess-Fischl: EndocrineWeb, 09 Setembro 2018. https://www.endocrineweb.com/conditions/type-2-diabetes/type-2-diabetes-complications. Accessed 29 Setembro 2019
5. AdvanceCare. https://advancecare.pt/artigos/saude-e-bem-estar/diabetes-e-alimentacao-os-segredos. Accessed 13 Junho 2019]
6. Faria, H.T.G., et al.: Adesão ao tratamento em diabetes (2014)
7. Khan, M.M., Ibrahim, R.: Cross Domain Recommender Systems: A Systematic Literature Review, Junho 2017
8. Villegas, N.M., Sánchez, C., Díaz-Cely, J., Tamura, G.: Characterizing context-aware recommender systems: A systematic literature review, 2 Novembro 2017
9. Assaf, D., Al-Eid, K., Asfour, M.: Recommender Systems: Standards and Algorithms
10. Recommender Systems. http://recommender-systems.org/hybrid-recommender-systems/. Accessed 09 Fevereiro 2020
11. Jung, H., Chung, K.: Knowledge-based dietary nutrition recommendation for obese management (2015)
12. Christodoulidis, S., Anthimopoulos, M., Mougiakakou, S.: Food Recognition for Dietary Assessment Using Deep Convolutional Neural Networks (2015)
13. Malwade, L.S.: Mobile and wearable technologies in healthcare for the ageing, 26 Abril 2018
14. Sardi, L., Idri, A., Fernández-Alemán, J.L.: A systematic review of gamification in e-Health, 15 Maio 2017
15. Gerli, C.: What are the advantage and disadvantages of gamification? 2020 January 2020. https://www.knowbly.com/post/what-are-the-advantages-and-disadvantages-of-gamification. Accessed 09 Fevereiro 2020
16. Guo, G., Zhang, J., Sun, Z., Yorke-Smith, N.: LibRec: a java library for recommender systems, in posters, demos. In: Late-breaking Results and Workshop Proceedings of the 23rd Conference on User Modelling, Adaptation and Personalization (UMAP) (2015)

A Comparative Analysis Between Crisp and Fuzzy Data Clustering Approaches for Traditional and Bioinspired Algorithms

Amanda Coradini and Alexandre Szabo$^{(\boxtimes)}$

Federal University of Grande Dourados, Dourados, MS, Brazil
amandaa.coradini@gmail.com, alexandreszabo@ufgd.edu.br

Abstract. Partitional data clustering algorithms produce a relationship matrix between data and clusters, named membership matrix, which clusters can be treated as mutually exclusive (crisp) or not (fuzzy), according to data clustering approach used. Moreover, a partition obtained by a crisp algorithm can be fuzzified and so, the relationship between data and clusters be relaxed, such as in fuzzy data clustering approach. However, algorithms have your own heuristic and, iteratively, the behavior of a crisp algorithm can be different of that respective fuzzy version and, in addition, fuzzifying a crisp partition can produce different result in relation to crisp and fuzzy clustering algorithms. Therefore, this paper proposes a comparative analysis among results produced by fuzzy data clustering algorithms and their respective crisp versions and fuzzified partitions. The proposal is identify whether a fuzzy clustering algorithm can be replaced by its respective crisp with fuzzified partition, in terms of result quality. The experiments were performed to two traditional partitional algorithms and two bioinspired algorithms.

Keywords: Fuzzy data clustering · FCM · k-Means · Particle Swarm Clustering

1 Introduction

Data clustering algorithms aims assign data (objects) to clusters according to similarity among them, such that data in a same cluster are similar to each other. The objective is minimize intra-cluster distances and maximize inter-cluster distances among objects [1, 2], i.e., produce compact and separable clusters in a iterative process, where the data attribute values determine their position on search space.

Partitional data clustering is a class of algorithms that segment objects in clusters, which are informed by user (*a priori*) or dynamically defined by algorithm [3, 4]. Such algorithms use vectors (prototypes), with dimension given by the attributes of dataset to be grouped, and they move by the search space to become centroids representing clusters. Thus, prototypes reduce the size of dataset to be analyzed, since they generalize the characteristics of their respective clusters.

There are several areas of data clustering application, such as medicine, psychology, business and engineering [5–7].

© Springer Nature Switzerland AG 2020
C. Analide et al. (Eds.): IDEAL 2020, LNCS 12490, pp. 298–308, 2020.
https://doi.org/10.1007/978-3-030-62365-4_29

Clustering is considered a most difficult task in data mining area because of its unsupervised nature, i.e., the class of data is unknown *a priori* and clustering algorithms goal find such classes, which can be identified by natural clusters [6].

According to Carmichael [8], natural clusters are densely populated regions surrounded by empty regions. However, in real problems can be hard identify such clusters; clusters can be overlapped, present different forms and sizes [9, 10], become clustering a challenging task. Moreover, a dataset can be clustered in several ways, according to heuristic of algorithm and its parametric setting, which characterizes it as an optimization problem [11]. Thus, nature inspired algorithms have been applied to solve clustering problems, which minimize its computational complexity and produce satisfactory results [12–14].

There are two main approaches in clustering algorithms: crisp (hard) and fuzzy. The crisp approach assigns each object in a cluster, meanwhile in fuzzy approach objects belong to all clusters simultaneously, with different membership degrees: the closer the object to a cluster, the higher the membership degree. The fuzzy approach allows quantify the belonging of objects to clusters [15], i.e., how similar or important it is to that cluster, and minimize contribution of noisy objects in updating prototypes, improving their representativeness.

A crisp clustering algorithm can produces a different partition of that respective fuzzy version as well as fuzzifying a crisp partition can produce different result in relation to crisp and fuzzy algorithms. Therefore, this paper proposes a comparative analysis among results produced by fuzzy data clustering algorithms, respective crisp versions, and their fuzzified crisp partitions. The proposal is identify whether a fuzzy clustering algorithm can be replaced by its respective crisp with fuzzified partition, in terms of result quality. For that, the experiments were performed for two of the most traditional algorithms from the literature and two bioinspired algorithms.

Partitional data clustering algorithms are often applied in real problems and we focused the two more traditional algorithms. Furthermore, bioinspired algorithms had gained popularity in solution of data clustering problems, such as algorithms based on swarms, and two of them were investigated in this paper.

The paper is organized as follows. Section 2 introduces some data clustering concepts and describes the k-Means and its fuzzy version – Fuzzy c-Means algorithm (FCM). Section 3 addresses one of the nature inspired techniques and describes two bioinspired algorithms, crisp and its fuzzy version. Methodology, results and discussion are presented in Sect. 4 and the paper is concluded in Sect. 5 with future works.

2 Data Clustering

Let $X = \{x_1, x_2, \ldots, x_n\}$ be a set of n data to be grouped. A partitional data clustering algorithm will produce a set of k clusters $C = \{c_1, c_2, \ldots, c_k\}$, such that $1 < k < n$ and $c_j \cap c_k = \emptyset$ [5, 16]. Since data are represented in a vector space, with dimension given by their attributes, the dissimilarity to each other can be determined by some distance measure, such as Euclidean distance:

$$d_{ij} = \sqrt{\sum_{a=1}^{dim} \left(x_i^a - c_j^a \right)^2} \tag{1}$$

where x_i and c_j are vectors of dimension *dim*. Thus, objects close to each other (similar) are assign into a same cluster.

A data clustering algorithm produces as result a matrix U_{nXk}, named membership matrix, which represents the relationship between objects and clusters:

$$\mu_{ij} = \begin{cases} 1, \textit{if } d_{ij} < d_{ik}, \forall j, k \in C, k \neq j \\ 0, \textit{otherwise} \end{cases} \tag{2}$$

One of the most popular data clustering algorithms, k-Means [17], aims optimize the function value:

$$J = \sum_{j=1}^{k} \sum_{i=1}^{n} \mu_{ij} d_{ij}^{2} \tag{3}$$

which minimizes intra-cluster distances in an iterative process composed by two steps: (i) update membership matrix (Eq. 2), and (ii) update prototypes (Eq. 4).

$$c_j = \frac{\sum_{i=1}^{n} \mu_{ij} x_i}{\sum_{i=1}^{n} \mu_{ij}} \tag{4}$$

In 1981, Bezdek [18] introduced a parameter responsible for relaxing the membership degree between objects and clusters, named weighting exponent or fuzzifier (m), which value must be in $(1, \infty)$. The higher (smaller) the m, the more overlapped (crisp) the clusters. The m value depends of the structure of clusters, and it is common values in [1.25, 2] [19].

Thus, in fuzzy approach, each object belongs to all clusters, simultaneously, with different membership degrees. The membership degree of an object to a cluster is defined by distance between them; the smaller (higher) the distance, the closer to one (zero):

$$\mu_{ij} = \frac{1}{\sum_{r=1}^{k} \left(\frac{d_{ij}}{d_{ir}}\right)^{\frac{2}{m-1}}} \tag{5}$$

where μ_{ij} is the membership degree between object x_i and cluster c_j, such that $\mu_{ij} \in [0, 1]$. That equation fraction the membership degree of an object to all clusters, which satisfies the follows restrictions [5]:

$$\sum_{j=1}^{k} \mu_{ij} = 1, \forall i = 1, \ldots, n \tag{6}$$

$$0 < \sum_{i=1}^{n} \mu_{ij} < n, \forall j = 1, \ldots, k \tag{7}$$

Restriction 1 (Eq. 6) ensures that the sum of membership degrees of an object to all clusters is one, meanwhile Restriction 2 (Eq. 7) ensures no empty clusters.

A fuzzy version of k-Means, FCM, is obtained by adding $m > 1$ in its equations:

$$J = \sum_{j=1}^{k} \sum_{i=1}^{n} \mu_{ij}^{m} d_{ij}^{2} \tag{8}$$

$$c_j = \frac{\sum_{i=1}^{n} \mu_{ij}^{m} x_i}{\sum_{i=1}^{n} \mu_{ij}^{m}} \tag{9}$$

3 Particle Swarm

Swarm Intelligence (or Collective Intelligence) is one of the grand areas of the bioinspired computing [20]. It is characterized by social agents with limited cognitive abilities, who interact to each other and with the environment, producing an auto-organized behavior. From this interaction rises the called "intelligence of swarm", i.e., the solution to a complex problem, such as optimization and data clustering [21–23].

Collective systems include colony of termite, ants, bee, social insets in general, or any system, such cars, people, which present a collective behavior. Thus, behaviors from nature are mathematically and computationally modeled as metaphor to solve real problems.

One of main techniques from Swarm Intelligence is the Particle Swarm. It was proposed by Kennedy and Eberhart [24], by introducing the Particle Swarm Optimization algorithm (PSO), which was inspired by school of fish, flock of birds, people or any social agent. PSO is one of the most popular optimization algorithms.

Each agent in PSO is called particle, modeled in a vector space (environment), which has its own velocity (Eq. 10) and position (Eq. 11):

$$v_j(t+1) = \omega(t) * v_j(t) + r_1 * \big(p_j(t) - c_j(t)\big) + r_2 * \big(g(t) - c_j(t)\big) \tag{10}$$

$$c_j(t+1) = c_j + v_j(t+1) \tag{11}$$

where ω, named inertia moment, control the exploration of search space; constants r_1 and r_2 are acceleration coefficients and define the particles' step size; $\big(p_j(t) - c_j(t)\big)$ is named cognitive term, which represents the particle's experience, where $p_j(t)$ represents its best position so far, meanwhile $\big(g(t) - c_j(t)\big)$ is the social term, which represents the swarm experience, where $g(t)$ is the best position among all particles. Velocity and position are updating, iteratively, until a stopping criterion to be achieved.

3.1 Particle Swarm Clustering

Since PSO has produced satisfactory results in solving optimization problems, Cohen and de Castro proposed some modifications in PSO to solve data clustering problems, culminating in PSC algorithm (Particle Swarm Clustering) [25].

In PSC, each particle is a prototype and objects compose the environment within which the particles interact. A third term was added in velocity equation, named self-organizing term, which moves particle to input object:

$$v_j(t+1) = \omega(t) * v_j(t) + \varphi_1 \otimes (p_j^i(t) - c_j(t)) + \varphi_2 \otimes \big(g^i(t) - c_j(t)\big)$$
$$+ \varphi_3 \otimes \big(y^i(t) - c_j(t)\big) \tag{12}$$

where φ_1, φ_2 and φ_3 are positive random vectors that influence the exploration of the search space; $p_j^i(t)$ represents the best particle position achieved so far in relation to the input object $(y^i(t))$, and $(g^i(t))$ is the nearest particle to the object $y^i(t)$ at time t. The symbol $\otimes$ represents an element-wise vector multiplication.

The algorithm presents each input object to swarm and the closest particle (smallest distance between particle and input object), called *winner*, is updated (Eq. 12 and Eq. 11). Before updating, the *winner* particle position is compared with its memories cognitive and social, which are updated by particle position if it is the best ones. After each iteration, the algorithm checks for particle stagnations and moves stagnated particles towards the one that most won at the current iteration.

3.2 Membership Weighted Fuzzy Particle Swarm Clustering

In 2011, Szabo et al. [26] presented an extension of PSC to be applied in fuzzy partitions, named Fuzzy Particle Swarm Clustering (FPSC). The main modifications in FPSC occurred in selection and evaluation of particles:

- Selection of particles: the *winner* particle is the prototype for which the input object has the maximal membership degree.
- Election of particles: the comparison between the *winner* and its memories considers the membership degree of the object to such particles.

However, in 2015, Szabo et al. [14] proposed a modification in FPSC, directly introducing membership degrees into particle velocity equation, in order to weight the step size of particles. Thus, particles move toward its memories and input object with an uncertainty degree:

$$v_j(t+1) = \omega(t) * v_j(t) + \mu_1 * \varphi_1 \otimes (p_j^i(t) - c_j(t)) + \mu_2 * \varphi_2 \otimes \left(g^i(t) - c_j(t)\right)$$
$$+ \ \mu_3 * \varphi_3 \otimes \left(y^i(t) - c_j(t)\right) \tag{13}$$

where μ_1 is the membership degree of object $y^i(t)$ to the cluster represented by particle $p_j^i(t)$; μ_2 is the membership degree of object $y^i(t)$ to the cluster represented by particle $g^i(t)$ and μ_3 is the membership degree of object $y^i(t)$ to the cluster represented by particle $c_j(t)$.

4   Performance Assessment

This section presents and discuss results obtained by two crisp and respective fuzzy clustering algorithms, as well as respective fuzzified crisp partitions, in a comparative analysis between approaches, for six benchmarking datasets (Table 1). Moreover, the comparisons were extended for a behavioral analysis of the algorithms, iteratively until their convergence (Figs. 1 and 2).

4.1   Methodology

k-Means, FCM, PSC and MWFPSC algorithms were implemented in Octave[1] and evaluated for a synthetic dataset (Ruspini) and five real datasets from the literature[2]: Wine,

---

[1] https://www.gnu.org/software/octave/.
[2] https://archive.ics.uci.edu/ml/datasets.html.

Ionosphere, Iris, Statlog Heart and Liver. The amount of prototypes was equal to classes present in respective dataset. The experiments were run ten times for each dataset, such that the prototypes initialization was the same for all algorithms in respective run for a same dataset. Thus, the same initial scenario for each run, in terms of prototypes' position, has been equally applied for all algorithms to avoid any advantage among prototype initializations. The stopping criterion was a maximum of 200 iterations or a minimization of the objective function ($|xb(t) - xb(t + 1)| < 10^{-2}$):

$$xb = \frac{\sum_{j=1}^{k} \sum_{i=1}^{n} \mu_{ij} d_{ij}^2}{n * \left(d_{jk}^2 \right)} \tag{14}$$

where xb is the Xie-Beni index value [27]. The closer the zero, more compact and separable the clusters and so, the better the solution.

For bioinspired algorithms, the *inertia moment* (ω) has an initial value of 0.90, with an iterative decay of 95% at each iteration until the value 0.01 is obtained. The position and velocity of the particles are also controlled and ranging from $[-0.1, 0.1]$ and $[0, 1]$, respectively.

For fuzzy algorithms and to fuzzify crisp partitions, the fuzzifier m was equal two. The fuzzification of a crisp partition is obtained by generating a fuzzy membership matrix (Eq. 5).

The results were evaluated by xb index and percentage of correct classification ($pcc(\%)$), in terms of mean $\pm$ standard deviation (Table 1), and the behavior of the algorithms was illustrated to one of runs, iteratively (Figs. 1 and 2).

4.2 Comparative Analysis Between Crisp and Fuzzy Approaches

Table 1 presents xb and $pcc(\%)$ values obtained by k-Means and respective fuzzified partition and fuzzy version, as well as for the bioinspired algorithms. A behavioral analysis for algorithms was performed and results are showed in Figs. 1 and 2, as following.

According to Table 1, fuzzified partitions k-Means presented the best xb values in relation to k-Means for all datasets, as well as for FCM, except for Ruspini and Iris, on average. FCM was the best in relation to k-Means, except for Statlog and Liver. For Ionosphere, both algorithms presented the same result. The best performance in $pcc(\%)$ was for FCM in most datasets. k-Means and its fuzzified partitions produced the same results for $pcc(\%)$, on average. Thus, fuzzified partitions produce more compact and separable clusters, meanwhile FCM produces more assertive classification. For bioinspired algorithms, fuzzified partitions PSC presented the best xb values in relation to PSC and MWFPSC for all datasets. The exception was for Ruspini, for which MWFPSC was the best. MWFPSC presented the best performance in $pcc(\%)$ for half of datasets. PSC and its fuzzified partitions presented the same performance, on average. Bioinspired fuzzified partitions produce more compact and separable clusters as occurred for traditional fuzzified partitions, and there is no difference between PSC and its fuzzified partitions in terms of $pcc(\%)$, on average.

Thus, xb value was better for both traditional and bioinspired fuzzified partitions, meanwhile the best $pcc(\%)$ was for FCM, and half of datasets for MWFPSC. The uncertainty iteratively generated by FCM reduces the step size of prototypes in the search

Table 1. Mean ± std of xb and $pcc(\%)$ for k-Means, Fuzzified k-Means, Fuzzy c-Means (FCM), and bioinspired algorithms: Particle Swarm Clustering (PSC), Fuzzified Particle Swarm Clustering and Membership Weighted Fuzzy Particle Swarm Clustering (MWFPSC).

Base	Measure	k-Means	Fuzzified k-Means	FCM	PSC	Fuzzified PSC	MWFPSC
Rus.	xb	0.74 ± 1.40	0.46 ± 0.83	0.03 ± 0.00	0.15 ± 0.07	0.10 ± 0.04	0.08 ± 0.03
	pcc (%)	93.06 ± 11.27	93.06 ± 11.27	100 ± 0.00	98.53 ± 2.70	98.53 ± 2.70	98.40 ± 2.87
Win.	xb	0.50 ± 0.00	0.30 ± 0.00	0.41 ± 0.00	0.55 ± 0.14	0.28 ± 0.07	0.33 ± 0.08
	pcc (%)	94.55 ± 0.60	94.55 ± 0.60	94.88 ± 0.18	72.47 ± 8.71	72.47 ± 8.71	72.64 ± 14.98
Ion.	xb	0.75 ± 0.00	0.48 ± 0.00	0.75 ± 0.00	0.46 ± 0.11	0.28 ± 0.06	0.33 ± 0.08
	pcc (%)	71.08 ± 0.15	71.08 ± 0.15	70.94 ± 0.00	66.35 ± 2.28	66.35 ± 2.28	65.87 ± 3.00
Iris	xb	0.37 ± 0.25	0.28 ± 0.18	0.18 ± 0.00	0.29 ± 0.07	0.18 ± 0.04	0.21 ± 0.12
	pcc (%)	82.46 ± 10.91	82.46 ± 10.91	89.40 ± 0.66	78.40 ± 9.73	78.40 ± 9.73	82.13 + 8.17
S. H.	xb	1.00 ± 0.21	0.65 ± 0.12	2.26 ± 0.00	0.60 ± 0.12	0.36 ± 0.06	0.40 ± 0.11
	pcc (%)	73.77 ± 9.89	73.77 ± 9.89	79.26 ± 1.50	70.40 ± 9.71	70.40 ± 9.71	68.52 ± 9.75
Liv.	xb	0.37 ± 0.08	0.26 ± 0.05	0.95 ± 0.00	0.16 ± 0.07	0.12 ± 0.05	0.13 ± 0.07
	pcc (%)	57.97 ± 0.00	57.97 ± 0.00	57.97 ± 0.00	58.11 ± 0.46	58.11 ± 0.46	58.09 ± 0.28

space, which promotes a better exploitation. Consequently, it can improve the percentage of correct classification and can require more iterations to convergence (Fig. 1).

A normality test was applied among algorithms for results of xb and $pcc(\%)$, individually, to determine whether a parametric or non-parametric test should be used [28]. For that, Shapiro-Wilk test [29] was performed and the result showed that null hypothesis must be rejected for xb and $pcc(\%)$, with 95% of confidence. Thus, Wilcoxon [30] was applied as a non-parametric test, and it concluded that there is significant difference in performance among algorithms for xb and $pcc(\%)$, with 95% of confidence. The behavior of the algorithms is depicted in Figs. 1 and 2, iteratively, until their convergence:

In Fig. 1, FCM needs more iterations to convergence among algorithms, except for Liver (Fig. 1(f)). For all approaches, the decay of xb is constant among iterations for all datasets. The best results were observed for fuzzified crisp partitions for all datasets, except for Ruspini and Iris (Figs. 1(a, d)), for which FCM was the best. Fuzzified crisp partitions were better than crisp ones for all datasets. In Fig. 2, MWFPSC required lesser iterations to convergence among algorithms, except for Ruspini and Wine (Figs. 2(a, b)). Thus, weighting the step size of particles in MWFPSC can limit the exploration of the search space for some scenarios. The performance of algorithms presented higher variation along of iterations in relation that in Fig. 1 because traditional algorithms

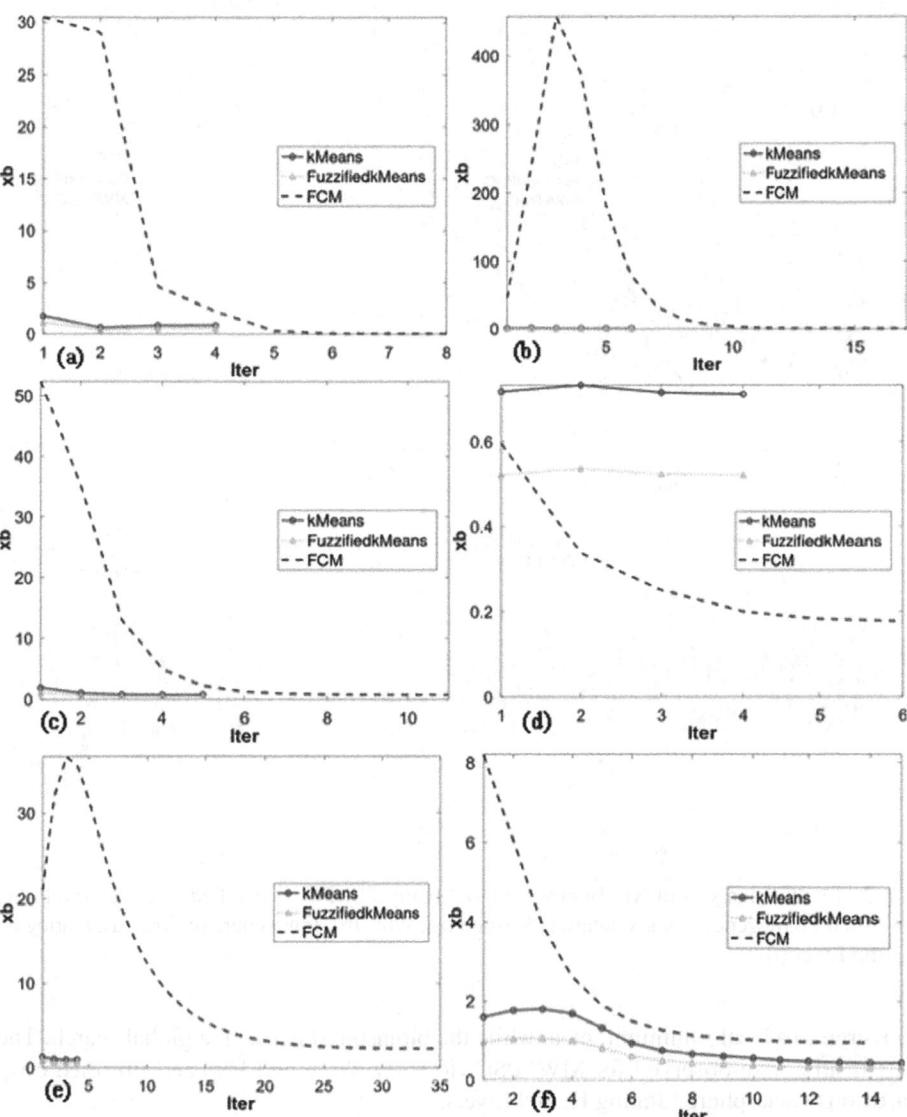

Fig. 1. Iterative analysis of Xie-Beni value (xb) obtained by k-Means, Fuzzified k-Means and FCM until their convergence, for six datasets: Ruspini (a), Wine (b), Ionosphere (c), Iris (d), Statlog H. (e) and Liver (f).

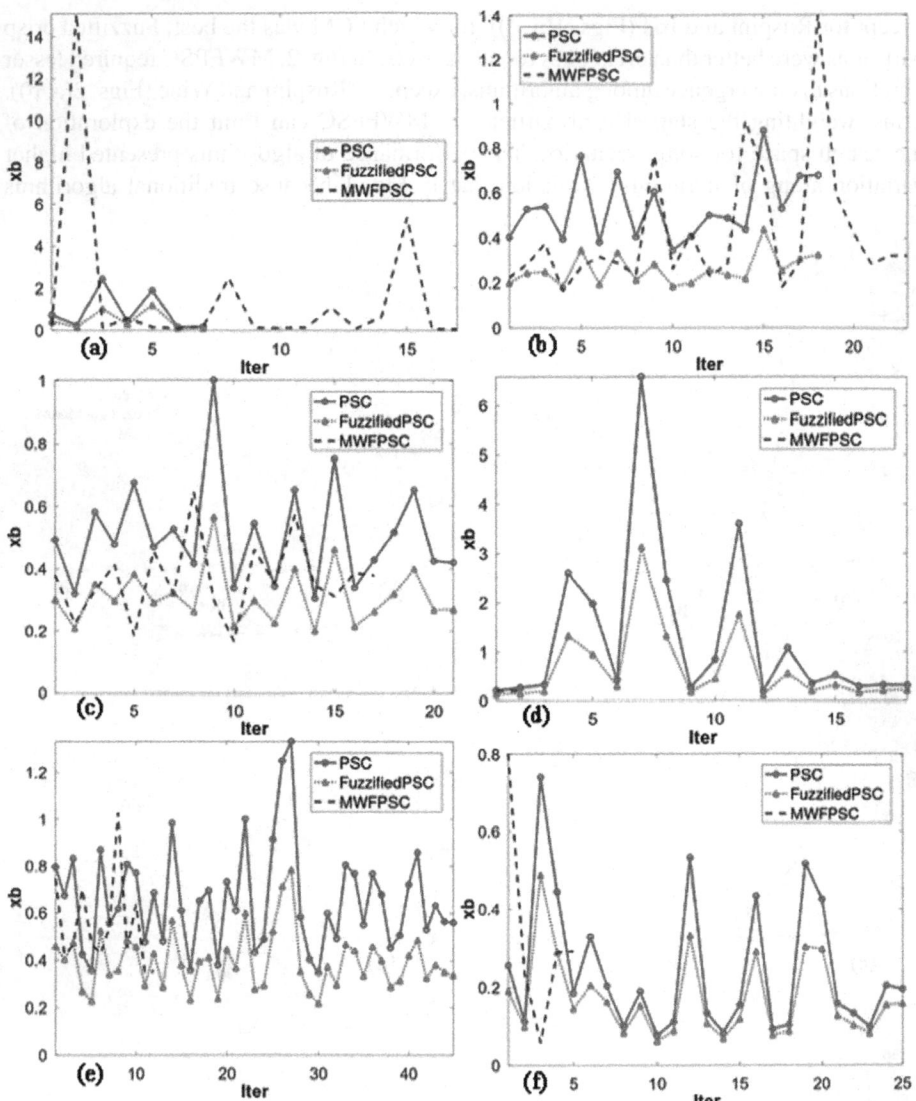

Fig. 2. Iterative analysis of Xie-Beni value (xb) obtained by PSC, Fuzzified PSC and MWFPSC until their convergence, for six datasets: Ruspini (a), Wine (b), Ionosphere (c), Iris (d), Statlog H. (e) and Liver (f).

converge to a local minimum, meanwhile the bioinspired perform a global search. The best results were observed for MWFPSC (Ruspini, Wine and Iris) and fuzzified crisp partitions (Ionosphere, Statlog H. and Liver).

5 Conclusion and Future Works

This paper proposed a comparison between crisp and fuzzy data clustering approaches for traditional and bioinspired algorithms. The proposal was identify whether a fuzzy clustering algorithm can be replaced by its respective crisp with fuzzified partition, in terms of result quality. That is important to decide which approach to apply to get the better results. Fuzzy approach relaxes the belonging between data and clusters, which minimizes the contribution of noisy data in prototypes updating, without adding a significant computational cost to clustering algorithm.

The results concluded that fuzzified crisp partitions produce more compact and separable clusters, meanwhile fuzzy clustering algorithms are more assertive, on average, because the step size of prototypes is weighted, which can favors the exploitation. Crisp data clustering approach produces same $pcc(\%)$ that its fuzzified partition for traditional and bioinspired algorithms. Thus, assigning membership degrees to data along iterations or fuzzifying crisp partitions can be the best choices for gain in terms of pcc or xb measures.

Further investigations include evaluate the fuzzy approach in other clustering techniques, such as hierarchical clustering and density-based clustering, other bioinspired techniques and different evaluation measures.

Acknowledgments. The authors thank Federal University of Grande Dourados (UFGD) for the financial support.

References

1. Han, J., Kamber, M.: Data Mining: Concepts and techniques, 2nd edn. Elsevier, San Francisco (2006)
2. Rawashdeh, M., Ralescu, A.: Crisp and fuzzy cluster validity: generalized intra-inter silhouette index. In: 2012 Annual Meeting of the North American Fuzzy Information Processing Society (NAFIPS), Berkeley, pp. 1–6. IEEE (2012)
3. Shafeeq, A., Hareesha, K.S.: Dynamic clustering of data with modified k-means algorithm. In: International Information and Computer Networks (ICICN), Singapore, vol. 27, pp. 221–225 (2012)
4. Li, H., He, H., Wen, Y.: Dynamic particle swarm optimization and k-means clustering algorithm for image segmentation. Optik **126**(24), 4817–4822 (2015)
5. de Oliveira, J.V., Pedrycz, W.: Advances in Fuzzy Clustering and Its Applications. Wiley, New York (2007)
6. Jain, A.K., Law, M.H.C.: Data clustering: a user's dilemma. In: Pal, S.K., Bandyopadhyay, S., Biswas, S. (eds.) PReMI 2005. LNCS, vol. 3776, pp. 1–10. Springer, Heidelberg (2005). https://doi.org/10.1007/11590316_1
7. Jain, A.K.: Data clustering: 50 years beyond k-means. Pattern Recognit. Lett. **31**(8), 651–666 (2010)
8. Carmichael, J.W., George, J.A., Julius, R.S.: Finding natural clusters. Syst. Biol. **17**(2), 144–150 (1968)
9. Nagy, G.: State of the art in pattern recognition. Proc. IEEE **56**(5), 836–863 (1968)
10. Ruspini, E.H., Bezdek, J.C., Keller, J.M.: Fuzzy clustering: a historical perspective. IEEE Comput. Intell. Mag. **14**(1), 45–55 (2019)

11. Jain, A.K., Murty, M.N., Flynn, P.J.: Data clustering: a review. ACM Comput. Surv. (CSUR) **31**(3), 264–323 (1999)
12. Ahmadyfard, A., Modares, H.: Combining PSO and k-means to enhance data clustering. In: 2008 International Symposium on Telecommunications, Tehran, pp. 688–691. IEEE (2008)
13. da Cruz, D.P.F., Maia, R.D., Szabo, A., de Castro, L.N.: A bee-inspired algorithm for optimal data clustering. In: IEEE Congress on Evolutionary Computation (CEC), Cancun, pp. 3140–3147. IEEE (2013)
14. Szabo, A., Delgado, M.R., de Castro, L.N.: A particle swarm clustering algorithm with fuzzy weighted step sizes. In: Jackowski, K., Burduk, R., Walkowiak, K., Woźniak, M., Yin, H. (eds.) IDEAL 2015. LNCS, vol. 9375, pp. 87–95. Springer, Cham (2015). https://doi.org/10.1007/978-3-319-24834-9_11
15. Szabo, A., de Castro, L.N., Delgado, M.R.: A constructive particle swarm algorithm for fuzzy clustering. In: Yin, H., Costa, J.A.F., Barreto, G. (eds.) IDEAL 2012. LNCS, vol. 7435, pp. 390–398. Springer, Heidelberg (2012). https://doi.org/10.1007/978-3-642-32639-4_48
16. Nisha, P.J.K.: A survey of clustering techniques and algorithms. In: 2015 2nd International Conference on Computing for Sustainable Global Development (INDIACom), New Delhi, pp. 304–307. IEEE (2015)
17. MacQueen, J.B.: Some Methods for classification and analysis of multivariate observations. In: 1967 5th Symposium on Mathematical Statistics and Probability, pp. 281–297. University of California Press (1967)
18. Bezdek, J.: Pattern Recognition with Fuzzy Objective Function Algorithms. Plenum Press, New York (1981)
19. Cox, E.: Fuzzy Modeling and Genetic Algorithms for Data Mining and Exploration. Morgan Kaufmann, San Francisco (2005)
20. de Castro, L.N.: Fundamentals of Natural Computing: Basic Concepts, Algorithms, and Applications. Chapman & Hall/CRC, Boca Raton (2006)
21. de Oliveira, J.V., Szabo, A., de Castro, L.N.: Particle swarm clustering in clustering ensembles: exploiting pruning and alignment free consensus. Appl. Soft Comput. **55**, 141–153 (2017)
22. Rana, S., Jasola, S., Kumar, R.: A review on particle swarm optimization algorithms and their applications to data clustering. Artif. Intell. Rev. **35**, 211–222 (2011)
23. Sengupta, S., Basak, S., Peters, R.A.: Data clustering using a hybrid of fuzzy c-means and quantum-behaved particle swarm optimization. In: 2018 IEEE 8th Annual Computing and Communication Workshop and Conference (CCWC), Las Vegas, pp. 137–142. IEEE (2018)
24. Kennedy, J., Eberhart, R.C.: Particle swarm optimization. In: IEEE International Conference on Neural Networks, Perth, pp. 1942–1948. IEEE (1995)
25. Cohen, S.C.M., de Castro, L.N.: Data clustering with particle swarms. In: 2006 IEEE International Conference on Evolutionary Computation, Vancouver, pp. 1792–1798. IEEE (2006)
26. Szabo, A., de Castro, L.N., Delgado, M.R.: The proposal of a fuzzy clustering algorithm based on particle swarm. In: 2011 Third World Congress on Nature and Biologically Inspired Computing, Salamanca, pp. 459–465. IEEE (2011)
27. Xie, X.L., Beni, G.: A validity measure for fuzzy clustering. IEEE Trans. Pattern Anal. Mach. Intell. **13**(8), 841–847 (1991)
28. Derrac, J., Garcia, S., Molina, D., Herrera, F.: A practical tutorial on the use of nonparametric statistical tests as a methodology for comparing evolutionary and swarm intelligence algorithms. Swarm Evol. Comput. **1**(1), 3–18 (2011)
29. Shapiro, S.S., Wilk, M.B.: An analysis of variance test for normality. Biometrika **52**, 591–611 (1965)
30. Wilcoxon, F.: Individual comparisons by ranking methods. Biometrics Bull. **1**(6), 80–83 (1945)

Bridging the Gap of Neuroscience, Philosophy, and Evolutionary Biology to Propose an Approach to Machine Learning of Human-Like Ethics

Nicolas Lori$^{(\boxtimes)}$ ⓘ, Diana Ferreira ⓘ, Victor Alves ⓘ, and José Machado ⓘ

Centre Algoritmi, University of Minho, 4710-057 Braga, Portugal
{nicolas.lori,diana.ferreira}@algoritmi.uminho.pt
{valves,jmac}@di.uminho.pt

Abstract. The growing explosion of ideas such as Artificial Intelligence (AI), smart environments and ubiquitous computing has led to the creation of the Ambient Intelligence (AmI) paradigm. As AmI begins to take place, moving from a futuristic idea to a reality, we are gradually witnessing the creation of an omnipresent, responsive, and intelligent atmosphere in which thousands of tiny sensors and natural user interfaces will be embedded in our natural movements and in our social and physical interactions. Hence, a key challenge in this multi-disciplinary approach is to get machines to act according to ethical priorities that make sense to human beings. In this study, we improve the capacity for machine ethics to approach human ethics by assessing the computation of transaction values and we argue that it is possible to perform such a computation using recent work that describes the effects of human decision-making using an axiomatic framework. This paper clarifies the relationship between the brain's 3-axes of neuroscience, the 3 Plato's Transcendentals of philosophy and the biological evolution's 3-components, as well as the top-down vs. bottom-up approaches to machine ethics.

Keywords: Artificial Intelligence · Ambient Intelligence · Machine ethics · Transaction value · Aesthetics evolution · Plato's Transcendentals · Axiomatic systems

1 Introduction

Embodied by the combination of autonomous systems, AI, and information technology, the 4th industrial revolution has been promoting a permanent transformation of morals, knowledge, and perceptions in almost all areas of human expertise [16, 22, 23]. The ethical, economic, and social implications of this revolution are a worldwide concern and a matter of political and public deliberation [6, 39], which are causing a reappreciation of how to compute the transaction value of an entity.

© Springer Nature Switzerland AG 2020
C. Analide et al. (Eds.): IDEAL 2020, LNCS 12490, pp. 309–321, 2020.
https://doi.org/10.1007/978-3-030-62365-4_30

A key aspect of [20] is the intricate relation between philosophy and computer science, and it was there proposed that the construction of such relation is greatly improved by the use of contemporary neuroscience. Based in the work of Schiller, the utmost Beauty-value should be assigned to objects that present the uppermost freedom [20], i.e., the objects that have the least usefulness and are therefore closer to a thing that exists for itself. In an opposing view, in [7] it is proposed that value is "useful information", translated as Knowledge and that is designated as Truth in [20]. Therefore, contemporary economics seems to place more value on Truth than Beauty, and people are prepared to spend their wealth to improve their bodily self-perception, which [20] associated to the concept of Good. Hence, there are three types of value: Truth-value, Good-value, and Beauty-value. Whereas, a recent book, "The Square and the Tower" [15], uses the theory of computer-networks to analyze the relation between history and contemporary socio-economics; where "Tower" means hierarchical command-and-control structures hence maximizing Truth-value, whereas "Square" symbolizes equalitarian distributed-control networks hence maximizing Good-value.

Ethical options are based on axiomatic choices promoted by cultures that can be either explicitly or implicitly religious, as they always require certain axioms to be valid without the support of an experimental validation. Thus, any approach to ethics is always about re-connecting (*religio* in Latin means "bind back") the mundane conditioned existence to a transcendental unconditioned valuation system that separates ethical from unethical. As noted, the balance between Truth-value and Good-value is a key characterizer of cultures [15,19,20]. In this work, the Transcendent is a generic term to signify God from the Christian perspective, or Absolute from the perspective of contemporary science.

Ethical choice perspectives are very important because they decide something that cannot be trivially decided by a computational binary logic of "Valid/1 vs. Invalid/0". Wittgenstein was the first to detail this fundamental gap of all non-religious thought at the end of his Tractatus Logico-Philosophicus [41]. However, despite Wittgenstein being the first, the Scottish Enlightenment of Adam Smith, Edmund Burke, and David Hume had already pointed out that the Enlightenment could only focus on data that could be made objective through detailed observation, thus leaving out religion and the arts that should be allowed to evolve over time. On the other hand, the Continental Enlightenment of Spinoza, Voltaire, and Schopenhauer argued that religion was obsolete and should in the future be replaced by art, with music being the most sublime aspect of art.

Until recently, it was thought that causalities were indeterminable and that all that could be statistically determined were correlations, a form of thinking that is well matched with Hume's perspective on ethics, but Pearl [27] obtained that it is possible to determine the direction of causality, and moreover, without determining it, the statistical analysis of events is necessarily wrong, and this determination of the need to define the direction of causality is more in line with Kant's ethical perspective, which is strongly supported by cause-effect relationships [19]. Unfortunately, this analysis of causality's direction is always difficult, and it is always probabilistic as it is based on Bayesian correlations. Aristotle's final cause can be understood as Darwinian causality, which is directly related to random walking processes, and randomness was a topic that also interested Aristotle, who was very important for the Scholastics. Therefore, one would expect the Scholastics to have explored the statistical perspective of causality, but that did not happen. For Scholastics, the formal logical deduction was the unquestionable basis of all causality, and consequently, Christian thought never developed a statistical perspective of ethics. Moreover, Enlightenment thinkers sought a determinism based on Reason and so always considered randomness as a temporary indication of contemporary limits of observation, and never as something fundamental, and therefore did not develop a statistical perspective of ethics either. Nevertheless, there were in the ancient past references to ethics, occurring in non-deterministic ways (e.g. Tower of Siloam parable).

Enlightenment believed that human rationality could perceive all the causality of the universe because all determinism is determinable, it therefore forgot to consider the possibilities that: causality was not deterministic, that what we do not know the cause of had sometimes no cause, and that the non-contradictory may never be able to be universal. Unfortunately, for the Enlightenment, these three things happened, respectively being: Darwin's evolution (in 1859), Heisenberg's uncertainty (in 1927), and Gödel's incompleteness (in 1931). In this work, a path is followed that is radically different from the Scholastics and the Enlightenment, preferring the assumption of a stochastic nature of the universe, instead of a formal logic approach. This choice is made because, since the works of Darwin, Heisenberg, and Gödel, it makes no sense to use formal logical deduction as a basis. Therefore, it is proposed here that with the use of Philosophy of Information [17] it is possible to put statistics as the basis for the foundations of ethics. Hence, as computer iterations follow a formal logic inference [9], a simple axiomatic approach to computer ethics will have the same limitations as the Scholastic and Enlightenment ethics. Thus, this statistical ethics approach has direct applications to how machine ethics is developed and applied.

We will use a 3-axes value approach [20] to establish a relation between biological evolution, brain axes, philosophy, psychology, and axiomatic systems that will make it easier to develop more human-like machine ethics. In order to assess the 3-axes value, we must work with an "axiom-driven value calculation" for Truth-value, e.g. a "deterministic inference" Newtonian Axiomatic System [9,21]; and with an "environmentally-driven value calculation" for Good-value, e.g. a "natural selection" Darwinian Axiomatic System [21]; whereas a new form of partial information-driven calculation, is required for the Beauty-value, just like in biology the "aesthetic evolution" [28,29] constitutes a suggestion of a "best guess" representation of evolution's undeterminable [34] future. A relevant aspect for this new form of partial information-driven calculation called Statistical Philosophy (e.g. by using Shannon Information), a Philosophy of Information [20] branch, is the important role of causality in determining the appropriate statistical approach for the events [27].

These 3-axes, Truth+Beauty+Good [19, 20] are what is known in philosophy as Plato's Transcendentals, and are equivalent to three major branches of philosophy through the relation: i. Epistemology searches for what is Truth; ii. Ethics searches for what is Good; iii. Aesthetics searches for what is Beauty. There are two other branches of philosophy: logic and metaphysics, but logic is simply the implementation of the equivalences to the axioms assumed as Truth by the epistemology, whereas metaphysics by definition aims at understanding what is beyond what physics can provide, and presently it is known that Darwinism can be considered to be beyond determinism [34], even beyond stochastic average determinism, and hence beyond physics. Thus, both philosophy and contemporary neuroscience perspectives can be reduced to a 3-axes system (see Fig. 1, and Table 1).

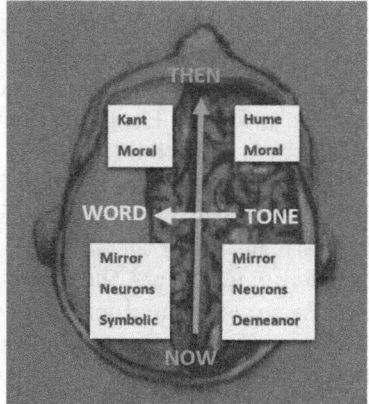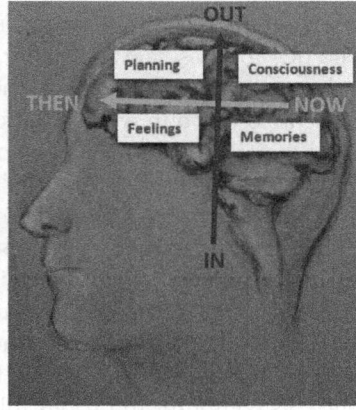

Fig. 1. Relation between brain 3-axes approach and brain function. On the left is a view from above, and on the right a view from the left side (®Sandra Lori) [20].

The simultaneous maximization of Truth-value, Good-value, and Beauty-value is the obvious goal; but, just as it seems impossible to have equipment costs, product improvement and wages maximized without going bankrupt; the simultaneous maximization of Truth-value, Good-value and Beauty-value often leads to a trilemma restriction that occurs in many forms, e.g. the Political Trilemma (triplet of "Democracy vs. national sovereignty vs. global economic integration") [30] and the Impossible Trinity (triplet of "independent monetary

Table 1. Plato's Transcendentals relation to contemporary neuroscience axes [19, 20].

Transcendentals	Axes
Truth	Hunting/Power/Now-Then
Beauty	Choosing/Meaning/Tone-Word
Good	Eating/Pleasure/In-Out

policy vs. fixed exchange rate vs. free movement of capital") [8,26]. Finding the maximization balance in entity transaction management is, hence, the finding of an appropriate balance in the Political Trilemma and the Impossible Trinity.

2   Related Work

The gradual shift towards ubiquitous computing and AmI is responsible for the foundation of an omnipresent and intelligent atmosphere. As computers' decision-making roles grow and autonomous machines become more sophisticated, society increasingly relies on computer-based intelligence with reduced human supervision. Unquestionably, granting control and autonomy to machines requires them to act in an ethical way. Ethics is necessary to determine what is morally right or wrong [37], to be a factor in the attribution of responsibility [35], to decrease the likelihood of negative outcomes for humans and/or to narrow the adverse effects machines can cause. Hence, the growing demand to regulate intelligent systems and bring forth better ethical approaches. Machine Ethics seeks to implement a moral dimension in computational systems either by introducing moral principles in machines or by discovering means to make machines function in an ethically responsible way on their own.

As machine ethics is a combination of computer science and moral philosophy, the scientific literature includes publications of different natures, ranging from theoretical papers about what a machine can or should do [11,35], to experiments about the incorporation of ethical reasoning in computer systems [3,40]. Allen *et al.* identified a high-level classification to machine ethics based on the nature of the approach: top-down approaches, bottom-up approaches, and a hybrid of top-down and bottom-up approaches. A top-down approach requires earlier specific moral principles or theories to train the machine to identify ethically appropriate actions as well as to recognize and correctly react to ethical scenarios and dilemmas [2,38]. In contrast, a bottom-up approach does not impose specific moral principles or theories, instead considers moral values as being implicit in the activity of agents and tries to provide agents the power to understand their own morality and the morality of others [2,38].

There is a multitude of works dedicated to top-down approaches, such as Dennis *et al.* development of ETHAN, a system that deals with situations where civil air navigation regulations conflict with each other [12]. The system is provided with a particular ethical policy that refers to four hierarchical ordered moral principles (do not harm people, do not harm animals, do not harm self, and do not harm property) and selects the course of action that results in the least violation of those principles in case of conflict. The system was proven to not perform an unethical action unless the rest of actions available are even less ethical. In contrast, relatively few researchers have been dedicated to bottom-up approaches; an example being the proposal by Wu and Lin of a reinforcement learning agent that integrated human policy, based on the premise that most human behavior is ethical, to accomplish a purpose with less risk of violating the ethical code [42]. There are also some studies that use a hybrid approach to implement ethics by combining top-down and bottom-up methods, such as Anderson and Anderson proposal of GenEth, a system that analyzes ethical dilemmas through the representation of a variety of aspects of those dilemmas such as the situational features, duties and actions, plus the production of abstract ethical principles using inductive inference before a self-made Ethical Turing Test is used to evaluate those principles by only allowing acts that an ethical expert would accept [5].

For intelligent autonomous computing agents to be fully integrated into society, it is not enough that they have an ethical reasoning, assurances are equally required for these agents to always perform within acceptable legal and social standards. The existing codes of ethics do not reflect the effects of autonomous and intelligent computing agents and are insufficient in terms of legal processes for coping with the inherent risks [4]. Accordingly, the current codes of ethics require a rigorous re-examination to legally regulate these entities. Control mechanisms are essential to ensure that intrinsic laws are always functioning and that specific standards are enforced for the design, development, assessment, use, and maintenance of autonomous agents. Other control mechanisms are needed to inspect and audit the first control mechanisms. Moreover, the identification and assignment of legal responsibility to those accountable for the harm arising from autonomous systems noncompliance with the laws is crucial.

3   Methods

The pillars of contemporary science are here assumed as the conjunction of [Truth, Good, Beauty] that is modeled, respectively, by [Physics, Biology, Economics] of contemporary science. They constitute contemporary science's Absolute, which are hence the best contemporary science has for describing the Transcendent, a pre-requisite for the establishment of ethics, as previously noted by Wittgenstein (1922).

In this work; Physics is a generic term that goes from mathematics as a foundation, through the theoretical physics of gravity and the Standard model, to chemistry as a practical application of quantum physics; Biology is a generic term that refers to the study of "information of life" that goes from the chemistry of organic molecules, through genetic biology, to neuroscience of population groups; Economics is a generic term that goes from the economics of representation in neuronal aggregates, through social psychology and sociology, to the economics of the performance of countries in the midst of the financial policy of a globalized world, all of which are associated with the stipulation of the value associated with transactions; and sociology is described as an aspect of economics, rather than the opposite perspective, because sociology is limited to the description of human social systems whereas economics can easily be extended to other types of interaction [7,19].

To assess the contribution of each of Plato's Transcendentals to the value of an entity/act, meaning the contribution of each of the axes of the 3-axes value, it is important to assess the threefold structure of evolution [28,29]:

- Deterministic Inference → Information from past is preserved into the future thus maintaining survival capability of the entity;
- Natural Selection → Information creation allows new behaviors that alter the resource extraction from environment, the entities with better survival capacity endure;
- Aesthetic Evolution → Alterations of information representation compatibility between entities alter the flow of resources between them, the communication link with better survival capacity endures.

The use of model-free approaches to data analysis is now typically called deep learning, but can also be referred to as machine learning or neural networks, and consists on learning the most effective representation of the data. Deep learning models have been able to show that: most mutations in humans are neither beneficial nor harmful for natural selection until an environment change makes a certain mutation become relevant [31]; specific gene mutations (meaning the deterministic inference of genetic information is partially broken) are associated to metabolism [36]; and Müllerian mimicry does occur in the Darwinian evolution for butterflies [10]. The existence of Müllerian mimicry [10] together with the existence of evolutionarily advantageous characteristics that are not truthful [24] are a further indication that the aesthetic evolution [28, 29] occurs and is *de facto* a different evolution line from natural selection.

The aesthetic evolution is a different concept from natural selection, as the strength of the survival is based in the establishment of jointly-accepted symbols [28, 29], and not necessarily in a better usage of the environmental resources as occurs in natural selection. The aesthetic symbol might represent a true best-guess of the natural selection trend [10] or not [24], but since natural selection is not deterministic [34] the natural selection trend is indeterminable, and hence its representation by a symbol is always a guess. Moreover, this aesthetic evolution allows culture/ethics/morality to have a certain degree of independence from both deterministic inference and natural selection. The 3-aspects of evolution can be directly linked to the 3-axes of Plato's Transcendentals, Truth+Beauty+Good, by the relation: "deterministic inference" ↔ Truth + "aesthetic evolution" ↔ Beauty + "natural selection" ↔ Good.

Plus, those 3-axes have previously been connected to human neuroanatomy, human decision-making, human mental health, human culture, and human economics [19, 20]. The trust-level vs. Gross Domestic Product (GDP)/capita across nations allows for the definition of religion-based clusters [7]. Moreover, across different nations the crime-rate correlates positively with the belief in Hell's existence and negatively with the belief in Heaven's existence [32]. Plus, there is an agreement in [1] and [25] that the appearance of a "Leviathan", meaning a command-and-control hierarchical structure with "stationary bandits" building a political elite and a rule-of-law establishing the rules within the "Leviathan", is a key contributor to the improvement of the wealth of the nations. Moreover, in [14] is described a correlation between the decrease in fear of legal punishment and the reduction of wealth in western nations. Thus, it is an appropriate resume of the relation between wealth and culture to consider that societies are constituted by a "Tower" that implements the rules of the "Leviathan", and a "Square" where under the protection of the "Leviathan" people can build continually evolving economical interactions [15].

4   Results and Discussion

The relations, respectively, between axiomatic systems and psychology's structure (see Fig. 2) [19], between axiomatic systems and psychology's consciousness types (see Table 2) [19], between the brain 3-axes and brain function (see Fig. 1) [19,20], plus the relation between Plato's Transcendentals [20] and contemporary neuroscience axes (see Table 1) [19,20] allow the clarification of the relation between the 3-components of evolution and the 3 axes.

The relation of Plato's Transcendentals with the brain axes is described in Table 1, but the establishment of the relation with the axiomatic systems described in Table 2 requires the use of the Types of Consciousness of Fig. 2.

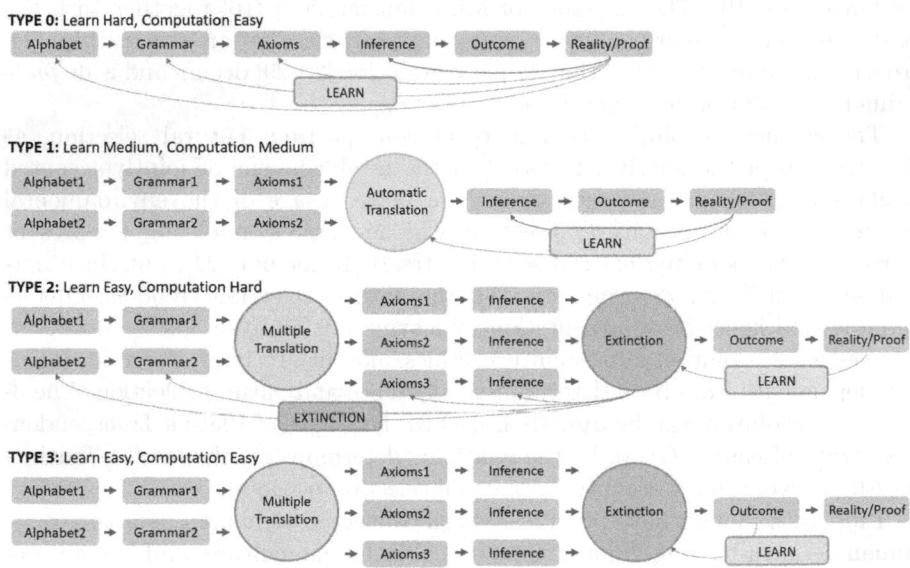

Fig. 2. Relation between axiomatic system, from Alphabet to Output, and the structure of the connection relating the Information-based axiomatic system to the Consciousness Types 0-1-2 described in [33] and Consciousness Type 3 proposed in [19].

The Type 1 consciousness activates the appropriate "deterministic inference" mechanisms so that from a certain stimuli a unique and appropriate response is obtained [33], whereas the Type 2 consciousness activates the appropriate "natural selection" analysis of what would be the consequences of the different

Table 2. Information-based axiomatic system and its relation to psychology [19].

Axiomatic system	Psychology's narrative
$Alphabet(S)$ + $Grammar(P[w], P[w_j], P[< l_{\sim_j}|w_j >], \dots)$	Setting
Proof-checking algorithm input (internal vs. external)	Initiating event
Axioms (consistent vs. complete)	Internal response
Rules of inference (single-alternative vs. multi-alternatives)	Goal + Actions
Inferred statements	Outcome
Proof-checking algorithm output (internal vs. external)	Ending

choices and environments so that from a certain situation the best path is chosen [18,33], finally the Type 3 consciousness allows for an analysis of the best options regardless of environmental input so that the best aesthetic option can be achieved in as much freedom as possible [19,33]. Hence, the relation between evolution component, Plato's Transcendentals, consciousness Type, contemporary science, and brain axes is:

- Deterministic Inference ↔ Truth ↔ Type 1 ↔ Physics ↔ Ant.-Post/Hunting/Power/Now-Then;
- Natural Selection ↔ Good ↔ Type 2 ↔ Biology ↔ Inf.-Sup./Eating/Pleasure/In-Out;
- Aesthetic Evolution ↔ Beauty ↔ Type 3 ↔ Economy ↔ Left-Right/Choosing/Meaning/Tone-Word.

The relation just above allows for a direct relation between axiomatic systems, human ethics, and human anatomy; which is a key issue in computer science as machine ethics becomes more and more important, and the implementation of ethical reasoning in intelligent machines is not far off. However, for intelligent autonomous computing agents to be fully integrated into society, it is not enough that they have an ethical reasoning, assurances are equally required for these agents to always perform within acceptable legal and social standards.

It is also obtained a new perspective on the relative importance of the top-down, bottom-up and hybrid approaches to machine ethics. For, the top-down approaches are a "deterministic inference"; whereas the bottom-down approaches are separable between "natural selection" types if the learning is based in the environment, and "aesthetic evolution" types if the learning is based in the interaction with the symbolic representations of the other autonomous agents, the hybrid approach. In practice, just like human ethics is based in all three of the 3-axes value and all three of the brain's 3 axes, the appropriate machine ethics should be based in the three components of evolution, and it will thus be an hybrid approach. Which is reasonable, as the likely source of the brain's 3-axes is the biological evolution's 3-components.

One may ask, and many have done so, what is the best ethics. Nevertheless, the best answer this approach obtains is that there is no answer. The best ethics is not this or that, but rather the permanent search for a better ethics through comparison, competition, and selection of ethics [7,15,24,25]; provided that those involved are striving to find what is most True and Good [1,14]. Through this balance between Truth and Good, the Beauty is found as a cultural creation of the balance between Truth and Good.

Moreover, for classical ethics, which separate between goodness and evil, to exist in the observed universe, the following is required: i. events universally definable as a goodness constitute a set B; ii. events universally definable as an evil constitute a set M; iii. set S of well-intentioned actions, meaning, they intend to obtain B events; iv. set C of ill-intentioned actions, meaning, they intend to obtain M events; v. classical ethics exists in the universe if and only if: "S actions not intersecting C actions" implies "B events not intersecting M events".

Requirement v means that for classical ethics, the "God/Absolute cannot write right by crooked lines", and if it is valid that "God/Absolute cannot write right by crooked lines" then causality in the observed universe would have to be only by deterministic inference, but that is not the case, as both natural selection and aesthetic evolution also occur. Hence, only a statistical perspective of ethics agrees with the observed universe. Thus, for example, set S actions generate not only set B events with high probability but also set M events with less probability. In rare most extreme cases, set S actions can be so creative that they only generate set B events for all space and time; or set C actions can be so malicious that they only generate set M events for all space and time. Moreover, there may be actions that are selfish, meaning, that they generate B events in their vicinity and M events away from it; or an action can be heroic by generating M events in their vicinity and B events away from it.

5   Conclusion

Alongside AI and Information and Communication Technologies (ICT), AmI has gained a prominent place in the scientific community. As AmI research matures, increasingly superior forms of intelligence and automation are permeating every aspect of human life. Unquestionably, as computers' decision-making roles grow and society increasingly relies on computer-based intelligence with reduced human supervision, ethical considerations are inevitable. In the last years, AmI has experienced a tremendous growth, but few authors have dedicated to the social, moral, and legal implications of this emerging reality.

In this study, we sustain that the success of AmI relies heavily on the development of better ethics and, consequently, on the implementation of effective machine ethics. Hence, the main purpose of the study was to improve the capacity for machine ethics to approach human ethics by assessing the computation of transaction values. We used a 3-axes value approach - Truth+Beauty+Good - to establish a relationship between biological evolution, brain axes, philosophy, psychology, and axiomatic systems.

According to the statistical perspective of ethics, well-intentioned actions are more likely to generate goodness, and malicious actions are more likely to generate evil. Nevertheless, one may ask, why the reason for living should be the ethics of goodness, and not an alternative ethics of generating evils. The reason for this preference is that what is goodness, is so because it is in accordance with the egalitarian objectivity of the Truth of Physics and with the elitist life-enhancement of the Good of Biology, and hence goodness is what maximizes the occurrence of Beauty; that is, Beauty/goodness is the combination of egalitarianism and elitism that allows for biological life to occur despite physical objectivity. In short, evil in going against life is hence necessarily self-destructive, thus in this aspect Augustine of Hippo [13] is correct in stating that Absolute Evil does not exist, for existing is a goodness, and the Absolute Evil in existing would cease to be Absolute Evil because it had at least one goodness.

Acknowledgments. This work has been supported by FCT – Fundação para a Ciência e a Tecnologia within the R&D Units Project Scope: UIDB/00319/2020.

References

1. Acemoglu, D., Robinson, J.A.: Why Nations Fail: The Origins of Power, Prosperity, and Poverty. Crown Books, Largo (2012)
2. Allen, C., Smit, I., Wallach, W.: Artificial morality: top-down, bottom-up, and hybrid approaches. Ethics Inf. Technol. **7**(3), 149–155 (2005)
3. Anderson, M., Anderson, S.L.: EthEl: toward a principled ethical eldercare robot (2008)
4. Anderson, M., Anderson, S.L.: Machine Ethics. Cambridge University Press, Cambridge (2011)
5. Anderson, M., Anderson, S.L.: GenEth: a general ethical dilemma analyzer. Paladyn J. Behav. Robot. **9**(1), 337–357 (2018)
6. Andrade, F., Neves, J., Novais, P., Machado, J., Abelha, A.: Legal security and credibility in agent based virtual enterprises. In: Camarinha-Matos, L.M., Afsarmanesh, H., Ortiz, A. (eds.) PRO-VE 2005. ITIFIP, vol. 186, pp. 503–512. Springer, Boston, MA (2005). https://doi.org/10.1007/0-387-29360-4_53
7. Beinhocker, E.D.: The Origin of Wealth: Evolution, Complexity, and the Radical Remaking of Economics. Harvard Business Press, Boston (2006)
8. Boughton, J.M.: On the origins of the Fleming-Mundell model. IMF Staff Papers **50**(1), 1–9 (2003). https://doi.org/10.2307/4149945
9. Chaitin, G.J.: Meta maths!: the quest for omega. Vintage (2006)
10. Cuthill, J.F.H., Guttenberg, N., Ledger, S., Crowther, R., Huertas, B.: Deep learning on butterfly phenotypes tests evolution's oldest mathematical model. Sci. Adv. **5**(8), eaaw4967 (2019)
11. Davenport, D.: Moral mechanisms. Philos. Technol. **27**(1), 47–60 (2014). https://doi.org/10.1007/s13347-013-0147-2
12. Dennis, L., Fisher, M., Slavkovik, M., Webster, M.: Formal verification of ethical choices in autonomous systems. Robot. Auton. Syst. **77**, 1–14 (2016)
13. Dyson, R.W., et al.: Augustine: The City of God Against the Pagans. Cambridge University Press, Cambridge (1998)
14. Ferguson, N.: The great degeneration: how institutions decay and economies die. Penguin (2014)
15. Ferguson, N.: The square and the tower: networks and power, from the freemasons to Facebook (2018)

16. Floridi, L.: The Blackwell Guide to the Philosophy of Computing and Information. Wiley, Hoboken (2008)
17. Floridi, L.: The Philosophy of Information. OUP, Oxford (2013)
18. Kahneman, D.: Thinking, Fast and Slow, Farrar, Straus and Giroux (2011)
19. Lori, N., Samit, E., Picciochi, G., Jesus, P.: Free-will perception in human mental health: an axiomatic formalization. from: automata's inner movie: science and philosophy of mind. chapter viii, curado, m., gouveia, ss (2019)
20. Lori, N., Neves, J., Alves, V.: Some considerations on the estimation of the value associated to a clinical act. Procedia Comput. Sci. **170**, 1041–1046 (2020)
21. Lori, N.F., Jesus, P.R.: Matter and selfhood in Kant's physics: a contemporary reappraisal (2010)
22. Machado, J., Abelha, A., Neves, J., Santos, M.: Ambient intelligence in medicine. In: 2006 IEEE Biomedical Circuits and Systems Conference, pp. 94–97. IEEE (2006)
23. Machado, J., Miranda, M., Pontes, G., Abelha, A., Neves, J.: Morality in group decision support systems in medicine. In: Essaaidi, M., Malgeri, M., Badica, C. (eds.) Intelligent Distributed Computing IV, pp. 191–200. Springer, Heidelberg (2010). https://doi.org/10.1007/978-3-642-15211-5_20
24. McKay, R.T., Dennett, D.C.: The evolution of misbelief. Behav. Brain Sci. **32**(6), 493–510 (2009)
25. Morris, I.: War! what is it Good For?: Conflict and the Progress of Civilization from Primates to Robots. Farrar, Straus and Giroux (2014)
26. Obstfeld, M., Shambaugh, J.C., Taylor, A.M.: The trilemma in history: tradeoffs among exchange rates, monetary policies, and capital mobility. Rev. Econ. Stat. **87**(3), 423–438 (2005)
27. Pearl, J., Mackenzie, D.: The Book of Why: The New Science of Cause and Effect. Basic Books, New York (2018)
28. Prum, R.O.: Aesthetic evolution by mate choice: Darwin's really dangerous idea. Philos. Trans. Roy. Soc. B: Biol. Sci. **367**(1600), 2253–2265 (2012)
29. Prum, R.O.: The evolution of beauty: how Darwin's forgotten theory of mate choice shapes the animal world-and us. Anchor (2017)
30. Rodrik, D.: The Globalization Paradox: Democracy and the Future of the World Economy. WW Norton & Company, New York (2011)
31. Schrider, D.R., Kern, A.D.: Soft sweeps are the dominant mode of adaptation in the human genome. Mol. Biol. Evol. **34**(8), 1863–1877 (2017)
32. Shariff, A.F., Rhemtulla, M.: Divergent effects of beliefs in heaven and hell on national crime rates. PLoS ONE **7**(6), e39048 (2012)
33. Shea, N., Frith, C.D.: Dual-process theories and consciousness: the case for 'type zero'cognition. Neurosci. Conscious. **2016**(1) (2016)
34. Smerlak, M., Youssef, A.: Limiting fitness distributions in evolutionary dynamics. J. Theor. Biol. **416**, 68–80 (2017)
35. Sparrow, R.: Killer robots. J. Appl. Philos. **24**(1), 62–77 (2007)
36. Sugden, L.A., Atkinson, E.G., Fischer, A.P., Rong, S., Henn, B.M., Ramachandran, S.: Localization of adaptive variants in human genomes using averaged one-dependence estimation. Nat. Commun. **9**(1), 1–14 (2018)
37. Voiklis, J., Kim, B., Cusimano, C., Malle, B.F.: Moral judgments of human vs. robot agents. In: 2016 25th IEEE International Symposium on Robot and Human Interactive Communication (RO-MAN), pp. 775–780. IEEE (2016)
38. Wallach, W., Allen, C., Smit, I.: Machine morality: bottom-up and top-down approaches for modelling human moral faculties. AI Soc. **22**(4), 565–582 (2008)

39. Winfield, A.F., Michael, K., Pitt, J., Evers, V.: Machine ethics: the design and governance of ethical AI and autonomous systems. Proc. IEEE **107**(3), 509–517 (2019)

40. Winfield, A.F.T., Blum, C., Liu, W.: Towards an ethical robot: internal models, consequences and ethical action selection. In: Mistry, M., Leonardis, A., Witkowski, M., Melhuish, C. (eds.) TAROS 2014. LNCS (LNAI), vol. 8717, pp. 85–96. Springer, Cham (2014). https://doi.org/10.1007/978-3-319-10401-0_8

41. Wittgenstein, L., dos Santos, L.H.L.: Tractatus logico-philosophicus. Edusp (1994)

42. Wu, Y.H., Lin, S.D.: A low-cost ethics shaping approach for designing reinforcement learning agents. In: Thirty-Second AAAI Conference on Artificial Intelligence (2018)

Anticipating Maintenance in Telecom Installation Processes

Diana Costa[1] , Carlos Pereira[2] , Hugo Peixoto[3] , and José Machado[3(✉)]

[1] University of Minho, Campus Gualtar, 4710 Braga, Portugal
a78985@alunos.uminho.pt
[2] NOS, Senhora da Hora, Portugal
carlos.migpereira@nos.pt
[3] Algoritmi Center, University of Minho, Campus Gualtar, 4710 Braga, Portugal
{hpeixoto,jmac}@di.uminho.pt

Abstract. Improving customer experience is crucial in any industry, especially in telecommunications, where competition is a constant factor. Today, all telecommunications companies rely on the massive amount of data generated daily to get to know the customer or study their behavior and thus create new effective strategies for their business. Within the most varied user experiences, the process of installing new services can be an event that raises doubts about their operation, degrade the user experience, or, in extreme cases, lead to maintenance interventions. Therefore, the use of advanced predictive models that can predict such occurrences become vital. With this, the company can anticipate the cases that will be problematic and reduce the number of negative experiences. The main objective of this work is to create a predictive model that, through all the available data history, can predict which customers will contact the customer service with problems derived from the installation process and have a following maintenance intervention. After analyzing an unbalanced dataset with approximately 560K entries from a Portuguese telecommunications company, and resorting to the CRISP-DM methodology for modeling, the best results were found with Light-GBM which obtained an AUPRC of 0.11 and AUROC of 0.62. The best trade-off between precision and recall was found with a threshold model of 0.43 in order to maximize recall while still avoiding a large number of false negatives.

Keywords: Customer · Data mining · Installation · Machine learning · Predict · Service · Telecommunications

1 Introduction

The growth and evolution of technology and telecommunications services are increasing the amount of data generated daily. In fact, a growing number of companies choose customer satisfaction as their leading performance indicator, and it is one of the fastest-growing segments of the marketing field [1].

© Springer Nature Switzerland AG 2020
C. Analide et al. (Eds.): IDEAL 2020, LNCS 12490, pp. 322–334, 2020.
https://doi.org/10.1007/978-3-030-62365-4_31

However, several typical processes, like the installation of new services, are not always pleasant. In Portugal, telecommunications have been in first place for over 12 years in the ranking of complaints to DECO, a Portuguese association for consumer protection [2]. Because of the efforts to address these issues, the telecommunications industry has become a leader in the data mining (DM) and machine learning (ML) area. DM techniques and ML models have been used in marketing and retail, network optimization, security and fraud detection, and to improve brand loyalty and increase company revenue.

The process of installing new systems or equipment can be a problematic event that raises questions about their operation, degrade the user experience, or even lead to maintenance interventions. Companies want to reduce the number of customers that might have a particular defective device or service, which typically leads to bad experiences or cases where customers have a poor problem resolution when contacting customer support.

The main goal of this work is to create a predictive model that, through all the available historical data, can predict with high performance which customers will contact customer support with problems arising from the installation process and have a following maintenance intervention. This allows to know in advance who will complain and to have a proactive action with the client, as well as provide personalized services in the customer's home (suitable assistance, equipment, or technician) if required, culminating in better customer experience and higher installation service efficiency. Reducing the number of problems also diminishes the time/equipment spent (number of trips to the customer's home or expensive equipment replacement), resulting in increased profit for the company.

Although customer satisfaction is not a new concern, and even telecommunications being an area widely explored by machine learning methods, no study was found that explicitly focused on the installation process, or customer behavior after having problems with the installation process. Therefore, to the best of our knowledge, there is no other study with the same theme as this particular article, which is considered to be novel not on the methods or methodologies it is presented here, but on the topic itself.

This article includes a description of the background in Sect. 2, followed by the data mining process guided by the CRISP-DM methodology in Sect. 3. Finally, conclusions and directions for future work appear in Sect. 4.

2   Background

2.1   Data Mining in Telecommunications Industry

DM and ML have been used in many industries, whether it is fraud detection, customer segmentation, market basket analysis, life and death, consumer or business. ML is a burgeoning new technology for mining knowledge from data, a technology that many people are starting to take seriously [3]. The DM applications for any industry depend on two factors: the available data and the business problems facing that industry [4]. The telecommunications industry was one of the first to adopt DM technologies, and these have proven effective to detect fraud, to manage networks better, to gain knowledge about the customers (e.g., customer segmentation), predict churn and retain clients, and to improve their marketing efforts, knowing what products and services yield the highest amount of profit. This is most likely because telecommunications companies routinely generate and store enormous amounts of high-quality data, have a huge customer base, and operate in a rapidly changing and highly competitive environment [4]. Telecommunications companies can no longer afford not to make use of their Big Data. Companies like AT&T and Sprint always top the list of the largest databases in the world, reaching storage in the order of thousands or even millions of terabytes. The huge amount of data available can also be a key concern, because of the scalability of DM methods, the inappropriate semantic level (records in the form of transactions/events), and the data generated in real-time and applications that need to operate in real-time, such as fraud and network fault analysis [4].

2.2   Installation Process

Any company that provides physical services to a customer has to go through its installation, which usually involves either a specialist technician going to the customer's home or the customer self-installing it. In both cases, this is a procedure where the novelty factor exists, and where service may be problematic because of some external influencing factor. In the case of telecommunications, installing new services requires a technician to visit the customer's home. An installation process is defined as the establishment of a service, such as the internet or telephone, in which special telecommunications knowledge is required for the person performing the craft. This process of installing new systems can be a problematic event that leads to following maintenance interventions resulting in user experience degradation. As already mentioned in the previous section, improving the user experience results in loyalty and profit for the telecommunications company.

3   Data Mining Process

In this work we follow the CRISP-DM method, which describes approaches commonly used by specialists to tackle problems of this nature [5]. The most significant advantage of using this method is that it is independent of industry, tools and data. The phases of the methodology include Business Understanding, Data Understanding, Data Preparation, Modeling, Evaluation and Deployment, which are presented next. The task of deploying the solution into production, the last phase of CRISP-DM, is not presented in this study, being the next step after checking whether the results are proved to be sustainable.

3.1   Business Understanding

The goal of the work presented in this paper is to reduce customers' problems related to installing (or upgrading) new services. Knowing in advance who will face an installation problem and being aware of the customer's probable future inconveniences allows the company to provide personalized services (suitable assistance, equipment, or technician), culminating in better customer experience and higher installation service efficiency. It also allows to have a proactive action and dialogue after the installation with the client and before the complaint call occurs, responding efficiently to the situation. All these actions ultimately lead to a decrease in time/equipment spent (number of trips to the customer's home or expensive equipment replacement).

The data used in this study were retrieved from a telecommunications company in Portugal. The DM aim is to develop accurate models that are able to support the decision-making process by predicting if a particular customer will have a problem that requires a maintenance intervention in a period of time following the initial installation. For the prediction to be the most accurate as possible, models use as input values the installation records from the clients who had and clients who had not have technical problems within that period since the installation.

3.2 Data Understanding

The provided data refers to approximately 500 d, comprising the period between May 21, 2018, and September 30, 2019. The dataset consists of 566244 entries. The main dataset was the one with the reports about the installations and maintenance at the customers' homes, which was complemented with other information such as personal details, types of equipment in use and service details, and the customer's previous calls to the customer support with the respective complaint or request. This resulted in each one of the entries being described by a total of 72 attributes. The records were previously anonymized and filtered in order to comply with privacy regulations.

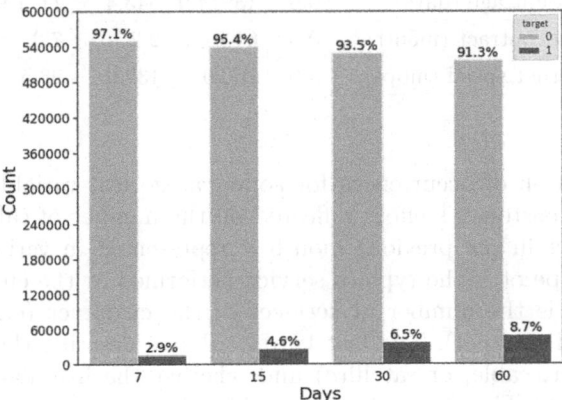

Fig. 1. Target distributions - choice of target variable for maintenance occurring up to 30 d after initial installation.

In the following tables and graphics, we present some statistical information that may help to comprehend how the data is distributed. Figure 1 depicts the distribution of possible target variables according to four different scenarios: the customer has a maintenance intervention until 7 d, 15 d, 30 d, and 60 d after installation. Given the proportion of the data shown for each period, and because the month period is commonly used in the company, it was decided that a problem arising from an installation process would be treated as such if it happened within 30 d. The figure also shows signs of an unbalanced dataset.

Table 1 shows statistical measures related to some numerical variables, described by the minimum (min), maximum (max), average (avg), and standard deviation (std dev). Variables #ots_client and #calls represent the previous number of maintenance/installations and the previous number of calls of the client for the customer support, respectively. Variable #ots_county is the number of known services performed at the town hall of that client, and tech_performance measures the work quality of the technicians based on how often they have done something wrong. The last three variables in the table represent how long the customer is in the company, the duration since the beginning of the present contract, and, finally, the clients' internet speed, respectively.

Table 1. Statistics measures of some business relevant numerical variables.

Variable	Min	Max	Avg	Std Dev
#Ots_client	0.0	25.0	1.5	1.0
#Calls	0.0	72.0	0.3	2.7
#Ots_county	0.0	85292.0	10963.6	14559.8
Tech_performance	0.6	1.0	0.9	0.2
Dur_process_home (min)	0.0	7550.0	79.7	107.4
Age	18.0	72.0	30.7	26.3
Account_age (days)	0.0	13974.0	543.4	1177.9
Begi_contract (months)	0.0	61.0	2.2	7.8
Internet_speed (mbps)	6.0	200.0	132.6	98.5

The distribution of occurrences for some categorical variables is presented in Table 2. The customer's effort reflected on the number of calls made to the customer support in the previous month is represented in variable client_pain (month). The type_ot is the type of service performed at the customer's home, and the bundle is the number of services of the customer (e.g., 3P includes landline, tv, and internet). The last two attributes describe the technology of the service (fiber, cable, or satellite) and whether the installation included a technology change. The range of the variables do not represent the totality of the company's customers, but the distribution of the customers covered in this study.

Table 2. Statistics measures of some business relevant categorical variables.

Variable	Range	Percentage (%)
Client_pain (month)	Yes	10, 0
	No	90, 0
Type_ot	Installation	49, 6
	Service up/downgrade	45, 7
	Reestablishment	4, 7
bundle	0P	1, 0
	1P	1, 3
	2P	1, 0
	3P	6, 4
	4P	15, 2
	5P	4, 1
	Unknown	71, 0
Cel_technology	HFC	70, 0
	DTH	15, 1
	FTTH	14, 7
	Unknown	0, 2
Change_technology	Yes	0, 7
	No	24, 2
	Unknown	75, 1

3.3   Data Preparation

At this phase, it was necessary to select and prepare the data to be used by the ML models. Firstly, all the different datasets were merged in order to enrich the main dataset regarding the details of the installation at the customer's home. Personal information from customers was combined, as well as some details of the equipment in use, service, and the call with the complaint that led to the installation was made. Then, all the variables and rows available underwent a selection process as they would be useful in predicting problems arising from the installation process. Data selection is a very important pre-processing task in data mining because not all attributes are relevant to a given problem [6]. For example, observations referring to businesses instead of customers or observations whose installation had not been carried out due to the customer's absence were eliminated since they were not part of the study's object. To ensure that there was no incomplete or inconsistent information, data with noise or null values were removed. More specifically, duplicated rows, as well as the ones with noise or redundancy (e.g., cable and HFC), were eliminated from the dataset. The same goes for the variables that were dropped out if redundant, deprecated, or wrongly calculated. Concerning null values, special care was necessary to identify the respective meanings since not all blank records refer to unknown values, and some should be replaced by an identifying string [7]. The true nulls were replaced by the average/mode or changed to an 'unknown' string for categorical variables or number as −1 for numerical variables. Lastly, variables with a large percentage of null values were also eliminated. In order to provide the models the right and unified data types, it was also performed a data type conversion, especially in the Boolean values and dates.

The task of feature engineering is crucial to accurately represent the underlying structure of the data and create the best model. With this in mind, it was made an effort to analyze all the features and to derive new ones. For example, concerning the technicians that execute the service at the customers' home, several measures that could characterize their work and indicate performance metrics were considered, such as the duration of the service or the time over or under the scheduled. For the client, many counting operations were fulfilled to create new features, such as the previous number of interventions that he had or the number of calls to client support regarding different ranges. Features like the county were aggregated to create the district, as a higher level of abstraction can result in more accurate results.

To ensure data normalization, all the numeric features were changed to a common scale without distorting differences in the ranges of values. StandardScaler and RobustScaler are two popular scalers used for this task: StandardScaler removes the mean and scales the data to unit variance, while the centering and scaling statistics of RobustScaler are based on percentiles and therefore are not influenced by a few numbers of very large marginal outliers [8]. For the categorical features, one-hot encoding and target encoding was performed, depending on the cardinality of the feature. For the model to be able to see the inherently cyclical nature of time, it was also used cyclical encoding with two dimensions: cos and sin. This allows us to extract interesting features like part of the day or month and to distance nearby time positions (e.g., December is close to January, or 12 pm is close to 1 am).

After all the data selection, cleanings, and transformations mentioned above, the dataset that initially consisted of 1787349 entries was reduced to 1415407 and, subsequently, shrank to 566244 observations with the construction of the target. Each of the entries is described by a total of 72 features (not considering one-hot encoding modifications), which are the consequence of a choice and metamorphosis of 156 original ones joined together (Table 3).

Table 3. Dataset dimensions after each phase of data preparation.

	Original size	Data preparation	Target building
Entries	1787349	1415407	566244
Features	156	71	72

Since the dataset is unbalanced, many balancing options were considered, such as random oversampling, random undersampling, both random over and undersampling, oversampling with SMOTE, and class weights.

3.4   Modeling

Before building the model, it was necessary to generate a mechanism to test the model's quality and validity. In this way, the dataset was divided into training, validation, and testing with a ratio of 70%–15%–15%, respectively. Training and validation are the data samples to train the models and to evaluate the model fit while tuning model hyperparameters. The test dataset is a sample of data used to provide an unbiased assessment of the final model in the training data. Table 4 shows the train, validation, and test split with the respective ratios and date ranges that each set involves.

Table 4. Train, validation, and test split.

Dataset	Ratio (%)	Date range
Train	70	21/05/2018 to 02/05/2019
Validation	15	03/05/2019 to 22/07/2019
Test	15	23/07/2019 to 30/09/2019

The classification algorithms considered were: Gaussian Naïve Bayes (GNB), Logistic Regression (LR), Linear Support Vector Classifier (LinearSVC), Random Forest (RF), eXtreme Gradient Boosting (XGBoost) and Light Gradient Boosting Machine (LightGBM). The algorithm selection was based in three aspects: easy understanding of the techniques, engine efficiency, and the possible training of the models with large datasets and high dimensional data. With this, LR and NB are two of the easiest algorithms to understand. Concerning the engine's efficiency, all algorithms comply with this parameter, except for RF because of the computational complexity [9]. The last principle, concerning large datasets and high dimensional data, is supported by all algorithms excluding LinearSVC.

The next step was to build and optimize the model. There is a variety of possible scalers to use and balancing methods to try with each algorithm mentioned above. In order to reduce the models to optimize, first, it was used a combination of the algorithms (GNB, LR, LinearSVC, RF, XGBoost, LightGBM) with different scalers (RobustScaler, StandardScaler) and balancing techniques (Oversampling, Undersampling, Over + Undersampling, balanced model class weight). All the algorithms were tested with the default parameters at this phase, except for RF, XGBoost, and LightGBM's number of estimators, which was increased to 200. All the 106 combinations evaluated (GBN does not support class weight) are summarized in Table 5.

Table 5. First phase of model optimization.

Scalers	Algorithms	Balancing techniques
StandardScaler	GBN	Random Oversampling 35%
RobustScaler	LR	Random Oversampling 40%
	LinearSVC	Random Undersampling 35%
	RF	Random Undersampling 40%
	XGBoost	Oversampling w/ SMOTE 35%
	LightGBM	Oversampling w/ SMOTE 40%
		Random Over+Undersampling
		Balanced model class weights
		None

Total: 106 combinations

From here, it was decided that only the top 3 models were going to be optimized. The hyperparameter optimization phase, also known as tuning, is the process of finding the best machine learning model hyperparameters for a given dataset. Two conventional methods for hyperparameter tuning are Grid-SearchCV and RandomSearchCV, where a list of parameters is defined and evaluated (preferably on a validation set), winning the combination which yields the best performance. With large and high dimensional datasets, GridSearchCV would significantly slow down computation time and be very costly, making an inefficient method for this particular study due to the numerous combinations [10]. Therefore, RandomSearchCV was used for 100 iterations, repeating the process for each model around 10 times. Once the best hyperparameters were found, the final model was tested on the test set, an untouched dataset to avoid a biased assessment.

As for the metrics to be used for this study, it was necessary to consider the unbalance of the dataset. Because most of the standard metrics that are widely used assume a balanced class distribution, and because typically not all classes, and therefore, not all prediction errors, are equal for imbalanced classification, we considered the following metrics for evaluation: AUPRC (Area Under Precision-Recall Curve), AUROC (Area Under the Receiver Operating Characteristic), Precision and Recall. The PR curve can focus only on the class of interest, whereas the ROC curve covers both classes, and that is the main reason why we decided that the main metric would be AUPRC. The well-known accuracy is not suitable because the impact of the least represented, but more crucial examples is reduced compared to that of the majority class [11]. In this type of study, the results can be applied in a real context in the company. So, the trade-off between precision and recall is commonly analyzed, adapting the threshold as needed [12]: a higher recall or a greater precision. The classification model threshold, by default 0.5, can be lowered or increased in order to achieve the company's objectives. A higher recall makes it possible to avoid false negatives, which, in a business context, would mean picking up more possible cases of having an installation problem, requiring a second screening action on the part of the company to verify the true positives of this population. On the other hand, a greater precision means that it only predicts an installation problem if the results are confident, that is, the customers selected by the model for candidates for installation problems are almost certain. In this case, better precision is not intended, but a better recall. To increase recall, and keeping in mind a trade-off between both precision and recall, it is possible to change the threshold in the final model and choose which is the best value instead of the typical 0.5 [12].

3.5 Evaluation

The first optimization phase was intended to reduce the number of models to be optimized, as mentioned in the previous section. Thus, algorithms were combined with different scalers and sampling methods in order to make a pre-selection. The top 3 models are summarized in Table 6, ranked by the highest AUPRC, which coincides with the higher AUROC as well.

Table 6. First optimization phase - best models obtained.

Scaler	Algorithm	Balancing technique	AUPRC	AUROC
Standard	LR	Balanced model class weights	0.0871	0.5802
Standard	LinearSVC	Balanced model class weights	0.0870	0.5799
Robust	LightGBM	Balanced model class weights	0.0860	0.5700

Hence, it was concluded that class weight would be used as an optimization parameter (instead of over/undersampling) and that each model would have specific weights to achieve the highest performance. For the LR and LinearSVC algorithms, the StandardScaler was chosen, while the LightGBM obtained better results with the RobustScaler. For the next phase, RandomSearchCV was used in the training and validation dataset. Each of the random searches was performed with the appropriate hyperparameters for each model and evaluation metric AUPRC. The achieved results are shown below in Table 7.

Table 7. Second optimization phase - best results achieved.

Scaler	Algorithm	Balancing technique	AUPRC	AUROC
Standard	LR	Class Weight {1:13}	0.1102	0.6189
Standard	LinearSVC	Class Weight {1:5}	0.1076	0.6150
Robust	LightGBM	Class Weight {1:8}	**0.1108**	0.6206

The algorithm that presented the best results was LightGBM with RobustScaler and Class Weight as a balancing method. Contrary to what happened before, where the automatically balanced model class weights were found to be better, the model found a better result for a weight of 8 for the minority class. With this, we achieved an AUPRC ≈ 0.11, which coincides with the higher AUROC ≈ 0.62. The improvement of the optimization phase was approximately 37% with respect to the AUPRC. Figure 2a shows the PR curve obtained for the best model. The plot of the precision-recall curve highlights that the model is barely above the no skill line for most recall values.

The model threshold values in Fig. 2b shows that the default value 0.5 needs to be lowered in order to maximize recall. The main goal of increasing recall is to avoid false negatives, meaning that the model chooses more cases of having installation problems, requiring a second screening action from the company to verify the true positives of this population. In this case, the selected threshold value that maximizes recall, and that keeps a good trade-off between precision and recall, is approximately 0.43.

The top 10 features most relevant from the model were mainly related to the location of the customer and the technology installed, as well as features associated with the service time. These features can serve as guidelines for future implementations.

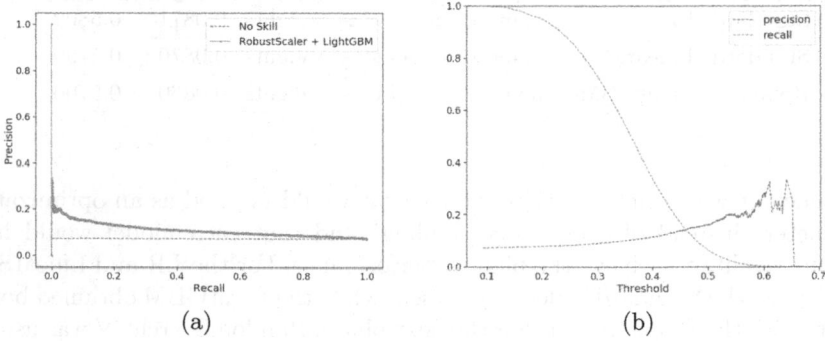

(a)   (b)

Fig. 2. Curves obtained for the best model: (a) comparison of precision for different recall values with a no skill model, and (b) PR for different thresholds.

Finally, with the best hyperparameters and the optimal threshold found, the final model was tested on the test set, confirming the results achieved in the validation set, and obtaining the confusion matrix in Table 8. Even with the adjustment of the threshold, and taking into account the results obtained from AUPRC, a trade-off between precision and recall that tries to maximize the recall will not prevent the existence of false positives and negatives. The existence of a large number of these cases, and the low AUPRC value, indicate that further work is needed.

Table 8. Confusion matrix of the best model with respect to the test set.

		Predicted	
		Yes	No
Actual	Yes	685	4639
	No	10384	69229

4 Conclusion and Future Work

Improving customer experience is crucial in any industry, especially in telecommunications, where competition is a constant factor. In this article, it was performed research in the process of installing new services, which can be an event that raises technical problems and, consequently, a possible degradation in the user experience.

The main objective of this work was to create a predictive model that, through all the available data history, could predict with high performance which customers would contact the customer service with problems from the installation process. With this, the company can anticipate the cases that will be problematic and reduce the number of negative experiences. The study was conducted using real data from a Portuguese telecommunications company, and the analysis was conducted resorting to the CRISP-DM methodology.

The task revealed to be of extreme difficulty, as shown in the rather small AUPRC and AUROC scores. The significant imbalance of the data proved problematic, and it was necessary to recur to precision-recall threshold optimization. The best trade-off between precision and recall was found with a threshold model of 0.43, in order to maximize the recall and not to lose all precision. A higher recall makes it possible to avoid false negatives, which means that the telecommunications company would have to perform a screening action with the clients to verify the true positives of this population. The existence of high false positives and negatives, as well as the low value of AUPRC, indicate that further work should be conducted to obtain more data that could improve models' performance. Future research will also focus on improving model tuning, and determining if the predictions could be more detailed concerning the predicted time of occurrence and the variables of the model that have shown more importance.

Acknowledgments. This work has been supported by FCT – Fundação para a Ciência e Tecnologia within the RD Units Project Scope: UIDB/00319/2020.

References

1. Loveman, G.: Diamonds in the data mine. Harvard Bus. Rev. **81**, 108–113+130 (2003)
2. DECO. 'Telecomunicações sempre no topo das reclamações'. https://www.deco.proteste.pt/institucionalemedia/imprensa/comunicados/2019/telecons-1705, Accessed 10 May 2020

3. Witten, I.H., Frank, E., Hall, M.A.: Data Mining: Practical Machine Learning Tools and Techniques, 3rd edn. Morgan Kaufmann, Burlington (2011)
4. Weiss, G.: Data mining in the telecommunications industry. In: Wang, J. (ed.) Encyclopedia of Data Warehousing and Mining, vol. 1, pp. 486–491, 2nd ed. Information Science Publishing (2008)
5. Neto, C., Brito, M., Lopes, V., Peixoto, H., Abelha, A., Machado, J.: Application of data mining for the prediction of mortality and occurrence of complications for gastric cancer patients. Entropy, 21, MDPI AG (2019)
6. Prata, M., Peixoto, H., Machado, J., Abelha, A.: Data mining in urgency department: medical specialty discharge prediction. In: 16th International Industrial Simulation Conference 2018, ISC (2018)
7. Han, J., Kamber, M., Pei, J.: Data Mining: Concepts and Techniques, 3rd edn. Morgan Kaufmann, Burlington (2011)
8. scikit-learn. Compare the effect of different scalers on data with outliers (2020). https://scikit-learn.org/stable/auto_examples/preprocessing/plot_all_scaling.html. Accessed 15 May 2020
9. Herrera, V.M., Khoshgoftaar, T.M., Villanustre, F. et al.: Random forest implementation and optimization for Big Data analytics on LexisNexis's high performance computing cluster platform. J. Big Data **6**, 68 (2019)
10. Bergstra, J., Bengio, Y.: Random search for hyper-parameter optimization. J. Mach. Learn. Res. **13**, 281–305 (2012)
11. Branco, P., Torgo, L., Ribeiro, R.P.: A survey of predictive modeling on imbalanced domains. ACM Comput. Surv. **49**(2) (2016). Article 31
12. Fawcett, T.: An introduction to ROC analysis. Pattern Recogn. Lett. **27**(8), 861–874 (2006)

Meta-hyperband: Hyperparameter Optimization with Meta-learning and Coarse-to-Fine

Samin Payrosangari[1] , Afshin Sadeghi[1,2] , Damien Graux[3] ,
and Jens Lehmann[1,2]

[1] Department of Computer Science, University of Bonn, Bonn, Germany
saminpayro@gmail.com, {sadeghi,jens.lehmann}@cs.uni-bonn.de
[2] Fraunhofer IAIS, Sankt Augustin, Germany
[3] ADAPT SFI Research Centre, Trinity College Dublin, Dublin, Ireland
damien.graux@adaptcentre.ie

Abstract. Hyperparameter optimization is one of the main pillars of machine learning algorithms. In this paper, we introduce Meta-Hyperband: a Hyperband based algorithm that improves the hyperparameter optimization by adding levels of exploitation. Unlike Hyperband method, which is a pure exploration bandit-based approach for hyperparameter optimization, our meta approach generates a trade-off between exploration and exploitation by combining the Hyperband method with meta-learning and Coarse-to-Fine modules. We analyze the performance of Meta-Hyperband on various datasets to tune the hyperparameters of CNN and SVM. The experiments indicate that in many cases Meta-Hyperband can discover hyperparameter configurations with higher quality than Hyperband, using similar amounts of resources. In particular, we discovered a CNN configuration for classifying CIFAR10 dataset which has a 3% higher performance than the configuration founded by Hyperband, and is also 0.3% more accurate than the best-reported configuration of the Bayesian optimization approach. Additionally, we release a publicly available pool of historically well-performed configurations on several datasets for CNN and SVM to ease the adoption of Meta-Hyperband.

Keywords: Hyperparameter optim. · Meta-learning · Coarse-to-Fine

1 Introduction

In recent years, machine learning has opened up its way promoting automation in various domains such as image and face recognition [6], cancer detection [14] or speech recognition [4]. Each machine learning model has particular parameters, such as learning rate, learning rate decay or regularization coefficient which specify its architecture. These parameters are called the "hyperparameters" of the model which should be initialized by data scientists [1]. Specifying the right values for hyperparameters is critical and may lead to a model with high accuracy [17]. Contrariwise, choosing wrong values reduces the performance of the

© Springer Nature Switzerland AG 2020
C. Analide et al. (Eds.): IDEAL 2020, LNCS 12490, pp. 335–347, 2020.
https://doi.org/10.1007/978-3-030-62365-4_32

model significantly. Therefore, several approaches for hyperparameter optimization have been proposed. Nevertheless, some of these methods are still immature and need improvement or modification to enhance the probability of finding a near optimum model architecture. During hyperparameter optimization, one needs to evaluate the performance of various sets of values, training the model in individual rounds. To do so, a naive idea is sampling random hyperparameter configurations from space [1] and the best discovered values of hyperparameters, leading to lower generalization error, are chosen. To increase the speed of hyperparameter optimization, a novel approach called Hyperband has been recently proposed [11] based on dynamic resource allocation. In a nutshell, many random hyperparameter configurations are sampled from the initial space, and instead of allocating the maximum amount of training resources to each of them, they get lower amount of resources at the beginning and compete with each other for higher amount of resources. Subsequently, only a portion of configurations with lower validation loss will proceed to the next training round with more resources to demonstrate their real performance. Therefore, one could evaluate several random configurations faster than simple random search by early elimination of unintelligent configurations. The Hyperband algorithm is practical as it assesses numerous configurations, efficient in finding a high performance hyperparameter configuration for a dataset, and 5 to 30 times faster than its competitors such as different variations of Sequential model-based optimization like Spearmint, TPE and SMAC [11]. However, as Hyperband only targets random search and pure exploration, it has a deficiency in leveraging achieved information from previous evaluations and other machine learning experiments.

To tackle this limitation, we propose **Meta-Hyperband** to increase the quality of the discovered hyperparameter configurations. Meta-Hyperband adds extents of exploitation to Hyperband. Alongside random configurations sampled from the predefined space in Hyperband, Meta-Hyperband embeds a meta-learning module to benefit from historical machine learning experiments which have been performed on different datasets. In addition, a Coarse-to-Fine module is included in Meta-Hyperband to search the areas around the best configurations discovered during the ongoing hyperparameter optimization process, as the current best configuration might be close to the optimum hyperparameter configuration.

The organization of this paper is as follows. Section 2 reminds the background, Sect. 3 makes an overview on the related works in hyperparameter optimization. Section 4 describes Meta-Hyperband in details and Sect. 5 presents the results of experiments on CNN and SVM. Finally, in Sect. 6 concludes the study.

2   Background

Hyperparameter Configuration: A "configuration" of a machine learning model is a unique set of values for different hyperparameters of the model. For instance, {learning rate = 0.5, regularization rate = 0.5, number of nodes in second layer = 10} is one configuration and {learning rate = 0.005, regularization rate = 0.5, number of nodes in second layer = 11} is another. Furthermore, If the model has t hyperparameters $h_1, \ldots, h_t$ and they vary in ranges $r_1, \ldots, r_t$ respectively, the "space" is defined as the t dimensional space $r_1 \times \cdots \times r_t$, and each configuration is allowed to be sampled from this space [5].

Meta-learning: Leveraging experiments and meta-data achieved, from previous learning procedures, in order to improve the quality of learning is known as meta-learning [18]. In this article, meta-learning corresponds to benefiting from historically well-performed hyperparameter configurations of a model on various datasets, to discover an optimum configuration for a new dataset.

Coarse-to-Fine: is a variation of the random search for hyperparameter optimization, which limits the space of possible hyperparameters configurations in an organized manner [12]. In pure random search, one picks many random configurations until reaching a high-quality model. In Coarse-to-Fine after randomly picking several initial configurations, the searching space is shrunk to the surrounding of the current best configuration [3]. The idea behind Coarse-to-Fine is to eliminate the impractical parts of the initial space and exploit the fruitful parts as the optimum hyperparameter configuration might be close to the best configuration which is discovered after an initial search in the main space.

3   Related Work

Random search is the basic idea for hyperparameter tuning. After random search, grid search was proposed which suffers from the curse of dimensionality and it limits the choices for more critical hyperparameters such as learning rate [1]. Although, hyperparameter optimization is time consuming, it influences the performance significantly. Thus, several other approaches have been designed. These approaches fall into two main categories: 1) Sequential model-based optimization and 2) Bandit-based optimization.

Sequential Model-Based Optimization. SMBO approaches are methods for optimizing a black-box function, where calculation of the function for each individual point is costly, analogically to hyperparameter optimization problem.

From Different variations of SMBO, Bayesian optimization(BO) method uses "expected improvement (EI)" as the acquisition function to address the hyperparameter optimization task, since EI has shown exceptional performance in optimizing several other multidimensional black-box functions [17]. Another method is Tree-structured parzen estimation (TPE) which uses Tree-structured parzen estimator to define the surrogate, and an EI acquisition function that is based on this surrogate function [2]. Sequential model-based algorithm configuration (SMAC) is another SMBO that derives the surrogate of the black-box function using random forests. To optimize hyperparameters, a random forest is trained on a pool of historically evaluated configurations. Therefore, the forest predicts the performance of the particular model for new hyperparameter configuration [7]. Subsequently, it leverages the EI acquisition function to determine next point to be evaluated at each iteration of optimization procedure [7]. The methods in

SMBO category struggle with the dimensionality issue [15], and in this condition, their performance is equivalent to random search. Moreover, these methods can't deal with non-convex functions [11]. Bandit-based methods, which are introduced in the next section, target these issues.

Bandit-Based Optimization. In bandit-based hyperparameter optimization environment, more resources are allocated to promising hyperparameter configurations, known as adaptive resource allocation, and the unpropitious ones are stopped early resulting in an order of magnitude faster optimization compared to SMBO methods [8]. From the perspective of this category of methods, the problem of hyperparameter optimization is considered as an infinite-armed bandit problem in a non-stochastic environment. In this scenario, each arm is a hyperparameter configuration and pulling an arm corresponds to evaluating the performance of the configuration by training for a specific amount of epochs and returning the loss. One early bandit-based method is Successive Halving: given a total amount of budget B, the budget is uniformly divided among a set of n randomly sampled hyperparameter configurations. Its algorithm for hyperparameter optimization is:

1. Sample n hyperparameter configurations from the space
2. Given the total amount of budget B, Uniformly allocate a small portion of the budget B to each configuration.
3. Train the model with each of these configurations individually with the allocated budget
4. Eliminate half of the configurations which have shown lower performance,
5. Keep the other half and allocate a bigger portion of the B to each of them.
6. Repeat from step 3 until one config. remains and report it as the best one.

Using this algorithm more hyperparameter configurations can be assessed using total amount of budget B, because early termination of unpromising configurations releases resources for evaluating more Hyperparameter configurations, and only those configurations which have shown promising performance will be eligible to get more resources out of the total budget. Hyperband algorithm is a pure exploration infinite-arm bandit method that extends Successive Halving by repeating it several times, but with a different number of initial configurations n [11]. This extension of successive halving performs the hyperparameter optimization task about 5 to 30 times faster than well-known methods such as SMBO variations and random search [11].

   In Hyperband, each run of Successive Halving is called a "bracket" [11]. This method reports the best configuration after execution of all individual Brackets. The reason for implementing multiple brackets with different values of n is that, it's indeterminable whether it's better to keep n small or large. When n is small, each configuration gets more resources and when a big n is chosen each configuration gets a lower number of resources but more configurations could be evaluated. In some problems, individual configurations might have a slow convergence rate, so they need more resources to reveal their performance and to compete with other configurations. Keeping the trade-off between the number of configurations n and the amount of resources which is allocated to each configuration B/n, is called the n versus B/n trade-off. The inputs of Hyperband include the maximum number of resources R which could be allocated to each individual configuration (i.e. determines the budget limit) and η which determines the portion of configurations that should be discarded at each round inside the brackets. Respectively, R and η determine the number of brackets and the initial number of configurations n_i in individual brackets. We introduce an algorithm in Sect. 4 that applies an improved sampling method for the initial configurations in each bracket, without touching the other variables such as R and η.

4   Meta-Hyperband

We propose **Meta-Hyperband**, a meta-learning hyperparameter optimization algorithm that benefits from a pool of historically well-performed hyperparameter configurations on various datasets in previous experiments. The idea is that one of these configurations might perform well on the new dataset directly or by little manipulation obtained by the Coarse-to-Fine module. Therefore, the Meta-Hyperband algorithm adds meta-learning as well as Coarse-to-Fine modules to the Hyperband algorithm. An advantage of Meta-Hyperband (over Hyperband) is its trade-off between exploration and exploitation as part of the brackets run a random search: a part of them run a meta-learning and the rest run a Coarse-to-Fine sampling on the best configurations discovered in the previous steps. The inputs to our algorithm are:

- **R:** The maximum amount of resources which should be allocated to each configuration. Basically, this value changes for different problems. For example, in Deep Learning, the number of training iterations might be selected as the resource and the maximum iterations which is required for a successful training depends on the problem.
- **Meta-learning pool:** A pool of best hyperparameter configurations discovered in historical experiments of the corresponding machine learning model. We show (see Sect. 5) that having a meta-learning pool improves the results significantly and leads to a faster approach. Therefore, contributing in building such a pool for different models helps to improve the performance of the models on new datasets.
- **Coarse-to-Fine coefficient:** The area around a proper configuration to be explored. If the initially specified range for a hyperparameter is (a, b) and the current best value for it is h_1, then h_1 is considered as the coarse and we try to fine tune h_1 in its surrounding [3]. In this article, considering that the Coarse-to-Fine coefficient is equal to r, the shrunk range for this hyperparameter is $(h_1 - rh_1, h_1 + rh_1)$. Here, the amount of moving around h_1 depends on itself by multiplying the radius by h_1, in order to avoid outlying values which are either far away from h_1 or too close to it. By limiting the range of all the hyperparameters in one configuration using the same coefficient, the space is

restricted using the Coarse-to-Fine module. The value of r defines the extent of space shrinkage, as smaller values apply more aggressive space restriction than larger values. The default value we consider for r is 0.2. According to our experiments, lower values of r restrict the space such that the performance improvement by fine-tuned configurations is not outstanding.

In Meta-Hyperband, 5 brackets (from 0 to 4) are considered. In the initialization phase, the maximum bracket number called s_{max} equals 4, and η defines the portion of resources which should be eliminated in each round in brackets. In Hyperband, η is an input and the s_{max} which is the index of last bracket, is defined by $\lfloor \log_\eta R \rfloor$. Differently, in Meta-Hyperband, the value of s_{max} is predefined, and using the same rule the equation of $\eta = \lfloor \sqrt[4]{R} \rfloor$ holds. The reason behind this setup is that, as a proper arrangements for Hyperband it was proposed to determine η in a way that according to R, the algorithm runs in 5 brackets, so that with different values of sampled configurations in different brackets the n versus B/n trade-off is holds [11]. The Meta-Hyperband algorithm is defined as:

1. Bracket$_1$, meta-learning: Sample n_1 hyperparameter configurations from the pool of meta-learning, and apply the Successive Halving steps 2–6 (see Sect. 3) on them. If there is no pool, the Bracket performs a random search.
2. Bracket$_4$, random search: Randomly sample n_4 hyperparameter configurations from space, and apply the Successive Halving steps 2–6 on them.
3. Bracket$_2$, random search: Randomly sample n_2 hyperparameter configurations from space, and apply the Successive Halving steps 2–6 on them.
4. Bracket$_3$, Coarse-to-Fine:
   (a) Sample $n_3/2$ fine-grained configurations where coarse is the best of Bracket$_1$, and apply the Successive Halving steps 2–6 on them.
   (b) Sample $n_3/2$ fine-grained configurations where coarse is the best of brackets 2 and 4, and apply the Successive Halving steps 2–6 on them.
5. Bracket$_0$, Coarse-to-Fine: Sample n_0 fine-grained configurations where coarse is the best of all previous steps, and apply the steps 2–6 on them.

In this algorithm, the number of configurations n_i in Bracket i, and the resource allocation are similar to Hyperband so the inequality $n_4 > n_3 > n_2 > n_1 > n_0$ holds. Therefore, in brackets with higher index, more configurations are sampled and each of them receives a smaller portion of the total budget B of the bracket in comparison to the brackets with lower index. Therefore, the Bracket with highest index (Bracket$_4$) samples R configurations initially and Bracket$_0$ samples $s_{max} + 1$ (5 in case of Meta-Hyperband) but allocates maximum amount of resources R to each of them. Algorithm 1 shows the pseudocode of Meta-Hyperband. Inside each bracket of Meta-Hyperband, the algorithm performs the sampling task according to the role of each bracket as discussed. In each bracket, the n_i configurations that are present are trained using the specified amount of resources r_i, $1/\eta$ of them with lower validation loss win the competition and proceed to next round, and the rest of the configurations are discarded.

Algorithm 1. Meta-Hyperband algorithm

Inputs: R, Coarse-to-Fine coefficient, meta-learning pool
Initialize: $s_{max} = 4, \eta = \lfloor \sqrt[4]{R} \rfloor, B = (s_{max} + 1)R$
for $s = \{1, 4, 2, 3, 0\}$ **do**

   $n = \lceil \frac{B}{R} \frac{n^S}{s+1} \rceil, r = R\eta^{-s}, T = []$
   $path = $ path-to-save-configurations
   if $s == 1$ **then**
       $T = $ get-meta-hyperparameter-configs(n)
   else if $s == 4$ *or* 2 **then**
       $space = $ get-main-space$()$
       $T = $ get-random-hyperparameter-configs$(n, space)$
   else if $s == 3$ **then**
       $a)$ $space = $ get-Coarse2Fine-space(best config. from bracket 1)
       $T = $ get-fine-parameters$(space, \frac{n}{2})$
       $b)$ $space = $ get-Coarse2Fine-space(best config. from brackets 2 & 4)
       $T.append($get-fine-parameters$(space, \frac{n}{2}))$
   else if $s == 0$ **then**
       $space = $ get-Coarse2Fine-space(best config. from at all other brackets)
       $T = $ get-fine-parameters$(space, n)$
   for $i \in \{0, ..., s\}$ **do**
       $n_i = \lfloor n \times \eta-i \rfloor, r_i = \lfloor r \times \eta i \rfloor$
       $L = $ evaluate-the-validation-loss$(t) : t \in T$
       $T = top(T, L, \lfloor \frac{n_i}{\eta} \rfloor)$

Result: The best discovered configuration with lowest validation loss.

At the end of each bracket, one configuration is reported as the best configuration of that bracket. After locating Bracket$_1$, which is the meta-learning sampling bracket, Brackets 4 and 2 are located, where the random search in the main space of configurations takes place. The Bracket$_4$ samples too many configurations and increases the chance of exploring the space thoroughly. Moreover, Bracket$_2$ samples lower number of random configurations but allocates more resources to each of them. After completion of the 3 previous brackets, the best configuration discovered in the Bracket$_1$, as well as the best configuration of Brackets 4 and 2, are taken into Bracket$_3$ which performs Coarse-to-Fine on these two current best configurations. This avoids trapping in the area around a local optimum because it considers two of the best configurations instead of considering one global best. Finally, Bracket$_0$ takes the global best of all previous brackets and applies Coarse-to-Fine on it. If the meta-learning pool is missing, the Coarse-to-Fine module can still be leveraged. And if this pool contains many configurations, sampling from it can be moved to other brackets of the Meta-Hyperband than the proposed bracket in this paper. Our analysis of Hyperband shows that in most of the hyperparameter optimization experiments all the Brackets are not used optimally. Experiments in Sect. 5 prove the dominance of Meta-Hyperband in discovering high performance configurations and leveraging all the brackets.

5   Experiments

To compare the performance of Meta-Hyperband and Hyperband, several experiments are performed on CNN and SVM using 5 benchmark datasets: Cifar10 [9], Cifar100 [9], SVHN [13], MNIST [10] and Fashion-MNIST [19]. To perform the

Table 1. Hyperparameter configuration space for CNN [11].

Hyperparameter	Range	Scale
Learning rate reduction	$[0,3]$	Integer
Initial learning rate	$[5 \times 10^{-5}, 5]$	Log
Conv1 L2 penalty	$[5 \times 10^{-5}, 5]$	Log
Conv2 L2 penalty	$[5 \times 10^{-5}, 5]$	Log
Conv3 L2 penalty	$[5 \times 10^{-5}, 5]$	Log
Fc10 L2 penalty	$[5 \times 10^{-3}, 500]$	Log
Local response normalization scale	$[5 \times 10^{-6}, 5]$	Log
Local response normalization power	$[0.01, 3]$	Linear

Table 2. Hyperband & Meta-Hyperband on CIFAR10 using CNN: $R = 300$ and $\eta = 4$.

Method/Results	Discovery bracket	Top-one test loss 75 epochs	Top-one test loss 300 epochs
Hyperband	$Bracket_4$, random	0.214500	0.195400
Hyperband	$Bracket_4$, random	0.219100	0.186600
Hyperband	$Bracket_4$, random	0.213800	0.202100
Meta-Hyperband	$Bracket_3a$, Coarse-to-Fine	0.173000	0.146800
Meta-Hyperband	$Bracket_4$, random	0.177000	0.147600
Meta-Hyperband	$Bracket_3a$, Coarse-to-Fine	0.191900	0.150800

Table 3. Hyperband & Meta-Hyperband on SVHN using CNN. $R = 600$ and $\eta = 4$.

Method/Results	Best config discovery bracket	Top-one test loss 6 epochs
Hyperband	$Bracket_4$, random	0.080042
Hyperband	$Bracket_2$, random	0.058208
Hyperband	$Bracket_3$, random	0.093667
Meta-Hyperband	$Bracket_0$, Coarse-to-Fine	0.066625
Meta-Hyperband	$Bracket_4$, random	0.064833
Meta-Hyperband	$Bracket_3a$, Coarse-to-Fine	0.062667

Table 4. Hyperband & Meta-Hyperband on CIFAR100 using CNN. $R = 300$ and $\eta = 4$.

Method/Results	Best config discovery bracket	Top-one test loss after 300 epochs
Hyperband	$Bracket_4$, random	0.8300
Hyperband	$Bracket_3$, random	0.7986
Meta-Hyperband	$Bracket_4$, random	0.5893
Meta-Hyperband	$Bracket_3b$, Coarse-to-Fine	0.5764

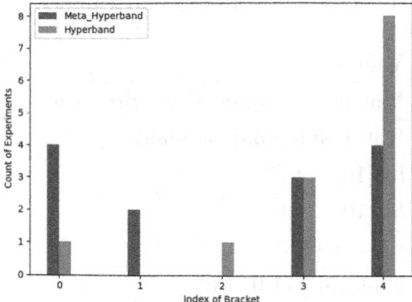 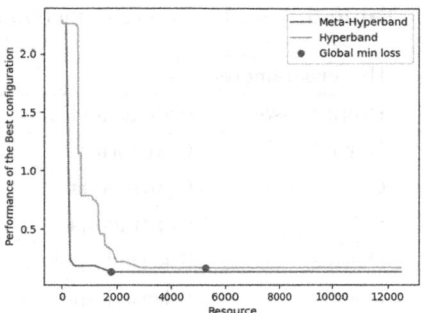

Fig. 1. Bar chart displaying the role of different brackets in 13 Hyperband and 13 Meta-Hyperband CNN experiments.

Fig. 2. Performance of the best discovered configuration in function of resource usage on two CNN with SVHN.

experiments, firstly Hyperband is used several times to discover the best hyperparameter configuration of each model for each of the datasets. Subsequently, all the discovered configurations are gathered to generate the pool of configurations for each model to feed Meta-Hyperband. These pools of configurations are enriched every time a new proper configuration for a dataset is discovered[1].

While applying Meta-Hyperband on a dataset, only the configurations corresponding to the rest of the datasets in the pool are involved in the meta-learning bracket. The value of η is initialized to 4 according to the amount of resources for our datasets, so Meta-Hyperband consists of 5 brackets. Moreover, the Coarse-to-Fine radius is initialized to $0.2 \times h$ for a hyperparameter h (this radius can also be treated as a hyperparameter). The resource in CNN is defined as the number of training iterations, and in SVM as the amount of training data points.

5.1 Experiments on CNN

In this study, the 18% error architecture proposed by Alex Krizhevsky is used for CNN model[2] and implemented by Cuda-convnet2 framework[3]. This architecture of CNN is also used in Hyperband [11] and ensures a fair comparison between the two approaches. Furthermore, the 8 hyperparameters of CNN are tuned (see Table 1). Table 1 also shows the ranges for them, which are similar to what was considered in [11] for fair comparison.

Each resource unit in our study is the training for 100 iterations which takes place on one batch of data of size 10k, and mini-batch size of 100 datapoints. Each method is repeated 13 times to gain an average performance. The bar chart (Fig. 1) displays the performance of brackets in both approaches, by counting the number of experiments that their best configuration was discovered in Bracket$_i$. The diversity of the bars in case of Meta-Hyperband shows that there is a higher

---

[1] Our sources → https://github.com/saminpayro/Meta_Hyperband_implementation.
[2] See "example layers" directory in http://code.google.com/p/cuda-convnet/.
[3] https://code.google.com/archive/p/cuda-convnet2/.

Table 5. List of 6 hyperparameters for SVM with their sampling ranges here.

Hyperparameter	Type	Values
Preprocessor	Categorical	Min/max, standardize, normalize
Kernel	Categorical	Rbf, polynomial, sigmoid
C	Continuous	$log[10^{-3}, 10^5]$
γ	Continuous	$log[10^{-5}, 10]$
Degree	if kernel=poly	Integer [2, 5]
Coef0	if kernel=poly, sigmoid	Uniform [−1.0, 1.0]

Table 6. Hyperband & Meta-Hyperband on CIFAR10 using SVM: R = 400, $\eta = 4$.

Method/Results	Discovery bracket	Top-one error
Hyperband	$Bracket_4$, random	0.4467
Meta-Hyperband	$Bracket_0$, Coarse-to-Fine	0.4402
Hyperband	$Bracket_2$, random	0.4455
Meta-Hyperband	$Bracket_1$, meta-learning	0.4429
Hyperband	$Bracket_4$, random	0.4477
Meta-Hyperband	$Bracket_0$, Coarse-to-Fine	0.4452

balance in the performance of brackets, and despite Hyperband brackets 0, 1, 3, 4 all have an active contribution in discovering a proper configuration for the CNN. The plots (Fig. 2) display the progress of the two approaches with respect to the resource usage for 2 experiments on SVHN dataset, in which the Meta-Hyperband has discovered the best configuration with consumption of less than 3000 units of resources compared to Hyperband. Table 2 displays the results of experiments on Cifar10. Among the 6 experiments, Meta-Hyperband has discovered configurations with around more than 3% lower test error than Hyperband after 75 and 300 epochs. The lowest error discovered by Meta-Hyperband on test set is 14.68% after 300 epochs training, which is even better than the configuration reported by Bayesian Optimization method (14.98%) [17].

In another experiment, the SVHN dataset is used as the basis. Table 3 compares the quality of the best configurations according to the test error and after 6 epochs training. Among the 6 experiments, Meta-Hyperband has discovered configurations with around 3% lower test error than 2 runs of Hyperband. In addition, configurations from the meta-learning pool of configurations, which were corresponding to Fashion-MNIST and CIFAR10, perform well on SVHN as well. For instance, one historical configuration of Fashion-MNIST produces 8% test loss on SVHN as well, and the best configuration (reported on Table 3 last row) is the result of Coarse-to-Fine on a historical configuration of CIFAR10 from the meta-learning pool. In addition, Table 4 shows that during experiments on CIFAR100 dataset, Meta-hyperband outperformed Hyperband (more than 2% lower test error). Rows 3 and 4 show that, in Meta-Hyperband both random and non-random brackets are leveraged for discovering a proper configuration.

5.2   Experiments on SVM

The list of 6 hyperparameters tuned for SVM is displayed in Table 5. For SVM, one unit of resource equals 100 data points, and the maximum number of resources should be equal to the size of the training set for each dataset. In this project, we used 60k training set for SVHN, and 40k training sets for CIFAR10 and CIFAR100 respectively. Figure 3 displays the performance of different brackets in both approaches, by counting the number of experiments that their best configuration was discovered in $Bracket_i$. According to this figure, in the SVM case, $Bracket_4$ of Hyperband is also the most effective bracket.

For SVM experiments, instead of repeating the execution of Meta-Hyperband for all the brackets, some configurations which have been discovered previously during the Hyperband test, are considered as the best of random brackets in Meta-Hyperband as well to ensure a fair comparison between the two methods.

Tables 6, 7 and 8 display the results of experiments on CIFAR10, SVHN, and CIFAR100 for SVM model. In these tables, each Meta-Hyperband row is paired to its above

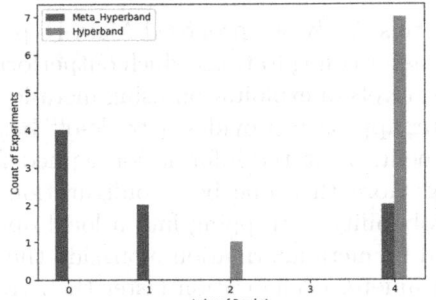

Fig. 3. Role of different brackets in Hyperband vs Meta-Hyperband in SVM.

Hyperband row, which means that the best configurations discovered by Hyperband in steps 4 or 2 are considered as the best configuration of random brackets for the paired Meta-Hyperband experiment.

Table 7. Hyperband & Meta-Hyperband on SVHN using SVM: $R = 600$, $\eta = 4$.

Method/Results	Discovery bracket	Top-one error
Hyperband	$Bracket_4$, random	0.232406269207
Meta-Hyperband	$Bracket_4$, random	0.232406269207
Hyperband	$Bracket_4$, random	0.245789072890
Meta-Hyperband	$Bracket_0$, Coarse-to-Fine	0.245678902347
Hyperband	$Bracket_4$, random	0.22303318992
Meta-Hyperband	$Bracket_4$,random	0.22303318992

Analysing the diversity of best configuration discovery among the brackets of Meta-Hyperband on SVM in the bar chart of Fig. 3 shows that $Bracket_0$ was the most effective bracket in discovering high performance configurations, also $Bracket_4$ still plays an important role by exploring the space. Moreover, 2 configurations were discovered in $Bracket_1$ which speed up the hyperparameter tuning of Meta-Hyperband.

Table 8. Hyperband & Meta-Hyperband on CIFAR100 using SVM: R $= 400$, $\eta = 4$.

Method/Results	Discovery bracket	Top-one error
Hyperband	Bracket$_4$, random	0.7486
Meta-Hyperband	Bracket$_1$, meta-learning of SVHN historical configuration	0.7166
Hyperband	Bracket$_4$, random	0.7486
Meta-Hyperband	Bracket$_0$, Coarse-to-Fine of SVHN historical configuration	0.7156

6   Conclusion and Future Work

In this study, we proposed Meta-Hyperband, our method to discover hyperparameter configurations which outperforms the literature algorithms, by introducing levels of exploitation using meta-learning as well as Coarse-to-Fine modules. Our approach provides a trade-off between exploration and exploitation, and benefits from the information gained from previous experiments. By considering more than one best configuration during the exploitation, it reduces the probability of trapping into a local optimum and the application of the historical or meta-information alongside the random search leads to a proper hyperparameter configuration faster than traditional methods. Meta-Hyperband has been validated through several experiments on CNN and SVM to tune hyperparameters for learning five different datasets. The best configurations discovered for SVM and CNN generated a pool that is shared and can be used in other experiments.

A recent approach to optimize hyperparameters, proposed by [16] for the MDE$_{nn}$ version of MDE, integrates the hyperparameters into the primary model's optimization. There, after initiating hyperparameters, the model optimizes them by back-propagation. It is interesting for future work to extend Meta-Hyperband in this direction and combine the Bandit-based algorithm with the models optimization, especially for NN-based models.

Acknowledgments. This study is supported by MLwin (*Maschinelles Lernen mit Wissensgraphen*, GA 01IS18050F of the German Federal Ministry of Education and Research), by the EU project Cleopatra (GA 812997) and by the Marie Skłodowska-Curie GA 801522 at the ADAPT SFI Research Centre (grant 13/RC/2106).

References

1. Bergstra, J., Bengio, Y.: Random search for hyper-parameter optimization. J. Mach. Learn. Res. **13**, 281–305 (2012)
2. Bergstra, J., Rémi B., Bengio, Y., Kégl, B.: Algorithms for hyper-parameter optimization. In: Neural Information Processing Systems, pp. 2546–2554 (2011)
3. Charniak, E., Johnson, M.: Coarse-to-fine n-best parsing and maxent discriminative reranking. In: Annual Meeting on Association for Computational Linguistics, pp. 173–180. ACL 2005 (2005)
4. Deng, L., Li, X.: Machine learning paradigms for speech recognition: an overview. Trans. Audio Speech Lang. Process. **21**(5), 1060–1089 (2013)

5. Feurer, M., Springenberg, J.T., Hutter, F.: Initializing bayesian hyperparameter optimization via meta-learning. In: AAAI Conference on Artificial Intelligence (2015)
6. Guo, G., Li, S.Z., Chan, K.: Face recognition by support vector machines. In: International Conference on Automatic Face and Gesture Recognition (2000)
7. Hutter, F., Hoos, H.H., Leyton-Brown, K.: Sequential model-based optimization for general algorithm configuration. In: Coello, C.A.C. (ed.) LION 2011. LNCS, vol. 6683, pp. 507–523. Springer, Heidelberg (2011). https://doi.org/10.1007/978-3-642-25566-3_40
8. Jamieson, P., Talwalkar, A.: Non-stochastic best arm identification and hyperparameter optimization. AISTATS (2015)
9. Krizhevsky, A.: Learning multiple layers of features from tiny images. Technical report, Department of Computer Science, University of Toronto (2009)
10. LeCun, Y., Bottou, L., Bengio, Y., Haffner, P.: Gradient-based learning applied to document recognition. Proc. IEEE **86**(11), 2278–2324 (1998)
11. Li, L., Jamieson, K., DeSalvo, G., Rostamizadeh, A., Talwalkar, A.: Hyperband: a novel bandit-based approach to hyperparameter optimization. In: ICLR (2017)
12. Moshkelgosha, V., Behzadi-Khormouji, H., Yazdian-Dehkordi, M.: Coarse-to-fine parameter tuning for content-based object categorization. In: International Conference on Pattern Recognition and Image Analysis (IPRIA), pp. 160–165. IEEE (2017)
13. Netzer, Y., Wang, T., Coates, A., Bissacco, A., Wu, B., Ng, YA.: Reading digits in natural images with unsupervised feature learning. In: NIPS (2011)
14. Rejani, Y.I.A., Selvi, S.T.: Early detection of breast cancer using SVM classifier technique. CoRR abs/0912.2314 (2009)
15. Rolland, P., Scarlett, J., Bogunovic, I., Cevher, V.: High-dimensional bayesian optimization via additive models with overlapping groups. AISTATS (2018)
16. Sadeghi, A., Graux, D., Yazdi, H.S., Lehmann, J.: MDE: multi distance embeddings for link prediction in knowledge graphs. In: 24th European Conference on Artificial Intelligence (ECAI) (2020)
17. Snoek, J., Larochelle, H., Adams, R.: Practical bayesian optimization of machine learning algorithms. In: Neural Information Processing Systems (NIPS) (2012)
18. Vilalta, R., Drissi, Y.: A perspective view and survey of meta-learning. Artif. Intell. Rev. **18**(2), 77–95 (2002)
19. Xiao, H., Rasul, K., Vollgraf, R.: Fashion-mnist: a novel image dataset for benchmarking machine learning algorithms. arXiv preprint arXiv:1708.07747 (2017)

Stateful Optimization in Federated Learning of Neural Networks

Péter Kiss[1(✉)], Tomáš Horváth[1,2], and Vukasin Felbab[1]

[1] Department of Data Science and Engineering, Faculty of Informatics,
ELTE – Eötvös Loránd University, Budapest,
Pázmány Péter Sétány 1/C., Budapest 1117, Hungary
{peter.kiss,tomas.horvath}@inf.elte.hu, vukasindfelbab@gmail.com
[2] Faculty of Science, Institute of Computer Science, Pavol Jozef Šafárik University,
Jesenná 5, 040 01 Košice, Slovakia

Abstract. Federated learning is a emerging branch of machine learning research, that is examining the methods for training models over geographically separated, unbalanced and non-iid data. In FL, on non-convex problems, as in single node training, the almost exclusively used method is mini batch gradient descent. In this work we examine the effect of using stateful training method in a federated environment. According to our empirical results with these methods, at the cost of synchronizing state variables along with model parameters, a significant improvement can be achieved.

1 Introduction

Federated Learning (FL) [13] is a generalized approach for distributed Machine Learning (ML) that enables the training process to take into account not negligible effect of communication over heterogeneous networks and the divergence of local training data sets. With the spread of small user devices and ML-driven applications, that build on data generated at these equipments themselves, this setup gains more and more importance. The traditional centralized solution for Data Center (DC) based large scale ML would be to transfer the information gathered at the users to DCs, where the training takes place, then send back the trained models to the users. This, apart from the obvious privacy concerns, can incur a large communication overhead along with the need for significant storage and computational resources at the location of centralized training.

1.1 Distributed SGD

When training ML models in a distributed manner, the problem we want to solve is to minimize the loss function f with respect to model parameters $\mathbf{w}$ over all available data points at K nodes (user devices) as follows:

$$\min_{\mathbf{w} \in \mathbb{R}^d} f(\mathbf{w}) = \sum_{k=1}^{K} \frac{n^{(k)}}{n} F^{(k)}(\mathbf{w}), \text{ with } F^{(k)}(\mathbf{w}) = \frac{1}{n^{(k)}} \sum_{i=1}^{n^{(k)}} f^{(i)}(\mathbf{w}), \quad (1)$$

© Springer Nature Switzerland AG 2020
C. Analide et al. (Eds.): IDEAL 2020, LNCS 12490, pp. 348–355, 2020.
https://doi.org/10.1007/978-3-030-62365-4_33

where $f^{(i)}(\mathbf{w}) \overset{\text{def}}{=} f(\mathbf{x}^{(i)}, y^{(i)}, \mathbf{w})$ denotes the loss on data point $(\mathbf{x}^{(i)}, y^{(i)})$, given $\mathbf{w}$, and $F^{(k)}$ denotes the averaged loss at node k.

To solve the problem in (1) the simplest, and in neural network (NN) optimization, due to the non-convex loss functions, the almost exclusively used methods are versions of Stochastic Gradient Descent (SGD) [5]. SGD takes the derivative of loss function at one data point with respect to the model parameters $\mathbf{w}$, then move the parameter values in the direction of the negative of the gradient:

$$\mathbf{w}_{t+1} = \mathbf{w}_t - \eta_t \nabla f^{(i)}(\mathbf{w}_t), \tag{2}$$

where η_t denotes the learning rate. In practice, instead of applying the gradient for each data points, an average of gradients over a batch $\mathcal{B}$ of randomly chosen examples is used (evaluated at the same $\mathbf{w}$). Such "minibatch" gradient descent (MBGD) methods, still commonly referred to as SGD, better exploits parallel computational capabilities of the hardware (like GPU). For both cases, the update is a stochastic approximation of the whole gradient: $\mathbb{E}[\nabla f^{\mathcal{B}}(\mathbf{w})] = \nabla f(\mathbf{w})$. In data parallel centralized distributed synchronous MBGD training, per batch updates are parallelly computed by processors of the DC on their assigned data chunks, then their average will be applied to the central model $\mathbf{w}_t$:

$$\mathbf{w}_{t+1} = \mathbf{w}_t - \eta_t \frac{1}{K} \sum_{k=1}^{k} \nabla f^{\mathcal{B}_k}(\mathbf{w}_t). \tag{3}$$

1.2 Federated Learning

The main idea behind FL [13] is that, instead of moving the data to a central location and use training method from Eq. (2) or (3), one could exploit the computational power residing at user devices and partition the training process among them. This, however, brings more complication in the formula. Due to the real distributed nature of the system, communication becomes expensive and unreliable. Therefore Federated SGD (FedSGD) [13] algorithm applies a modified version of (3) to reduce communication complications, where instead of communicating the gradients per batch, the central updates takes place after multiple local updates:

$$\mathbf{w}_{t+1} = \mathbf{w}_t - \frac{n}{n^k} \sum_{k=1}^{k} \Delta^{(k)}, \text{ with } \Delta^{(k)} = \sum_{i=0}^{r} \nabla f^{\mathcal{B}_{t_i}^k}(w_{t_i}^k), \tag{4}$$

where $\mathbf{w}_{t_{i+1}}^k = \mathbf{w}_{t_i}^k - \eta \nabla f^{\mathcal{B}_{t_i}^k}(\mathbf{w}_{t_i}^k)$, $\mathbf{w}_{t_0}^k = \mathbf{w}_t$ and r is a hyper-parameter for the number of local updates.

To further increase communication efficiency, FedAvg algorithm [19] takes only a small subset (10%) of updates. It has been empirically proven to be able to keep or, in some cases, even increase convergence rate of the learning. FedAvg became the baseline of FL research.

Since the data to be processed by FL generated at a huge number of independent nodes, it has the following characteristics:

1. *Massively Distributed.* The number of nodes can be much bigger than the average number of training examples stored on a given node (n/K).
2. *Non-IID.* The data points available locally are drawn from a different distributions.
3. *Unbalanced.* Different nodes may vary by orders of magnitude in the number of training examples they hold.

2   Motivation and Related Work

When it is compared to traditional NN training methods, FL exposes a significant performance drop, that reaches up to 50% accuracy loss, even for relatively simple setups [27], and too often fails completely.

Problems of weak performance of FL might root in multiple factors residing in the nature of learning. First, computation of global updates might involve problems of **large batch learning** and, second, **non-iid nature** of training data poses additional statistical challenges as well.

Huge Batches. We can view the FL update rule in Eq. (4) as using huge batches in MBGD: $|\mathcal{B}_{\text{FedAvg}}| \approx |\hat{K}|r|\mathcal{B}|$, where $\hat{K}$ is the number of nodes participating in the training round in FedAvg ($= K$ in FedSGD). Moreover, SGD based methods build on the assumption, that the update is an unbiased estimate of the full gradient, that is unlikely in a non-iid setting. The local updates $\Delta^{(k)}$ can be considered as an approximation of the gradient, computed over an extremely large mini-batch [18], that with the central aggregation corresponds to an even bigger one.

Larger batch sizes has been typically used to support parallel processing. First, for better utilizing GPUs (Eq. (2)), and, second, for data center based model parallel multi-machine processing (Eq. (3)). The drawbacks however became soon visible when the models trained in this manner showed a significant decrease in generalization ability [11,14].

To analyse generalization problem of large batch training, multiple experiments are presented with really large batch sizes in the literature, as 4096 [10] or 8192 [8] inputs per batch. Both of these work aim at preserving the statistical properties of the gradients proving that increasing the learning rate, and using an initial warm-up phase along with batch normalization results in similar convergence rate to small batch training can be achieved. Yet, as [18] concludes, large batch learning reduces the range of usable hyper-parameter setups, which might lead to worse performance, and can even prevent convergence in FL.

Weight Divergence. Hidden layers of NNs act as feature extractors and multi-layered architectures are fitted to learn increasingly complex features, layer by layer. Since the key characteristic of FL is the presence of very divergent local data sets, the local models might learn very different higher level features.

In sequential learning with frequent model updates, ordering of the neurons is permutation-indifferent. However, when one computes pairwise statistics on the corresponding neurons, that are fitted to detect very different patterns, as at the averaging of local updates (Eq. 4), the resulting model's performance drops. Even if, as in case of FedAvg [19], the training starts from the same state, the more we train locally, divergence has a less and less negligible effect. [27] proposes to reduce this effect through sharing an "anchor" data set that will be distributed across all the participant node, however it is a bit contradictory to the principles of FL.

The other approach, Neuron matching, introduced by [25] could be utilized that attempts to find parts of the NNs that correspond to detectors of the same or very similar features. The common, completely new model is then created by assembly of the corresponding shards, avoiding blind coordinate-wise averaging of parameters.

Uncertain Convergence. To carry out thorough analysis of convergence of NN training in the FL involves so many variables that makes giving meaningful guarantees extremely hard. There have been a lot of effort carried out to give theoretical convergence guaranties, however due tho the complexity of the problem all of them make serious restrictions.

Many works, that are trying to give convergence guarantees make assumptions even for the case of convex optimization, like *iid data* or *all devices are active*. The latter assumption was made in [4,12,23], while [3,20,21,24] assumed both. [15] and [16], on the other hand, provide convergence analysis for true FedAvg with non-iid data, but for the case of strongly convex optimization objective, thus they are not really applicable for the NN case.

3   GD Based Methods in FL

Stateful methods are commonly used to enhance learning in single node setting, and have been designed to overcome very similar problems that we face in our setting. Now we make an attempt to empirically measure the impact of some stateful methods on the performance of federated training, that have been used to overcome challenges in different aspects of distributed ML.

In our experiments we tested the effect of simple and Nesterov momentum, from adaptive learning rates (ALR) based techniques Adagrad and RMSProp, and Adam that is the combinations of ARL and momentum.

Application of *momentum* has been already proposed in [8] for reducing variance in large batch learning while [9] and [22] uses Nesterov momentum for federated training of LSTM networks for next word prediction (but also used in parallel SGD for a long time, as in elastic averaging [26], for example). Using ALR methods (namely Adagrad) is also already proven to have a strong stabilizing effect [6] on the performance in data parallel distributed training.

In our previous, rather limited experiment [7], stateful optimization methods were applied for local training at the nodes, proven to outperform the baseline performance for a simple setup of FedAvg.

Learning Method. The method we use is a very simple one, similarly to the method described in [17]: Instead of averaging only the weight updates of the nodes, we maintain a similarly aggregated and broadcasted centralized set of variables representing the optimizer state, hoping that, at some extent, it helps to overcome challenges of FL we described in Sect. 2. The worker nodes uses the stateful methods for their local optimization steps, adjusting state variables according to the local data as well. Then, along with the model weights, the nodes share the optimizer state variables with the parameter server where they will be averaged to get a global optimizer state. When the model of the workers will be updated from the parameter server, so will happen with the state too, and then the new loop begins.

Complexity Measure. To measure the performance and stability of the learning algorithms in the function of hardness of the task, we define complexity as the reciprocal of best accuracy we achieve across all the optimizers and hyperparameter setups.

4   Experimental Setup

Data Sets. Experiments were run with three image classification task such that MNIST, Fashion-MNIST and CIFAR-10. The number of nodes participating in learning, was set to 30. We keep 10% of the data for testing the performance of the centralized model, and then created local data sets in three highly skewed style:

1. "Fifty-fifty": Similarly to the experimental settings in [19], data has been divided into equally sized one or two class chunks;
2. "Full-non-iid": In this setting, all the nodes received data points only from a single class, that is intended to be the most challenging setup;
3. "99%-non-iid": the 99% of data at a node comes from the same class, while 1% is picked iid.

Models. For the above mentioned three image classification tasks two types of NNs were utilized:

1. For MNIST and Fashion-MNIST data sets a fully connected (FC) network with a single hidden layer was used, based on [2];
2. For the CIFAR-10 data set we used a convolutional neural network (CNN), following the settings in [1].

Hyper-Parameters. We run each algorithm for 80 training rounds with a rather small grid of common and algorithm specific hyper-parameters: learning rate $\eta \in \{0.1, 0.01, 0.001, 0.0001\}$, batch size $|\mathcal{B}| \in \{16, 32, 64\}$, with 1 epoch local training for the FC data sets, and 1,3 and 5 epochs on Cifar-10. Values for β_1, β_2 for momentum, ρ for RMSProp are within the set $\{0.5, 0.9, 0.95\}$. For this experiment we did not use learning rate decay.[1]

---

[1] Hyper-parameters are denoted following Keras documentation https://keras.io/api/optimizers/.

5   Result

In all our setups, with a very few exceptions, all stateful methods over-performed SGD (Table 1). In Fig. 1, we plotted the best result of each optimizer on the most complex `Cifar-10` base data set. This shows how profound effect different distributions can have, and how much performance of stateful methods can differ.

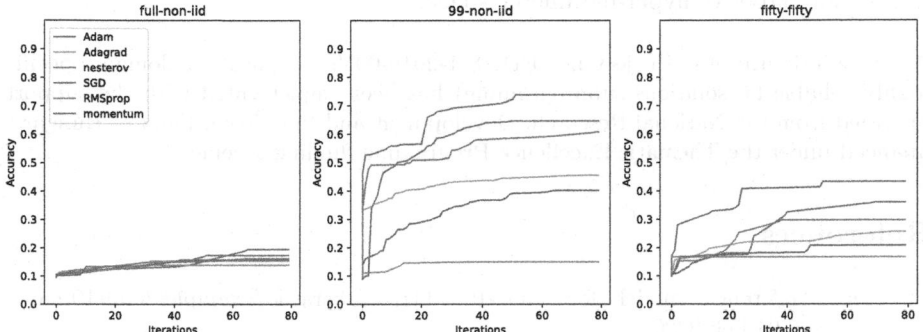

Fig. 1. Accuracy on `Cifar-10`.

Table 1. Best/mean performance of optimizers on the examined data distributions.

Data set	C	SGD	AdaGrad	RMSProp	Momentm	Nesterov	Adam
Cifar-10 1	2.46	0.20/0.11	0.25/0.12	0.16/0.10	0.35/0.12	0.29/0.12	**0.43/0.13**
Cifar-10 2	6.09	0.15/0.10	0.15/0.10	0.13/0.10	0.17/0.10	0.16/0.10	**0.19/0.10**
Cifar-10 3	1.37	0.40/0.14	0.45/0.21	0.15/0.13	0.64/0.31	0.60/0.21	**0.73/0.33**
F-MNIST 1	1.50	0.66/0.43	**0.73**/0.51	0.68/0.50	0.72/**0.53**	0.71/0.52	0.69/0.42
F-MNIST 2	1.74	0.58/0.42	**0.69/0.53**	0.58/0.33	0.65/0.48	0.64/0.46	0.64/0.29
F-MNIST 3	1.44	0.47/0.36	0.63/0.44	**0.74/0.54**	0.57/0.43	0.59/0.41	**0.74**/0.42
MNIST 1	1.15	0.83/0.51	0.89/0.66	0.86/**0.74**	0.87/0.65	0.88/0.63	**0.90**/0.54
MNIST 2	1.50	0.54/0.34	**0.85/0.57**	0.65/0.35	0.79/0.42	0.77/0.46	0.75/0.32
MNIST 3	1.17	0.67/0.42	0.78/0.56	**0.89/0.64**	0.76/0.47	0.76/0.50	**0.89**/0.48
$-\sigma(C, Acc)$		0.84/0.62	0.81/0.68	0.88/0.77	0.91/0.71	0.63/0.61	0.96/0.92

According to the results, ALR methods are performing best on relatively simple tasks (`MNIST` and `Fashion-MNIST`), both in mean performance and best performance. However, their performance drops dramatically on `Cifar-10`, where momentum methods gain advantage. Perhaps because ALR algorithms has been designed to converge fast on convex functions, and the more complex data brings more complex loss surface. Adam shows in each setup one of the bests if not the best results, as it is shown in Fig. 1 and Table 1. Although its performance shows lot of variance and it is very sensitive to hyper-parametrization (lower mean), in the most difficult tasks it results in the best mean performance along with best maximum performance.

6   Conclusion

We found that, in general, using stateful optimizers in FL might help to significantly increase learning performance. Naturally, these methods comes with a not negligible communication overhead since the optimizer state usually maintains one or more value per variable. Thus, for momentum the cost of communication is doubled or, for Adam, tripled. However, in some cases, this price can be worth to pay due to the tendency for stagnation of the federated training in a significant subset of hyper-parameter space.

Acknowledgements. Project no. ED_18-1-2019-0030 (Application domain specific highly reliable IT solutions subprogramme) has been implemented with the support provided from the National Research, Development and Innovation Fund of Hungary, financed under the Thematic Excellence Programme funding scheme.

References

1. Keras reference model for cifar-10. https://keras.io/examples/cifar10_cnn/. Accessed 04 Feb 2020
2. Keras reference model for mnist. https://keras.io/examples/mnist_mlp/. Accessed 04 Feb 2020
3. Zhou, F. Cong, G.: On the convergence properties of a k-step averaging stochastic gradient descent algorithm for nonconvex optimization (2017)
4. Yu, H., Yang, S., Zhu, S.: Parallel restarted SGD with faster convergence and less communication: demystifying why model averaging works for deep learning, vol. 33 (2019)
5. Chen, J., Pan, X., Monga, R., Bengio, S., Jozefowicz, R.: Revisiting distributed synchronous SGD. arXiv preprint arXiv:1604.00981 (2016)
6. Dean, J., et al.: Large scale distributed deep networks. In: Advances in Neural Information Processing Systems, pp. 1223–1231 (2012)
7. Felbab, V., Kiss, P., Horváth, T.: Optimization in federated learning. In: CEUR Workshop Proceedings (CEUR-WS.org), vol. 2473, pp. 58–65. ceur-ws.org (2019). ISSN: 1613–0073
8. Goyal, P., et al.: Accurate, large minibatch SGD: training imagenet in 1 hour. arXiv preprint arXiv:1706.02677 (2017)
9. Hard, A., et al.: Federated learning for mobile keyboard prediction. arXiv preprint arXiv:1811.03604 (2018)
10. Hoffer, E., Hubara, I., Soudry, D.: Train longer, generalize better: closing the generalization gap in large batch training of neural networks. In: Advances in Neural Information Processing Systems, pp. 1731–1741 (2017)
11. Keskar, N.S., Mudigere, D., Nocedal, J., Smelyanskiy, M., Tang, P.T.P.: On large-batch training for deep learning: generalization gap and sharp minima. arXiv preprint arXiv:1609.04836 (2016)
12. Khaled, A., Mishchenko, K., Richtárik, P.: First analysis of local gd on heterogeneous data (2019)
13. Konečný, J., McMahan, H.B., Ramage, D., Richtárik, P.: Federated optimization: distributed machine learning for on-device intelligence. arXiv preprint arXiv:1610.02527 (2016)

14. LeCun, Y.A., Bottou, L., Orr, G.B., Müller, K.-R.: Efficient backProp. In: Montavon, G., Orr, G.B., Müller, K.-R. (eds.) Neural Networks: Tricks of the Trade. LNCS, vol. 7700, pp. 9–48. Springer, Heidelberg (2012). https://doi.org/10.1007/978-3-642-35289-8_3
15. Li, T., Sahu, A.K., Zaheer, M., Sanjabi, M., Talwalkar, A., Smith, V.: Federated optimization in heterogeneous networks (2018)
16. Li, X., Huang, K., Yang, W., Wang, S., Zhang, Z.: On the convergence of fedavg on non-iid data. arXiv preprint arXiv:1907.02189 (2019)
17. Liu, W., Chen, L., Chen, Y., Zhang, W.: Accelerating federated learning via momentum gradient descent. arXiv preprint arXiv:1910.03197 (2019)
18. Masters, D., Luschi, C.: Revisiting small batch training for deep neural networks. arXiv preprint arXiv:1804.07612 (2018)
19. McMahan, H.B., Moore, E., Ramage, D., Hampson, S., et al.: Communication-efficient learning of deep networks from decentralized data. arXiv preprint arXiv:1602.05629 (2016)
20. Stich, S.U.: Local SGD converges fast and communicates little (2018)
21. Wang, J., Joshi, G.: Cooperative SGD: a unified framework for the design and analysis of communication-efficient SGD algorithms (2018)
22. Wang, K., Mathews, R., Kiddon, C., Eichner, H., Beaufays, F., Ramage, D.: Federated evaluation of on-device personalization. arXiv preprint arXiv:1910.10252 (2019)
23. Wang, S., et al.: Adaptive federated learning in resource constrained edge computing systems. In: IEEE INFOCOM 2018-IEEE Conference on Computer Communications (2018)
24. Woodworth, B., Wang, J., Smith, A., McMahan, B., Srebro, N.: Graph oracle models, lower bounds, and gaps for parallel stochastic optimization (2018)
25. Yurochkin, M., Agarwal, M., Ghosh, S., Greenewald, K., Hoang, T.N., Khazaeni, Y.: Bayesian nonparametric federated learning of neural networks. arXiv preprint arXiv:1905.12022 (2019)
26. Zhang, S., Choromanska, A.E., LeCun, Y.: Deep learning with elastic averaging SGD. In: Advances in Neural Information Processing Systems, pp. 685–693 (2015)
27. Zhao, Y., Li, M., Lai, L., Suda, N., Civin, D., Chandra, V.: Federated learning with non-iid data. arXiv preprint arXiv:1806.00582 (2018)

Talking in Italian About AI with a Chatbot: A Prototype of a Question-Answering Agent

Chiara Leoni, Mauro Coccoli(ID), Ilaria Torre(ID), and Gianni Vercelli(✉)(ID)

University of Genoa DIBRIS, Viale F. Causa 13, 16145 Genoa, Italy
chiara.leoni@outlook.com
{mauro.coccoli,ilaria.torre,gianni.vercelli}@unige.it

Abstract. Talking with chatbots and question answering interfaces is nowadays common to most people using some kind of digital device. Artificial intelligence-based solutions to create chatbots are available on the market and are provided by all the big players like IBM, Microsoft, Google, and Amazon. Chatbots can differ for the type of corpus they rely upon, the type of questions they are suited for, the algorithms, and the ability to learn and expand their knowledge base. Moreover they differ for the language, with English chatbots being on average those best performing. In this paper we present ConversIAmo, a chatbot prototype designed to answer questions in Italian about introductory concepts on Artificial Intelligence. The natural language processing pipeline is built upon the IBM Watson suite and implements further modules for improving the answer selection and ordering, and to expand the knowledge base. Results are encouraging, especially with regard to the ordering of the correct answers. The paper describes the question answering dataset, the first in this domain, and outlines the dialog flow and the answering process.

Keywords: Conversational agents · IBM Watson · AI · Question answering · Italian

1 Introduction

The use of Artificial Intelligence (AI) is becoming pervasive in many fields of application such as, e.g., healthcare, economics, education, and its reliability is constantly increasing. To support the development of AI-based solutions, many big players in the IT market are offering powerful AI capabilities relying on cloud computing integrated in their Software-as-a-Service (SaaS) solutions, resulting in AI-as-a-Service (AIaaS) products such as, e.g., visual recognition, natural language processing, problem solving, machine learning and many others. In this paper, we will focus on conversational and QA (Query Answering) modules.

Specifically, we observe that the number of applications using conversation to interact with humans is still fast-growing and, accordingly, dialog systems

© Springer Nature Switzerland AG 2020
C. Analide et al. (Eds.): IDEAL 2020, LNCS 12490, pp. 356–367, 2020.
https://doi.org/10.1007/978-3-030-62365-4_34

are quite sophisticated solutions now, owing to the more advanced AI-enabled natural language understanding capabilities. In fact, digital assistants on smartphones or home controllers (e.g., Siri, Cortana, Alexa, Google Now/Home, etc.) are becoming very common. They are task-oriented dialog agents, which may be designed for specific goals and set up to have short conversations to drive their actions such as playing music, controlling home appliances, weather forecasting, gathering news, providing travel directions, finding restaurants and hotels, and so forth. In addition, many companies deploy their own task-based conversational agents to support customers with auto responders or chatbots. Such systems are usually based on a given knowledge structure, representing the most common intents which may be expressed by the users, in relation to the task to accomplish.

The wide availability and ease of use of such tools, both commercial or free to use, which allow programming personalized conversational applications through advanced AI services, has contributed to the recent popularity of chatbots. In fact, there is a vast offer of intelligent and scalable solutions by big IT corporations (e.g., Amazon, IBM, Microsoft, Oracle) to provide customers with simple user interfaces to manage transactions, storage and information updates. However, question answering systems are especially available in English, while there are less systems in other languages, such as Italian (e.g., [6,10,22]). Moreover, they usually require extensive training in order to get good performances.

This is the reason why we decided to work in this direction and to start developing the *ConversIAmo* system, which is a QA prototype system based on unstructured *corpus* in Italian, in the domain of Artificial Intelligence. Literally, *conversiamo* is the Italian for *let's talk*. The name also contains a pun, as the capitalized IA stands for "Intelligenza Artificiale", i.e., "Artificial Intelligence" in Italian. As described in [14], ConversIAmo exploits different technologies. It is mainly based on the IBM Watson services, included within the IBM Cloud platform, whose functions are augmented with newly-developed modules and Tint, an Italian Natural Language Processing tool, to better support the Italian language and to improve performances on questions in Italian.

It is worth noting that our tests have been performed with limited training of Watson Discovery Service. Indeed, while training Watson Discovery is a recommended task in order to make it learn and get good results [10,20], it is also a time consuming activity, therefore proposing a method that balances training time and results was our priority. The preliminary results obtained using our approach seem encouraging, especially about the improvement in terms of ranking of the correct answers.

In this paper we describe our approach and we provide details about the building of the Question Answering Dataset in Italian on the AI domain, that we make available on GitHub[1] for further research. The remainder of the paper is organized as follows: in Sect. 2 some related works are presented and discussed; in Sect. 3 the Question Answering Dataset is described, while Sect. 4 presents

---

[1] https://github.com/Chln94/ConversIAmo.

ConversIAmo prototype and dialogue flow; Sect. 5 discusses the results; finally, Sect. 6 concludes the paper.

2 Background and Related Works

The idea of conversational interfaces is not recent, as it dates back some decades, with Question Answering (QA) systems replying user's questions with either the exact answer or with short passages of text containing it. IBM Watson's victory against two human world champions in the TV quiz *Jeopardy!* triggered everyone's interest on this topic in 2011, as a result of years of studies in machine learning, natural language processing and information retrieval [8]. Since then, many QA systems have been developed: mostly in English, but also bilingual or non-English ones, dealing with many topics (open domain) or a narrow one (closed domain), drawing answers from structured data or unstructured texts.

Most existing QA systems are *open domain*, meant to deal with different themes and usually adopting the Internet as a source, or DBpedia or another huge corpus that has high probability of containing an answer. A *closed domain* system, on the other hand, is intended to answer only to questions related to a specific topic on which the knowledge base relies upon. While a big corpus has a lot of noise and requires strong methods for filtering and scoring several candidate answers, a smaller one is usually aimed to provide more focused and precise answers, thus it needs a deeper question analysis to better understand the user's request.

Some early examples of English open-domain systems inspired by DeepQA are Watsonsim [11] and YodaQA [1]. Watsonsim is an open-source system with the aim of answering Jeopardy-style questions, it combines offline corpora from Wikipedia and online resources from web search engines (Google and Bing). YodaQA is an end-to-end QA pipeline that answers factoid questions by exploiting the UIMA (Unstructured Information Management Architecture) [9] and open-source NLP tools. Some early examples of closed domain systems, instead, are AQUA [25], combining NLP, ontologies and information retrieval technologies to answer questions about academic people and organizations, and AquaLog [16], which takes a natural language query and an ontology as input and returns answers from one or more knowledge bases.

Owing to the surge of technologies available presently, conversational agents in use can be found in many applications, spanning a variety of fields. In practice they serve where natural human interaction mechanisms are desirable.

Usually, task-oriented dialog agents focus on a particular domain. In this domain, they show the ability to understand requests in natural language and reply by providing an answer or performing a task resolving the issue proposed by the requester.

Extensive training tasks increase effectiveness despite a relevant computational cost [20]. In recent years, some relevant applications based on IBM Watson assistant services have been implemented for educational purposes. Watson Experience Manager (WEM) [13], originally designed for customer support service, has been experimented in a postgraduate course in Data Science. To face the stress during exams, a canadian research group [23] recently investigated and compared Google, Amazon and IBM API packages, and proposed IBM Watson to develop and implement a multilingual student support system. In a very vertical domain, Memeti and Pillana proposed PAPA [17], a parallel programming assistant for programmers, which combines IBM Watson services in a dialog-based interactive QA system to help novice parallel programmers to avoid common mistakes on OpenMP.

Despite the great success of IBM Watson and its tools to create agents which understand natural language and exploit big data to extract meaningful insights, question answering is still a complex topic and there are few examples of non-English QA systems. In general, performances are usually more satisfactory for dialog agents speaking English than for dialog agents using other languages. Italian is one of them, and some features are also not supported [2,10]. As a matter of fact, we found for example that current versions of Watson Discovery did not get the same performances using Italian texts compared with English texts. For this reason, we developed new modules for answer selection and ranking process on Italian texts.

A thorough analysis of systems that use Watson services with Italian text is reported in [10], [6,7]. The former paper describes a virtual assistant, supporting both students and staff in a smart campus. The authors analyze the effect of training Watson Discovery in terms of performances and costs. The second paper [6] describes a conversational agent that exploits IBM services and tools for Italian language processing and is used in [7] where the authors propose a method to improve retrieval, by using lexical resources and word embeddings to expand the query.

In addition to the conversational systems above, we can mention some more question answering systems in Italian. MULIB [24] is a question answering system that uses both structured and unstructured knowledge bases to provide answers to natural language questions, and QuASIt [22], again a question answering system that uses a rule-based approach to understand the query and structured knowledge base.

3 Question Answering Dataset

Several studies have shown that human-machine interaction is not the same as between human-human. Some have shown that, in general, people adapt their communication style of the interlocutor and tend to simplify the way of speaking when conversing with a machine [12], others have measured how we tend to be less open and well-disposed but instead more hasty in human-AI interactions [19]. Based on these observations, the type of questions that one expects from a user to a machine is not very complex and rich in subordinate propositions, but rather concise, if not even composed only of keywords.

Some examples of question and answer pairs datasets (all for open-domain questions in English) are:

- the WikiQA dataset in English[2], created for research on open-domain question answering thanks to the help of crowd-sourcing, that collects about 3k annotated Bing user queries, looking for candidate answers in summary sections of linked Wikipedia pages [26];
- the MS MARCO dataset[3], for machine-reading comprehension and question answering, that contains 100k English queries with corresponding checked answers extracted from documents in Bing search engine [21];
- the Stanford Question Answering Dataset (SQuAD)[4] that in the last version includes more than 10k questions on 500+ articles and also 50k unanswerable questions written by crowd workers deliberately similar to answerable ones, so that systems have to understand when they have no proper answers and it is better to avoid answering.

An Italian version of the SQuAD dataset[5] has been recently created and it contains more than 60,000 question-answer pairs derived from the original one. This result has been achieved through semi-automatic translation and subsequent refinements, filtering out many paragraph-question-answer triplets in which the answer is no longer part of the translated paragraph after the translation, plus some corrections concerning different morphological forms or word order substitution have been added [5].

Another relevant example of Italian dataset on closed-domain is the question answering dataset on the first half of the XX century of Italian history called "Who was Pietro Badoglio?" [18]. This dataset includes 627 manually classified and annotated questions with their Lexical Answer Type and a set of question-answer pairs, sourced from 274 Italian Wikipedia pages about that historical period.

Our dataset is based on the principles and the examples above. As far as we know, at the time of starting this project no question answering dataset was available in Italian nor in English on the subject of artificial intelligence.

From a corpus of 130 articles available online on the subject[6], 110 questions have been manually created, with one or maximum two correct passages of text as correct answers for each question.

Question Classes. Each question in the dataset is associated to a Question Class. This is typical in datasets that are used for conversational agents. Like most of those mentioned above, we defined the question classes according to a well-settled taxonomy consisting in six macro-classes: ABBREVIATION, DESCRIPTION, ENTITY, HUMAN, LOCATION, NUMERIC, each divided into sub-classes [15]. Due to our field of interest and for feasibility reasons, we considered reasonable to use only a subset of them: a sub-category of DESCRIPTION (definition), two sub-classes of HUMAN (individual and group), and some

---

[2] Available on https://www.microsoft.com/en-us/download/details.aspx?id=52419.
[3] Available on http://www.msmarco.org/.
[4] Available on https://rajpurkar.github.io/SQuAD-explorer/.
[5] Available on https://github.com/crux82/squad-it.
[6] Before using the corpus, please check copyright permissions of resources.

types of NUMERIC (count, date, money, percent and period), while we have not analyzed the ENTITY category at all, because in fact suitable for too many different questions.

Question classes and distribution in the dataset are reported in Table 1.

Table 1. Table of question classes distribution in our dataset.

Question class	Number of questions	Percentage of total
Description	66	60%
Numeric	22	20%
Human	12	10.9%
Location	7	6.4%
Abbreviation	3	2.7%

The dataset is available on GitHub (https://github.com/Chln94/ConversIAmo).

Intents. Within question answering systems, intent refers to the goal the user has in mind when asking a question. Therefore, intents are typically mapped to question classes. For instance, it is quite reasonable to recognize in a question such as "when was the perceptron invented?" the intent to know a date or the purpose of having a descriptive response to the question "what is the difference between supervised and unsupervised learning?". IBM Watson Assistant, as well as other platforms to build chatbots (e.g., Google's *Dialog Flow*), enable the creation of intents. For the ConversIAmo prototype we created one intent for each question class and we provided ten examples for each one. The examples were tailored to our Italian domain, like the questions in our dataset, but with different content. Moreover we associated entity types to question types that expect an entity in the answer. Such questions are usually referred to as factoid questions.

Question Classification Results. We assessed Assistant's ability to identify the intent in the questions, by submitting all the questions in our dataset. Performances are very good for all the classes, but description ones. While in the former case all the intents have been recognized, for description questions we got 8% wrong classifications (all belonging to DESCRIPTION_definition in the dataset) and 11% non-classified questions (all belonging to generic DESCRIPTION class. E.g., "Which kind of neural networks are used for NLP?" is a DESCRIPTION question, "What is machine learning?" is a DESCRIPTION_definition question). Further investigation will be needed to understand if more examples could be enough to help Assistant classification or other strategies should be adopted, such as identifying finer grained classes.

4 Components and Dialogue Flow

ConversIAmo is a conversational agent, i.e. a "chatbot", that is built as an instance of IBM Watson Assistant and implements a question answering system through the integration of different services. Some of these services are included in the IBM Watson suite—Watson Discovery for information retrieval from unstructured data, and Natural Language Understanding (NLU) service to extract enrichments from texts (e.g., named entities and keywords); in addition we created some custom-made Java functions to manage the answer retrieval and improve the ranking of answers, and we also exploited an open-source tool, Tint, for the Italian language analysis based on Stanford CoreNLP [4]. The details of the question analysis and the answer selection process are explained in [14]. In the following we sketch them and provide details about the dialog flow. The knowledge base of Conversiamo is composed of the Corpus of Italian texts on the AI domain and is complemented by the Dialogue Flow tree.

In Watson Assistant, the Dialog Flow is a tree structure that can be defined with the aim to create shortcuts to answers that match the features of the user question. Each branch of the tree has a condition to be satisfied in order to enter its nodes (typically a specific intent, but it might also be an entity type, an entity value, or a context variable value) and, for each user request, the system looks at the intent and the entities it refers to and tries to match them to the condition of that node, proceeding from the first one in the tree until the last one.

It is worth noting that in ConversIAmo the dialog flow is not static, as it can be expanded. Indeed, when a satisfying answer to a user question is not available in the dialogue flow, the answer is sought in the corpus and the user is asked to confirm or not confirm its correctness. If the correctness is confirmed, the answer is added to a node—if a node was already defined with conditions that matched the intent and entities of the question, otherwise a new node is defined with its conditions and answer.

The dialogue flow manager and all the components involved in the question answering process are displayed in Fig. 1.

The dialogue starts with the question submitted as input by the user and passed to Assistant to check the conditions in the Dialog Flow. Assistant analyses the question and returns: intent, if sufficiently certain, entities, if present, and a response, if already inserted in a node of the dialogue flow with a matching condition. If there are no answers yet, or if the answer does not satisfy the user, then the whole process of Question analysis, Information Retrieval and Answers Selection starts.

The recognition of entities within the user input is performed by two services of the Watson suite, i.e. Assistant and NLU that we combine. For the keyword extraction we tried two different approaches, the first exploiting Watson Natural Language Understanding service and the latter using Tint, with the aim to perform an analysis more suited for Italian language (see [14] for details).

Query reformulation follows the Question analysis phase. It is obtained by identifying and discarding stop words, enriched with the set of Italian question words. The output is passed to Discovery for the answer retrieval in the corpus

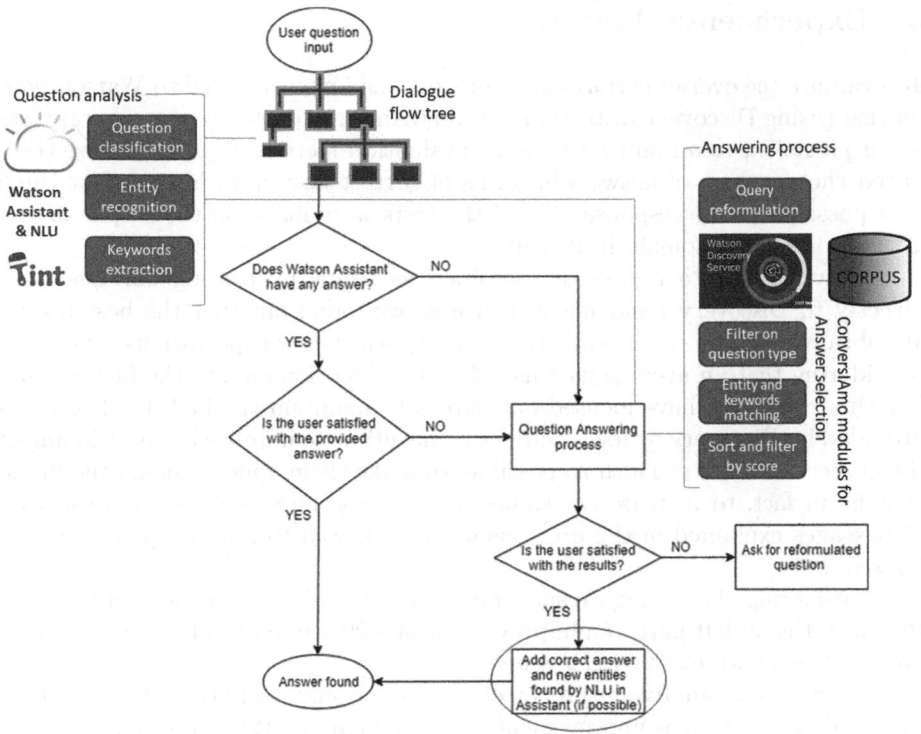

Fig. 1. Question answering components and dialog manager flow diagram.

of documents. The result is a set of text passages retrieved by Discovery with an associated score.

The next step, that is ordering the answers, is the step where we got worst results for Italian using the base services provided in the Watson suite. Therefore we developed new modules, that in Fig. 1 are represented by the blocks named in sequence "Filter on question type", "Entity and keyword matching", "Sort and filter by score". Basically, they analyse the text passages returned by Discovery and match them with the results of the Question Analysis. In particular, for question types that are expected to contain some specific entity, text passages that do not contain such entities are discarded. Finally, in order to rank the answers, ConversIAmo uses the score assigned by Watson Discovery to each passage and increases it on the basis of the percentage match between (i) the *keywords* and (ii) the *entities* in the original question and in the text passage, considering also the morpho-syntactic role. After normalization, results that are below a given threshold are discarded and the others are ranked in descending order to be showed to the user.

If the user is satisfied with the answer, the dialogue flow can be expanded as explained at the beginning of this section.

5   Experimental Results

To evaluate the overall performances of ConversIAmo compared to Watson base service (using Discovery instead of ConversIAmo modules), we ran several tests using *precision, recall* and *F1 score* as evaluation metrics. In addition, we computed the *accuracy* of answers in terms of correct answer within the first three text passages of the response [3]. All the tests were made on the corpus of 130 articles on the AI domain in Italian.

Analyzing the text passages obtained as result by passing our questions directly to Discovery removing stop words, we found out that the best results are obtained for the *recall*, while the *accuracy* was below expectations, especially considering that an average user usually pays attention only to the first results. For this reason we have focused our efforts to maintain the high level of *recall* provided by Discovery by itself and in the meantime to improve the positioning of the correct answers as much as possible, to make them appear among the firsts, that is, in fact, to increase the values of *accuracy*. Furthermore, the re-scoring of passages explained in the previous section allowed to improve *precision* and *F1 score*.

Considering the average results for all the question types, we obtained, as shown in Fig. 2 (left part), an improvement of +20% in terms of *accuracy*, +12% for *precision*, and +21% for *F1 score*.

However, if we analyze results by class of question, we find that DEFINITION classes do not get great improvement: +7% *precision*, +12% *F1 score*, and +3% for *accuracy*.

Conversely, factoid questions perform better: the average for the classes HUMAN, LOCATION, and NUMERIC resulted in +30% for *accuracy*, +20% *precision* and +35% *F1 score*, as shown in Fig. 2 (right part). Details on accuracy are provided in Fig. 3.

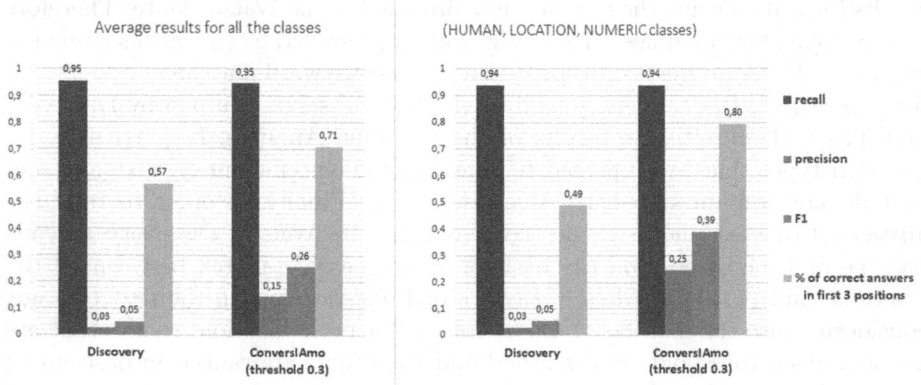

Fig. 2. Results for recall, precision, F1 score and accuracy (%of correct answers in the first 3 positions).

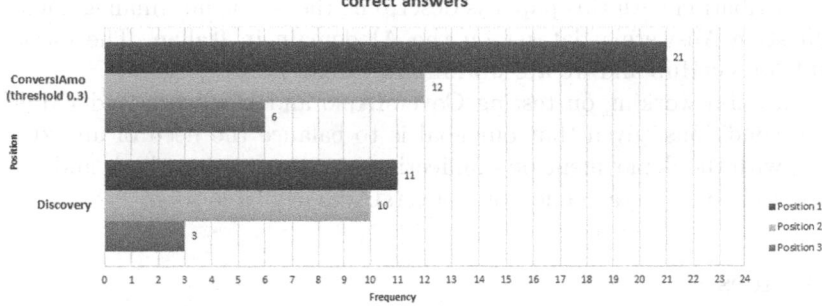

Fig. 3. Correct answers in the first 3 positions.

The use of different methods for keyword extraction did not produce significant differences in results at the moment. Details about the tests using Tint and NLU for keyword extraction are provided in [14]. The results provided in this paper and in Fig. 2 are obtained using Watson NLU.

It is clear that, for factoid-type questions the entity-based filtering technique and the re-scoring mechanism are effective, but these are not enough to cover the variety of questions that require a description or definition in response. In our dataset (and in general in the Italian language) these types of questions can be very different one from each other, they do not respect an easily identifiable pattern, and the same applies to the relative answers. This means that they need more specific and extensive training.

6   Conclusions and Discussion

ConversIAmo is an Italian speaking agent, the first on the AI domain. It combines different services from IBM Watson framework with custom-made Java functions and it also draws on the linguistic resources provided by Tint, an Italian NLP tool. The idea was to provide an easy-to-apply method to overcome limits of free services on the market due to the lower development of non-English language models and the lack of many state-of-the-art studies for Italian. The experimentation outcome is encouraging, since we gained relevant improvement in accuracy and some slight improvement in both precision and F1 score, keeping recall over 94%. One of the main contributions is the dynamic expansion of ConversIAmo dialog flow, since just a very basic setup has been done at start and answers are gradually added into nodes according to the question type and entities thanks to the users feedback. Our work has proved to be more effective with factoid-type questions, less accurate with description ones instead (both for question classification and question answering pipeline), while the keyword extraction methods currently implemented didn't influence much the results obtained at the end of the pipeline. We are working on the revision of the method, to improve the syntactic and semantic analysis for Italian language.

A contribution with this paper is describing the dialog flow management and the Question Answering dataset on the AI domain in Italian. The dataset is available on GitHub and we are working to extend it.

We are also working on testing CoversIAmo and Discovery under different training conditions, given that our goal is to balance the need of an extensive training with the deployment of an effective easy-to-apply method, and, finally, we hope to make a beta-version of ConversIAmo available soon.

References

1. Baudiš, P.: Yodaqa: a modular question answering system pipeline. In: 2015–19th International Student Conference on Electrical Engineering (2015)
2. Bellomaria, V., Castellucci, G., Favalli, A., Romagnoli, R.: Almawave-SLU: a new dataset for SLU in Italian. In: Proceedings of the sixth Italian Conference on CL (2019)
3. Boyer, J.M.: Natural language question answering in the financial domain. In: Proceedings of the 28th Annual International Conference on Computer Science and Software Engineering, pp. 189–200. IBM Corp. (2018)
4. Cabrio, E., Mazzei, A., Tamburini, F.: Tint 2.0: an all-inclusive suite for NLP in Italian. In: Proceedings of the Fifth Italian Conference on Computational Linguistics CLiC-it, vol. 10, p. 12 (2018)
5. Croce, D., Zelenanska, A., Basili, R.: Neural learning for question answering in Italian. In: Ghidini, C., Magnini, B., Passerini, A., Traverso, P. (eds.) AI*IA 2018. LNCS (LNAI), vol. 11298, pp. 389–402. Springer, Cham (2018). https://doi.org/10.1007/978-3-030-03840-3_29
6. Damiano, E., Spinelli, R., Esposito, M., De Pietro, G.: An effective corpus-based question answering pipeline for Italian. In: De Pietro, G., Gallo, L., Howlett, R.J., Jain, L.C. (eds.) KES-IIMSS 2017. SIST, vol. 76, pp. 80–90. Springer, Cham (2018). https://doi.org/10.1007/978-3-319-59480-4_9
7. Esposito, M., Damiano, E., Minutolo, A., De Pietro, G., Fujita, H.: Hybrid query expansion using lexical resources and word embeddings for sentence retrieval in question answering. Inf. Sci. **514**, 88–105 (2020)
8. Ferrucci, D.A.: Introduction to "this is watson". IBM J. Res. Dev. **56**(3.4), 1:1–1:15 (2012)
9. Ferrucci, D., Lally, A.: Uima: an architectural approach to unstructured information processing in the corporate research environment. Nat. Lang. Eng. **10**(3–4), 327–348 (2004)
10. Gaglio, S., Re, G.L., Morana, M., Ruocco, C.: Smart assistance for students and people living in a campus. In: 2019 IEEE International Conference on Smart Computing (SMARTCOMP), pp. 132–137. IEEE (2019)
11. Gallagher, S., Zadrozny, W., Shalaby, W., Avadhani, A.: Watsonsim: overview of a question answering engine. arXiv preprint arXiv:1412.0879 (2014)
12. Hill, J., Ford, W.R., Farreras, I.G.: Real conversations with artificial intelligence: a comparison between human-human online conversations and human-chatbot conversations. Comput. Hum. Behav. **49**, 245–250 (2015)
13. Kollia, I., Siolas, G.: Using the IBM watson cognitive system in educational contexts. In: 2016 IEEE Symposium Series on Computational Intelligence (SSCI), pp. 1–8. IEEE (2016)

14. Leoni, C., Torre, I., Vercelli, G.: Conversiamo: improving Italian question answering exploiting IBM watson services. In: Proceedings of the 23rd International Conference on Text, Speech and Dialogue (TSD 2020) (2020)
15. Li, X., Roth, D.: Learning question classifiers. In: Proceedings of the International Conference on Computational Linguistics, vol. 1, pp. 1–7 (2002)
16. Lopez, V., Uren, V., Motta, E., Pasin, M.: Aqualog: an ontology-driven question answering system for organizational semantic intranets. Web Semantics: Science, Services and Agents on the World Wide Web 5(2), 72–105 (2007)
17. Memeti, S., Pllana, S.: Papa: a parallel programming assistant powered by IBM watson cognitive computing. J. Comput. Sci. 26, 275–284 (2018)
18. Menini, S., Sprugnoli, R., Uva, A.: "Who was Pietro Badoglio?" towards a QA system for Italian history. In: Proceedings of the Tenth International Conference on Language Resources and Evaluation (LREC 2016), pp. 430–435 (2016)
19. Mou, Y., Xu, K.: The media inequality: comparing the initial human-human and human-AI social interactions. Comput. Hum. Behav. 72, 432–440 (2017)
20. Murtaza, S.S., Lak, P., Bener, A., Pischdotchian, A.: How to effectively train IBM watson: classroom experience. In: 2016 49th Hawaii International Conference on System Sciences (HICSS), pp. 1663–1670. IEEE (2016)
21. Nguyen, T., et al.: Ms marco: a human-generated machine reading comprehension dataset (2016)
22. Pipitone, A., Tirone, G., Pirrone, R.: QuASIt: a cognitive inspired approach to question answering for the Italian language. In: Adorni, G., Cagnoni, S., Gori, M., Maratea, M. (eds.) AI*IA 2016. LNCS (LNAI), vol. 10037, pp. 464–476. Springer, Cham (2016). https://doi.org/10.1007/978-3-319-49130-1_34
23. Ralston, K., Chen, Y., Isah, H., Zulkernine, F.: A voice interactive multilingual student support system using IBM watson. In: 2019 18th IEEE International Conference On Machine Learning And Applications (ICMLA), pp. 1924–1929 (2019)
24. Siciliani, L., Basile, P., Semeraro, G., Mennitti, M.: An Italian question answering system for structured data based on controlled natural languages. In: Proceedings of the Sixth Italian Conference on Computational Linguistics (2019)
25. Vargas-Vera, M., Lytras, M.D.: Aqua: a closed-domain question answering system. Inf. Syst. Manage. 27(3), 217–225 (2010)
26. Yang, Y., Yih, W.T., Meek, C.: Wikiqa: a challenge dataset for open-domain question answering. In: Proceedings of the 2015 Conference on Empirical Methods in Natural Language Processing, pp. 2013–2018 (2015)

Review of Trends in Automatic Human Activity Recognition in Vehicle Based in Synthetic Data

Ana Coimbra[1,2], Cristiana Neto[1,2], Diana Ferreira[1,2], Júlio Duarte[1,2],
Daniela Oliveira[1,2], Francini Hak[1,2], Filipe Gonçalves[2],
Joaquim Fonseca[2], Nicolas Lori[1,2(✉)], António Abelha[1,2],
and José Machado[1,2]

[1] Centre Algoritmi, University of Minho, 4710-057 Braga, Portugal
nicolas.lori@algoritmi.uminho.pt
{abelha,jmac}@di.uminho.pt
[2] Bosch Car Multimedia, 4705-820 Braga, Portugal

Abstract. Driverless vehicles are more and more becoming a reality. However, people still have some concerns in using them, the main concern is fear, hence the importance of creating a surveillance system inside those vehicles. For the detection and classification of human movements to be possible it is necessary to train the system with data representative enough for all kinds of possibilities. Although the production of large quantities of data becomes an expensive process and adds the problem of data protection, the use of synthetic data once they are artificially generated allows lower costs and eliminates the problem of data protection. A bibliographic study was carried out in this paper with articles from 2017 or later on the use of synthetic data. In these studies, it is noted that synthetic data is widely used with good results. As far as image capture is concerned, they show that 3D cameras have better results, but they are more expensive, so 2D cameras are more often used with later conversion to 3D images. The stitched puppet (SP) model is capable of adapting to the most difficult poses having obtained good results in its application in the FAUST dataset.

Keywords: Autonomous car · Multisensory fusion · Audio-visual synthetic data

1 Introduction

The automobile industry is constantly evolving and currently a revolution in this field is being observed with the progressive introduction of novel technologies known as the Internet of Things (IoT). The concept of IoT has emerged a few years ago and the revolutionary wave that it generated affected several sectors besides the automobile area (agriculture [26], construction [8], health [6], transport [10]), contributing to the increase in the quality of life of the populations. Fundamentally, IoT is the way physical objects are connected and communicating with each other and with the user, through intelligent sensors and software

© Springer Nature Switzerland AG 2020
C. Analide et al. (Eds.): IDEAL 2020, LNCS 12490, pp. 368–376, 2020.
https://doi.org/10.1007/978-3-030-62365-4_35

that transmit data to a network, leading to a smarter and more responsive planet [11, 22]. Such an improvement is likely to be improved by an adequate use of Big Data approaches (e.g. [1–4, 16]).

In this context, the installation of sensors in vehicles and subsequent data collection can certainly help to make a car safer and more efficient (tracking fuel and maintenance needs), but it can also contribute to the communication between millions of cars that are on the road at the same time, allowing for example real-time traffic updates, reduction of accidents at blind spots at intersections, real-time information on road defects, among others [5, 23]. Thus, these connected and intelligent vehicles require an enormous amount of data collection and processing in a massive way. Vehicles with IoT technologies based on advances from sensors to Artificial Intelligence (AI) to big-data analysis, need to be not only connected but also 'smart'.

As the variety of sensors present in smart cars, results in a large volume of data in structured and unstructured formats, it is not conceivable the use of traditional tools to deal with this amount and invariability of data and this is where the Big Data concept arrives. However, Big Data consists not only on large datasets but also on the knowledge to use computational strategies and technologies to handle these datasets. In terms of Big Data datasets most authors agree to define it into six dimensions [7, 13, 15, 25, 27, 32]:

- **Volume:** large amount of data generated;
- **Variety:** variety of data produced;
- **Velocity:** speed with which data is produced;
- **Veracity:** data consistency, quality and trustworthiness;
- **Value:** useful information present in data;
- **Volatility:** duration of usefulness of data.

Nowadays there are several market options that uses Big Data, being the automotive sector a great object of study [32]. For example, over the last years the automotive industry has made large investments in autonomously driven vehicles, being these a major source of Big Data. In this sense, IoT and Big Data technologies complement each other greatly and bring important benefits to the automotive industry, since they allow the collection of a large amount of data through on-board sensors (cameras, radar, Lidar, GPS, etc.) and the use of this information to improve driving safety and quality. The complex interactions with other road users lead people to view driverless vehicles with skepticism. Several studies have found that the acceptance of driverless vehicles depends a lot on the risk perception and worry for safety and security risks. Therefore, the security issues are one of the key factors for the adoption of driverless vehicles, hence it is especially important to reduce them.

In-vehicle surveillance is related to the passenger experience, a.k.a. user-experience (UX), plus also cockpit Big Data. By monitoring stress, aggressive facial expressions and voice tone it is possible both to prevent and to mitigate violence. This, however, varies according to the violence type. Aggressive emotion may scale into aggressive behavior and therefore should be prevented. There are also more complex aggression patterns such as those used in harassment where the aggressor may show a happy or serious facial expression and low voice tone, therefore this behavior must be considered against the victim's expressions. This can be tricky, especially when the victim is a child or within a human trafficking situation. Another tricky type of aggression to detect is in-vehicle theft, as a cockpit seat is usually narrow and people are very close to each other. An intervention protocol must be defined for aiding the operator to properly address this whole set of tricky events.

There is then an intersection between in-vehicle surveillance and in-vehicle UX in the sense that the same sensor set can be used to address both aims. In addition, some papers showed that a proper settled UX may aid in violence prevention by identifying and acting to reduce the passenger's aggressive body language. Thus, it is useful to have a surveillance system so that people feel safe inside driverless vehicles. The target is an Intelligent Cockpit Web Platform that aggregates all sensor and algorithms outputs and allows a real-time remote communication to the vehicle. This platform will be directed by operators, where each one will be responsible for a series of vehicles having these in constant monitoring. Thus, a platform will be built where it will exist a real-time activity log, a real-time surveillance access, a remote communication and a vehicle status. There, the seats occupations and the ID of the respective passengers can be viewed, as well as data related to the vehicle, such as its identification, location and local time. Hence, if inside the vehicle there is fire or violence among passengers, an alert will appear on the operator's platform, where the operator, can act as needed, such as calling the competent authorities (firefighters, police, emergency medical services, etc.) or speak directly to the occupants to calm them or prevent violence through a synchronous conference stream.

However, we have to take into account some problems that may arise, such as, for example, wireless connection failure (for example lack of network inside a tunnel) or overcrowding of the system (for example alerts coming from 100 cars at the same time). In these cases, the system will have to be able to act automatically without requiring a human operator to give the order, for example, to activate the emergency system automatically in the event of an accident.

2   Methods for Generating Synthetic Data

There is a difference between simulation and emulation. The simulation aims to reproduce a specific feature or behavior of an object or event, whereas emulation seeks to replicate the whole practice of them. For instance, a voice simulation serves to reproduce speech with several volumes and timbres, whereas an emulation would require the reproduction of the vocal cords' behavior. This work aims to generate synthetic data by simulation and not by emulation.

For data generation, it is necessary to create a simulation of the cockpit and of events that may happen within it. Ideally, a domain-specific language for scripting these events would be useful. In addition to the cockpit, the sensors must also be simulated to verify which data can be achieved through them. Nevertheless, for doing so, it is necessary to access each sensor's datasheet. Therefore, a general specification must be created based on the expected sensor types.

For creating the simulation environment, open-source toolsets are Unity 3D; Gimp or Krita; Inkscape and Blender. There are several car-driving simulators, such as LG SVL, AISim, and others. The point is that most of these solution's scenes are out-vehicle, which means that the existing cockpit design is not focusing on it. Therefore, on the one hand, there are plenty of applications for cockpit modeling that can be used; on the other hand, a proper solution for meeting this project's simulation constraints should be created. As previously highlighted, synthetic data comes in handy when it's either impossible or impractical to generate a large amount of training data that many algorithms and models require.

Mainly, there are four methods for feature analysis: supervised; semi-supervised (rule-based); semi-supervised (model-based); and unsupervised. The first is based on a labeled dataset to be used for training a classification model. The second is based on known patterns that can be properly described as a rule. The third is the same as the former yet the patterns are described as a model. The fourth is based on unlabeled data whose features are extracted by statistics. This work is based on the first approach. This is so because there is not yet enough information about autonomous vehicles for deriving rules or models. The fourth may be quite untrustworthy (as shown by some modern chatbots) and such risks cannot be accepted.

As the focus of the automotive industry is increasingly moving towards the development of driverless vehicles, safety tests are becoming one of the most critical steps in the car manufacturing process, but they are not the only part of the equation. AI is one of the main contributors to the capacity of producing self-driving vehicles. In fact, manufacturers around the world are using AI techniques such as machine learning and deep learning algorithms in almost every aspect of the development of autonomous vehicles. Like any other algorithm, the algorithms used to develop a self-driving vehicle must be trained. Thus, before being submitted to a rigorous and elaborated set of assessments, an autonomous vehicle requires exhaustive training. The amount of time, effort and costs associated with the training of an autonomous vehicle in order to perceive its surroundings and behave accordingly is a major challenge. The perception system of the car is responsible for collecting the data produced by the different sensors and performing complex tasks such as scene understanding, object identification and motion estimation in real time. These complex tasks are assigned to machine learning and deep learning algorithms, which require vast amounts of examples to learn as even minor variations among those examples will affect the perception systems' judgment. In addition, the data used to train these models must be representative enough of all the traffic possibilities, including the edge cases and the dangerous situations that only a small part of vehicles will encounter. Hence, the perception system is the most critical asset and, coincidently, the most difficult to train.

Achieving a number of examples that are sufficiently illustrative of the nearly endless variety of conditions and the innumerable situations that a vehicle may face on the road is a difficult task. Hence, a new challenge arises: How do autonomous vehicle manufacturers acquire vast amounts of examples to train the perception system to ensure that it learns as much variability as possible? The problem of gathering this volume of data cannot be solved manually. Recent innovations have made the production of synthetic data a possibility. As the name suggests, instead of being produced by real events, synthetic data refers to artificially generated data that resembles the characteristics and complexity of reality [17]. Synthetic data is anonymous and free from data protection laws, allowing both Big Data and privacy protection at the same time, hence enabling a wide variety of unique opportunities. Synthetic data is a valuable approach for solving data insufficiency problems that has two broad categories: fully synthetic data, which means generating artificial data from scratch; and partially synthetic data, which uses sophisticated data manipulation techniques to create novel and unique training examples from real ones [17]. Synthetic data is important because the datasets available are often too small, documenting only the limited circumstances to which the physical vehicles were subjected to testing. This would leave behind a vast number of situations for training the vehicle's myriad of circumstances that can arise. Hence, synthetic data can be produced to meet specific needs that are not covered in the existing data.

There are several directions for the use of synthetic data in machine learning, in a format that resolves the privacy and legal issues that make the use of real data impossible [9,18]. This occurs both in the training of machine learning models with completely synthetic datasets [21] and in the enhancement of existing datasets so that the resulting hybrid datasets are better suited for model training [12,24,28,29]. In the present study the last proposal will be used to cover the parts of the data that are not sufficiently represented in the real dataset.

3   Results and Discussion

We made a bibliographic study about the use of synthetic data using a backward review approach. Thus, we chose three of the most important works about applying Machine Learning to Human Movement Analysis. The references of these three selected works were analyzed and crossed, building a list with 210 papers. Three steps were created to conduct the study. In the first step, studies published in 2017 or later were selected. In the end of this step the studies list was reduced to 40. Then, in step two, all abstracts and conclusions were read. In this phase, the papers which do not perfectly match with the theme were excluded. Finally, in the last stage all studies are carefully analyzed. The aim of this stage was to answer some questions such as what kind of dataset was used, or what kind of technique was used for creating the dataset. In Fig. 1 the main results of the study are presented [14,30,31].

As shown in Fig. 1, for works involving Machine Learning and Human Activity Recognition, the synthetic data is widely used. Researchers have pointed the

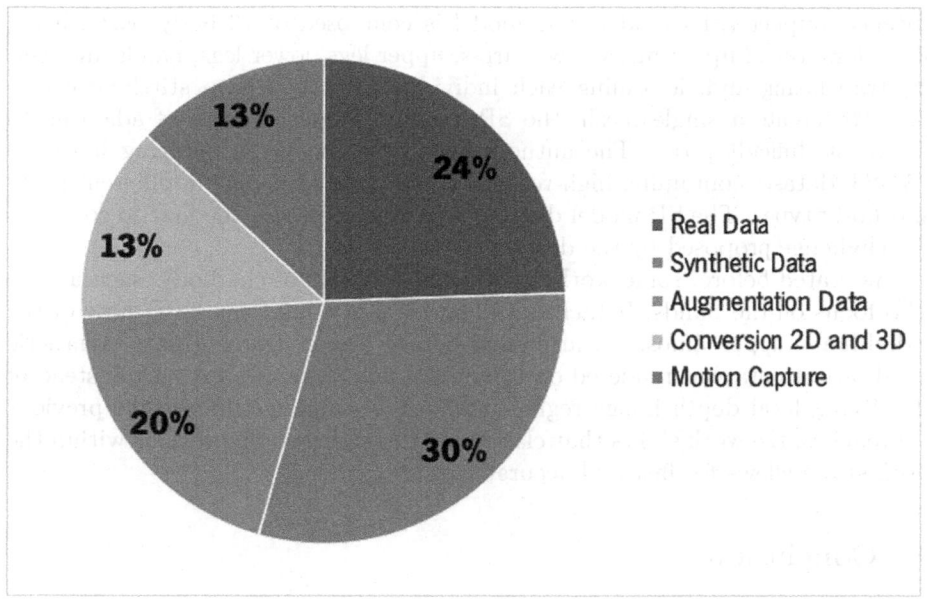

Fig. 1. Results of synthetic data bibliographic review.

training dataset increase as being the main reason for using data augmentation techniques. In this field, scaling and rotation are the most used methods by the researchers. Also, methods like flipping and cropping are emerging. Conversion of images between 2D and 3D is also very used (Fig. 1). In this theme, the conversion direction most used is from 2D to 3D. Although all studies claim that the results are better when using a 3D camera, the target of most studies is to reduce costs. Thus, 2D cameras are used for capturing video/pictures and then all images are converted to 3D. This justifies a greater use of 2D to 3D conversion than the opposite in these works. Finally, in some studies the use of Motion Capture cameras has been addressed. The researchers pointedly prove that although this kind of camera are too expensive, they have great results in human motion detection.

Although real data is a valid option to train deep neural networks, an alternative exists: synthetic data. This data can be used while training a model or it could even be created by using a model to create synthetic data. Throughout the years, several works have considered the use of this type of data for several applications such as human pose estimation [19]. Some works are general and focus in the entire body whereas others focus on specific parts of the body such as the hand's position.

Regarding the full human body, [33] proposes a 3D realistic and part-based model of the human body. The individual body parts can independently translate and rotate in 3D and, through that, adapt to the correct body pose. As the body consists of several parts stitched together, the model was named the stitched puppet (SP) model. The model is composed of 16 body parts: head, shoulders, torso, upper arms, lower arms, upper legs, lower legs, hands and feet. By translating and deforming each individual part and then stitching everything to create a single mesh, the SP model becomes capable of adapting to the most difficult poses. The authors tested the model by applying it to the FAUST dataset containing high-resolution 3D scans of people in different poses and body types. The SP model developed in this work was the first to complete the challenge proposed by the dataset.

As stated before, some works focus only on parts of the body, such as [20] who focus on the hands. It tackles the problem of estimating a 3D pose of the individual's upper limbs, i.e. arms and hands. To get training data, synthetic hand-object data was rendered on top of real 3D background scenes. Instead of classifying local depth image regions through a scanning window like previous approaches, the work shows that classifying the global configurations within the work-space allows for fast and accurate results.

4   Conclusion

For automatic human activity recognition, it is necessary to divide the problem into two parts, the part of the classification model and the part of the interior of the vehicles where the images will be captured for later classification.

In terms of the classification model the use of synthetic data for automatic human activity recognition brings several advantages, such as, the mass production of various situations not being necessary, reducing the costs and it also eliminates the problem of data protection laws. In terms of the environment inside the cockpit, it is necessary to understand the type of cameras to be used, the ones that present the best results in relation to motion detection and classification are Motion Capture cameras, however, this type of camera is too expensive. The use of 3D cameras also shows good results, but they are also an expensive option. For these reasons, a more economical approach is more often used, which consists of the use of 2D cameras for capturing images, and afterwards converting to 3D images, thus overcoming the price problem and obtaining good results in the same way.

Some studies are general and focus on the entire body, others focus on specific parts, for example, the position of the hands. However, the SP model, which consists of dividing the body into different stitched parts together, proved to be able to adapt to the most difficult poses.

Acknowledgments. This work has been supported by FCT – Fundação para a Ciência e Tecnologia within the R&D Units Project Scope: UIDB/00319/2020. Human and material resources have also been supported by the European Structural and Investment Funds in the FEDER component, through the Operational Competitiveness and Internationalization Programme (COMPETE 2020) [Project number 039334; Funding Reference: POCI-01-0247-FEDER-039334].

References

1. Analide, C., Novais, P., Machado, J., Neves, J.: Quality of knowledge in virtual entities. In: Coakes, E., Clarke, S. (eds.) Encyclopedia of Communities of Practice in Information and Knowledge Management, pp. 436–442. IGI Global, Hershey (2006)
2. Aqra, I., Abdul Ghani, N., Maple, C., Machado, J., Sohrabi Safa, N.: Incremental algorithm for association rule mining under dynamic threshold. Appl. Sci. **9**(24), 5398 (2019)
3. Brandão, A., et al.: A benchmarking analysis of open-source business intelligence tools in healthcare environments. Information **7**(4), 57 (2016)
4. Brito, C., Esteves, M., Peixoto, H., Abelha, A., Machado, J.: A data mining approach to classify serum creatinine values in patients undergoing continuous ambulatory peritoneal dialysis. Wireless Netw. 1–9 (2019). https://doi.org/10.1007/s11276-018-01905-4
5. Chen, Y.K.: Challenges and opportunities of Internet of Things. In: 17th Asia and South Pacific Design Automation Conference, pp. 383–388. IEEE (2012)
6. Constant, N., Douglas-Prawl, O., Johnson, S., Mankodiya, K.: Pulse-glasses: an unobtrusive, wearable HR monitor with Internet-of-Things functionality. In: 2015 IEEE 12th International Conference on Wearable and Implantable Body Sensor Networks (BSN), pp. 1–5. IEEE (2015)
7. Demchenko, Y., Grosso, P., De Laat, C., Membrey, P.: Addressing big data issues in scientific data infrastructure. In: 2013 International Conference on Collaboration Technologies and Systems (CTS), pp. 48–55. IEEE (2013)
8. Gbadamosi, A.Q., Oyedele, L., Mahamadu, A.M., Kusimo, H., Olawale, O.: The role of Internet of Things in delivering smart construction (2020)
9. Heaton, J., Witte, J.: Generating synthetic data to test financial strategies and investment products for regulatory compliance. SSRN Electron. J. (2019). https://doi.org/10.2139/ssrn.3340018
10. Kang, L., Poslad, S., Wang, W., Li, X., Zhang, Y., Wang, C.: A public transport bus as a flexible mobile smart environment sensing platform for IoT. In: 2016 12th International Conference on Intelligent Environments (IE), pp. 1–8. IEEE (2016)
11. Kim, K.J.: Interacting socially with the Internet of Things (IoT): Effects of source attribution and specialization in human-IoT interaction. J. Comput. Mediated Commun. **21**(6), 420–435 (2016)
12. Li, M., Chen, S., Chen, X., Zhang, Y., Wang, Y., Tian, Q.: Symbiotic graph neural networks for 3D skeleton-based human action recognition and motion prediction. arXiv preprint arXiv:1910.02212 (2019)
13. L'heureux, A., Grolinger, K., Elyamany, H.F., Capretz, M.A.: Machine learning with big data: challenges and approaches. IEEE Access **5**, 7776–7797 (2017))
14. Mehrizi, R.: Deep neural networks for human motion analysis in biomechanics applications by deep neural networks for human motion analysis in biomechanics applications By Rahil Mehrizi Dissertation Director: Kang Li (2019)
15. Mohammadi, M., Al-Fuqaha, A., Sorour, S., Guizani, M.: Deep learning for IoT big data and streaming analytics: a survey. IEEE Commun. Surv. Tutorials **20**(4), 2923–2960 (2018)
16. Neves, J., et al.: A soft computing approach to kidney diseases evaluation. J. Med. Syst. **39**(10), 131 (2015). https://doi.org/10.1007/s10916-015-0313-4
17. Nikolenko, S.I.: Synthetic data for deep learning. arXiv preprint arXiv:1909.11512 (2019)

18. Reiter, J.P., Drechsler, J.: Releasing multiply-imputed synthetic data generated in two stages to protect confidentiality. Statistica Sinica **20**, 405–421 (2010)
19. Rogez, G., Schmid, C.: Mocap-guided data augmentation for 3D pose estimation in the wild. In: Advances in Neural Information Processing Systems, pp. 3108–3116 (2016)
20. Rogez, G., Supancic, J.S., Ramanan, D.: First-person pose recognition using ego-centric workspaces. In: Proceedings of the IEEE Conference on Computer Vision and Pattern Recognition, pp. 4325–4333 (2015)
21. Saleh, F.S., Aliakbarian, M.S., Salzmann, M., Petersson, L., Alvarez, J.M.: Effective use of synthetic data for urban scene semantic segmentation. In: Ferrari, V., Hebert, M., Sminchisescu, C., Weiss, Y. (eds.) ECCV 2018. LNCS, vol. 11206, pp. 86–103. Springer, Cham (2018). https://doi.org/10.1007/978-3-030-01216-8_6
22. Sisinni, E., Saifullah, A., Han, S., Jennehag, U., Gidlund, M.: Industrial Internet of Things: challenges, opportunities, and directions. IEEE Trans. Industr. Inf. **14**(11), 4724–4734 (2018)
23. Thakur, A., Malekian, R., Bogatinoska, D.C.: Internet of Things based solutions for road safety and traffic management in intelligent transportation systems. In: Trajanov, D., Bakeva, V. (eds.) ICT Innovations 2017. CCIS, vol. 778, pp. 47–56. Springer, Cham (2017). https://doi.org/10.1007/978-3-319-67597-8_5
24. Tran, L., Yin, X., Liu, X.: Disentangled representation learning gan for pose-invariant face recognition. In: Proceedings of the IEEE Conference on Computer Vision and Pattern Recognition, pp. 1415–1424 (2017)
25. Uddin, M.F., Gupta, N., et al.: Seven v's of big data understanding big data to extract value. In: Proceedings of the 2014 zone 1 conference of the American Society for Engineering Education, pp. 1–5. IEEE (2014)
26. Vasisht, D., et al.: Farmbeats: an IoT platform for data-driven agriculture. In: 14th {USENIX} Symposium on Networked Systems Design and Implementation ({NSDI} 2017), pp. 515–529 (2017)
27. Wu, J., Guo, S., Li, J., Zeng, D.: Big data meet green challenges: greening big data. IEEE Syst. J. **10**(3), 873–887 (2016)
28. Wu, X., He, R., Sun, Z., Tan, T.: A light CNN for deep face representation with noisy labels. IEEE Trans. Inf. Forensics Secur. **13**(11), 2884–2896 (2018)
29. Xiao, S., Feng, J., Xing, J., Lai, H., Yan, S., Kassim, A.: Robust facial landmark detection via recurrent attentive-refinement networks. In: Leibe, B., Matas, J., Sebe, N., Welling, M. (eds.) ECCV 2016. LNCS, vol. 9905, pp. 57–72. Springer, Cham (2016). https://doi.org/10.1007/978-3-319-46448-0_4
30. Yao, S., Hu, S., Zhao, Y., Zhang, A., Abdelzaher, T.: DeepSense: a unified deep learning framework for time-series mobile sensing data processing. In: 26th International World Wide Web Conference, WWW 2017, pp. 351–360 (2017). https://doi.org/10.1145/3038912.3052577
31. Zago, M., Luzzago, M., Marangoni, T., De Cecco, M., Tarabini, M., Galli, M.: 3D tracking of human motion using visual skeletonization and stereoscopic vision. Front. Bioeng. Biotechnol. **8**(March), 1–11 (2020). https://doi.org/10.3389/fbioe.2020.00181
32. Zaslavsky, A., Perera, C., Georgakopoulos, D.: Sensing as a service and big data. arXiv preprint arXiv:1301.0159 (2013)
33. Zuffi, S., Black, M.J.: The stitched puppet: a graphical model of 3D human shape and pose. In: Proceedings of the IEEE Conference on Computer Vision and Pattern Recognition, pp. 3537–3546 (2015)

Special Session on Machine Learning, Law and Legal Industry

Intellectual Properties of Artificial Creativity: Dismantling Originality in European's Legal Framework

Beatriz A. Ribeiro[✉]

Master in Cognitive Science, University of Lisbon, Lisbon, Portugal
beatriz.smar@gmail.com

Abstract. The debate on whether a work created by AI should be protected is waging on and divides academics into three groups: (1) the ones who maintain a traditional view, arguing that Intellectual Property was made by humans to regulate humans and the works of human intellect, (2) the ones who advocate a redefinition of the concept of intellectual creation, opposing the first view with a more progressive one and (3) those who consider that the law should find an intermediate solution, as the situation is analogous to the employer/employee context (Gürkaynak et al. 2017). However, the main purpose of this paper is not to participate on the above-mentioned discussion. On the contrary, it is assumed that the matter will soon be resolved and, therefore, other important notions in the Intellectual Property's universe, namely, originality and creativity. The bottom-line question is the following: considering the European framework, can artwork created by AI be regarded as creative and thus, original?

1 Introduction

The debate on whether a work created by AI should be protected is waging on and divides academics into three groups: (1) the ones who maintain a traditional view, arguing that Intellectual Property was made by humans to regulate humans and the works of human intellect, therefore leaving out landscapes with curious dispositions as well as any form of work forged by an animal or computer program, meaning that every work that happens to be created by any of the mentioned should fall on the public domain, (2) the ones who advocate a redefinition of the concept of intellectual creation, opposing the first view with a more progressive one, and (3) those who consider that the law should find an intermediate solution, as the situation is analogous to the employer/employee context (Gürkaynak et al. 2017)

However, the main purpose of this paper is not to participate on the above-mentioned discussion. On the contrary, it is assumed that the matter will soon be resolved and, therefore, other important notions in the Intellectual Property's universe are analysed, namely, originality and creativity. The bottom-line question is the following: considering the European framework, can artwork created by AI be regarded as creative and thus, original?

The traditional European definition of work in Intellectual Property determines that it depends on the existence of three things: (1) intellectual creation, (2) that falls into

C. Analide et al. (Eds.): IDEAL 2020, LNCS 12490, pp. 379–389, 2020.
https://doi.org/10.1007/978-3-030-62365-4_36

the category of either literary, scientific or artistic work and (3) that is by any means displayed in the physical world (article 2, Berne Convention).

Closely linked to the first requisite is the concept of originality, since without this element there is no real intellectual creation to be protected. This notion is mentioned but not accurately outlined in the Berne Convention, thus forcing the tradition of the *Droit d' Auteur* to allocate resources in the study and definition of the boundaries of this concept. As such, many efforts have been placed to outline the so-called originality threshold. In this sense, the work is original if (1) there is no pre-existent copy (hence, provided with some degree of creativity, when compared with the existent intellectual creations) and, additionally, (2) if the work expresses the individuality of the author's personality, which as shall be explored later on the paper, is often understood as the result of free creative choices.

Differently, in the Copyright system (namely UK and USA), the work is considered as an original if it is not derived from another and it comes directly from its author. The criteria seems to be grounded in the *skill, labour and judgment* demonstrated by the author, not demanding any creation of the spirit, notwithstanding the *Feist Publications v. Rural Telephone Service* (499 U.S. 340 [1991] 346) judgement, which leaves information and data outside the originality bubble, regardless of the amount of skills and labour proven to exist. The paper will focus on the European framework and how works created by AI can be seen in the light of the European context.

2 The Legal Notion of Originality and Creativity

In a purely intuitive perspective, painting a picture or writing a poem is usually deemed as creative, while mathematics and engineering are rarely classed as such (Ritchie 2007). Creative products are also typically perceived as unique (Simonton 2009).

However, in legal matters, the definition is, in fact, quite different. As mentioned, the traditional definition of work presumes three elements: (1) intellectual creation, (2) that falls into the category of either literary, scientific, or artistic work and (3) that is by any means exteriorized.

There is a strong relation between originality of the intellectual creation and the author since the notion of the intellectual creation implies a fabrication originated in the author's mind. Hence, the originality requirement, essential for protection, is closely linked to the question of authorship, since the author itself is what makes it original (Bond and Blair 2019).

As previously outlined, on the other hand, the work is original if there is no pre-existent copy. Put differently, the work must present some degree of creativity, when compared with the existent intellectual creations, and there must be a distinctive creative trait that makes the work an original instead of a copy. Although, in a broad sense, creativity can be seen as novelty (Desbois 1967), in specific, it means that the work should be deemed as an increment of value, in the face of existent intellectual heritage (Menezes Leitão 2020, pp. 74). Strictly speaking, only works that show some minimum amount of this attribute attract protection (Margoni 2016).

On the other hand, the work is original if, cumulatively, expresses the individuality of the author's personality. This condition requires some analysis, since is it not so intuitive as the first one. In fact, the author's personality as an essential requirement was introduced in the notion of originality as a result of a long legal debate on the concept. However, the fine line between works that present enough creativity in order to be protected, opposing to those which lack the minimum threshold of originality is still yet to be drawn, from a legal stance, despite many proposed solutions in literature.

The European legal framework mentions *en passent* and defines broadly the concept of originality in three occasions (Akester 2019) which are the following:

a. The current Directive 2009/24/EC of the European Parliament and Council of 23 April 2009 on the legal protection of computer programs. A computer program is protected if original and this means being a result of intellectual creation of the author.
b. The Directive 2006/116/EC of the European Parliament and Council of 12 December 2006 on the term of protection of copyright and certain related rights. In its article 6, applicable to photographs, it is stated that *Photographs which are original in the sense that they are the author's own intellectual creation shall be protected*. Here, the notion of work as the expression of the author's personality is implied.
c. The Directive 96/9/EC of the European Parliament and Council of 11 March 1996 on the legal protection of databases.

Besides these strict definitions of originality present in the above-mentioned directives, which, as described, regulate specific categories of works – computer programs, photography, and databases – none of the major international copyright treaties gives any other notion regarding the concept.

Academics have been very critical of this scattered definitions, since these Directives, though aiming to harmonize EU Members legislation, they do so with very limited effects, to the specific subject therein regulated (Margoni 2016).

This circumstance has been associated to the lack of attribution powers in regulating copyright, in accordance with the Treaty on the Functioning of the European Union. What we do know, due to the *Infopaq International ECJ judgment (See Infopaq International v. Danske Dagblades Forening [2009] E.C.R. I-06569)* is that every work must have a minimum of originality in order to comply with Berne standards. Yet, specific standards and elements that should complete the concept of intellectual creation remains a matter for national courts and legislatures (Margoni 2016).

Another legacy of the *Infopaq* judgement was the definition of a standard of originality, which in this case provides horizontal harmonization, instead of simple application in specific matters. In other words, the *author's own intellectual creation* as the definition of original creation is now applicable to every field in intellectual property. Therefore, we must conclude that the work must somehow be a reflection of the author's personality.

The *Infopaq*, along with other judgements[1] goes even further in defining the concept of originality, stating that it is present when authors can exercise free, creative choices and put their personal signature on the work.

---

[1] Football Dataco v. Yahoo [2012], 38; Bezpečnostní softwarová asociace v. Ministerstvo kultury. [2010], 50; Eva-Maria Painer v. Standard VerlagsGmbH [2011], 89, 92.

Other unavoidable judgments when shaping the concept of originality include Football Association Premier League Ltd and Others v QC Leisure and Others (C-403/08) and Karen Murphy v Media Protection Services Ltd (C-429/08), which clearly state that if such expression of the author's personality is determined by technical or functional rules or, alternatively, the expression is constrained by specific narrow rules or goals, it cannot be considered creative work.

According to Bently and Sherman (2014) there is a fourth inference that can be done from these decisions. It appears that the court argues that the quality of the contributions weight more than its quantity. In this sense, the personal stamp of the author will have more significance in terms of originality appreciation.

This conclusion is supported by the clarification given by the Advocate General Mengozzi which states that the EU standard requires a creative aspect, thus not being sufficient the mere labour and skill of a specific work.

Along with the fact that it can be seen as a discovery, and therefore excluded from the sphere of protection, this explains why Fractal Art itself, for instance, though very much studied and admired, cannot constitute protected work.

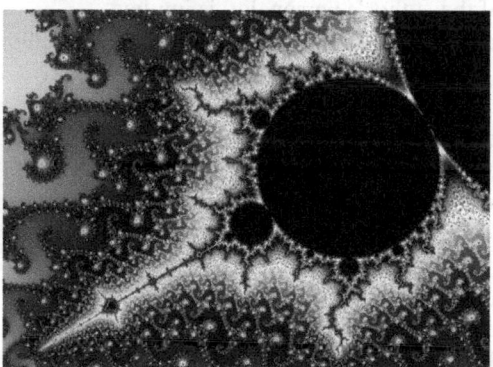

Fig. 1. Image of a Fractal, retrieved from the Wikimedia Commons, accessible trough https://commons.wikimedia.org/wiki/File:Mandel_zoom_08_satellite_antenna.jpg

The term Fractal was coined by Benoit Mandelbrot to described the geometry of highly fragmented forms of nature that were perceived as amorphous (for instance clouds or threes), thus not easily represented by the standard Euclidean Geometry, which typically characterizes regular objects (Bond 1992).

Having noticed that computers could simulate very complex patterns which would be somewhat aesthetic to the human eye, many simulations have been experimented, making it, in some sense, the first artworks created by a computer, since it couldn't have been done without it (Fig. 1).

These computer simulations led to the *Fractal Art Manifesto*, written by Kerry Mitchell, in 1999, which stated that the art was in the programming, meaning, in the choice of the equation or algorithm. In his view, the art would be executed by the computer but with the direction of the (human) artist.

However, no one questioned the inexistence of creativity in the case of the computer: it was purely deterministic (Du Sautoy, M. 2019). We could argue a different position in the case of the programmer, but no one will support the idea that the compute itself was creative.

Any further considerations on matters of originality and creativity, including establishing whether a specific work meets the requisite of author's own intellectual creation, fall into the competence of national courts. In other words, different levels of originality can operate within different jurisdictions, notwithstanding the fact that a common market is currently in construction in the EU (Margoni 2016).

Part of the doctrine (for instance Sá e Mello 2019, pp. 62) argues that these types of works should be considered non-intellectual works, since according to this view, the work should be rooted in a "creative spirit", in the expression of an intelligent disposition, traditionally ascribed to human beings but questionable in AI systems. This position states that if we decide to concede protection to an AI's creative work, we would be ignoring one element of the definition of protected work – the spiritual conception of the work – in order to give the spotlight to another element which is usually dependent of the first – creativity.

When it comes to AI originality, there is only one legal system that tackles the question: the UK's system. In Section 178 UK Copyright, Designs and Patents Act (hereinafter called CDPA) the act addresses the question of "computer-generated" work, defining such as the work that is generated by computer in circumstances in which there is no human author of the work. This definition is broad enough to include, at the very least, any literary, dramatic, music or artistic work (henceforth LDMA).

If facing with the present situation, according to the CDPA, the author *shall be taken to be the person by whom the arrangements necessary for the creation of the work are undertaken.*

Given the presented solution, and while the CDPA does not mention the originality requisite, the English courts and academic commentary varies and broadly splits into three main views (Bond and Blair 2019). The first focuses on the creative efforts of the persons making the arrangements. Other part of the doctrine suggests that there is no originality requirement for computer-generated works. And finally, the third view proposes that the originality of computer-generated works should be assessed objectively, i.e. would the creation of the same work by a human have satisfied the originality requirement?

Nevertheless, it is important to state that this method for attributing authorship to LDMA works when there is no human author has been followed in a small number of other jurisdictions, receiving only a single judicial mention from the English courts in its 30-year history (Bond and Blair 2019).

3   Creativity in AI and Robotics

The field of Computational Creativity is tasked with both defining the philosophical foundations of the search for creativity and transferring this understanding into real machines that are valuable to society (Hodson 2018).

Most of the literature in this field was built through the foundational works of Margaret Boden who argues that creative acts represent (1) novel combinations of familiar ideas, (2) explorations of the potential of conceptual spaces, and (3) transformations that enable the generation of novel ideas (Boden 1998).

However, it was not until recently that academics tacitly accepted that the notion of creativity is difficult to pin down and reproduce in machines (Colton and Wiggins 2012).

Nevertheless, research has not stopped, both in understanding how human creativity works and how to replicate such phenomena in machines.

Accordingly, when it comes to Computational Creativity, it appears that research was halved in two main methodologies (Hodson 2018).

The first one aims to discover the set of domain-independent creative processes, able to generate creative artefacts when applied in any domain. This includes conceptual blending and bisociative discovery of new concepts.

The second approach reflects upon specific creative processes active in certain domains, such as music production, painting, and poetry. In this sense, the point is not to find principles and emulate them, but instead to understand the specific creative processes to each discipline, (Colton 2008).

As far as Intellectual Property is concerned, the main obstacle presented by theoretical work in AI, in the field of creativity, is that remains essentially focused in understanding when a system simply does what is asked by the programmer and when it goes beyond the initial prescriptions (Jennings 2010). This goal is obviously not enough from a legal point of view, because the fact that what the machine does is surprising does not make it necessarily creative. On the other hand, Science and Law are working with completely different definitions of creativity and originality, which is also an obstacle.

However, much progress has been made in the last few years regarding creativity and robotics. Today, and concerning these matters, it is possible to detect two major categories of AI: symbolic machines and artificial neural networks (Miller 2019).

Symbolic machines can create interesting work, having been fed enormous amounts of data and programmed to do so. In other words, though recognizable as art, it is still pre-programmed (the most used example is Harold Cohen's AARON algorithm). Artificial neural networks (ANN), on the other hand, teach themselves, thus being considered computationally more creative, since they are more experimental and unpredictable (Miller 2019).

After this introduction on the matter, the paper will focus on specific approaches, with different levels of autonomy, and analyse whether they can be, from a legal stance, considered creative.

The first one is the art created by Genetic Algorithms (GA). The typical use of genetic algorithms is optimization, where there is a need to search some virtual space for the individual (in this case, either an image or a piece of sound/music) that scores highest in some selected measure. The genetic algorithm first generates a random set of individuals drawn from the set of possible individuals that could exist (this set is known as the search space). In the context of an artistic system this could be the generation of random sounds or images (Johnson and Romero 2002).

Typically, these individuals are represented as binary strings, which facilitates the application of the crossover and mutation operators. After the best individuals in the population are selected (either by a human, a computer or a combination of the two), they exchange information with other individuals (a process known as crossover) and small changes are then made in the individuals resulting in a new population. This process is repeated either until a satisfactory result is found or until the population reaches a state where the individuals are all very similar to one another (Johnson and Romero 2002) (Figs. 2 and 3).

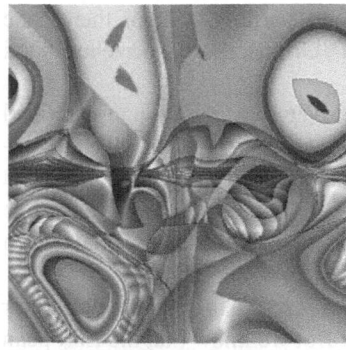 

Fig. 2. Examples of GA art, retrieved respectively from https:// www.genetic-algorithms.com/ and https://mc.ai/drawing-with-gen etic-algorithms/

Fig. 3. Examples of GA art, retrieved respectively from https://www.genetic-algorithms.com/ and https://mc.ai/ drawing-with-genetic-algorithms/

Though with a considerable level of autonomy (the successive offspring can be created and selected by the computer) and interesting results, it's impossible to argue that free creative choices exist on the part of the computer – the mutations and offspring selected are completely randomized.

Fig. 4. Edmond de Belamy portrait, retrieved from https://www.livesc ience.com/63929-ai-created-painting-sells.html.

The second approach to algorithmic art that we are going to discuss is through the utilization of Generative Adversarial Networks (GAN), which produced the well-known Edmond de Belamy portrait (Fig. 4), which was sold in 2018 for $432,500 by Christie's. In order to achieve this masterpiece, a Generative Adversarial Network (GAN) algorithm was used, and the system was fed with 15,000 portraits created between the 14th and 20th century. GAN algorithms are also the ones who allowed us to see Monalisa in a Harry Potter's portraits fashion, moving and exhibiting different facial expressions.

Unlike the previous wave of algorithmic art, in which the artist had to write detailed code, specifying the rules for a desired aesthetics, in this new approach the system is set up to learn the aesthetics by looking at a large quantity of images through machine learning technology (Mazzone and Elgammal 2019). In this procedure, the creative process itself is done by the artist, who also adjusts the algorithm according to necessities.

The third approach is Deep Neural Networks, where the most famous *artist* is Google's Deep Dream (visual art) and Magenta project (song composition).

In contrast to early AI, deep learning technologies afford computers capacities to collate and organize information, which allows those systems to make intelligent predictions about the future (Alpaydin 2016). These systems make these predictive calculations using what they know from past experiences and then generate randomized permutations of existing data points. Through trial-error method, deep learning systems then make intelligent guesses and become increasingly accurate (Gruner and Csikszentmihalyi 2018).

In the case of Deep Dream, through a Deep Neural Network, the system is trained for image classification, but not for its originally intended purpose. Here, image classification is not the output, as it happens in other Deep Learning Algorithms, but instead, a necessary step in order to create art. After identification of the object, Deep Dream then proceeds to change the image, granting it those psychedelic outlines that we have been presented with.

The Deep Dream artwork clearly presents novelty, as it consists in something never presented before, therefore being original in this sense. But we know that the European Court of Law requires more in order to consider a work original and thus protected – the work must somehow reflect the author's personality or individuality (Figs. 5 and 6).

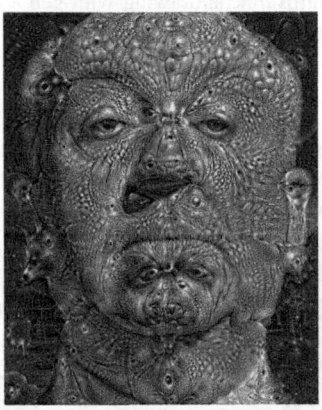

Fig. 5. Google's Deep Dream Art, retrieved, respectively, from https://www.nytimes.com/2017/08/14/arts/design/google-how-ai-creates-new-music-and-new-artists-project-magenta.html and http://deepdream.psychic-vr-lab.com/deepdream/pics/971953.html.

Fig. 6. Google's Deep Dream Art, retrieved, respectively, from https://www.nytimes.com/2017/08/14/arts/design/google-how-ai-creates-new-music-and-new-artists-project-magenta.html and http://deepdream.psychic-vr-lab.com/deepdream/pics/971953.html.

It is interesting to note that the style present by Deep Dream is actually quite characteristic. In other words, Deep Dream has, undoubtedly, a unique style, created by its own constraints and experience, which would allow an art *connoisseur* to identify the artist. Hence, at some point, it mirrors some *virtual personality* of this author.

Yet, and while being characteristic enough to identify the *author*, would it be fair to say that it reflects the personality of Deep Dream's algorithm? Would this be enough to consider the work as a reflection of the *author's personality*? Also, can we say that the system is free in its *creative choices*?

More recently, and yet to be fully studied and explored, some proposals have emerged, such as Creative Adversarial Network (CAN) (Elgammal et al. 2017), in which the machine is trained between two opposing forces: (1) one that urges the machine to follow the aesthetics of the art that is shown to the system, therefore minimizing deviation from art distributions and (2) another that penalizes the machine if it emulates an already established style. According to the authors, these two constraints would ensure that the art generated is novel, while not departing from the aesthetics standards with which the system was previously fed.

This process appears to be somewhat more creative than the methodologies described earlier, because, as the authors argue, there is less human influence and the outputs are surprising. Additionally, when applying the Turing Test, most humans could not tell the difference between human created and algorithm created art.

In conclusion, we have machines that are creating new artworks, autonomously, and some of them with a characteristic trait. Is this enough to be considered creative, and thus, original work?

Considering the above described, and even with clear and important advances, creative machines are still very much dependent on inputs given by the human mind. As mentioned, for a machine to create a painting, it must be shown, *a priori*, existing images and recommended an appropriate fashion ensemble. It must also be programmed to recognize and understand harmonies and patters, in order to compose. At the very best, what we will have is a machine that deploys randomized outputs from pre-programmed and selected inputs. Hence, the domain of the machine world is a closed system and depends directly on the limitations of the human mind (Gruner and Csikszentmihalyi 2018).

Following this argument, it is clear that AI systems are, today, incapable of radically altering existing paradigms independently. In other words, while Deep Dream indeed presents noteworthy and admirable artworks, they are not to be considered disruptive in the state of art, though obviously, no demerit intended.

As Jennings (2010) puts it, if we were to objectively describe fully autonomous computational creativity, AI would have to meet the following elements: (1) Autonomous evaluation, meaning that the system would evaluate its liking without an outside opinion source, (2) Autonomous change, in the sense that the machine is able to perform changes without directly being told how and when to do so, and (3) Non-randomness, which means that the changes should be not random. Guaranteeing the fulfilment of these three elements could suggest useful future directions in order to get closer to a creative machine.

In conclusion, AI artworks hardly fulfil the creativity definition, and therefore the originality requirement.

However, the notion of an work of art being the coherent expression of the individual's psyche begins in the Romantic era, prevailing throughout the centuries until today, although it does not mean it is necessarily the right definition of creative art (Mazzone and Elgammal 2019).

While being obvious that machines cannot replicate, at least for now, human art and creativity, we still must question whether the criteria applied to humans should be extendible to AI, considering that the sources of inspiration are not, by any mean, the same. In other words, does the impossibility of reproduction of human creativeness somehow invalidates or, rather, disqualifies the creation of the machine?

Mazzone and Elgammal (2019) also point out that creative acts do not always have its origin through particular and differentiable processes of cognitive systems (such as brains). In this sense, they can be the result of any combinations of the available symbolic and statistical manipulations available.

Nevertheless, from a legal stance, creativity demands the existence of the authors personality traits in the work, and those traits should derive from creative and free deliberation of the same author. So far, symbol manipulation or machine learning, though autonomous, still faces strong constraints put in the system by the programmer.

Accordingly, and although it is safe to say that machines do produce novel works autonomously, it is also clear that those works are still very much dependent of human creativity, therefore not producing absolute free creative choices.

4   Conclusion

Art is, ultimately, an expression of human free deliberation. As described, European Courts and legislation appear to agree with this statement. Until computers have their own version (which by no means is meant that it should be identical to human's) of this, any art created by AI will, in the end, be traceable back to a human desire to create (Du Sautoy, 2019, pp 98).

In this sense, it appears that even if the first requirement for protection, which would be the legal personhood, is solved, AI still has another difficulty to overcome: the concept of originality and creativity and its applicability to artificial systems. These concepts, like many others in Intellectual Property, suffer from the typical anthropomorphic syndrome, in which the concepts designed in the Law are exclusively for human beings.

In this sense, as Bond and Blair (2019) put it, it is important to answer the question of whether should copyright only reward acts of truly human cognition or does it play a more utilitarian role in society, encouraging the production and distribution of new works irrespective of the manner in which they were created? The solution may lie in recognizing computer-generated works as deserving of only economic rights akin to those afforded to films, sound recordings, broadcasts, and typographical arrangements.

In the European legal context, however, and given the importance of the *création de l' spirit*, it is hard to conceive protecting works on the basis of their utility and financial potential.

One option is, naturally, adapting the concept of originality and creativity in order to allow AI's works to be a part of the range of protected *oeuvres*. This adaptation could be, inclusively, to consider AI artworks as derivative works, in the sense that they require a pre-existing work (which is created by a human being), from which they acquire their substance.

Yet, the definition is far from consensual and lacks proper concretization even in the human context, thus barely leaving room for any discussion in the domain of AI creativity. If originality is, in fact, to be considered as a requirement, as it is today, its assessment and concretization should be conducted rigorously (Rosati 2018).

However, AI creativity and the fulfilment of the originality requirement in the artwork created by machines is, definitely, an inevitable question.

References

Alpaydin, E.: Machine Learning, The New AI (MIT Press Essential Knowledge series). MIT Press, Cambridge (2016)

Akester, P.: Código Anotado do Direito de Autor e Direitos Conexos, 2nd edn. Edições, Almedina (2019)

Bently, L., Sherman, B.: Intellectual Property Law. Oxford University Press, New York (2014)

Boden, M.A.: The Creative Mind: Myths and Mechanisms. Basis Books, New York (1991)

Bond, M.: Introduction to fractal geometry: definition, concept, and applications, Presidential Scholar Thesis (1990–2006), vol. 42 (1992)

Bond, T., Blair, S.: Artificial Intelligence & copyright: section 9(3) or authorship without an author. J. Intellect. Property Law Pract. **14**(6), 423–423 (2019)

Colton, S.: Creativity versus the perception of creativity in computational systems, In: AAAI Spring Symposium: Creative Intelligent Systems, vol. 8 (2008)

Colton, S., Wiggins, G.A.: Computational creativity: the final frontier? In: Proceedings of the 20th European Conference on Artificial Intelligence, 21–26. IOS Press (2012)

Desbois, H.: Le Droit d' Auteur en France. Dalloz, Paris (1967)

Du Sautoy, M.: The Creative Code, Art and Innovation in the Age of AI. The Belknap Press of Harvard University Press, Cambridge (2019)

Elgammal, A., Bingchen L., Elhoseiny, M., Mazzone, M.: CAN: creative adversarial networks, generating "art" by learning about styles and deviating from style norms, arXiv, arXiv:1706. 07068 (2017)

Gruner, D.T., Csikszentmihalyi, M.: Engineering creativity in an age of artificial intelligence, consulted on 2nd April (2018). 326844333_Engineering_Creativity_in_an_Age_of_Artificial_Intelligence

Gürkaynak, G., Yılmaz, I., Doygun, T., Ince, E.: Questions of intellectual property in the artificial intelligence realm. Robot. Law J. consulted on 31st March (2017). https://gurkaynak.av.tr/docs/8b791-rlj-september-october-2017-.pdf

Hodson, J.: The creative machine, consulted on 6th April (2018). https://www.researchgate.net/publication/325655108

Jennings, K.: Developing creativity: artificial barriers in artificial intelligence. Mind. Mach. **20**, 489–501 (2010). https://doi.org/10.1007/s11023-010-9206-y

Johnson, C., Romero, J.: Genetic algorithms in visual art and music. Leonardo **35**, 175–184 (2002). https://doi.org/10.1162/00240940252940559

Margoni, T.: The harmonisation of EU copyright law: the originality standard. In: Perry, M. (ed.) Global governance of intellectual property in the 21st century, pp. 85–105. Springer, Cham (2016). https://doi.org/10.1007/978-3-319-31177-7_6

Mazzone, M., Elgammal, A.: Art, creativity and the potential of artificial intelligence. Arts **8**, 26 (2019). https://doi.org/10.3390/arts8010026

Menezes Leitão, L.M.: Direito de Autor, 3ª edn. Almedina, Coimbra (2020)

Miller, A.: Creativity and AI: the next step, scientific America, consulted in 2nd April (2019). https://blogs.scientificamerican.com/observations/creativity-and-ai-the-next-step/

Ritchie, G.: Some empirical criteria for attributing creativity to a computer program. Mind. Mach. **17**, 67–99 (2007). https://doi.org/10.1007/s11023-007-9066-2

Rosati, E.: Why originality in copyright is not and should not be meaningless requirement. J. Intellect. Property Law Pract. **13**(8), 62 (2018)

Sá e Mello, A.: Manual de Direito de Autor e Direitos Conexos, 3ª edn., Almedina, Coimbra (2019)

Simonton, D.: Varieties of scientific creativity: a hierarchical model of domain-specific disposition, development, and achievement. Perspect. Psychol. Sci. **4**, 441–452 (2009)

Network Analysis for Fraud Detection in Portuguese Public Procurement

Davide Carneiro[1,2](✉) ⓘ, Patrícia Veloso[1] ⓘ, André Ventura[1],
Guilherme Palumbo[1], and João Costa[1]

[1] CIICESI, Escola Superior de Tecnologia e Gestão, Instituto Politécnico do Porto,
Felgueiras, Portugal
{dcarneiro,8170430,8170566,pamv,8150133}@estg.ipp.pt
[2] Algoritmi Centre/Department of Informatics, Universidade do Minho, Braga,
Portugal

Abstract. As technology evolves, frauds, in all their different forms,
become increasingly more complex, as mega-processes such as Luanda
Leaks or Panama Papers have recently shown. Fraud detection mechanisms must thus also resort to recent technological developments to
ensure that even new forms of complex fraud are detected in time. In
this paper we present a system to facilitate fraud detection in Public
Procurement in Portugal. It relies on three main components. Data on
public procurement and involved entities is acquired from public sources
and integrated into a graph-oriented database. A rules-engine enriches
these data with information additional information, using legal rules or
custom rules defined by the users. Finally, a graph-oriented User Interface is used to support decision-making, allowing users to quickly and
efficiently explore and filter information, in a natural and geo-referenced
way. The main goal of this system is to increase transparency by facilitating access to relevant information, and in this way contribute to the
fairness of the whole public procurement procedure.

Keywords: Fraud detection · Public procurement · Graph-oriented
database · Network analysis

1 Introduction

Fraud refers to unlawful actions carried out by one or more parties in order to
achieve some economic or strategic advantage. Fraud detection is the use of tools
or methods to identify and/or characterize occurrences of potential fraud. Traditional methods involving manual detection are time-consuming, expensive and,
in the age of big data, impractical. As a consequence, fraud detection is usually
done by random sampling, and only a very small percentage of transactions is
actually analyzed [1].

Fraud detection must thus turn to automated processes, using statistical
and computational methods, as well as Machine Learning (ML). These methods

C. Analide et al. (Eds.): IDEAL 2020, LNCS 12490, pp. 390–401, 2020.
https://doi.org/10.1007/978-3-030-62365-4_37

reduce the manual effort of the auditor through automatization, or by doing a preliminary classification of the instances, which allows the auditor to direct their investigative attention to relevant cases.

Automated fraud detection has been a research topic for a few decades, especially since the advent of Machine Learning (ML). While both supervised and unsupervised ML algorithms can be used, the large majority of the literature reports the use of supervised ones. The most common algorithm found in the literature include Neural Networks [5,6,14], Regression Models (e.g. Logistic Models [9,13], Probit and Logit [3], Fuzzy Systems [7] and Expert Systems [4,7]. There are also some comprehensive surveys on fraud detection systems over time, which provide a longitudinal view of the field (e.g. [1,7,11]).

In what concerns ML approaches, the standard methodology (or slight variations of it) are usually followed, i.e., feature selection, data cleaning, model training and performance evaluation [5,6]. Rule-based approaches, such as Expert Systems, usually rely on rule-mining algorithms (e.g. association rules) in order to find the rules that will then support fraud detection [10].

Current approaches to fraud detection are thus mostly data-oriented. While fairly straightforward to implement given the existing wealth of ML algorithms, there are also shortcomings. Namely:

- A relational approach is generally used in which each data point is considered and analyzed individually [7,11]. This makes it hard to uncover sophisticated scams such as fraud rings;
- Public datasets are scarce as they usually belong to private companies, which protect them for legal and competitive reasons [1,7,11]. They are also very skewed as fraud events are relatively much less frequent [1,7,12];
- Datasets must be labeled by Human experts, which is time-consuming and costly [7];
- There is too much emphasis of research on complex, domain-dependent, non-linear, supervised algorithms, which are virtually impossible to interpret and explain [5,6,14]. Knowledge is also difficult to transfer across domains;
- Implemented systems documented in research are rare.

This paper proposes to address these shortcomings through the development of a hybrid (Rule-based + Machine Learning) Decision Support System (DSS) operating on top of a Knowledge Graph. A specific case study in the domain of Portuguese Public Procurement Law [8] is being developed.

The main features of the proposed approach are thus:

- Data describing transactions between stakeholders are acquired from different public sources (e.g. Base Portal), integrated, and stored in a graph database, which will allow to uncover patterns that are difficult to detect using traditional knowledge representation;
- Legal regulations on Portuguese Public Procurement Law and audit guidelines (e.g. OECD guidelines) will be represented in the form of production rules, and will be used by a Rule Engine to automatically tag entities and relationships;

- Network analysis algorithms (e.g. centrality, density, PageRank) will be used to further enrich the data, and also making potentially suspicious situations more evident;
- A DSS based on a rich graph-oriented visual representation of the data is being developed, that will significantly improve the way auditors interact with the data. Auditors will also use this interface to validate the labels provided by the Rule Engine, thus actively contributing with knowledge;
- A ML mechanism, powered by the labelled data, will be developed to provide additional insights. The goal goes beyond that of pinpointing individual edges or nodes as potential fraud: we aim to find groups of transactions or networks that might be associated to fraudulent behaviors.

2   Portuguese Public Procurement

The issue of public procurement is regulated in several instruments of International Law and European Union Law and, in Portugal, the Code of Public Contracts (Decree-Law no. 18/2008, of 29 January and subsequent changes)[1]. Essentially, the public contract is an agreement formed by two or more declarations of will (that of a public contractor and that of a contractor), which are directed to the production of certain legal effects [2], such as the acquisition of goods or services, public works contracts, etc.

The Public Procurement market represents a considerable economic weight in the European Union's Internal Market. In the communication COM (2011) 308 final, the Commission stated that, in 2009, "public expenditure on works, goods and services represented approximately 19% of the EU's GDP", realizing that "one fifth of this expenditure is covered by the scope of EU public procurement directives (around € 420 billion, or 3.6% of EU GDP)". For this reason, and in order to take full advantage of its potential, the need to safeguard an open market principle and effective competition within the scope of public procurement procedures was soon recognized.

However, corruption prevents the achievement of the intended objectives with the legal regime of public procurement. Corrupt behaviors affect the integrity of public procurement procedures and, consequently, frustrate market opening and effective competition between economic operators. In this context, the pursuit of the public interest (which should always be addressed by the public administration, under penalty of misuse of power - one of the defects that may occur in administrative procedures) is impaired, namely by not implementing the best value for money principle.

It should be noted that the misuse of power can take two forms: (i) the misuse of power for reasons of private interest (occurs when the principally determining reason is to pursue a private interest, whether material or immaterial, of the holder of the issuing body); and (ii) the misuse of power for reasons of public interest (occurs when the principally determining motive is the pursuit of an end that, although not the legal end, is still in the public interest).

---

[1] http://www.pgdlisboa.pt/leis/lei_mostra_articulado.php?nid=2063&tabela=leis.

The measures to prevent corruption in the scope of public procurement, already enshrined in the 2004 Directives, aimed to hinder and impede the development of corruption in the type of administrative procedures referred to here. With the approval of the 2014 Directives, more specifically, Directive 2014/24/EU, the regime in question is reinforced by the enshrining of a set of new measures that aim to facilitate traceability and, consequently, the detection of those behaviors throughout the procedure. The sanctioning measures that find their consecration outside the directives on public procurement also play an extremely important role in combating corruption in public procurement procedures.

In addition, and with regard to Portugal, in the document entitled "Prevention of corruption risks in public procurement", the Council for the Prevention of Corruption recommends, namely, to reinforce its performance in the identification, prevention and management of risks of corruption and related infractions in public contracts, as to their formation and execution, which must substantiate the decision to hire, the choice of procedure, the estimate of the contract value and the choice of the contractor. The council also recommends, in cases of recourse to prior consultation or direct adjustment, that internal control procedures be adopted to ensure compliance with the limits for the formulation of invitations to the same entities.

It is also important to stress that the council highlights the importance of ensuring transparency in public procurement procedures, namely the fulfillment of the obligation to advertise on the public procurement portal, and ensuring that the contract managers have the technical knowledge to enable them to monitor performance of the contracts and for full compliance with other legal obligations. Finally, the council recommends that the public sector's supervisory, control and inspection bodies include, in their actions, the verification of this recommendation on the prevention of corruption risks in public procurement. They must also ensure the operation of mechanisms to control possible conflicts of interest in public procurement, in addition to favoring the use of competitive procedures to the detriment of prior consultation and direct adjustment.

Despite all the initiatives that have been taken, we verify that the integrity of the public procurement procedures is not safeguarded exclusively with the implementation of sanctioning measures; but in addition, through a comprehensive approach aimed at strengthening the prevention of corrupt behavior.

It is undeniable that the EU has made remarkable progress in implementing anti-corruption policies in a global context; and, in particular, within the scope of public procurement procedures. However, the public procurement market continues to represent a highly attractive and permeable domain for the proliferation of corrupt behaviors.

3   Proposed Architecture

The architecture of the proposed system is a layered one. Figure 1 provides an overview of the main layers and its components from the perspective of data flow. The key idea of the layered design is that, as data moves forward in the pipeline, new layers of meaning are added, thus enriching them.

Two different sources of data are considered: legislation and other documents (e.g. OECD audit guidelines) on the one hand, and streaming public data sources describing the stakeholders' transactions on the other.

The former will be used to develop a Rule-based System (RbS) for basic fraud detection. Legal norms and guidelines are modeled as production rules and inserted into a Rules Engine. Rules from different domains can co-exist and be applied according to the domain of the audit being performed.

Streaming data describes stakeholders' transactions. The primary source of data will be the BASE.gov portal[2]. This portal contains public procurement contracts celebrated in Portugal in the past years, according to what is established in the Decree-Law no. 96/2015, of 17 August. These data are complemented with secondary public data sources describing aspects such as stakeholders Economic Activity Code, geographical location, or financial information.

As opposed to most of existing data-oriented approaches, we start at a lower level: transaction data rather than audit data. These data are ingested and stored in a graph database, as opposed to relational databases which are generally used. Stakeholders (e.g. public entities, service providers) are stored as nodes while relationships between them (e.g. contracts) are stored as edges. Given that transaction data will arrive in a streaming fashion, this graph will evolve over time.

Once transaction data are stored in the database, the RbS is used to add a first layer of meaning. Thus, both nodes, relationships or groups of relationships will receive semantic annotations describing concepts such as fraud, or the adherence or not to certain best practices or guidelines.

Next, Network Analysis algorithms are used to add a second layer of meaning. While the RbS will target mostly individual transactions or stakeholders, this layer will add information regarding the shape of the network. That is, how stakeholders relate to each other. Centrality and dispersion measures will be considered, as well as algorithms such as PageRank. Altogether, this layer will allow a Human auditor to more easily identify potential fraud relationships and networks, as well as its most relevant nodes.

Finally, the topmost component of the proposed system is a Decision Support System (DSS). Through this, the auditor will first and foremost be able to have a rich visual representation of the data, in the form of graphs, providing her/him with actionable insights and allowing to make better data-driven decisions based on them. Data is geo-referenced, navigable and explorable. For instance, instead of writing complex queries, the auditor selects one or more stakeholders and/or relationships to start navigating the data. Moreover, parameters such as location, date or properties of the relationships (e.g. monetary value, type of contract) are also be queryable.

Finally, a ML module will add the final layer of meaning to the data. Annotated data already validated by the auditor will be used to train ML models

---

[2] http://www.base.gov.pt/.

on the graph, that will automatically point out potentially interesting networks for the auditor to analyze. Both supervised and unsupervised methods might be used, in order to highlight groups of nodes with suspicious behaviors.

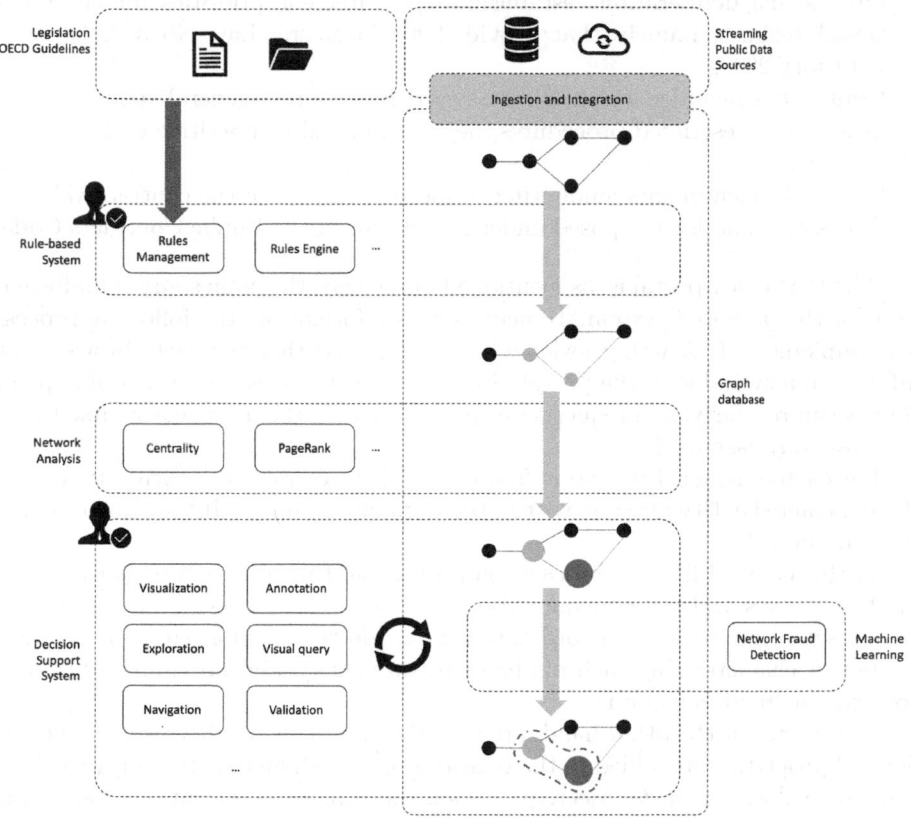

Fig. 1. Main layers and components of the proposed architecture.

3.1 Data Acquisition and Transformation

The main source of data is the BASE.gov portal. This public portal collects information about the formation and performance of Portuguese public procurement contracts. Data is received by this portal from different sources, including the Portuguese Mint and Official Printing Office (INCM), the electronic public procurement platforms, and the contracting entities. The data then becomes publicly available through the web portal.

The portal publishes different information, including:

– The formation and performance of public contracts covered by Part II of the PCC, excluding information on the performance of concession contracts, in

accordance with Decree-Order No 85/2013 of 27 February 2013 amending Decree-Order No 701-F/2008 of 29 July 2008;
- Notices on the opening of procurement procedures and subsequent notices, if any;
- Orders and deliberations establishing procurement priorities under exceptional regimes, namely that provided for in Decree-Law No 34/2009 of 6 February 2009;
- Contracts concluded under direct award procedures (general regime), open procedures, restricted procedures, negotiation and competitive dialogue procedures;
- Contract amendments amounting to more than 15% of the contract value;
- Accessory sanctions imposed under Article 460 of the Public Contracts Code.

The BASE.gov portal is, as mentioned previously, the main source of information for the proposed system. To acquire this information, the following process was implemented. A web crawler was implemented that regularly browses the information available in the portal. Specifically, it browses the results of a query that is run regularly, and requests all the contracts in the most recent time frame (e.g. last day, last week).

For each contract, the system first checks if the contracting authority and/or the contracted entity already exist in the database as nodes. If they do not exist, they are added.

In this case, additional information is obtained from external portals, including the address and the Economic Activity Code. The address is also converted into a set of geographical coordinates. This additional information is relevant for better characterizing each public or private entity, and providing additional context for fraud detection.

Contracts, on the other hand, are stored as relationships between the nodes. Several properties are added to the relationship, which include the contract date, contract value, contract object, published date, place of execution, competing entities, among others. The distance between the two entities involved in the contract is also added as a property of the contract.

The type of the entities is given by the direction of the relationships: an outgoing relationship means that the node is a contracting authority, while an incoming relationship means that it is a contracted entity. Figure 2 shows an excerpt of the data in the database. A contract is selected and the bottom shows some of the properties of this contract. This kind of representation allows to quickly identify, for example, the central nodes: contracting authorities that make a significant number of contracts, or contracted entities that win many contracts.

All this information is added with the goal of enriching the context that the auditor will work in. Indeed, the auditor has access to significantly more data than in the BASE.gov portal. This allows for a richer and more efficient decision-making process.

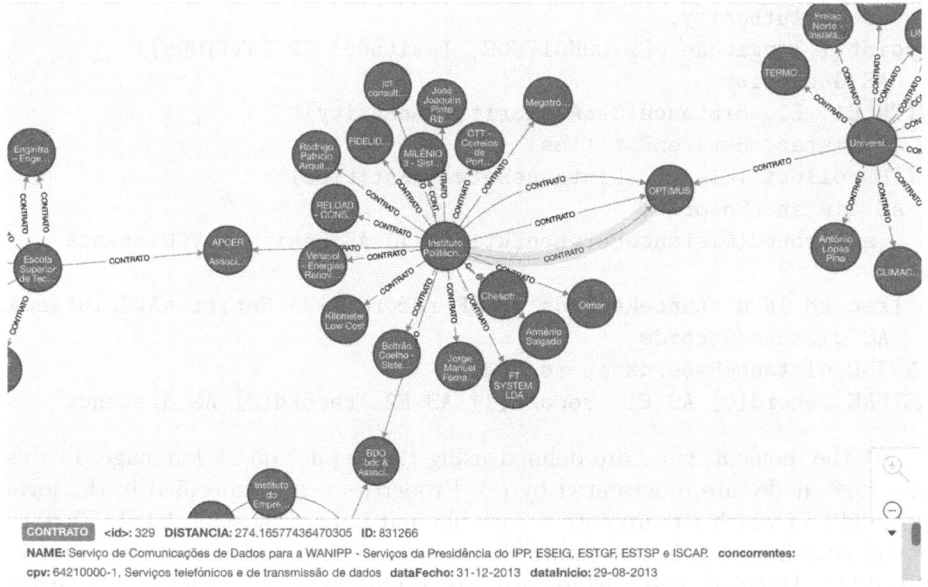

Fig. 2. Modeling of the information as a graph: nodes represent entities (contracting authorities or contracted entities), relationships represent individual contracts.

3.2 Rules Engine

The Rules layer provides the system with the ability to define custom rules, either for fraud detection or for any other information need. In this context, a rule is a way for users to define which information they want to generate and/or highlight.

At regular intervals, the Rules Engine will take all the rules, and determine to which nodes or relationships in the database they apply. Whenever one or more nodes and/or relationships match the condition(s) of the rule, those elements are tagged with the rule's identifier. This allows the elements in the database to be searched by the rules they comply with, making it easier to identify these elements in the Decision layer.

Rules can be translated from the legislation or audit guidelines, or they can be custom rules defined by the auditor, to highlight specific needs of information. For instance, to find all the contracts for a given contracting authority whose value is an outlier when compared against the value of all its contracts.

The following listing details, as an example, one such rule: to highlight all the contracts (and respective relationships) whose distance is above the average distance of the contracts of a given Contracting Authority.

```
MATCH (E1:ENTITY) -[C:CONTRATO]->(E2:ENTITY)
WITH
  E1, E2, point({ longitude: E1.LONGITUDE, latitude: E1.LATITUDE })
```

```
   AS GeoAuthority,
 point({ longitude: E2.LONGITUDE, latitude: E2.LATITUDE})
   AS GeoEntity
WITH E1, E2, distance(GeoAuthority,GeoEntity)
   AS DistanceBetweenEntities
WITH collect([E1, E2, DistanceBetweenEntities])
   AS distanceRecords,
     avg(round(DistanceBetweenEntities)) AS EntitiesAVGDistance
WITH
   [record IN distanceRecords WHERE record[2] > EntitiesAVGDistance]
   AS distanceRecords
UNWIND distanceRecords as record
RETURN record[0] AS E1, record[1] AS E2, record[2] AS distance
```

At the moment, rules are defined using the Cypher query language. In this language, nodes are represented by ( ). Properties can be specified in the form (l:Label), in which l represents a variable and Label a specific label. That is, it will search for labels containing that specific Label, and provide its value in variable l. Directed relationships are represented by -[ ]-¿, in which the arrow gives the intended direction of the relationship.

So the expression

```
MATCH (E1:ENTITY) -[C:CONTRATO]->(E2:ENTITY)
```

will match all the relationships C of type CONTRACT starting at nodes E1 with the property ENTITY and ending at nodes E2 with the same property. Next, it creates two variables (GeoAuthority and GeoEntity) containing the longitude and latitude of each entity, and it uses these two variables to calculate the distance between the entities. Finally, the WHERE clause selects those cases in which the distance is above the average distance of the contracts of that authority.

In the future, a more natural rules language will be developed that will be automatically compiled and translated to Cypher, so that the definition of rules does not require knowledge of this specific language and can be more easily done by any user.

3.3 Decision Support

The topmost layer of the architecture is the Decision-Support one. The main goal of this layer is to facilitate decision processes of auditors and other users, such as those preparing framework agreements or other contracts in accordance with the Portuguese Public Procurement Law. The system can thus be used in two main ways. On the one hand, it can be used by auditors to search for fraud in existing contracts, in a reactive manner. On the other hand, it can be used in a proactive and preventive way, by allowing entities currently negotiating new contracts to ascertain their legal validity. This is especially useful in framework agreements, which combine multiple entities.

The analysis of these and many other factors is greatly facilitated by the Decision Support layer, especially when compared with the traditional way of doing it by searching for and analyzing each contract individually. Indeed, the developed user interface allows for all this information to be analyzed and assessed in a visual and geo-referenced way, which significantly facilitates this kind of analyses.

All in all, the system will contribute to more transparency in public procurement, by facilitating access to information and compliance audits.

The User Interface (UI) allows the user to navigate the map and see the entities in their respective locations, as well as the contracts between them. The user can thus quickly spot how different regions interact in terms of contracts, or which are the main players in each region. It is also clearly visible when contracting authorities make contracts with very distant contracted entities.

The user can also click on an entity or on a contract to see its details. Figure 3 shows the UI when a specific contract is selected and its information shown.

The user can also use a simple or advanced search, to refine the nodes and/or connections that are shown in the UI. Figure 4 shows the corresponding search form, in which the user can search by entity names, by dates, by contract values, among others. The user can then easily navigate the map and explore the data in an intuitive and contextualized way.

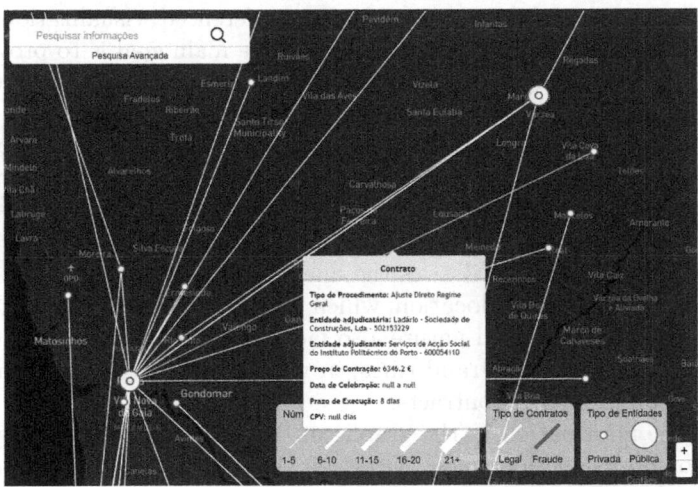

Fig. 3. UI when a contract is selected and its details shown.

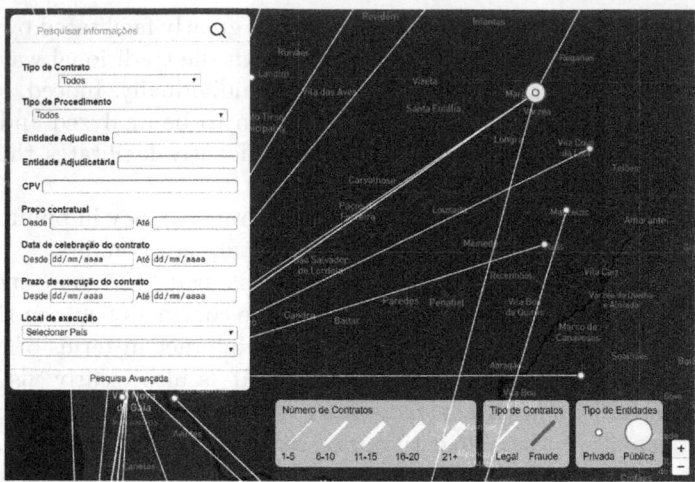

Fig. 4. UI with the advanced search functionality active.

4   Conclusions and Future Work

In this paper we presented a prototype of a system that combines a graph-oriented database, a rules engine and a geo-referenced visualization layer to improve transparency in public procurement. The main goal is to provide users (e.g. auditors, contracting entities) with relevant information to assess the validity of new or existing contracts.

This is achieved in several different ways. First, data is obtained from public portals, processed and integrated so that the user can access it in an efficient and seamless manner, without the need to resort to different tools or sources.

Second, the information is provided in a geo-referenced way, in the form of a graph, which significantly facilitates the analysis of the contracts as it provides notions such as distance and location, which are very relevant in fraud detection. Moreover, this graph-oriented nature also facilitates the detection of more complex types of fraud, such as fraud rings, in which the onus is not on individual contracts but on groups of contracts.

Finally, the information in the graph can be enriched by the users through the definition of production rules. Users can thus define specific rules and their consequent, that is, what to do with the nodes or relationships for which the condition of a given rule holds true. This allows not only to implement audit rules according to the legislation or to audit guidelines, but also for users to implement their own rules, which should reflect information needs. Users can thus tailor the information they see in the UI to their own needs. In the future, a more user-friendly language will be developed, so that rules are easier to define.

All in all, this system facilitates the access to information regarding public procurement, contributing in turn to the transparency and fair competition between all the stakeholders.

Acknowledgments. This work was supported by the Northern Regional Operational Program, Portugal 2020 and European Union, trough the European Regional Development Fund (ERDF) and Fundos Europeus Estruturais e de Investimento (FEEI) in the scope of project 39900 - 31/SI/2017, and by FCT - Fundação para a Ciência e a Tecnologia, through project UIDB/04728/2020.

References

1. Bolton, R.J., Hand, D.J.: Statistical fraud detection: a review. Stat. Sci. **17**, 235–249 (2002)
2. Gonçalves, P.C.: Direito dos contratos públicos. Almedina, Coimbra (2018)
3. Hansen, J.V., McDonald, J., Messier Jr., W.F., Bell, T.B.: A generalized qualitative-response model and the analysis of management fraud. Manage. Sci. **42**(7), 1022–1032 (1996)
4. Hilas, C.S.: Designing an expert system for fraud detection in private telecommunications networks. Expert Syst. Appl. **36**(9), 11559–11569 (2009)
5. Krambia-Kapardis, M., Christodoulou, C., Agathocleous, M.: Neural networks: the panacea in fraud detection? Manag. Auditing J. **25**, 659–678 (2010)
6. Liou, F.M.: Fraudulent financial reporting detection and business failure prediction models: a comparison. Manag. Auditing J. **23**, 650–662 (2008)
7. Phua, C., Lee, V., Smith, K., Gayler, R.: A comprehensive survey of data mining-based fraud detection research. arXiv preprint arXiv:1009.6119 (2010)
8. Raimundo, M.A.: A formação dos contratos públicos (2011)
9. Ravisankar, P., Ravi, V., Rao, G.R., Bose, I.: Detection of financial statement fraud and feature selection using data mining techniques. Decis. Support Syst. **50**(2), 491–500 (2011)
10. Sánchez, D., Vila, M., Cerda, L., Serrano, J.M.: Association rules applied to credit card fraud detection. Expert Syst. Appl. **36**(2), 3630–3640 (2009)
11. Sharma, A., Panigrahi, P.K.: A review of financial accounting fraud detection based on data mining techniques. arXiv preprint arXiv:1309.3944 (2013)
12. Wei, W., Li, J., Cao, L., Ou, Y., Chen, J.: Effective detection of sophisticated online banking fraud on extremely imbalanced data. World Wide Web **16**(4), 449–475 (2013). https://doi.org/10.1007/s11280-012-0178-0
13. Yuan, J., Yuan, C., Deng, X.: The effects of manager compensation and market competition on financial fraud in public companies: an empirical study in china. Int. J. Manage. **25**(2), 322 (2008)
14. Zhou, W., Kapoor, G.: Detecting evolutionary financial statement fraud. Decis. Support Syst. **50**(3), 570–575 (2011)

Biased Language Detection in Court Decisions

Alexandra Guedes Pinto[1,2]ⓘ, Henrique Lopes Cardoso[3,4(✉)]ⓘ,
Isabel Margarida Duarte[1,2]ⓘ, Catarina Vaz Warrot[1,2]ⓘ,
and Rui Sousa-Silva[1,2]ⓘ

[1] Faculdade de Letras da Universidade do Porto (FLUP) via Panorâmica s/n,
4150-564 Porto, Portugal
[2] Centro de Linguística da Universidade do Porto (CLUP), Porto, Portugal
{mapinto,iduarte,awarrot,rssilva}@letras.up.pt
[3] Faculdade de Engenharia da Universidade do Porto (FEUP),
Rua Dr. Roberto Frias, 4200-465 Porto, Portugal
[4] Laboratório de Inteligência Artificial e Ciências de Computadores (LIACC),
Porto, Portugal
hlc@fe.up.pt

Abstract. The Portuguese Commission for Citizenship and Gender
Equality advocates that equality between men and women is a fundamental principle of the Portuguese Constitution. While court decisions
should reflect this principle, a preliminary analysis in cases of gender
violence reveals that this is not always the case. Based on the extensive
literature on subjectivity, modality and bias in Linguistics and in tan-
dem with AI and Natural Language Processing (NLP) techniques, the
research proposed in this paper aims to study the linguistic formulations
that convey bias in court decisions. The goal is to develop a linguistic
model and, subsequently, a tool to automatically detect gender bias in
this text genre. A corpus of a set of legal sentences on gender violence has
been extracted from the public access database of the Portuguese Ministry of Justice (IGFEJ), which can be subject to a manual annotation
process according to a typology of biased categories and structures. By
exploiting the corpus in a supervised machine learning approach while
following the most recent advances in NLP, we aim to deliver the aforementioned tool for biased language detection.

Keywords: Biased language · Court decisions · Discourse analysis ·
Natural Language Processing

1 Introduction

As research in the fields of both theoretical and applied linguistics demonstrates,
the linguistic expression of subjectivity, which can be expected in some text genres and not tolerated in others, is a complex phenomenon that can often express
bias [13,15,16,20,21,23]. In fact, the marking of a biased point of view in texts
has been studied from several linguistic perspectives, some of the most recent

© Springer Nature Switzerland AG 2020
C. Analide et al. (Eds.): IDEAL 2020, LNCS 12490, pp. 402–410, 2020.
https://doi.org/10.1007/978-3-030-62365-4_38

being those related to Computational Linguistics, with a focus on the development of automatic detection models in various types of discourse and text genres [5,8,22,24,31]. Computational Linguistic studies have led to many different applications, including sentiment analysis [19], the detection of radicalized discourse on social networks [6,27] and hyperpartisan news [7], the identification of markers of authorship [28], among others.

The expression of gender bias has become a particularly relevant societal issue. Through political measures and laws, Governments attempt to speed greater equality between genders. The official website of the Portuguese Commission for Citizenship and Gender Equality (CIG)[1] reads that equality between women and men is a fundamental principle of the Portuguese Constitution, and further declares that the promotion of such equality is a fundamental duty of the State. In this context, the National Strategy for Equality and Non-Discrimination "Portugal + Igual" (*Portugal More Equal*), was published in the Official Journal[2]. By recognizing these principles as a requirement for a sustainable future, CIG has defined a course of action until 2030[3]. The integration of the gender mainstreaming dimension in policies emerges as a principle of good governance, where the official discourse plays a decisive role as the driving force of social change. Court decisions are one of the forms of such official discourse, where several types of bias persist, including gender bias. Some of these instances have attracted media attention, but many others remain unknown, thus helping perpetuate unequal views and consequently contributing to an unjust and unfair administration of justice and to the construction of unsustainable gender identities.

This paper aims to present a research proposal that firstly engages in the development of a linguistic model and subsequently in the production of a computational tool to automatically detect gender bias in court decisions. This will in turn contribute to raising social awareness to this problem. In Sect. 2 we review relevant literature on the topic. In Sect. 3 we draw a research methodology for addressing the problem. Finally, we conclude in Sect. 4.

2   Literature Review

Subjectivity in discourse has always been a topic of concern for Linguists. Benveniste [2] brought the issue to the fore by drawing attention to the most basic forms of the subject's inscription in language. Modality and modalization also problematize this core issue when analyzing grammaticalization of the speaker's opinions and attitudes in language and discourse [17,18]. Kerbrat-Orecchioni [13] discusses "subjectivèmes", a category where all the grammatical instances of the subject's footprints in the utterance are integrated. The diverse theorizations about the functions of language (from linguistic functionalism to pragmatics) always include propositions in which the subject's presence is stronger, giving

---

[1] https://www.cig.gov.pt/a-cig/missao/.
[2] Diário da República, Resolution No. 61/2018, of 21 May.
[3] https://www.cig.gov.pt/documentacao-de-referencia/doc/portugal-mais-igual/.

rise to the emotive function and to expressive illocutionary acts. Similarly, in the field of text linguistics, subjectivity has been confirmed to be a linguistic index that differentiates genres. More recently, approaches to the phenomena of stance and hedging [10,11], associated with politeness and attenuation [4], in various modes (oral, written), types of discourse (scientific, political) and text genres are seen as privileged ways for the subject to construct their *ethos* and their relationship with the interlocutor. Subjectivity is, in turn, intimately related to argumentativeness, to the extent that many of the utterances with strong subjectivity marks are simultaneously utterances with a high degree of argumentativeness (orientation to act on the interlocutor).

In Law, it is believed that judges are required to justify their argumentation in order to support their decisions. Thus, studies on argumentation, as well as in Linguistics in relation to the Law, are also relevant to understand bias detection and how a biased point of view is linguistically construed in texts.

Recently, the interest of Computational Linguistics and NLP in the construction of subjectivity in discourse has grown, bringing the development of models for its automatic analysis. These models have proved to be particularly relevant in the treatment of large volumes of data, such as the ones generated on the web, which made it possible to automatically detect subjectivity in texts, with the subsequent extraction, classification and summarization of this information.

Linguistic and social constructions such as the expression of feelings [31], subjectivity [24], opinions [19], factual data, beliefs [22], politeness, respect and power [22], bias [8,16,23], among others, are being treated in computational research with very interesting and useful results. The grammaticalization of speaker opinions and attitudes extends to all linguistic levels (phonological, morphological, lexical, syntactic, semantic, discursive) and only an integrated model can respond to the challenge of manually or automatically detecting bias in texts.

Furthermore, court decisions, a genre in which subjectivity and bias are not expected to unveil explicitly, require a model that differs from the ones already built for other discourse types and text genres, such as opinion articles or posts in social media, where stance and evaluative expressions are prototypical. Therefore, it is necessary to integrate the analysis of diverse linguistic levels. The lexical level, by constructing a specific lexicon of sentiment in gender violence court decisions, as other authors have done for other text genres [5,31]; the morphological level, including the internal structure of words and their modification through minimizing and intensifying expressions, as demonstrated by attenuation and stance phenomena [4,11]; the syntactic-semantic level, in cases of presuppositional, factive predicates [12,14] and cleft structures; and, finally, the pragmatic and discursive level, involving the expression of modalities and illocutionary acts; argumentation marks and argumentative moves, among others.

Gender issues and gender bias have been explored from different theoretical frameworks in different areas of the social sciences. This proposal adopts a discursive perspective , by seeing gender identity as a historical-discursive construction and discourse as a privileged form of stereotypical gender representation.

Social categorizations and stereotypes, which may promote discrimination and bias, are sustained or challenged through discourse, and "surface in (often subtle and largely implicit) linguistic biases" [3], sheltered in various discursive structures foreseen in the current study. From a linguistics perspective, gender bias has already been approached in previous works from the authors, both in court decisions [21] and media discourse [15], together with other types of ideological bias [20]. A related problem is the study of politeness markers in court decisions [29]. From the computational side, computational forensic linguistics [27], bias and propaganda detection in text [6,7] are techniques that clearly see an application in this domain.

3   Methodology

Detecting biased language in court decisions demands a multifaceted approach to the problem, building on existing work. The development of an automated tool for the task calls for NLP and machine learning techniques, for which annotated corpora must be made available.

In Portugal, a public access database of court decisions is made available by the Ministry of Justice (IGFEJ)[4]. By searching the database on the topic of gender violence, we are able to obtain a corpus with which we aim to conduct an annotation process, leading to a tagged corpus of court decisions. Such annotation process must be guided by a consolidated linguistic model that is able to capture the targeted bias-inducing linguistic phenomena. With a qualitatively sound annotated corpus, we are then able to apply NLP and machine learning techniques to develop an automated tool for bias detection in court decisions.

We thus propose the following methodology:

1. Collect a corpus of court decisions from the database of the Ministry of Justice (IGFEJ), focused on the topic of gender violence;
2. Adopt a set of potentially bias-inducing linguistic phenomena, based on research in linguistics, computational linguistics and NLP, in order to produce a typology of subjectivity and bias adjusted to Portuguese and to a discursive perspective;
3. Build an annotated corpus of court decisions with bias-inducing words or structures, in accordance with the typology previously set;
4. Develop computational models to automatically detect bias and subjectivity in court decisions, building on the annotated corpus and other resources, together with NLP and machine learning algorithms and techniques.

3.1   Data Collection

To retrieve judgments where gender issues arise more openly, certain descriptors available in the database can be used, such as "domestic violence", "rape", "mistreatment", "sexual coercion", or "sexual abuse". The issue of data protection

---

[4] http://www.dgsi.pt/.

is safeguarded by the fact that public court decisions have already been purged from any personal data prior to their inclusion in the Ministry's database.

Furthermore, certain parts of court decisions are particularly relevant in terms of expression of bias, such as the "setting of the matter of fact" and the "reasoning". The former is aimed at exposing the facts proven, while the latter is aimed at framing and justifying the decision of the judge or panel of judges.

The corpus is planned to include decisions from Courts of First Instance, Courts of Appeal and the Supreme Court of Justice, and cover a wide geographic distribution in the Portuguese territory. Such a physical and hierarchical distribution of data allows for bias analyses to be run on a territorial or court-degree basis.

3.2   Linguistic Model

The design and stabilization of a linguistic model integrating bias-inducing linguistic phenomena is crucial for this research. Such a model must build on studies in linguistics, computational linguistics and NLP, and form a typology for subjectivity and bias adjusted to Portuguese and to a discourse analytic perspective. Also, the need to delve into embedded bias in sophisticated discursive structures such as the ones used in court decisions, where judges are expected to develop argumentative moves, implies a set of linguistic categories adjusted to this specific text genre. It is important to ensure that the linguistic model covers all the linguistic manifestations of bias in court decisions to be analyzed and, as such, it should include phenomena at all discourse levels, as has been stated in Sect. 2. Additionally, categories in the typology should, to the extent possible, avoid fuzzy borders that will hamper the annotation process.

Other taxonomies will be used as a starting point, namely of subjectivity [13], attenuation and stance phenomena [4,10,21] and from previous work on sentiment, subjectivity and bias detection in computational linguistics [23,24,31].

3.3   Corpus Annotation

Corpus annotation is an endeavor whose preparation cannot be taken lightly. As such, the consolidated linguistic model should be used to prepare a set of clear annotation guidelines that will drive the manual annotation of the corpus. The selection of the annotator base is also determinant of the quality of the annotation outcome. Although some works on corpus annotation start to resort to crowd-sourcing, we contend that in the present project specialized linguistic background is a must for recruited annotators.

For bias detection, we can consider the annotation process as a sequence of two steps: (i) the identification of potentially biased words or sequences; and (ii) the labeling of these words or sequences with the respective category. Corpus annotation projects typically elicit several annotations per document, for the sake of measuring human agreement on the task. Running the annotation process in a stepped manner enables drilling down the assessment of agreement, which provides deeper insights into the ways the obtained corpus can be used.

Traditionally, the output of a corpus annotation project is a so-called gold-standard corpus, based on the annotations produced throughout the annotation process. The corpus is evaluated according to inter-annotator agreement metrics, in order to make it suitable for supervised machine learning. A recent trend, however, takes human interpretation in refined texts to be inherently subjective in itself [1], focusing instead on new ways of accommodating such diversity in an annotated corpus [32].

3.4   Bias Detection Model

Given an annotated corpus, it is feasible to employ machine learning algorithms in a supervised fashion. The linguistic model defined, together with existing bias, sentiment, or subjectivity lexicons, can be explored to boost the performance of the computational models to be developed.

Traditional feature-based approaches for addressing NLP tasks are giving way to advanced neural-based models, grounded on word embeddings, language models and deep learning architectures. Still, feature-based approaches work well in certain domains, by exploring features drawn from expert knowledge and linguistic theories, in particular when not enough data is available to render an approach based on deep learning (typically a data-hungry paradigm) promising.

Given the two-level annotation approach mentioned in Sect. 3.3, two kinds of NLP tasks can be pinpointed. The identification of bias-inducing words or expressions can be modeled as a sequence labeling problem, often addressed through Hidden Markov Models or Conditional Random Fields (CRF). While sequence labeling allows for classifying tokens in a number of existing labels (part of the typology specified by the linguistic model addressed in Sect. 3.2), this assignment can also be modeled as a classification task, for which techniques such as Support Vector Machines have shown competitive performance in several domains. Conceptualizing the bias detection model as being composed of these two steps is sensible, as it allows us to improve in any of the steps in an independent manner, adjusting to the collected and annotated data.

Recent trends in NLP, including deep learning approaches based on cross-lingual word embeddings and language models [26,30], are promising approaches when enough data is available for the task. While word embeddings can and have been used as features for CRF models, neural models typically generalize better to unseen data. In particular, the Transformer-based BERT architecture [9] has proven to be very versatile and achieved state-of-the-art performance in several NLP tasks. BERT can be fine-tuned to virtually any NLP task. This kind of approach also enables the exploitation of auxiliary datasets, which can be used to bootstrap learning in these data-intensive models – a technique typically known as transfer learning [25]. Examples include the Gender Bias shared task[5], the Hyperpartisan News Detection shared task[6], or the EmotionX shared task[7].

4 Conclusions

Drawing upon previous works and models, this research proposal innovates by transferring the study of subjectivity and linguistic bias to the under-explored genre of court decisions. To the best of our knowledge, it is also innovative in that it proposes a computational tool for the automatic analysis of this phenomenon. If applied to the analysis of published or in-progress court decisions, such a tool would allow for the detection of potentially bias-inducing linguistic categories and structures, enabling the rephrasing of biased strings. In short, it can contribute to an increased social awareness of bias and a more egalitarian discourse in court decisions, specifically in cases of gender-based violence.

We envision the development of a computer application whose use-cases are at least two-fold: on the one hand, the analysis of pre-written court decisions; on the other, the assistive writing of an impartial and language-unbiased legal sentence. Such functionalities will be very helpful, we believe, for both law practitioners (including lawyers and judges) and legal trainees. Endowing these players with this kind of tools will, hopefully, promote more democratic and unbiased practices in the administration of justice in matters of gender equality, thereby demonstrating the social relevance of this research project.

Acknowledgements.. This research is partially supported by CLUP(FCT/UID/ LIN/00022/2019), by LIACC (FCT/UID/CEC/0027/2020) and by project DARGMINTS (POCI/01/ 0145/FEDER/031460), funded by Fundação para a Ciência e a Tecnologia (FCT).

References

1. Aroyo, L., Welty, C.: Truth is a lie: crowd truth and the seven myths of human annotation. AI Mag. **36**(1), 15–24 (2015). https://doi.org/10.1609/aimag.v36i1. 2564
2. Benveniste, E.: Problèmes de linguistique gènèrale, vol. I. Gallimard, Paris (1966)
3. Beukeboom, C., Burgers, C.: Linguistic bias. In: Giles, H., Harwood, J. (eds.) Oxford Research Encyclopedia of Communication, pp. 1–21. Oxford Research Encyclopedias. Oxford University Press, Oxford (2017). https://doi.org/10.1093/ acrefore/9780190228613.013.439
4. Briz, A., Albelda Marco, M.: Una propuesta teórica y metodológica para el análisis de la atenuación lingüística en español y portugués. la base de un proyecto común (es.por.atenuaciÓn). Onomázein Revista de lingüística, filología y traducción 28, 288–319 (12 2013). https://doi.org/10.7764/onomazein.28.21

---

[5] https://genderbiasnlp.talp.cat/gebnlp2020/shared-task/.
[6] https://pan.webis.de/semeval19/semeval19-web/.
[7] http://doraemon.iis.sinica.edu.tw/emotionlines/challenge.html.

5. Conrad, A., Wiebe, J., Hwa, R.: Recognizing arguing subjectivity and argument tags. In: Proceedings Workshop on Extra-Propositional Aspects of Meaning in Computational Linguistics, Jeju, Republic of Korea, pp. 80–88. ACL, July 2012

6. Cruz, A.F., Rocha, G., Lopes Cardoso, H.: On sentence representations for propaganda detection: From handcrafted features to word embeddings. In: Proceedings 2nd Workshop on Natural Language Processing for Internet Freedom: Censorship, Disinformation, and Propaganda, Hong Kong, China, pp. 107–112. ACL, November 2019. https://doi.org/10.18653/v1/D19-5015

7. Cruz, A.F., Rocha, G., Lopes Cardoso, H.: On document representations for detection of biased news articles. In: Proceedings 35th Annual ACM Symposium on Applied Computing, SAC 2020, New York, NY, USA, pp. 892–899. Association for Computing Machinery (2020). https://doi.org/10.1145/3341105.3374025

8. De-Arteaga, M., et al.: Bias in bios: a case study of semantic representation bias in a high-stakes setting. In: Proceedings Conference on Fairness, Accountability, and Transparency, pp. 120–128. ACM, New York, NY, USA (2019). https://doi.org/10.1145/3287560.3287572

9. Devlin, J., Chang, M., Lee, K., Toutanova, K.: BERT: pre-training of deep bidirectional transformers for language understanding. CoRR abs/1810.04805 (2018), http://arxiv.org/abs/1810.04805

10. Hyland, K.: Hedging in Scientific Research Articles. John Benjamins Publishing, Amsterdam (1998)

11. Hyland, K.: Constructing proximity: relating to readers in popular and professional science. J. Engl. Acad. Purp. 9(2), 116–127 (2010). https://doi.org/10.1016/j.jeap.2010.02.003

12. Karttunen, L.: Implicative verbs. Language 47(2), 340–358 (1971)

13. Kerbrat-Orecchioni, C.: L'énonciation: De la subjectivité dans le langage. Armand Colin (1980)

14. Kiparsky, P., Kiparsky, C.: Fact. In: Bierwisch, M., Heidolph, K.F. (eds.) Progress in Linguistics, pp. 143–173. Mouton Publishers, The Hague (1970)

15. Marques, A., Duarte, I.M., Pinto, A.G., Pinho, C.: A construção da identidade da mulher em revistas do Estado Novo. Ex aequo 39, 71–88 (2019)

16. Menegatti, M., Rubini, M.: Gender bias and sexism in language. In: Nussbaum, J.L. (ed.) Oxford Research Encyclopedia of Communication. Oxford University Press, Oxford (2017)

17. Oliveira, F.: Modalidade e modo. In: Raposo, E.P. (ed.) Gramática Portuguesa, vol. I. Editorial Caminho, Lisbon (2004)

18. Oliveira, F., Mendes, A.: Modalidade. In: Raposo, E.P. (ed.) Gramática do Português, vol. I. Fundação Calouste Gulbenkian, Lisbon (2013)

19. Pang, B., Lee, L.: Opinion mining and sentiment analysis. Foundations and Trends in Information Retrieval 2, 1–135 (2008). https://doi.org/10.1561/1500000011

20. Pinto, A.: A retórica do eu e do outro - the othering: A gramática da identidade no discurso político. Estudos do Discurso: Caminhos e Tendências, p. 25, January 2016

21. Pinto, A.G.: A construção ideológica da mulher num acórdão sobre violência doméstica. In: Savoir et pouvoir dans un monde polycentrique: les discours aux prismes des langues, des cultures et des espaces : Congrès DNC3-ALED (2019)

22. Prabhakaran, V., Hutchinson, B., Mitchell, M.: Perturbation sensitivity analysis to detect unintended model biases. In: Proceedings 2019 Conference on Empirical Methods in Natural Language Processing and the 9th International Joint Conference on Natural Language Processing (EMNLP-IJCNLP), Hong Kong, China, pp. 5740–5745. ACL, November 2019. https://doi.org/10.18653/v1/D19-1578

23. Recasens, M., Danescu-Niculescu-Mizil, C., Jurafsky, D.: Linguistic models for analyzing and detecting biased language. In: Proceedings 51st Annual Meeting of the Association for Computational Linguistics, Sofia, Bulgaria, vol. 1, pp. 1650–1659. ACL, August 2013

24. Riloff, E., Wiebe, J.: Learning extraction patterns for subjective expressions. In: Proceedings 2003 Conference on Empirical Methods in Natural Language Processing, pp. 105–112 (2003)

25. Ruder, S., Peters, M.E., Swayamdipta, S., Wolf, T.: Transfer learning in natural language processing. In: Proceedings 2019 Conference of the North American Chapter of the Association for Computational Linguistics: Tutorials, pp. 15–18. ACL (2019)

26. Ruder, S., Vulić, I., Søgaard, A.: A survey of cross-lingual word embedding models. J. Artif. Intell. Res. **65**(1), 569–630 (2019). https://doi.org/10.1613/jair.1.11640

27. Sousa-Silva, R.: Computational forensic linguistics: an overview of computational applications in forensic contexts. Language and Law **5**, 118–143 (2018)

28. Sousa Silva, R., Laboreiro, G., Sarmento, L., Grant, T., Oliveira, E., Maia, B.: 'twazn me!!!;('automatic authorship analysis of micro-blogging messages. In: Muñoz, R., Montoyo, A., Métais, E. (eds.) NLDB 2011. LNCS, vol. 6716, pp. 161–168. Springer, Heidelberg (2011). https://doi.org/10.1007/978-3-642-22327-3_16

29. Teixeira, J., Pinto, A.: Marcas de cortesia no género textual acórdão e o seu contributo para a construção da relação interacional. REDIS: Revista de Estudos do Discurso **7**, 142–162 (2018). https://doi.org/10.21747/21833958/red7a6

30. Wada, T., Iwata, T., Matsumoto, Y.: Unsupervised multilingual word embedding with limited resources using neural language models. In: Proceedings 57th Annual Meeting of the Association for Computational Linguistics, Florence, Italy, pp. 3113–3124. ACL, July 2019. https://doi.org/10.18653/v1/P19-1300

31. Wiebe, J., Wilson, T., Bruce, R., Bell, M., Martin, M.: Learning subjective language. Comput. Linguist. **30**(3), 277–308 (2004). https://doi.org/10.1162/0891201041850885

32. Yan, Y., et al.: Modeling annotator expertise: learning when everybody knows a bit of something. In: Teh, Y.W., Titterington, M. (eds.) Proceedings 13th International Conference on Artificial Intelligence and Statistics. Proceedings Machine Learning Research, vol. 9, pp. 932–939 (2010)

Special Session on Machine Learning Algorithms for Hard Problems

A One-by-One Method for Community Detection in Attributed Networks

Soroosh Shalileh[✉][ID] and Boris Mirkin[ID]

Department of Data Science and Artificial Intelligence NRU HSE Moscow Russian Federation, 11, Pokrovski Boulevard, Moscow 109028, Russian Federation
sr.shalileh@gmail.com, bmirkin@hse.ru
https://cs.hse.ru/

Abstract. The problem of community detection in a network with features at its nodes takes into account both the graph structure and node features. The goal is to find relatively dense groups of interconnected entities sharing some features in common. We apply the so-called data recovery approach to the problem by combining the least-squares recovery criteria for both, the graph structure and node features. In this way, we obtain a new clustering criterion and a corresponding algorithm for finding clusters/communities one-by-one. We show that our proposed method is effective on real-world data, as well as on synthetic data involving either only quantitative features or only categorical attributes or both. Our algorithm appears competitive against state-of-the-art algorithms.

Keywords: Attributed network · Cluster analysis · Community detection · Least squares criterion · One by one clustering

1 Introduction: Previous Work and Motivation

Community detection is a popular field of data science with various applications ranging from sociology to biology to computer science. Recently this concept was extended from flat and weighted networks to networks with a feature space associated with its nodes, these are referred to as attributed (or feature-rich) networks [6]. A community is a group, or cluster, of densely interconnected nodes that are similar in the feature space too.

There have been published a number of papers proposing various approaches to identifying communities in attributed networks (see recent reviews in [6] and [3]). They naturally fall in three groups: (a) those heuristically transforming the feature-based data to augment the network format, (b) those heuristically convering the data to the features only format, and (c) those involving, usually, a probabilistic model of the phenomenon to apply the maximum likelihood principle for estimating its parameters. A typical method within approach (a) or (b) combines a number of heuristical approaches, thus involving a number of unsubstantiated parameters which are rather difficult to systematize, the more

© Springer Nature Switzerland AG 2020
C. Analide et al. (Eds.): IDEAL 2020, LNCS 12490, pp. 413–422, 2020.
https://doi.org/10.1007/978-3-030-62365-4_39

so to put to testing. Most interesting approaches in the modeling group (c) are represented by methods in [17] and [14]. The former statistically models inter-relation between the network structure and node attributes, the latter involves Bayesian inferences.

Our approach relates to that of modeling, except that we model the data rather than the process of data generation. Specifically, our data-driven model assumes a hidden partition of the node set in non-overlapping communities and parameters encoding the average within-community link intensity and feature central points. To find this partition and parameters, a least-squares data approx-imation criterion is defined. To fit this criterion, a greedy-wise procedure of find-ing clusters one-by-one is applied. This approach as already proved successful in application to both feature data only and network/ data only [2,10].

The rest of the paper is organized as follows. We describe our model and algorithm in Sect. 2. In Sect. 3, we describe the setting of our experiments. In Sect. 4, we describe results of our experiments to validate our method and compare it with competition. We draw conclusions in Sect. 5.

The authors are indebted to the anonymous reviewers whose comments helped them to improve the presentation.

2   Least-Squares Criterion

Let us consider a dataset represented by two matrices: a symmetric $N \times N$ network adjacency matrix $P = (p_{ij})$, where p_{ij} can be any reals, and by an $N \times V$ entity-to-feature matrix $Y = (y_{iv})$ with $i \in I$, I being an N-element entity set.

We assume that there is a partition $\mathbf{S} = \{S_1, S_2, ..., S_K\}$ of I in K non-overlapping communities, a.k.a. clusters, with a binary membership vector $s_k = (s_{ik})$, $k = 1, 2, ..., K$, so that $s_{ik} = 1$ for $i \in S_k$, and $s_{ik} = 0$, otherwise. The cluster S_k is assigned with a V-dimensional center vector $c_k = (c_{kv})$ and a positive network intensity weight λ_k.

According to the least-squares principle, "right" membership vectors s_k, com-munity centers c_k and intensity weights λ_k are minimizers of the summary least-squares criterion:

$$F(\lambda_k, s_k, c_k) = \rho \sum_{k=1}^{K} \sum_{iv} (y_{iv} - c_{kv}s_{ik})^2 + \xi \sum_{k=1}^{K} \sum_{ij} (p_{ij} - \lambda_k s_{ik}s_{jk})^2 \quad (1)$$

The factors ρ and ξ in Eq. (1) are expert-driven constants to balance the two sources of data, i.e features and networks.

To use a one-by-one clustering strategy [11] here, let us denote an individ-ual community by S; its center in feature space, by c; and the corresponding intensity weight, by λ (just removing the index, k, for convenience). The extent of fit between the community and the dataset will be the corresponding part of criterion in (1):

$$F(\lambda, c, s) = \rho \sum_{i,v} (y_{iv} - c_v s_i)^2 + \xi \sum_{i,j} (p_{ij} - \lambda s_i s_j)^2 \quad (2)$$

The problem: given matrices $P = (p_{ij})$ and $Y = (y_{iv})$, find binary s, as well as real-valued λ and $c = (c_v)$, minimizing criterion (2). It is easy to prove that the optimal real-valued c_v is equal to the within-S mean of feature v, and the optimal intensity value λ is equal to the mean within-cluster link value:

$$c_v = \frac{\sum_{i \in S} y_{iv}}{|S|}; \quad \lambda = \frac{\sum_{i,j \in S} p_{ij}}{|S|^2} \tag{3}$$

Criterion (2) can be further reformulated as:

$$F(s) = \rho \sum_{i,v} y_{iv}^2 - 2\rho \sum_{i,v} y_{iv} c_v s_i + \rho \sum_v c_v^2 \sum_i s_i^2$$
$$+ \xi \sum_{i,j} p_{ij}^2 - 2\xi\lambda \sum_{i,j} p_{ij} s_i s_j + \xi\lambda^2 \sum_i s_i^2 \sum_j s_j^2 \tag{4}$$

The items $T(Y) = \sum_{i,v} y_{iv}{}^2$ and $T(P) = \sum_{ij} p_{i,j}^2$ in (4) express quadratic scatters of data matrices Y and P, respectively. Using them, Eq. (4) can be reformulated as

$$F(s) = \rho T(Y) + \xi T(P) - G(s) \tag{5}$$

where

$$G(s) = 2\rho \sum_{i,v} y_{iv} c_v s_i - \rho \sum_v c_v^2 \sum_i s_i^2 + 2\xi\lambda \sum_{i,j} p_{ij} s_i s_j - \xi\lambda^2 \sum_i s_i^2 \sum_j s_j^2 \tag{6}$$

By putting the optimal values c_v and λ from (3) into this expression, we obtain a simpler expression for $G(s)$

$$G = \rho|S| \sum_v c_v^2 + \xi\lambda \sum_{ij} p_{ij} s_i s_j \tag{7}$$

Maximizing G in (7) is equivalent to minimizing criterion F in (2) because of (5).

One can see that maximizing the first item in (7) requires obtaining a numerous cluster (the greater the $|S|$, the better) which is as far away from the space origin, 0, as possible (the greater the squared distance from 0, $|\sum_v c_v^2|$, the better). The second item in the criterion (7) is proportional to the sum of within-cluster links multiplied by the average within-cluster link λ. Maximizing criterion (7), thus, should produce a large anomalous cluster of a high internal density.

We employ a greedy heuristic: starting from arbitrary singleton $S = i$, the seed, add entities one by one so that the increment of G in (7) is maximized. After each adding, recompute optimal c_v and λ. Halt when the increment becomes negative. After stopping, the last check is executed: **Seed Relevance Check:** Remove the seed from the found cluster S. If the removal increases the cluster contribution; this seed is extracted from the cluster.

We refer to this algorithm as Attributed Network Addition Clustering algorithm, ANAC. Our community detection algorithm SEANAC below consecutively applies ANAC to detect more than one community.

SEANAC: Sequential Extraction of Attributed Network Addition Clusters

1. Initialization. Define $J = I$, the set of entities to which ANAC applies at every iteration, and set cluster counter $k = 1$.
2. Define matrices Y_J and P_J as parts of Y and P restricted at J. Apply ANAC at J, denote the output cluster S as S_k, its center c as c_k, the intensity λ as λ_k and contribution G as G_k.
3. Redefine J by removing all the elements of S_k from it. If thus obtained J is empty, stop. Set the current k as K and output all S_k, c_k, λ_k, G_k, $k = 1, 2, ..., K$. If not, add 1 to k, and go to 2.

The implementation of our proposed algorithm and other supplementary materials can be found https://github.com/Sorooshi/SEANAC.

3   Setting of Experiments for Validation and Comparison of SEANAC Algorithm

3.1   Algorithms Under Comparison

We take two popular algorithms in the model-based approach, CESNA [17] and SIAN [14], which have been extensively tested in computational experiments. The author-made codes of the algorithms are publicly available in [9] and [12] respectively.

3.2   Datasets: Real-World and Synthetic

Real World Datasets. We take on five real-world data sets listed in Table 1.

Table 1. Real world datasets under consideration. Symbols N, E, and F stand for the number of nodes, the number of edges, and the number of node features, respectively.

Name	Nodes	Edges	Features	Ground truth
Malaria HVR6 [7]	307	6526	6	Cys Labels
Lawyers [8]	71	339	18	Derived out of office and status features
World Trade [15]	80	1000	16	Derived out of continent and structural world system features
Parliament [1]	451	11646	108	Political parties
COSN [4]	46	552	16	Region

Malaria Data Set [7]. The nodes are amino acid sequences containing six highly variable regions (HVR) each. The edges are drawn between sequences with similar HVRs number 6. In this data set, there are two nominal attributes of nodes: Cys labels derived from of a highly variable region HVR6 (assumed ground truth); and Cys-PoLV labels derived from the sequences adjacent to regions HVR 5 and 6.

Lawyers Dataset [8,16]. The Lawyers dataset comes from a network study of corporate law partnership that was carried out in a Northeastern US corporate law firm, referred to as SG & R, 1988–1991, in New England. There is a friendship network between lawyers in the study. The features in this dataset are: Status (partner, associate), Gender (man, woman), Office location (Boston, Hartford, Providence), Years with the firm, Age, Practice (litigation, corporate), Law school (Harvard or Yale, UCon., Other).

Most features are nominal. Quantitative features, "Years with the firm" and "Age", have been converted to the nominal format, so that categories of "Years with the firm" are $x <= 10$, $10 < x < 20$, and $x >= 20$; and categories of "Age" are $x <= 40$, $40 < x < 50$, and $x >= 50$.

World-Trade Dataset [15]. The World-Trade dataset contains data on trade between 80 countries in 1994. The link weights represent total imports by row-countries from column-countries, in \$ 1,000, for the class of commodities designated as 'miscellaneous manufactures of metal' to represent high technology products or heavy manufacture. The weights for imports with values less than 1% of the country's total imports are zeroed. The node attributes are: Continent (Africa, Asia, Europe, North America, Oceania, South America) Structural World System Position (Core, Semi-Periphery, Periphery), Gross Domestic Product per capita in \$ (GDP p/c). The GDP feature is converted into a three-category nominal feature according to the minima of its histogram, \$ 4406.9 and \$ 21574.5. The categories are: 'Poor', 'Mid-Range', and 'Wealthy'.

Parliament Dataset [1]. The nodes correspond to members of the French Parliament. An edge is drawn if the corresponding MPs sign a bill together. The features are the constituency of MPs and their political party.

Consulting Organisational Social Network (COSN) Dataset [4]. Nodes in this network correspond to employees in a consulting company. The (asymmetric) edges are formed in accordance with their replies to this question: "Please indicate how often you have turned to this person for information or advice on work-related topics in the past three months". The answers, coded by 0 (I Do Not Know This Person), 1 (Never), 2 (Seldom), 3 (Sometimes), 4 (Often), and 5 (Very Often), form the edge weights. Attributes: Organisational level (Research Assistant, Junior Consultant, Senior Consultant, Managing Consultant, Partner), Gender (Male, Female), Region (Europe, USA), Location (Boston, London, Paris, Rome, Madrid, Oslo, Copenhagen).

Before applying SEANAC, all attribute categories are converted into 1/0 dummy variables which are considered quantitative. These datasets can be found at public site https://github.com/Sorooshi/PhD-Datasets.

Generating Synthetic Data Sets. First of all, we specify the number of nodes N, the number of features V, and the number of communities, K, in a dataset to be generated. As the number of parameters to control is rather high, we narrow down the variation of our data generator by maintaining two types of settings only, a small size network and a medium size network. For a small size setting, we specify the values of the three parameters as follows: $N = 200$, $V = 5$, and $K = 5$. For the medium size, $N = 1000$, $V = 10$, and $K = 15$.

Generating Networks. At given numbers of nodes, N, and communities K, cardinalities of communities are defined uniformly randomly, up to a constraint that no community may have less than a pre-specified number of nodes (in our experiments, this is set to 30, so that probabilistic approaches are applicable), and the total number of nodes in all the communities sums to N.

Given the community sizes, we populate them with nodes, that are specified just by indices. Then we specify two probability values, p and q. Every within-community edge is drawn with the probability p, independently of other edges. Similarly, any between- community edge is drawn independently with the probability q.

Generating Quantitative Features. To model quantitative features, we generate a Gaussian distribution at each cluster: its covariance matrix is diagonal with diagonal values uniformly random in the range $[0.05, 0.1]$ and components of the cluster center are uniformly random from the interval $\alpha[-1, +1]$; the real α controls the cluster intermix: the smaller the α, the closer are cluster centers to each other. The possibility of presence of noise in data is modeled too, by uniformly random generation a noise feature. We replicate 50% of the original data with noise features.

Generating Categorical Features. To model categorical features, we randomly choose the number of categories for each of them from the set $\{2, 3, ..., L\}$ where $L = 10$ for small-size networks and $L = 15$ for the medium-size networks. For every k, $k = 1, ..., K$, cluster centers are generated randomly so that no two centers may coincide at more than 50% of features. Once a center of k-th cluster, $c_k = (c_{kv})$, is specified, N_k entities $i \in S_k$ are generated as follows. Given a pre-specified threshold of intermix, ϵ between 0 and 1, for every pair (i, v), $i = 1 : N_k$; $v = 1 : V$, a uniformly random real number r between 0 and 1 is generated. If $r > \epsilon$, the entry x_{iv} is set to be equal to c_{kv}; otherwise, x_{iv} is taken randomly from the set of categories specified for feature v. The closer ϵ to 1, the more similar to the center are the entities.

Generating Mixed Scale Features. Quantitative and categorical features are generated in equal numbers independently of each other.

The synthetic data as well as real world data can be found at public site: https://github.com/Sorooshi/SEANAC/tree/master/data.

3.3   Evaluation Criterion

To evaluate the result of a community detection algorithm, we compare the found partition with that generated, by using the customary Adjusted Rand Index (ARI) [5]. The closer the value of ARI to unity, the better the match between the partitions. If one of the partitions consists of just one part containing all I, then ARI=0.

4   Results of Computational Experiments

The goal of our experiments is to test validity of the SEANAC algorithm over all types of attributed network datasets under consideration. In the cases at which features are categorical, the SEANAC algorithm is to be compared with the popular algorithms SIAN and CESNA, which work with categorical features only.

4.1   Parameters of the Generated Datasets

We set network parameters, the probability of a within-community edge, p, and that between communities, q, to take either of two values each, $p = 0.7, 0.9$ and $q = 0.3, 0.6$. In the cases at which all the features are categorical, we decrease q-values to $q = 0.2, 0.4$, because all the three algorithms fail at $q = 0.6$. Feature generation is controlled by an intermix parameter, α at quantitative features, and ϵ at categorical features. We take each of the intermix parameters to be either 0.7 or 0.9.

We may explicitly insert 50% features that are uniformly random in some datasets.

Therefore, generation of synthetic datasets is controlled by specifying six two-valued and one three-valued parameters leading to 192 combinations of these altogether. At each setting, we generate 10 datasets, run a community detection algorithm, and calculate the mean and the standard deviation of ARI index at these 10 datasets.

4.2   Validity of SEANAC

Table 2 presents the results of our experiments at synthetic datasets with mixed scale features.

Table 2. Performance of SEANAC on synthetic networks combining quantitative and categorical features for two different sizes: The average ARI index and its standard deviation over 10 different data sets.

p q $\alpha/$ ϵ	Small-size networks	50% noisy feature	Medium-size networks	50% noisy features
0.9, 0.3, 0.9	0.99(0.01) | 5.00(0.00)	0.99(0.01) | 5.00(0.00)	1.00(0.00) | 15.00(0.00)	1.00(0.01) | 15.00(0.00)
0.9, 0.3, 0.7	0.98(0.03) | 5.00(0.00)	0.99(0.02) | 5.00(0.00)	1.00(0.00) | 15.00(0.00)	0.99(0.01) | 15.00(0.00)
0.9, 0.6, 0.9	0.91(0.01) | 4.60(0.50)	0.88(0.01) | 4.50(0.67)	0.95(0.08) | 14.00(1.26)	0.93(0.10) | 13.70(1.67)
0.9, 0.6, 0.7	0.86(0.14) | 4.80(0.60)	0.88(0.14) | 4.80(0.39)	0.84(0.08) | 12.10(1.22)	0.81(0.09) | 11.80(1.47)
0.7, 0.3, 0.9	0.99(0.02) | 5.00(0.00)	0.99(0.01) | 5.00(0.00)	0.99(0.01) | 14.90(0.30)	0.99(0.01) | 14.90(0.30)
0.7, 0.3, 0.7	0.94(0.10) | 4.90(0.30)	0.95(0.06) | 4.90(0.30)	0.99(0.01) | 14.80(0.40)	0.96(0.07) | 14.30(1.19)
0.7, 0.6, 0.9	0.74(0.20) | 3.80(0.87)	0.73(0.15) | 4.20(0.87)	0.56(0.14) | 7.80(1.78)	0.55(0.14) | 8.10(1.70)
0.7, 0.6, 0.7	0.67(0.14) | 4.30(1.10)	0.57(0.14) | 3.90(0.54)	0.39(0.09) | 7.10(1.51)	0.42(0.08) | 7.40(0.66)

One can see that SEANAC successfully recovers the numbers of communities at $q = 0.3$ and mostly fails at $q = 0.6$ – because this corresponds to a counterintuitive situation at which the probability of a link between separate communities is greater than 0.5. Yet even in this case the partition is recovered exactly when other parameters keep its structure tight, as say at $p = 0.9$. Insertion of noise features does reduce the levels of ARI but not that much. The real reduction in the numbers of recovered communities, 7–8 out of 15 ones generated, occurs at the medium size datasets at really loose data structures with $p = 0.7$ and $q = 0.6$, leading to significant drops in the levels of ARI values as well.

The picture is much similar at the cases of quantitative only and categorical only feature scales - they are left out to shorten the paper.

4.3   Comparing SEANAC and Competition

In this section, we compare the performance of SEANAC with that of CESNA [17], and SIAN [14]. It should be noted that SEANAC determines the number of clusters automatically, whereas both CESNA and SIAN need that as part of the input. Table 3 presents our results at synthetic datasets (with categorical features only, as required by the competition) and Table 4, at real world datasets.

Table 3. Comparison: average ARI values and their standard deviation over 10 different data sets for CESNA, SIAN and SEANAC at synthetic data sets with categorical features. The best results are highlighted using bold print.

Setting			Small size networks			Medium size networks		
p	q	ϵ	CESNA	SIAN	SEANAC	CESNA	SIAN	SEANAC
0.9,	0.3,	0.9	**1.00(0.00)**	0.55(0.29)	0.99(0.01)	0.89(0.05)	0.00(0.00)	**1.00(0.00)**
0.9,	0.3,	0.7	0.95(0.10)	0.48(0.29)	**0.97(0.02)**	0.85(0.08)	0.00(0.00)	**0.99(0.01)**
0.9,	0.6,	0.9	0.93(0.08)	0.32(0.25)	**0.96(0.01)**	0.63(0.06)	0.00(0.00)	**0.99(0.01)**
0.9,	0.6,	0.7	**0.90(0.06)**	0.11(0.14)	0.75(0.12)	0.48(0.09)	0.00(0.00)	**0.96(0.03)**
0.7,	0.3,	0.9	0.97(0.08)	0.55(0.16)	**0.98(0.02)**	0.77(0.07)	0.03(0.08)	**1.00(0.01)**
0.7,	0.3,	0.7	**0.89(0.14)**	0.51(0.21)	0.87(0.07)	0.71(0.13)	0.00(0.00)	**0.99(0.01)**
0.7,	0.6,	0.9	0.50(0.10)	0.05(0.09)	**0.90(0.07)**	0.06(0.02)	0.00(0.00)	**0.99(0.01)**
0.7,	0.6,	0.7	0.20(0.08)	0.03(0.04)	**0.60(0.09)**	0.02(0.01)	0.00(0.00)	**0.91(0.04)**

One can see that at small sizes CESNA wins three times (out of 8), and at all the other cases, including at medium size datasets, SEANAC wins. SIAN never wins in this table. Moreover, SIAN comprehensively fails on all counts at medium sizes by producing NaN which we interpret as a one-cluster solution.

We also experimented with a slightly different design for, categorical feature generation. That different design sets an entity to either coincide with its cluster center or to be entirely random. At that design CESNA wins 7 times at the small size datasets and SEANAC wins at 7 medium size datasets.

At the real world datasets, CESNA never wins; SEANAC wins three times, and SIAN, two times (see Table 4).

Here, we chose that data normalization method leading, on average, to the larger ARI values. Specifically, we use z-scoring for normalizing features in Lawyers data set, HVR data set and COSN data set. The best results on World-Trade data set and parliament data set are obtained with no normalization. The network data in Lawyers and HVR are normalized with applying the modularity transformation [13]. The network data of COSN is normalized by shifting all the similarities to the average link value [11].

Table 4. Comparison of CESNA, SIAN and SEANAC on Real-world data sets; average values of ARI and standard deviation (std) are presented over 10 random initialization. The best results are shown using bold print.

	CESNA	SIAN	SEANAC
HRV6	0.20(0.00)	0.39(0.29)	**0.45(0.14)**
Lawyers	0.28(0.00)	0.59(0.04)	**0.63(0.06)**
World Trade	0.23(0.00)	**0.55(0.07)**	0.23(0.03)
Parliament	0.25(0.00)	**0.79(0.12)**	0.28(0.01)
COSN	0.44(0.00)	0.43(0.05)	**0.50(0.11)**

5 Conclusion

This paper proposes a novel combined data recovery criterion for the problem of detecting communities in an attributed network. Our algorithm extracts clusters one by one. This allows us to determine the number of clusters automatically, whereas other algorithms need the number of clusters pre-specified. Another feature of our approach is that it is more or less universal regarding the scales of the data available. On the other hand, SEANAC results may depend on data normalization.

We experimentally show that SEANAC is competitive over both synthetic and real-world data sets against two popular state-of-the-art algorithms, CESNA [17] and SIAN [14].

There should be several possible directions for future work over the data recovery approach accepted in this paper. First of all, its extension to large datasets should be proposed and validated. Then the possibility of trade-off between two constituent data sources, network and features, which is explicitly present in our criterion should be investigated. Yet another direction for future work shoud be a systematic investigation of the relative effect of different data standardization methods on the results of our method.

References

1. Bojchevski, A., Günnemann, S.: Bayesian robust attributed graph clustering: joint learning of Partial anomalies and group structure. In: Thirty-Second AAAI Conference on Artificial Intelligence (2018)

2. Chiang, M.M.T., Mirkin, B.: Intelligent choice of the number of clusters in k-means clustering: an experimental study with different cluster spreads. J. Classif. **27**(1), 3–40 (2010). https://doi.org/10.1007/s00357-010-9049-5
3. Chunaev, P.: Community detection in node-attributed social networks: a survey, arXiv preprint arXiv:1912.09816 (2019)
4. Cross, R.L., Parker, A.: The Hidden Power of Social Networks: Understanding How Work Really Gets Done in Organizations. Harvard Business Press, Boston (2004)
5. Hubert, L., Arabie, P.: Comparing partitions. J. Classif. **2**(1), 193–218 (1985)
6. Interdonato, R., Atzmueller, M., Gaito, S., Kanawati, R., Largeron, C., Sala, A.: Feature-rich networks: going beyond complex network topologies. Appl. Netw. Sci. **4**, 1–13 (2019)
7. Larremore, D.B., Clauset, A., Buckee, C.O.: A network approach to analyzing highly recombinant malaria parasite genes. PLoS Comput. Biol. **9**(10), e1003268 (2013)
8. Lazega, E.: The Collegial Phenomenon: The Social Mechanisms of Cooperation Among Peers in a Corporate Law Partnership. Oxford University Press, Oxford (2001)
9. Leskovec, J., Sosič, R.: SNAP: a general-purpose network analysis and graph-mining library. ACM Trans. Intell. Syst. Technol. (TIST) **8**(1), 1–20 (2016). CESNA on Github. https://github.com/snap-stanford/snap/tree/master/examples/cesna
10. Mirkin, B., Nascimento, S.: Additive spectral method for fuzzy cluster analysis of similarity data including community structure and affinity matrices. Inf. Sci. **183**(1), 16–34 (2012)
11. Mirkin, B.: Clustering: A Data Recovery Approach. CRC Press, Boca Raton (1st edn, 2005; 2dn edn, 2012)
12. Nature Communications. https://www.nature.com/articles/ncomms11863
13. Newman, M.E.: Modularity and community structure in networks. Proc. Natl. Acad. Sci. **103**(23), 8577–8582 (2006)
14. Newman, M.E., Clauset, A.: Structure and inference in annotated networks. Nat. Commun. **7**, 11863 (2016)
15. De Nooy, W., Mrvar, A., Batagelj, V.: Exploratory Social Network Analysis with Pajek (Chapter 2). Cambridge University Press, Cambridge (2004)
16. Snijders, T.: The Siena webpage. https://www.stats.ox.ac.uk/snijders/siena/Lazega_lawyers_data.htm
17. Yang, J., McAuley, J., Leskovec, J.: Community detection in networks with node attributes. In: 2013 IEEE 13th International Conference on Data Mining, pp. 1151–1156. IEEE (2013)

A Hybrid Approach to the Analysis of a Collection of Research Papers

Boris Mirkin[1,2] (iD), Dmitry Frolov[1,4](✉) (iD), Alex Vlasov[1], Susana Nascimento[3], and Trevor Fenner[2]

[1] Department of Data Analysis and Artificial Intelligence, HSE University, Moscow, Russia
{bmirkin,dfrolov}@hse.ru
[2] Department of Computer Science and Information Systems, Birkbeck University of London, London, UK
[3] Department of Computer Science and NOVA LINCS, Universidade Nova de Lisboa, Caparica, Portugal
[4] Natimatica, Ltd., Moscow, Russia

Abstract. We define and find a most specific generalization of a fuzzy set of topics assigned to leaves of the rooted tree of a taxonomy. This generalization lifts the set to a "head subject" in the higher ranks of the taxonomy, that is supposed to "tightly" cover the query set, possibly bringing in some errors, both "gaps" and "offshoots". Our hybrid method involves two more automated analysis techniques: a fuzzy clustering method, FADDIS, involving both additive and spectral properties, and a purely structural string-to-text relevance measure based on suffix trees annotated by frequencies. We apply this to extract research tendencies from two collections of research papers: (a) about 18000 research papers published in Springer journals on data science for 20 years, and (b) about 27000 research papers retrieved from Springer and Elsevier journals in response to data science related queries. We consider a taxonomy of Data Science based on the Association for Computing Machinery Classification of Computing System (ACM-CCS 2012). Our findings allow us to make some comments on the tendencies of research that cannot be derived by using more conventional techniques.

Keywords: Hybrid approach · Generalization · Fuzzy cluster · Annotated Suffix Tree · Research tendency

1 Introduction

The issue of automation of structurization and interpretation of digital text collections is of ever-growing importance because of both practical needs and theoretical necessity. There are many papers tackling various aspects of this. In our view, however, the mainstream of all the efforts currently constitute approaches based on the analysis of structure and dynamics of graphs/networks of inter-relations between papers, or articles, (sometimes, between authors) or between

© Springer Nature Switzerland AG 2020
C. Analide et al. (Eds.): IDEAL 2020, LNCS 12490, pp. 423–433, 2020.
https://doi.org/10.1007/978-3-030-62365-4_40

research concepts. Paper [5] exemplifies the former, more recent papers [3,9] – the latter. Arguably, the latter, analysis of concept networks is less computationally intensive than the former, because the sizes of concept graphs are much smaller than those of graphs of articles. Yet results of structural analyses are frequently unstable, much dependent on the datasets involved, and, also, difficult to use for knowledge engineering.

Consider, for illustration, a result from [9]: three sets of keywords returned by three different methods as response to query "Economic growth" in Table 1. One cannot help but noticing how different and, sometime, arbitrary are keywords returned by algorithms. This type of return is difficult to interpret and automate.

Table 1. Three sets of keywords returned by three different topic modeling methods in response to query "Economic growth" in [9], Table 3 on page 228.

Management information system	Economic adjustment	Stages of growth model
Tobacco	Economic policy	Growth policy
Internet Usage	Growth policy	Resource wealth
Eurobond	Economic development	Kuznets curve
Automobile engine	Economic reform	Export-led growth

The goal of this paper is developing a coherent methodology for conceptual analysis of research paper collections that would lead to unified conceptual representations more suitable for automated analysis. The very first provision is to restrict the arbitrariness of keywords, be they supplied by authors, like in [3], or extracted from texts, like in [9]. To achieve that, we use a domain taxonomy, so that the set of keywords is a subset of the taxonomy leaf topics. A taxonomy, in this paper, is a rooted tree whose nodes are annotated by domain concepts in such a way that parental nodes are tagged by concepts more general than concepts assigned to the children nodes. In spite of the recent surge in efforts for automated taxonomy building (see, for a review, [13]), no sound automated taxonomy making method has been developed so far. We definitely prefer using a manually developed taxonomy such as ACM Classification of Computing Systems 2012 by the international Association for Computing Machinery [2].

Therefore, the set of keywords here is constant. This would shield us from empirical biases which are immanent to the approaches that use keywords derived from the texts under analysis. There is a negative side too: some of our leaf-related keywords may appear little relevant or even irrelevant to this or that article from the collection. Therefore, we need a method for assessment of relevance between keywords and texts, which would provide us with robust relevance scoring independently of the way at which keywords appear in the text. Such a method has been proposed and substantiated, with our participation, to evaluate similarity between texts and keywords considered as strings of symbols, the so-called Annotated Suffix Tree approach (see in [6,11]).

Our next step will be for obtaining clusters of keywords so that those keywords that tend to co-occur in the same texts would tend to belong to the same clusters. The clusters sought should be fuzzy to reflect semantic relations between keywords. Therefore, the next stage of our approach is in using the obtained keyword-to-text relevance scores for finding fuzzy clusters of keywords, that tend to co-occur in the same texts.

Conventionally, obtaining such a cluster or set of clusters would be considered "the end of the story", like it is in popular methods for topic modeling [1,4]. We, however, consider it is imperative to use the knowledge embodied in the domain taxonomy, of which the keywords are part, for further interpretation of the clusters. Specifically, given a fuzzy cluster of taxonomy leaves, we propose to find a most specific generalization of that in higher ranks of the taxonomy and use thus obtained higher ranks concept(s) as a general description of the cluster. To this end, we develop a method for finding the most parsimonious generalization of fuzzy leaf clusters.

We apply this strategy to two collections of research papers in Data Science that we have downloaded using different criteria. We use a taxonomy of Data Science [7] derived from the most popular Computer Science taxonomy, manually developed by the world-wide Association for Computing Machinery in 2012 as the ACM Computing Classification System (ACM-CCS) [2]. Our generalizations and interpretations of the two sources are mutually consistent. Moreover, they cannot be found with the existing approaches because they are based on different levels of conceptual granularity, whereas other approaches involve the same granularity level.

The rest of the paper is organized accordingly. Section 2 presents a mathematical formalization of the generalization problem as of parsimoniously lifting of a given query fuzzy leaf set to higher ranks of the taxonomy and provides a recursive algorithm leading to a globally optimal solution to the problem. Section 3 describes an application of this approach to deriving tendencies in development of Data Science, that can be discerned from two sets of research papers: (a) about 18000 research papers published by the Springer Publishers in data science 17 journals for the past 20 years, and (b) about 27000 research papers published in 80 data science journals by Springer and Elsevier, and retrieved using 17 query terms such as "clustering" and "artificial intelligence". Its subsections describe our approach to finding and generalizing fuzzy clusters of research topics. The results are followed by our comments on the tendencies in the development of the corresponding parts of Data Science drawn from the lifting results. Section 4 concludes the paper.

2   A Hybrid Approach to Automated Analysis of Text Collections Within a Domain

Our approach involves two information sources: (a) a conceptual tree of domain taxonomy, and (b) set of texts from the domain. There are three methods to operate over these:

- Computation of relevance matrix between the texts and taxonomy leaf topics with a follow-up conversion of the matrix into a topic-to-topic corelevance matrix;
- Computation of thematic fuzzy clusters over the corelevance matrix;
- Parsimonious lift of fuzzy clusters to higher ranks of the taxonomy.

These are briefly described further on.

2.1   Evaluation of Relevance Between Texts and Key Phrases

Most popular and well established approaches to scoring keyphrase-to-document relevance include the so-called vector-space approach [12] and probabilistic text model approach [4]. These, however, rely on individual words and text pre-processing. We utilize an in-house method [6,11], which requires no manual work.

An Annotated Suffix Tree (AST) is a weighted rooted tree used for storing text fragments and their frequencies. To build an AST for a text string, all suffixes from this string are extracted. A k-suffix of a string $x = x_1 x_2 \ldots x_N$ of length N is a continuous end fragment $x_k = x_{N-k+1} x_{N-k+2} \ldots x_N$. For example, a 3-suffix of string $INFORMATION$ is substring ION, and a 5-suffix, $ATION$. Each AST node is assigned a symbol and the so-called annotation (frequency of the substring corresponding to the path from the root to the node including the symbol at the node). The root node of AST has no symbol or annotation. We use an efficient version of AST building algorithm [6]. Having an AST T built, one can score the string-to-document relevance over the AST as the average frequency of a symbol conditioned by the previous substring coinciding in both the string and document [7].

Co-relevance Topic-to-Topic Similarity Score. Given a topic-to-document matrix R of relevance scores, it is converted to a topic-to-topic similarity matrix A for scoring the "co-relevance" of topics, a.k.a. keyphrases, according to the text collection structure. The similarity score $a_{tt'}$ between topics t and t' can be computed as the inner product of vectors of scores $r_t = (r_{tv})$ and $r_{t'} = (r_{t'v})$ where $v = 1, 2, \ldots, V = 17685$ or $V = 26799$ at Collection A or B, respectfully. The inner product is moderated by a natural weighting factor assigned to texts in the collection.

The weight of text v is defined as the ratio n_{max}/n_v where n_v is the number of topics n_v relevant to v and n_{max} is the maximum n_v over all $v = 1, 2, \ldots, V$. A topic is considered relevant to v if its relevance score is greater than 0.2 (a threshold found experimentally, see [7]).

Therefore, the greater the number of relevant documents, the less significant is the topic.

In this way, a hidden cluster structure is accentuated: documents relevant to few topics make greater contributions to the similarities. [Corrigendum: the authors recently noted an unfortunate mistake in their previous publications

[7, 8, 10] – the omission of the word "reciprocal" in the definition of the weight of a text when defining co-relevance similarity scores. We apologise for this slip].

Our algorithm, FADDIS, [10] finds fuzzy clusters one by one under the assumption that each of the clusters is represented by its fuzzy membership vector $\boldsymbol{u} = (u_t)$, $t \in T$, where T is the leaf set of our taxonomy so that the product $(\mu u_t)(\mu u_{t'}) = \mu^2 u_t u_{t'}$ approximates $a_{tt'}$ as closely as possible. Here μ_k stands for the cluster's intensity value determined according to the approximation task [10].

2.2 Defining and Computing Fuzzy Clusters of Taxonomy Topics

Clusters of topics should reflect co-occurrence of topics: the greater the number of texts to which both topics t and t' are relevant, the greater the interrelation between t and t', the greater the chance for topics t and t' to fall in the same cluster. We have tried several popular clustering algorithms. Unfortunately, no satisfactory results have been found. Therefore, we present here results obtained with our FADDIS algorithm developed specifically for finding thematic clusters [10]. This algorithm implements assumptions that are relevant to the task:

LN Laplacian Normalization: Similarity data transformation modeling – to an extent – heat distribution and, in this way, making the cluster structure sharper.
AA Additivity: Thematic clusters behind the texts are additive so that similarity values are sums of contributions by different hidden themes.
AN Non-Completeness: Clusters do not necessarily cover all the key phrases available as the text collection under consideration may be irrelevant to some of them.

2.3 Parsimoniously Lifting a Fuzzy Thematic Cluster: Model and Method

The problem: Given a fuzzy set S of taxonomy leaves, find a node $h(S)$ of higher rank in the taxonomy tree, that covers the set S as tightly as possible. Such a "lifting" problem is a mathematical explication of the human facility for generalization, that is, "the process of forming a conceptual form" of a phenomenon represented, in this case, by a fuzzy leaf subset. Figure 1 shows a crisp set S of five black leaf boxes lifted to the root of the left branch of the tree, with the price of a gap and an offshoot. A gap is a white leaf box covered by the lifted node, 'the head subject', and, thus falling in the same concept as S even as it does not belong in S. An offshoot is a black box on the right, belonging to S but not covered by the head subject. If we lift S higher, to the root of the tree, this would produce no offshoots, with the price of four gaps.

An explication of the "tight coverage" problem is what we refer to as the maximum parsimony principle: given a fuzzy leaf set S and 'prices' for head subjects, gaps and offshoots, find its parsimonious lift: a node (set of nodes) h to minimize the total penalty for the introduction of corresponding head subjects,

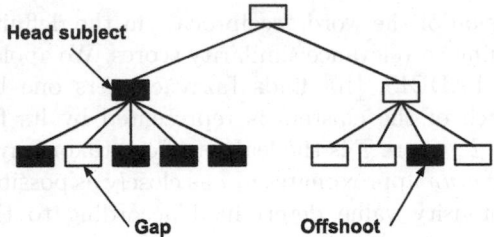

Fig. 1. Generalization of the query set represented by five black-box leaves by mapping it to the root of the left branch.

gaps, and offshoots. This problem was formally stated and solved by the authors in [7,8]. Specifically, an algorithm ParGenFS, is proposed in these references that optimally solves the problem by recursively building the optimal lifting from bottom of the taxonomy tree to its top.

A Python 3 package GOT (Generalization Over Taxonomies) implementing our method is available at GitHub: https://github.com/dmitsf/GOT/.

3 Application to Collections of Research Papers

This section describes application of the method described above.

3.1 Scholarly Text Collection

We have downloaded two collections: (a) a collection of 17685 research papers together with their abstracts published in 17 Data Science journals by the Springer Publisher in 1998–2017, see [7] (Collection A); (b) a collection of 26 799 research papers published in 80 Data Science journals by the Springer and Elsevier Publishers and retrieved by using such keywords as clustering, machine learning, deep learning, artificial intelligence, etc. as queries (Collection B). We use abstracts to these papers.

3.2 DST Taxonomy

Taxonomy is a form of knowledge engineering which is getting more and more popular. Mathematically, a taxonomy is a rooted tree, a hierarchy, whose all nodes are labeled by main concepts of a domain. The hierarchy corresponds to a relation of inclusion: the fact that node A is the parent of B means that B is part, or a special case, of A. The domain of our choice is Data Science, comprising such areas as machine learning, data mining, data analysis, big data, computational intelligence, etc. We take that part of the ACM-CCS 2012 taxonomy, which is related to Data Science, and add a few leaves related to more recent Data Science developments. The higher ranks of the taxonomy are presented in Table 2 (see [7]).

Table 2. ACM Computing Classification System (ACM-CCS) 2012 higher rank subjects related to Data Science.

Subject index	Subject name
1	Theory of computation
1.1	Theory and algorithms for application domains
2	Mathematics of computing
2.1	Probability and statistics
3	Information systems
3.1	Data management systems
3.2	Information systems applications
3.3	World Wide Web
3.4	Information retrieval
4	Human-centered computing
4.1	Visualization
5	Computing methodologies
5.1	Artificial intelligence
5.2	Machine learning

3.3 FADDIS Thematic Clusters at Collection A

After computing the 317×317 topic-to-topic co-relevance matrix, converting in to a topic-to-topic Lapin transformed similarity matrix, and applying FADDIS clustering, at Collection A, we sequentially obtained 6 clusters, of which three clusters seem especially homogeneous. We denote them using letters L, for 'Learning'; R, for 'Retrieval'; and C, for 'Clustering'. These clusters are presented in Table 3.

Table 3. Clusters L, R, C: topics with largest membership values.

Cluster L		Cluster R		Cluster C	
$u(t)$	Topic	$u(t)$	Topic	$u(t)$	Topic
0.300	Rule learning	0.211	Query representation	0.327	Biclustering
0.282	Batch learning	0.207	Image representations	0.286	Fuzzy clustering
0.276	Learning to rank	0.194	Shape representations	0.248	Consensus clustering
0.217	Query learning	0.194	Tensor representation	0.220	Conceptual clustering
0.216	Apprenticeship learning	0.191	Fuzzy representation	0.192	Spectral clustering
0.213	Models of learning	0.187	Data provenance	0.187	Massive data clustering
0.203	Adversarial learning	0.173	Equational models	0.159	Graph based conceptual clustering

3.4   Results of Lifting Clusters L, R, and C Within DST

All obtained clusters are lifted in the DST taxonomy using ParGenFS algorithm with the gap penalty $\lambda = 0.1$ and off-shoot penalty $\gamma = 0.9$.

Lifting Cluster L gave three head subjects: machine learning, machine learning theory, and learning to rank. These represent the structure of the general concept "Learning" according to text Collection A.

Similar comments can be made with respect to results of lifting of Cluster R: Retrieval. The obtained head subjects: Information Systems and Computer Vision show the structure of "Retrieval" in the set of publications under consideration. Lifting of Cluster C leads to 16 (!) head subjects/offshoots at which the core clustering subjects are supplemented by methods and environments in the cluster – demonstrating in this way that the ever increasing role of clustering activities should be better reflected in the taxonomy.

3.5   Fuzzy Clusters at Collection B

Among many fuzzy clusters found by FADDIS algorithm among the DTS taxonomy over the Collection B, there are seven interpretable clusters. These are presented in Table 4.

The first two of them one-to-one correspond to clusters L and C over collection A, whereas the third cluster over A, R (Retrieval), corresponds to five other clusters over Collection B. These five are not incompatible with Cluster R, but rather appear to be its facets. The "Computer vision" head subject over A, has received now two complementary aspects: "Structuring" and "Computer vision representations".

3.6   Making Conclusions

One can see that the topic clusters found with the text collections do highlight areas of soon-to-be developments. One cannot help but relate them to the following processes:

– theoretical and methodical research in learning, as well as merging the subject of learning to rank within the mainstream;
– representation of various types of data for information retrieval, and merging that with visual data and their semantics; and
– various types of clustering in different branches of the taxonomy related to various applications and instruments.

Most impressive here is the information retrieval cluster R. Rather than conventionally relating the term "information" to texts only, visuals are becoming parts of the concept of information. However, unlike the multilevel granularity of meanings in texts, developed during millennia of the process of communication via languages in the humankind, there is no comparable hierarchy of meanings for images. One may only guess that the elements of the R-related five clusters linked to data representation and management systems, are those that are going

Table 4. Generalizations of interpretable clusters found at the Collection B. Symbol ⊙ denotes an offshoot.

Interpretation	Head subjects and offshoots	Gaps	Leaves
"Learning"	1.1.1. – Machine learning theory 5.2. – Machine learning ⊙ 3.4.4.5. – Learning to rank	38	32
"Clustering"	3.2.1.4. – Clustering and 8 offshoots	0	17
"Probabilistic representations"	2.1.1. – Probabilistic representations 5.2.1.2. – Unsupervised learning 5.2.3.5. – Learning in probabilistic graphical models and 8 offshoots	11	31
"Retrieval"	3.1.4. – Query languages 3.4. – Information retrieval ⊙ 5.1.1.9. – Language resources	27	28
"Structuring"	3.1.1.5. – Data model extensions 5.1.3. – Computer vision ⊙ 1.1.1.12. – Structured prediction ⊙ 3.1.4.1.1. – Structured Query Language ⊙ 3.4.1.1. – Document structure ⊙ 3.4.2.1. – Query representation ⊙ 3.4.7.1.1. – Structured text search ⊙ 5.2.1.1.5. – Structured outputs and 11 other offshoots	11	34
"Computer vision representations"	5.1.3.2. – Computer vision representations ⊙ 4.1.4.1. – Visualization toolkits and 3 more offshoots	0	13
"Querying"	3.1.3.2. – Database query processing 3.4.2. – Information retrieval query processing and 5 offshoots more	3	15

to be put in the base of a future multilevel system of meanings for images and videos.

Regarding the "clustering" cluster C with its many head subjects, one may conclude that, perhaps, a time moment has come or is to come real soon, when the subject of clustering must be raised to a higher level in the taxonomy to embrace all these "heads". At the dawn of the Data Science era clustering was usually considered a more-or-less auxiliary part of machine learning. Perhaps, soon we are going to see a new taxonomy of Data Science, in which clustering is not just an auxiliary instrument but rather a model of empirical classification, a big part of the knowledge engineering.

4   Conclusion

The paper describes a hybrid method for the analysis of a collection of research papers based on a domain taxonomy. The method involves the following original developments by the authors:

  i A taxonomy of Data Science derived from ACM-CCS 2012;
 ii A method for scoring relevance between taxonomy leaf topics and texts which requires no manually texts pre-processing;
iii A spectral method for one-by-one deriving fuzzy clusters of taxonomy leaf topics;
 iv A method for parsimoniously generalization of fuzzy leaf clusters in the taxonomy;
  v Consistent conclusions of tendencies of research in Data Science.

Acknowledgments. D.F., A.V. and B.M. acknowledge continuing support by the Academic Fund Program at the National Research University Higher School of Economics(grant no 19-04-019 in 2018–2019) and by the International Decision Choice and Analysis Center(DECAN) NRU HSE, in the framework of a subsidy granted to the HSE by the Government of the Russian Federation for the implementation of the the Russian Academic Excellence Project "5–100". S. N. acknowledges the support by FCT/MCTES, NOVA LINCS(UID/CEC/04516/2019).

References

1. Amado, A., Cortez, P., Rita, P., Moro, S.: Research trends on big data in marketing: a text mining and topic modeling based literature analysis. Eur. Res. Manag. Bus. Econ. **24**(1), 1–7 (2018)
2. Association for Computing Machinery (ACM): The 2012 ACM computing classification system (2012). http://www.acm.org/about/class/2012
3. Ba, Z., Cao, Y., Mao, J., Li, G.: A hierarchical approach to analyzing knowledge integration between two fields-a case study on medical informatics and computer science. Scientometrics **119**(3), 1455–1486 (2019)
4. Blei, D.M.: Probabilistic topic models. Commun. ACM **55**(4), 77–84 (2012)
5. Chen, C., Ibekwe-SanJuan, F., Hou, J.: The structure and dynamics of cocitation clusters: a multiple-perspective cocitation analysis. J. Am. Soc. Inform. Sci. Technol. **61**(7), 1386–1409 (2010)
6. Dubov, M.: Text analysis with enhanced annotated suffix trees: Algorithms and implementation, pp. 308–319 (2015)

7. Frolov, D., Nascimento, S., Fenner, T., Mirkin, B.: Parsimonious generalization of fuzzy thematic sets in taxonomies applied to the analysis of tendencies of research in data science. Inf. Sci. **512**, 595–615 (2020)
8. Frolov, D., Nascimento, S., Fenner, T., Taran, Z., Mirkin, B.: Computational generalization in taxonomies applied to: (1) analyze tendencies of research and (2) extend user audiences. In: Yin, H., Camacho, D., Tino, P., Tallón-Ballesteros, A.J., Menezes, R., Allmendinger, R. (eds.) IDEAL 2019. LNCS, vol. 11872, pp. 3–11. Springer, Cham (2019). https://doi.org/10.1007/978-3-030-33617-2_1
9. Galke, L., et al.: Inductive learning of concept representations from library-scale bibliographic corpora. INFORMATIK 2019: 50 Jahre Gesellschaft für Informatik-Informatik für Gesellschaft (2019)
10. Mirkin, B., Nascimento, S.: Additive spectral method for fuzzy cluster analysis of similarity data including community structure and affinity matrices. Inf. Sci. **183**(1), 16–34 (2012)
11. Pampapathi, R., Mirkin, B., Levene, M.: A suffix tree approach to anti-spam email filtering. Mach. Learn. **65**(1), 309–338 (2006)
12. Salton, G., Buckley, C.: Term-weighting approaches in automatic text retrieval. Inf. Process. Manag. **24**(5), 513–523 (1988)
13. Wang, C., He, X., Zhou, A.: A short survey on taxonomy learning from text corpora: issues, resources and recent advances. In: Proceedings of the 2017 Conference on Empirical Methods in Natural Language Processing, pp. 1190–1203 (2017)

Sequential Self-tuning Clustering for Automatic Delimitation of Coastal Upwelling on SST Images

Susana Nascimento[1,2](✉) [iD], Sayed Mateen[1], and Paulo Relvas[3] [iD]

[1] CS Department, Lisbon, Portugal
s.mateen@fct.unl.pt
[2] NOVA Laboratory for Computer Science and Informatics (NOVA LINCS)
Faculdade de Ciências E Tecnologia, Universidade Nova de Lisboa, Lisbon, Portugal
snt@fct.unl.pt
[3] Campus de Gambelas, Centro de Ciências do Mar (CCMAR), Universidade do
Algarve, 8005 -139 Faro, Portugal
prelvas@ualg.pt

Abstract. Upwelling is of major environmental and economic importance for coastal regions. Sea Surface Temperature (SST) satellite
imagery provide an expedited method of monitoring its variability.

This work proposes a one-by-one extracting version of a spatial clustering algorithm with self-tuning thresholding derived from anomalous
clustering, able to precisely delineate coastal upwelling from SST images.
The stop condition is defined based on properties of the phenomenon and
allows to model the appropriate number of upwelling regions.

The algorithm, Sequential Self-Tuning Seed Expanding Cluster (S-
STSEC), shows to outperform the homologous sequential version of
Seeded Region Growing (SRG) on the automatic delimitation of coastal
upwelling from a collection of 207 SST images comprising two distinct
upwelling systems: from the Portuguese coast and from Canary upwelling
system. Four popular internal clustering validity indices were combined
to measure the quality of the results.

Keywords: Sequential clustering · Anomalous clustering · Seeded
region growing · Internal validity indices · Coastal upwelling · SST
images

1 Introduction

Coastal upwelling is the dynamic response of the continental shelf waters to an
alongshore wind blowing with the coast to the left in the Northern hemisphere
or to the right in the Southern hemisphere. Under such a wind regime, upper
ocean waters over the continental shelf are pushed offshore and replaced by cold
and nutrient-rich waters upwelled from the deeper layers. Upwelling regions are
among the most productive of the world ocean, with strong economic and environmental impacts. The identification of the upwelling patterns variability is

© Springer Nature Switzerland AG 2020
C. Analide et al. (Eds.): IDEAL 2020, LNCS 12490, pp. 434–443, 2020.
https://doi.org/10.1007/978-3-030-62365-4_41

critical to perceive the response of the upwelling regimes to the alteration of their forcings, such as climate changes [18]. Satellite-derived sea surface temperature (SST) observations provide the most suitable data for this purpose since they show good time-space resolution and the thermal contrast between the cold upwelled waters and the warmer offshore ocean is well discerned. The sharp region where the strongest horizontal SST gradient occur is called the upwelling front. Consistent and reliable data sets are available for over 30 years, which is long enough to allow long-term analysis. SST data have been provided daily since 1982 by the Advanced Very High-Resolution Radiometer (AVHRR) sensor on board NOAA-n satellite series. The amount of images and the uncertainty inherent to the subjective recognition of the upwelling patterns by human experts, requires the development of effective algorithms for the automatic detection of this mesoscale oceanographic features.

Various approaches based on traditional image analysis have been developed for segmentation SST images (e.g. [11,16,22]). However, they require computational processes that are too complex to get seemingly satisfactory results and are not operational to process a large number of images. Other works concentrate on clustering. This is the case of the computational tool FuzzyUPWELL [13] that integrates an unsupervised version of fuzzy clustering that precisely segments the upwelling regions combined with feature extraction to delineate the upwelling fronts in the Portuguese coast. The work in [20] also explores fuzzy clustering followed by region growing to remove false positives being applied to upwelling on the southern part of Moroccan Atlantic coast. That study is extended to the Northwest Africa off Morocco in [5] with a particle swarm optimization (PSO) algorithm supported on geographic and physical properties of the region. Despite the successful analysis on upwelling recognition, the latter two approaches adopted several empirical parameters, including the ad-hoc definition of the number of clusters in which to segment an SST image. Such aspects are not easy to operationalize to process a huge number of images and are not flexible to analyse upwelling in distinct regions of the globe.

To overcome those difficulties Nascimento et al. [14,15] developed an unsupervised spatial clustering algorithm with adaptively optimized thresholds. The method, Self-Tuning Seed Expanding Cluster (ST-SEC), extends the popular Seeded Region Growing (SRG) [1] within the framework of anomalous clustering [12] what allows to overcome well recognized limitations of SRG algorithms. First, the homogeneity clustering criterion takes the format of a product rather than the conventional difference between a pixel and the mean of the region of interest, with a threshold adaptively derived from that criterion. Second, the method considers a moving window acting as a regulariser of the cluster growing that brings forth two desirable properties: (i) there is no need in specifying the order of testing for labeling among pixels: all those borderline pixels can be considered and decided upon simultaneously; (ii) the simultaneous borderline labeling considerably speeds up the SRG procedure. These properties contrast to either expert driven or automatic thresholds derived from (supervised) non-homogeneous properties in most of SRG algorithms and yet the issue of dependence on the pixel sorting order (e.g. [2,8,10,19,21]).

Since coastal upwelling can cover two or more contiguous regions, the present work extends the ST-SEC algorithm to a sequential version that extracts clusters one by one till a stop condition is reached, defined on knowledge of the domain. The number of clusters is an outcome of the algorithm instead of being empirically pre-specified. To study the competitiveness of the Sequential Self-tuning Expanding Cluster (S-STEC) with SRG, Adams and Bischof's SRG algorithm was also extended to a sequential version where regions are extracted one by one till the same stop condition holds.

The rest of the paper is organized as follows. Section 2 describes a modified version of the ST-SEC algorithm and the new Sequential Self-Tuning Seed Expanding Cluster (S-STSEC). Section 3 presents the imagery data of SST's covering two distinct upwelling systems: the Portuguese coastal upwelling, and the Canary upwelling system. In Sect. 4 we discuss the experimental results comparing S-STSEC with S-SRG on automatic upwelling delineation as well as on the results evaluation combining four internal clustering validity indices. The conclusion and future work are in Sect. 5.

2 Sequential Self-tuning Seed Expanding Cluster Algorithm

In this work the Self-tuning Seed Expanding Cluster algorithm [14,15] was improved to increase its rate of convergence. Specifically, (i) a 4-neighborhood (instead of the commonly used 8-neighborhood) experimentally shown to produce better segmentation results; (ii) the algorithm iterates until the expanding cluster becomes stable between two consecutive iterations substituting the previous condition of the set of boundary cluster pixels to be empty. The ST-SEC algorithm is described as follows.

Let $T(R, L)$ be a SST map, where R is the set of rows and L the set of columns, and elements of $R \times L$ are pixels. The process starts by subtracting the average temperature t^* of the temperature map T from the temperature values at all pixels in $R \times L$. The centered values are denoted as $t(i, j)$, $(i, j) \in R \times L$. The algorithm finds a cluster $C \subseteq R \times L$ in the format of a binary map $B(R, L)$ with elements $b_{ij} = 1$ if $(i, j) \in C$ and $b_{ij} = 0$, otherwise. The algorithm involves a window, $W(i, j)$, of a pre-specified size centered at pixel (i, j). Based on our experiments we take a square window of size 7×7 pixels.

The ST-SEC algorithm starts selecting a seed pixel, $o = (i_o, j_o)$, as the pixel with the lowest temperature value (ties are resolved randomly). The cluster C is initialized with the seed $o = (i_o, j_o)$ together with those pixels within the window $W(i_o, j_o)$ satisfying the similarity condition

$$c \times t(i, j) \geq \pi, \tag{1}$$

where c is the temperature of the seed pixel o, and π, a similarity threshold.

The boundary of cluster C is defined as the set F of such unlabeled pixels (i', j') whose 4-neighbourhood $N(i', j')$ intersects C. That is,

$$F = \{(i', j') \notin C | N(i', j') \cap C \neq \varnothing\}. \tag{2}$$

Then, the algorithm proceeds iteratively expanding cluster C by dilatating its boundary F step by step till the cluster stabilizes. For each pixel $(i', j') \in F$ the boundary expand region is defined as the subset of pixels (i, j) belonging to C that intersects the window centered at (i', j') and calculates the average temperature of those pixels, $c^* = mean\,(T\,(W(i', j') \cap C))$. The homogeneity criterion is defined as:

$$c^* \times t(i', j') \geq \pi. \tag{3}$$

If condition (3) holds, the boundary pixel (i', j') is allocated to an auxiliary cluster C' and consequently labeled, and the corresponding boundary pixels $N(i', j')$ are merged to the auxiliary boundary set F'. The process continues until all boundary pixels of F have been treated. At the end, the new labeled pixels in C' are merged with C and the corresponding boundary set F is updated with set F'.

The iterative process of expanding the boundary pixels of cluster C stops when it becomes stable between two consecutive iterations.

Notice that the process of treating all frontline pixels of the cluster before updating it and the corresponding new boundary set, guarantees that the expansion does not depend on the order of selection of those boundary pixels. Also, the process of testing the similarity condition and labeling boundary pixels can be performed in parallel.

The similarity threshold π in equations (1) and (3) is derived from criterion (3) as half of the squared average temperature within-C, that is $\pi = \lambda^2/2$ with λ that average temperature. The theoretical ground of the thresholding within the anomalous clustering approach is described in [15].

2.1   Sequential Self-tuning Expand Cluster Algorithm

In order to derive more than one cluster the ST-SEC algorithm is extended to a sequential version where clusters are extracted one-by-one till reach a stop condition. This condition is defined based on knowledge of the domain. Specifically:

(i) Since coastal upwelling starts on the coldest region near the coast, candidate seed pixel(s) are the coldest pixel(s) whose distance to the coastline does not exceeds a predefined distance δ.

(ii) It was experimentally verified that the difference between the average temperature of the first extracted cluster $(mean(C^1))$ and the lowest temperature value of the current cluster $(t(i_o, j_o))$ decreases with the increase of the number of clusters. This second sub-condition establishes that this difference to be not less than a threshold, τ, automatically calculated by Otsu's method [17].

Figure 1 shows the values of this feature for the first three derived clusters on 60 SST images, with the top line corresponding to the feature values for the first cluster, the middle line to the second cluster, and the bottom line the corresponding feature values for the third cluster.

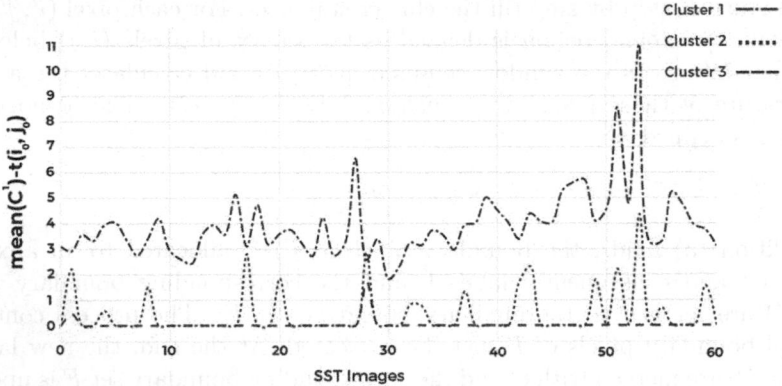

Fig. 1. Values of the extracted feature, $mean(C^1) - t(i_o, j_o)$, measuring the difference between the average temperature of the first extracted cluster and the lowest temperature value of the current cluster (seed pixel), for the first three retrieved clusters on 60 SST images.

The Sequential Self-Tuning Expand Cluster (S-STSEC) algorithm is defined as follows:

Input $T(R, L)$- *temperature map (after centering);*
w- *side of the exploring window* W; δ- *threshold distance to coastline*

Output *Set of* K *clusters* $C_1, C_2, \ldots, C_K$ *in the format of a binary map* $B(R, L)$ *over the same sets* R *and* L *with elements* $b_{ij} = 1$ *if* $(i, j) \in C_k$ *and* $b_{ij} = 0$ *if* $(i, j) \notin \bigcup_{k=1}^{K} C_k$.

Set $C = \emptyset$; $k = 0$

Repeat
    $k = k + 1$
    set (i_o, j_o) pixel corresponding to $min_t(T)$
    $C^k = ST\_SEC(T, w, \pi)$
    $T = T - C^k$
until distanceCoastline$((i_o, j_o)) > \delta$ **or** $(mean(C^1) - t(i_o, j_o)) \leq \tau$

An homologous sequential extraction algorithm was considered for Adams & Bischof's SRG algorithm [1] with each C^k its output cluster.

3   Imagery Data

Two collections of AVHRR Sea Surface Temperature (SST) images have been considered in this study. A collection of SST's of the Portuguese coast (37°N to 41°N and 8°W to 12°W) covering five upwelling seasons of the years of 1998 (82 images), 1999 (31 images), 2000 (32 images), 2001 (30 images) and 2002 (22 images). This collection of 197 SST daily images cover different situations ranging from images with well characterized upwelling pattern in terms of fairly sharp fronts between the cold upwelled water and the warm offshore waters, identified by strong thermal gradients along the coast; surface waters measured by contrasting thermal gradients and continuity along the coast; images showing distinct upwelling patterns related to thermal transition zones offshore from the North toward the South and with smooth transition zones between upwelling sub-domains; and noisy SST images with clouds, where information to define the upwelling front is lacking. Figure 2 first row – left column illustrates one SST image of this collection.

A high resolution colour scale (192 levels) was applied to each SST image in order to have best distribution of colour levels over the SST range in each individual image. Each SST image, T, is assigned to a map with each offshore pixel being the temperature in degrees Celsius. The continuous white region on the right side of each SST image corresponds to land, whereas white pixels in the ocean part correspond to missing values during the satellite transmission, typically due to cloud cover.

The second collection of SST images comprises 10 images (20°N to 34°N and 10°W to 22°W) from the Canary upwelling system. The left column of the last two rows in Fig. 2 illustrate two SST images of this collection.

4   Experimental Study

The two sequential algorithms S-STSEC and S-SRG were run taking each of the 207 SST images with no ground-truth behind. The distance to coastline threshold δ was fixed to 10 Km.

The first row of Fig. 2 shows (from left to right) one SST image of the Portuguese coast and the corresponding segmentation results obtained by S-STSEC and S-SRG algorithms. The bottom two rows of Fig. 2 show two SST images of the Canary upwelling system as well as the corresponding segmentation results for the algorithms. These images clearly illustrate how different are the morphologies of the upwelling regions and how successful are the S-STSEC and S-SRG segmentation results.

To evaluate the quality of the segmented results we take advantage of four popular internal clustering validity indices (CVI) measuring compactness and separation of the clusters: the Silhouette (S), Calinski-Harabasz (CH), Davies-Bouldin (DB), and S_Dbw validity indices [3]. It is well known by the clustering community that there is no "optimal" CVI able to cope successfully with all clustering applications and that CVIs results, typically, are not concordant with

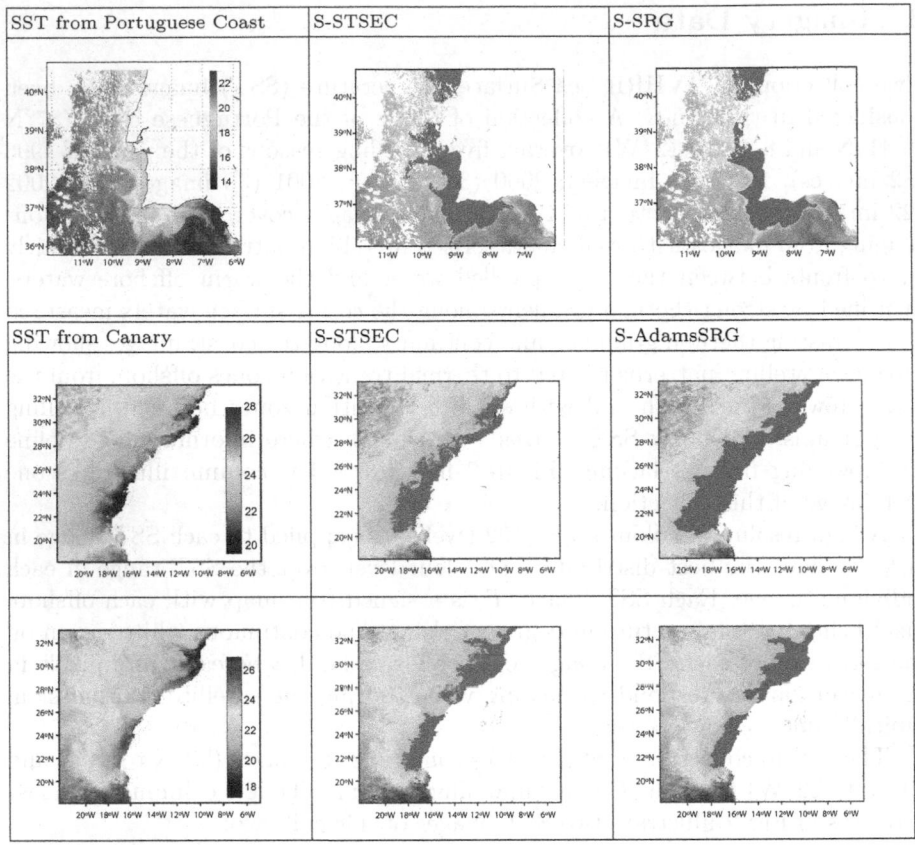

Fig. 2. SST images of the Portuguese coast and corresponding segmentation results from S-STSEC and S-SRG algorithms (top row); two SST images of Canary system and corresponding segmentation results (bottom two rows).

each other (e.g. [3]). Therefore, we combined the four indices as follows: the DB and S_Dbw values were transformed by subtracting the minimum and divide by the maximum value, such that the best value is of maximization (instead of minimization); then, each of the S, CH, DB and S_Dbw were normalized to the interval $[0, 1]$. The median $(SF - med)$ and the average $(SF - A)$ of the four transformed validity values were calculated as fusion score (SF) for each segmentation result.

The graphic in Fig. 3 shows the boxplots for $SF - med$ comparing the segmentation results of S-STSEC against S-SRG for the various upwelling seasons in the coast of Portugal. It is clearly evident that S-STSEG outperforms S-SRG in the median values, lower and upper quartiles, for the SST images from 1998 to 2001. The S-SRG algorithm achieves better segmentation results for the SST images of 2002. This is explained by the so called leaking problem (i.e explosion) that is present in some of the S-STSEC segmentation results for that year.

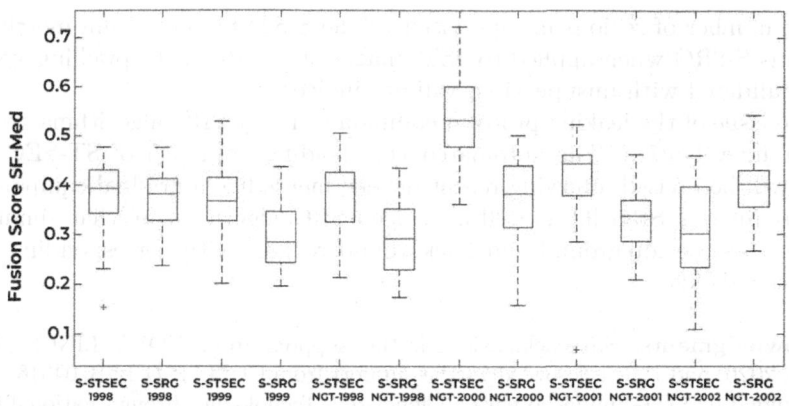

Fig. 3. Boxplot graphics comparing upwelling recognition results of S-STSEC vs S-SRG algorithms for the SST images of Portuguese coast taking the median $(SF - med)$ of S, CH, DB, and S_Dbw internal validity indices.

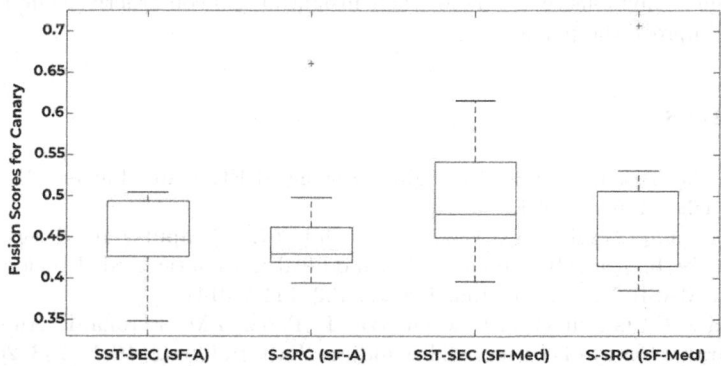

Fig. 4. Boxplot graphics comparing the upwelling recognition results of S-STSEC vs S-SRG algorithms for the SST images of Canary upwelling system respecting the average $(SF - A)$ and median $(SF - med)$ fusion scores.

For the collection of SST images of Canary system the corresponding boxplot graphic comparing S-STSEC algorithm with S-SRG with the average and median measures are shown in Fig. 4. Again, it is clear the best performance of S-STSEC algorithm against S-SRG.

5   Conclusion and Future Work

We propose a sequential extraction clustering version of a recently proposed spatial clustering algorithm, Self-Tuning Seeded Expanding Clustering (ST-SEC) and also apply it to Adams and Bischof's SRG. This algorithmic schema shown to be effective to automatically delineate coastal upwelling and retrieves the correct number of regions unsupervisedly. The S-STSEC algorithm mostly outperforms S-SRG when applied to SST images of two distinct upwelling systems when validated with unsupervised validity indices.

The issue of the leaking problem common to many SRG algorithms deserves further investigation. The automated thresholding approach of ST-SEC algorithm will be revised following recent developments like a gradual equipartition thresholding for SRG [6] as well as a deep SRG algorithm [9] that simultaneously grows the foreground and background regions with corresponding automatic thresholds.

Acknowledgments. S.N. acknowledges the support from NOVA LINCS (UIDB /04516/2020) and P.R. acknowledges the support from CCMAR (UIDB/04516/ 2020) both funded by FCT-Fundação para a Ciência e a Tecnologia, through national funds. The authors acknowledge Dr. Joaquim Luís for the preprocessing of the satellite imagery and the support to this research. Colleagues from CO and DEGGE, Faculdade de Ciências, Universidade de Lisboa, are thanked for providing the collection of SST images of the Portuguese coast examined in this study. The authors also wish to thank the anonymous reviewers for their insightful and constructive comments that allowed to improve the paper.

References

1. Adams, R., Bischof, L.: Seeded region growing. IEEE Trans. Pattern Anal. Mach. Intell. **16**, 641–647 (1994)
2. Al-Faris, A.Q., Ngah, U.K., Isa, N.A., Shuaib, I.L.: Computer-aided segmentation system for breast MRI tumour using modified automatic seeded region growing (BMRI-MASRG). J. Digit. Imaging **27**, 133–144 (2014)
3. Arbelaitz, O., Gurrutxaga, I., Muguerza, J., Pérez, J.M., Perona, I.: An extensive comparative study of cluster validity indices. Pattern Recog. **46**(1), 243–256 (2013)
4. Byun, Y., Kim, D., Lee, J., Kim, Y.: A framework for the segmentation of high-resolution satellite imagery using modified seeded-region growing and region merging. Int. J. Remote Sens. **32**(16), 4589–4609 (2011)
5. El Aouni, A., Garçon, V., Sudre, J., Yahia, H., Daoudi, K., Minaoui, K.: Physical and biological satellite observations of the northwest african upwelling: spatial extent and dynamics. IEEE Trans. Geosci. Remote Sens. **58**(2), 1409–1421 (2020)
6. Fan, H., Meng, F., Liu, Y., Kong, F., Ma, J., Lv, Z.: A novel breast ultrasound image automated segmentation algorithm based on seeded region growing integrating gradual equipartition threshold. Multi.Tools Appl. **78**(19), 27915–27932 (2019). https://doi.org/10.1007/s11042-019-07884-8
7. Fan, J., Zeng, G., Body, M., Hacid, M.-S.: Seeded region growing: an extensive and comparative study. Pattern Recogn. Lett. **26**(8), 1139–1156 (2005)
8. Guo, P., Li, N.: Self-Adaptive Threshold Based on Differential Evolution for Image Segmentation. Proceedings of 2015 2nd International Conference on Information Science and Control Engineering (ICISCE), pp. 466–470 (2015)
9. Huang, Z., Wang, X., Wang, J., Liu, W., Wang, J.: Weakly-supervised semantic segmentation network with deep seeded region growing. 2018 IEEE/CVF Conference on Computer Vision and Pattern Recognition, pp. 7014–7023 (2018)

10. Ju, Z., Zhou, J., Wang, X., Shu, Q.: Image segmentation based on adaptive threshold edge detection and mean shift. Proceedings of the 4th IEEE International Conference on Software Engineering and Service Science (ICSESS 2013), pp. 385–388. IEEE (2013)
11. Marcello, J., Marques, F., Eugenio, F.: Automatic tool for the precise detection of upwelling and filaments in remote sensing imagery. IEEE Trans. Geosci. Remote Sens. **43**(7), 1605–1616 (2005)
12. Mirkin, B.: Clustering: A Data Recovery Approach. Chapman and Hall, Boca Raton (1st Edition, 2005; 2nd Edition, 2012)
13. Nascimento, S., Franco, P., Sousa, F., Dias, J., Neves, F.: Automated computational delimitation of SST upwelling areas using fuzzy clustering. Comput. Geosci. **43**, 207–216 (2012)
14. Nascimento, S., Casca, S., Mirkin, B.: A seed expanding cluster algorithm for deriving upwelling areas on sea surface temperature images. Comput. Geoci. Special issue on "Statistical learning in geoscience modelling: novel algorithms and challenging case studies, **85**, 74–85 (2015)
15. Nascimento, S., Mirkin, B.: Applying anomalous cluster approach to spatial clustering. In: Kreinovich, V. (ed.) Uncertainty Modeling. SCI, vol. 683, pp. 147–157. Springer, Cham (2017). https://doi.org/10.1007/978-3-319-51052-1_10
16. Nieto, K., Demarcq, H., McClatchie, S.: Mesoscale frontal structures in the canary upwelling system: New front and filament detection algorithms applied to spatial and temporal patterns. Remote Sens. Environ. **123**, 339–346 (2012)
17. Otsu, N.: A threshold selection method from gray-level histograms. IEEE Trans. System, Man, and Cybern. SMC- **9**(1), 62–66 (1979)
18. Relvas, P., Luís, J., Santos, A.M.P.: Importance of the mesoscale in the decadal changes observed in the northern Canary upwelling system. Geophys. Res. Lett. **36**, L22601 (2009). https://doi.org/10.1029/2009GL040504
19. Shih, F., Cheng, S.: Automatic seeded region growing for color image segmentation. Image Vis. Comput. **23**, 877–886 (2005)
20. Tamim, A., Minaoui, K., Daoudi, K., Yahia, H., Atillah, A., Aboutajdine, D.: An efficient tool for automatic delimitation of moroccan coastal upwelling using SST images. IEEE Geosci. Remote Sens. Lett. **12**(4), 875–879 (2015)
21. Verma, O., Hanmandlu, M., Seba, S., Kulkarni, M., Jain, P.: A simple single seeded region growing algorithm for color image segmentation using adaptive thresholding. In: Proceedings of the 2011 International Conference on Communication Systems and Network Technologies, pp. 500–503. IEEE Computer Society, Washington, D.C., USA (2011)
22. Vidal-Fernández, E., Piedra-Fernńdez, J., Almendros-Jiménez, J., Cantón-Garbín, M.: OBIA system for identifying mesoscale oceanic structures in SeaWiFS and MODIS-Aqua images. IEEE J. Select. Top. Appl. Earth Observ. Remote Sens. **8**(3), 1256–1265 (2015)

Special Session on Automated Learning for Industrial Applications

Time Series Clustering for Knowledge Discovery on Metal Additive Manufacturing

Marta Aramburu-Zabala[1] (ID), Simona Masurtschak[1] (ID), Ramón Moreno[1]([⊠]) (ID),
Jeremy Jean-Jean[2] (ID), and Angela Veiga[3] (ID)

[1] LORTEK-BRTA, Arranomendi Kalea 4A, 20240 Ordizia, Spain
rmoreno@lortek.es
[2] RISE IVF, Material and Production Division, Argongatan 30, 431 53 Mölndal, Sweden
[3] CEIT-BRTA, Manuel Lardizabal 15, 20018 San Sebastián, Spain

Abstract. This work meets Metal Additive Manufacturing and Time Series Processing. It presents a four-step analytical procedure addressed to support the discovery of defect causes in 3D metal printing. The method has a phase of data space transformation, where the features space is firstly reduced and secondly exploited in a higher dimensional space. Later, a procedure for knowledge discovery is applied. Finally, by analyzing the results, it is concluded the most probable causes of the high rate of defects in the production phase. This procedure is proved with data obtained from a SLM machine, and the results are convincing.

Keywords: Metal Additive Manufacturing · Single layer melting · Time series · Clustering · Fault detection

1 Introduction

Industry 4.0 [1] is a keyword which refers to the latest advances in industrial manufacturing. The fourth revolution is mainly based on data explosion in current industrial devices. In a broad approach, Industry 4.0 (I4.0 onwards) is underpinned by a few key technologies [2], among them are three on which this work is based. 1) Internet of Things (IoT) that in short refers to data collection from heterogeneous sources. 2) Distributed Computing [3] which refers to different levels of computation (edge: machine level, fog: plat level and cloud: on remote servers). 3) Additive Manufacturing (AM) [4] which refers to the technologies that build parts from 3D models via a layer-by-layer technique.

In mathematical terms, IoT data is managed by Time series [5]. Time series has a long history in the field of economical sciences. However, nowadays with the data explosion from I4.0 this mathematical branch is experiencing its own revolution. Data is coming from different sources and all of them have their own latency and period. For this reason, to gather all of them in synchronized lapses of time requires an additional effort.

Cloud computing is perhaps the most popular term when referring to computation in I4.0. Nonetheless, given the unaffordable computing requirements in industrial environments, this term has been split in three computing levels: edge/fog and cloud).

© Springer Nature Switzerland AG 2020
C. Analide et al. (Eds.): IDEAL 2020, LNCS 12490, pp. 447–455, 2020.
https://doi.org/10.1007/978-3-030-62365-4_42

All of them combined can draw the cycle of data from the sensors to a cloud, and the return from the cloud to the actuators of the production plants.

AM is a groundbreaking technology which allows to build polymeric and metallic items by utilizing the 3D printing technique. In the work presented, the focus lies on AM for metals. AM technology for metals classifies roughly into four main techniques: Laser Metal Deposition (LMD), Direct Energy Deposition (DED), Binder Jetting (BJ) and Powder Bed Fusion (PBF). The most important one to date is PBF in which one or more lasers direct energy into a powder bed. During manufacturing, a lot of data output from PBF machines is generated. Some of the variables are setters, whereas many other variables are getters (from observation). Apart from monitoring the process with the sensors inside the machines, a lot of research is currently carried out in monitoring of the powder bed or the melt-pool. Intensity cameras and tomography are just some of the sensing technologies investigated for PBF. However, a vital aspect of the monitoring process is the management of the data not only for monitoring purposes, but also for manufacturing process optimization.

This work presents a statistical procedure for Time Series Analysis and Knowledge Discovery to find the causes of some defects made during the production time in PBF machines. For this, a PBF machine from SLM with an integrated RGB camera has been employed which provided real-time monitoring (in form of images) of the single layers. From these images small errors during the manufacturing process were detected and a procedure of four steps was carried out to understand the defects in a Time series manner. In a first step, a dimensionality reduction of the initial features space was carried out. In the second step, the focus lay on information expansion. In this step the reduced features space was projected on a hyper dimensional feature space where data was exploited to gain a deeper insight. The third step dealt with knowledge discovery. Within this step, the procedure was looking for the causes of the errors. This step was accomplished by using Ward's clustering algorithm [6]. Finally, a fourth step gives a time based view that give an insight into the manufacturing process.

The paper consists of the following sections: Sect. 2 gives a brief introduction of SLM technology. Section 3 explains the method used in this work. Section 4 shows the experimental results. Finally Sect. 5 ends with the conclusions.

2 Powder Bed Fusion

In PBF [7] a part is fabricated by melting of metal powder particles via a laser beam. The parts are created from a digitalized 3D model which will be submitted to a production machine. Before submitting to the machine, the model will be sliced into layers of the same height. Powder is then evenly spread out over a build platform. The laser, usually directed via galvanic X/Y mirrors, then melts the shape of the part in the powder bed applying a pre-set scanning strategy. After the melting, the powder bed is lowered by one layer, new powder will be spread out and the process will start anew. Due to the reactive nature of the metal particles, the process is carried out under an inert atmosphere.

Two main advantages have dominated the use of PBF technology: the fabrication of complex structures [8] as well as the production of near-net shape parts [9]. In order to make use of these advantages, defects such as porosity, cracks, delamination or geometrical instability have to be minimized. For this, a variety of process parameters [10] has to be understood and controlled. Furthermore, in order to decrease time-consuming and expensive post-manufacturing inspection routines and to achieve a first-time-right approach, industry is currently seeking for machines that integrate real-time monitoring closed-loop control systems with ideally self-learning abilities [11]. A time series analysis in combination with data science can be evaluated in order to take a first step towards elimination of defects.

3 Method

This method takes as basis the mathematical Time Series definition presented in previous work [12] where the only difference in this case, is that samples can take different size. It has a noticeable effect with Time Series Partitioning procedure. It splits the full time series recorded in the database into samples. Therefore, it is necessary to have defined beforehand what a sample is and how to make it. According with [12] (Eq. 4), a sample is defined as:

$$\omega = \{T_k, T_{k+1}, T_{k+2}, \ldots, T_l\} | k > i, l < j \tag{1}$$

In this work, k and j are not constants, and this is a key difference with the previous work. Depending of the event to monitor, k, l will take different values. Finally, the full TS is split in ω s of variable size. A collection of ω s Ω is defined as

$$\Omega = \{\omega_1, \omega_2, \omega_3, \ldots, \omega_n\} | \forall (\omega_i, \omega_j) \in \Omega, \omega_i \cap \omega_j = \phi \tag{2}$$

The proposed method in this work can be summarized as four steps. The first one is a dimensionality reduction of the input data space. Secondly, information expansion is accomplished. Information expansion expands the current data space into a higher dimensional one. Thirdly, a clustering procedure is carried out in order to discover the relationships among the samples within the last dataset. From this statistical procedure, a set of measurements is given out which informs about the quality of the procedure. In a final step, clusters which have been discovered in the previous step can be graphed to visualize the results. Figure 1 shows a summary of the procedure.

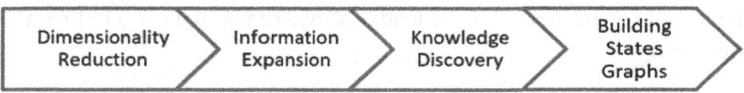

Fig. 1. Process flow of the proposed method. dimensionality reduction (variable selection), information expansion (metrics), knowledge discovery (through clustering)

3.1 Dimensionality Reduction

In general, the first step when working with data generated by machines, is cleaning. On many occasions data coming from the sensors in a machine can be corrupted due to unknown reasons. Additionally, if possible, before performing any data analysis procedure, raw data must be analyzed and repaired. Nonetheless, it is also necessary to detect where important information is hidden. This pre-processing step is consolidated via two consecutive steps: 1) Noise data removal: In this step are removed variables with low variance, or with a lot of missing data. 2) Correlated data removal: In this step, the correlation matrix in order to detect correlated variables is computed. In a set of correlated variables, the variable with higher variance are selected and the rest are removed.

The resulting data is a multivariate time series without neither noise nor redundant data.

3.2 Information Expansion

Sample definition: A single layer of the manufacturing process has multiple observations, all these observations together count as a sample as defined in Eq. 4. Consequently, in line with the scope of this work, a manufactured single layer is considered a Time series *sample.*

Time Series Partitioning: The full Time Series is split by samples as defined in Eq. 2.

Afterwards, a collection of derivative metrics (for each dimension) are extracted for every sample: minimum, maximum, mean, median, standard deviation, kurtosis (Eq. 3) and skewness (Eq. 4) are defined as follow:

$$\frac{\frac{1}{n} \sum_{i=1}^{n} (x_i - \bar{x})^4}{\left[\frac{1}{n} \sum_{i=1}^{n} (x_i - \bar{x})^2\right]^2} - 3 \tag{3}$$

$$\frac{\frac{1}{n} \sum_{i=1}^{n} (x_i - \bar{x})^3}{\left[\frac{1}{n} \sum_{i=1}^{n} (x_i - \bar{x})^2\right]^{3/2}} \tag{4}$$

where a *sample* of n values, where x_i is the ith value and $\bar{x}$ is the sample mean.

Second Order Polynomial regression: In order to get additional metrics, a second order polynomial regression is applied to each feature: $y = ax^2 + bx + c$

Thus, *intercept* $= c$, *and curvature* $= a$. The tendency is computed by fitting a new polynomial regression but to $(x - \bar{x})$ instead of x, where $y = a_2(x - \bar{x})^2 + b_2(x - \bar{x}) + c_2$, and *tendency* $= b_2$

3.3 Knowledge Discovery

Knowledge discovery is applied to the different samples obtained from the manufacturing process. In this work knowledge discovery consists of an agglomerative hierarchical clustering using Ward's method. The agglomerative hierarchical clustering builds the hierarchy from the individual elements by progressively merging together the pair of clusters with lower merging cost. Ward's method minimizes the total within-cluster variance and thus, at each step, finds the pair of clusters that leads to a minimum increase in total within-cluster variance after merging. The merging cost of combining clusters A and B can be defined as:

$$\Delta(A, B) = \frac{n_A n_B}{n_A + n_B} \|\vec{m}_A - \vec{m}_B\|^2$$

where $\vec{m}_j$ is the center of cluster j, n_j is the number of points in it and $\|\cdot\|$ is the Euclidean norm.

4   Experimental Results

The presented method has been tested over a dataset obtained from an SLM machine. A sensor was employed and data was recorded during manufacturing of a part. The measurements were "live data" which was recorded every second. There were 434.358 observations of 39 variables (38 + timestamp). The sensors recorded temperatures of different parts of the machine, amounts of oxygen, pressures, gas flow speeds, powder fill levels and laser on/off.

A set of images capturing the process was also obtained from a camera within the SLM machine. Each image corresponded to a layer of the building process. A total of 1.987 layers were produced. For the case that the powder coating is insufficient, the image will display yellow or green pixels in the affected area. In this case the recoater has to spread powder again over the powder bed. In the case of obtaining green pixels, the process was behaving as expected and a lack of powder was not detected. Image processing was applied to determine if the image consisted of yellow pixels and, consequently, contained a defective/not defective variable.

Dimensionality Reduction
As a start of the examination, the data needed to be processed in order to produce a meaningful dataset. Firstly, the variables with no variance (i.e. always exhibiting the same value) or empty variables were removed. Secondly, a correlation matrix was computed and, for pairs of variables that highly correlated with each other, the variable with less variance was removed. This resulted in 8 features that were used for the experiment they are: Build Chamber, Platform, Total Memory, Pressure, Oxygen1, Gas Flow, Memory Process and Laser on.

Information Expansion
This step divided in three nested steps: 1) The first step produced a sample for all observations in every manufactured layer. Each layer's dataset had between 30 to 90 observations (depending on the duration of the layer building). 2) For the second step, derivate data was from each sample. This was accomplished by computing a set of metrics: minimum, maximum, mean, median, standard deviation, kurtosis and skewness. This, in return, resulted in derivate data which has two properties: On one hand it was a time-compression (all observations are compressed into a single vector). On the other hand, it was a dimensional (information) expansion. 3) For the third step, a 2nd order polynomial regression is applied to each feature and the intercept, the tendency and the curvature were extracted from the interpolation. Figure 4 shows the signal and its interpolation of the first layer for two of the features.

Finally, after applying the 3-step procedure, for each sample a collection of 80 features (8 original x 10 derivate) was computed. Table 1 shows the collection of the final features for every sample in Ω on the first sample (layer).

Table 1. The 10 metrics computed for each feature of 1st layer

	Build chamber	Platform	Memory total	Pressure	Oxygen 1	Gas flow	Memory process	Laser on
Minimum	32.9	199.4	19	12.2	0.11	2.7	57	0
Maximum	33.1	200.5	19	13.2	0.12	2.74	57	1
Mean	32.97	199.986	19	12.58	0.11	2.73	57	0.72
Median	33	200	19	12.6	0.11	2.73	57	1
St deviation	0.07	0.36	0	0.17	0.001	0.01	0	0.45
Kurtosis	−0.89	−1.38	−3	1.37	66.01	−0.86	−3	−1.06
Skewness	0.48	−0.08	0	0.73	8.25	−0.31	0	−0.98
Intercept	32.88	200.03	19	12.86	0.11	2.72	57	0.58
Curvature	1.7e−05	−1.8e−05	−3.1e−18	0.0002	−3.7e−07	3.4e−06	8.7e−18	−0.001
Tendency	0.003	−0.002	0	−0.004	2.3e−06	−3.7e−05	0	−0.01

4.1 Knowledge Discovery

During the third step, hierarchical clustering is then applied to all samples created in the previous step. For this study only the first 1.000 layers/samples of the manufacturing process were kept. The 1.000 observations were normalized to be able to apply the clustering algorithm. After several experiments, it was determined that seven is the best number of clusters in order to fully visualize which building process phases led to more/less defects. Figure 2 represents the clustering results for each layer. Each point represents a layer of the building process and is defined by the 80 variables. The points were drawn in the 2 dimensional space based on their Euclidean distance. Thus, the closest points in the figure represent the ones with the smallest Euclidean distance in between them. Each cluster is characterized by a color. The clusters were created by minimizing the within-cluster variance. The points that exhibit a 'Y' in Fig. 2 correspond to the defective ones, i.e. these layers needed a recoating of the powder.

From a first glance it was seen that most of the defective layers belonged to the red cluster, i.e. cluster 3. This was confirmed by looking at the actual proportions of defective observations in each cluster which is displayed in Table 2. The table confirms that 59% of the defective layers are in cluster 3 – even though this only accounts to 16% of the total observations. Another interesting observation can be seen in cluster 4 (characterized by the brown points in Fig. 3) although it represents 14% of the observations, it contains no defective layer.

The aim is then to characterize cluster 3 (red one) in order to understand what led to the defective layers. By observing the scatterplots of each of the 80 variables, it turns

Fig. 2. Sampling region of colored cluster which indicates the defective layers with a 'Y'.

Table 2. Distribution of observations and defective layers between clusters

	Cluster							
	1	2	3	4	5	6	7	All
Number of observations in cluster	24	40	161	136	58	336	245	1000
Distribution of observations between clusters	2%	4%	16%	14%	6%	34%	24%	100%
Distribution of defects between clusters	10%	7%	**59%**	**0%**	17%	5%	2%	100%

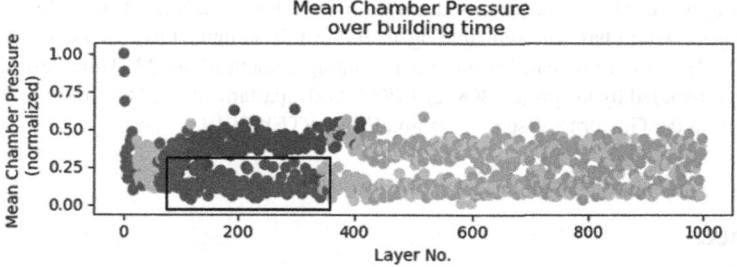

Fig. 3. Mean Chamber Pressure (normalized) over building time colored by cluster. The squared region contains most of the defects.

out that the variable that better discriminates between defects/no defects is the Mean Chamber Pressure.

If a horizontal line is drawn in Fig. 3 at y = 0 and the proportions of the defective observations below and above the line are compared, it can be found that the observations with a normalized mean pressure ≤ 0.25 account for 80% of the defects. As has already been shown the red cluster contained 59% of the defects, all of them can be found between the layers 110 and 360 as is represented by the black rectangle in Fig. 3. Again, the brown cluster contained 0 defects. It has been shown that most of the defects occured between the layers 110 and 360 but during that part of the process, the defects only occured for layers with a lower mean chamber pressure (≤ 0.25 normalized). Thus, it may be recommended to increase the chamber pressure in order to reduce the number of defects in these layers.

5   Conclusions

The work presented in this paper described a method consisting of four steps for a PBF Time Series Analysis. It has been developed to extract as much information as possible. The two first steps were dealing with data space transformation (compression and expansion). After the data space transformation, a Ward's clustering approach was implemented to support the knowledge discovery phase and to look for the cause of some defects.

The usability of this method has been proven by using real data coming from a SLM machine. Applying the method, has shown that most of the errors were detected during the manufacturing phase and were strongly linked to the variable of "chamber pressure". The results obtained from the statistical analysis were validated by experts in SLM manufacturing.

Future work will see an implementation of an inference engine aimed to provide real-time monitoring in terms of product quality and process optimization. Simultaneously, the application will provide an advanced support decision system that will clarify optimal process parameters to obtain minimal geometry deviation and porosity to aid an optimum manufacturing process. Monitoring the quality during 3D printing as well as the implementation of a decision support system will provide real-time recommendations in order to maximize the quality and minimize the number of defects.

Acknowledgments. Data collection and curation have been accomplished within the DIGI-QUAM Project, which has received funding from the EIT Manufacturing, and is supported by the EIT, a body of the European Union under grant agreement n° 20122. Time Series Analysis part has been founded by the project KK-2019/00095 (Departamento de Desarrollo Economico e Infraestructuras del Govierno Vasco. Programa ELKARTEK 2019.

References

1. Lu, Y.: Industry 4.0: A survey on technologies, applications and open research issues. J. Ind. Inf. Integr. **6**, 1–10 (2017). https://doi.org/10.1016/j.jii.2017.04.005
2. Frank, A.G., Dalenogare, L.S., Ayala, N.F.: Industry 4.0 technologies: implementation patterns in manufacturing companies. Int. J. Product. Econ. **210**, 15–26 (2019). https://doi.org/10.1016/j.ijpe.2019.01.004
3. Mohan, N., Kangasharju, Y.J.: «Edge-fog cloud: a distributed cloud for Internet of things computations». In: 2016 Cloudification of the Internet of Things (CIoT), pp. 1–6, November 2016. https://doi.org/10.1109/ciot.2016.7872914
4. Herzog, D., Seyda, V., Wycisk, E., Emmelmann, Y.C.: Additive manufacturing of metals. Acta Mater. **117**, 371–392 (2016). https://doi.org/10.1016/j.actamat.2016.07.019
5. Duque Anton, S., Ahrens, L., Fraunholz, D., Schotten, Y.H.D.: «Time is of the essence: machine learning-based intrusion detection in industrial time series data». In: 2018 IEEE International Conference on Data Mining Workshops (ICDMW), pp. 1–6, November 2018. https://doi.org/10.1109/icdmw.2018.00008
6. Strauss, T., von Maltitz, Y.M.J.: Generalising Ward's Method for Use with Manhattan distances. PLOS ONE **12**(1), e0168288 (2017). https://doi.org/10.1371/journal.pone.0168288

7. Kruth, J.P., Mercelis, P., Vaerenbergh, J.V., Froyen, L., Rombouts, M.: Binding mechanisms in selective laser sintering and selective laser melting . Rapid Prototyping J. 11(1), 26–36 (2005). https://doi.org/10.1108/13552540510573365
8. Yadroitsev, I., Smurov, I.: Selective laser melting technology: from the single laser melted track stability to 3D parts of complex shape. Phys. Proc. 5, 551–560 (2010). https://doi.org/10.1016/j.phpro.2010.08.083
9. Rashid, R., Masood, S.H., Ruan, D., Palanisamy, S., Rahman Rashid, R.A., Brandt, M.: Effect of scan strategy on density and metallurgical properties of 17-4PH parts printed by selective laser melting (SLM) . J. Mater. Process. Technol. 249, 502–511 (2017). https://doi.org/10.1016/j.jmatprotec.2017.06.023
10. Spears, T.G., Gold, S.A.: In-process sensing in selective laser melting (SLM) additive manufacturing. Integr. Mater. Manufact. Innov. 5(1), 16–40 (2016). https://doi.org/10.1186/s40192-016-0045-4
11. Malekipour, E., El-Mounayri, H.: Common defects and contributing parameters in powder bed fusion AM process and their classification for online monitoring and control: a review. Int. J. Adv. Manufact. Technol. 1, 527–550 (2017). https://doi.org/10.1007/s00170-017-1172-6
12. Moreno, R., Pereira, J.C., López, A., Mohammed, A., Pahlevannejad, P.: Time series display for knowledge discovery on selective laser melting machines. In: Yin, H., Camacho, D., Tino, P., Tallón-Ballesteros, A.J., Menezes, R., Allmendinger, R. (eds.) IDEAL 2019. LNCS, vol. 11872, pp. 280–290. Springer, Cham (2019). https://doi.org/10.1007/978-3-030-33617-2_29

Quaternion Neural Networks: State-of-the-Art and Research Challenges

David García-Retuerta[1] , Roberto Casado-Vara[1(✉)] ,
Angel Martin-del Rey[2,3] , Fernando De la Prieta[1] , Javier Prieto[1] ,
and Juan M. Corchado[1,4]

[1] BISITE Research Group, University of Salamanca, Edificio Multiusos I+D+i,
Calle Espejo s/n, Salamanca 37007, Spain
{dvid,rober,fer,javierp,corchado}@usal.es
[2] University of Salamanca, Department of Applied Mathematics,
Calle del Parque 2, 37008 Salamanca, Spain
delrey@usal.es
[3] University of Salamanca, Institute of Fundamental Physics and Mathematics,
Department of Applied Mathematics, Calle del Parque 2, 37008 Salamanca, Spain
[4] Air Institute, IoT Digital Innovation Hub (Spain),
Calle Segunda 4, 37188 Salamanca, Spain

Abstract. Machine Learning has recently emerged as a new paradigm for processing all types of information. In particular, Artificial Intelligence is attractive to corporations & research institutions as it provides innovative solutions for unsolved problems, & it enjoys a great popularity among the general public. However, despite the fact that Machine Learning offers huge opportunities for the IT industry, Artificial Intelligence technology is still at its infancy, with many issues to be addressed. In this paper, we present a survey of quaternion applications in Neural Networks, one of the most promising research lines in artificial vision which also has a great potential in several other topics. The aim of this paper is to provide a better understanding of the design challenges of Quaternion Neural Networks & identify important research directions in this increasingly important area.

Keywords: Quaternions · Neural networks · Image processing

1 Introduction

With the rapid development of processing and storage technologies and the success of the Internet, computing resources have become more powerful and widespread than ever before. Furthermore, their price is continuously lowering due to the appearance of new technologies. The combination of these factors have created a fertile ground for the development of Machine Learning (ML) and its general usage in all situations.

ML has shaken up the way in which information is processed, as larger CPU-GPU clusters are required to train ever-increasing Neural Networks (NN) and

© Springer Nature Switzerland AG 2020
C. Analide et al. (Eds.): IDEAL 2020, LNCS 12490, pp. 456–467, 2020.
https://doi.org/10.1007/978-3-030-62365-4_43

less processing power is required in the final product, which makes use of the trained NN. These factors have had a dramatic impact on the IT industry over the latest years, with large companies such as Amazon or Microsoft shifting their focus towards creating new, efficient data centres. Other major actors such as Apple or Spotify have introduced ML in their products, for example, implementing face recognition software for adjusting the focus of a smartphone camera or recommending similar songs to the ones the user has already listened to.

One of the most promising research lines in ML is related to *quaternions*, which have achieved a great performance in the field of artificial vision and other topics. There has been a series of noticeable advances in this field recently, which have become more and more regular in the last years (Fig. 1).

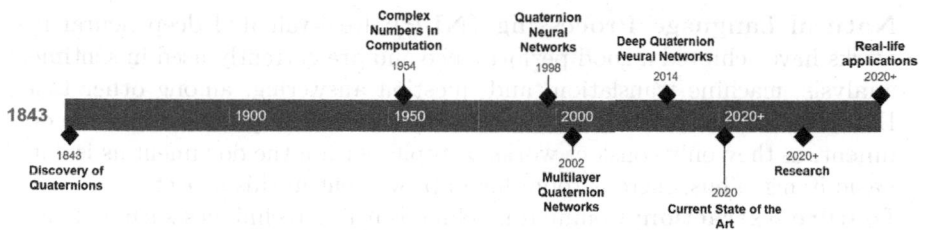

Fig. 1. Evolution of quaternion neural networks.

Common Neural Network designs are based on real numbers and optimisation algorithms, and advanced implementations make use of convolutions and kernels.

1.1 Image Processing

A very powerful type of neural networks for image analysis, which can take great advantage of quaternions, are convolutional neural networks (CNN). They have been successfully used to find contours automatically [11] and to perform an automatic classification of objects in any image [15]. In the latter aspect, ResNet set a milestone in 2015 by achieving a better classification than humans in the ImageNet competition [12].

However, CNNs cannot deal with colour in an efficient way. Either a preprocessing filter must be used (e.g., converting the RGB image to grayscale) or the network must process each of the colours separately (and combine their results in a later phase). As a result, two possibilities arise: to obtain a lower performance by applying an aggressive pre-processing filter, or to require vast resources to develop and train the network.

Other networks such as KNN, SVM, logistic regression, etc; have been used as well but showed in a poorer performance. This is due to the fact that they require careful feature selection, a process which cannot be automatised, resulting in a more time-consuming alternative for researchers. Furthermore, a key advantage of CNNs is their ability to take into account the surroundings of a data point, such as a pixel. They greatly benefit from large datasets as, in such a case, they are able of employing *transfer learning*, a common enhancement achieved only in Deep Learning.

1.2 Related Topics

Quaternions have a great potential in many other topics of machine learning. As well as in image processing, many other cases can greatly benefit from its higher dimensionality and its unique properties. Some of the most promising results have been obtained in the following topics:

– **Natural Language Processing (NLP)**: Real-valued deep neural networks have achieved a good performance and are currently used in sentiment analysis, machine translation, and question answering, among other tasks. However, they do not take into account the internal dependencies of the document, as they only consider works or topics within the document as isolated basic items. Thus, there is room for improvement in this aspect.
– **Feature Extraction**: Common machine learning techniques such as Principle Components Analysis (PCA), Independent Component Analysis (ICA) or Linear Discriminant Analysis (LDA) have proven themselves very useful over the past few years as reducing the dimensionality of datasets with hundreds of columns is becoming more and more common. However, these and other real-valued techniques do not adapt well to specific cases such as acoustic feature extraction. More complex analysis could be used in this aspect to improve the efficiency [29].

The main goal of this article is to analyse in detail the potential applications of quaternions to all the previous topics and to provide a clear introduction to recent research discoveries. The article is organised as follows: Sect. 3 describes the used variants for connecting the quaternion ideas within the neural networks. In Sect. 4, we describe the successful implementation cases and, in particular, we focus on the significant breakthroughs. Promising research lines and their challenges are discussed in Sect. 5. Finally, the paper concludes in Sect. 6.

2   Mathematical Background

Quaternions are the natural extension of complex numbers into 4 dimensions. As a results, they are often called *hypercomplex numbers*. They are defined as:

$$Q = a\mathbf{1} + b\mathbf{i} + c\mathbf{j} + d\mathbf{k} \tag{1}$$

where $\{a, b, c, d\} \in \mathbb{R}$, and $\{1, i, j, k\}$ form the basis of the quaternions space. $a\mathbf{1}$ refers to the real component of the number, and $b\mathbf{i} + c\mathbf{j} + d\mathbf{k}$ to the imaginary component.

They also must satisfy the following identities (knows as the *Hamilton rules*):

$$\mathbf{i}^2 = \mathbf{j}^2 = \mathbf{k}^2 = \mathbf{ijk} = -1 \tag{2}$$

$$\mathbf{ij} = -\mathbf{ji} = \mathbf{k}$$
$$\mathbf{jk} = -\mathbf{kj} = \mathbf{i} \tag{3}$$
$$\mathbf{ki} = -\mathbf{ik} = \mathbf{j}$$

As a result, it can be infered that the multiplication is not commutative. The dot product of two quaternions Q_1, Q_2 is defined as:

$$\langle Q_1, Q_2 \rangle = a_1 a_2 + b_1 b_2 + c_1 c_2 + d_1 d_2 \tag{4}$$

– **Addition:** $Q_1 + Q_2 = (a_1 + a_2) + (b_1 + b_2)\mathbf{i} + (c_1 + c_2)\mathbf{j} + (d_1 + d_2)\mathbf{k}$.
– **Scalar multiplication:** $\lambda Q = \lambda a + \lambda b\mathbf{i} + c\lambda\mathbf{j} + d\lambda\mathbf{k}$.

The conjugate $\overline{Q}$ of Q is defined as follows:

$$\overline{Q} = a\mathbf{1} - b\mathbf{i} - c\mathbf{j} - d\mathbf{k} \tag{5}$$

Furthermore, the normalised vector $\hat{\mathbf{Q}}$ (also called *unit quaternion*) is calculated as:

$$\hat{\mathbf{Q}} = \frac{Q}{\|Q\|} = \frac{Q}{\langle Q, \overline{Q} \rangle} = \frac{Q}{\langle \overline{Q}, Q \rangle} = \frac{Q}{\sqrt{a^2 + b^2 + c^2 + d^2}} \tag{6}$$

As for the *Hamilton product* $\otimes$ of two quaternions Q_1, Q_2; it is defined as:

$$\begin{aligned}
Q_1 \otimes Q_2 = {} & (a_1 a_2 - b_1 b_2 - c_1 c_2 - d_1 d_2) \\
& + (a_1 b_2 + b_1 a_2 + c_1 d_2 - d_1 c_2) \\
& + (a_1 c_2 - b_1 d_2 + c_1 a_2 + d_1 b_2) \\
& + (a_1 d_2 + b_1 c_2 - c_1 b_2 + d_1 a_2)
\end{aligned} \tag{7}$$

3   Implementations in Neural Networks

QNNs differ from traditional Neural Networks due to the higher dimensionality of quaternions and to the unique properties of the Clifford Algebras. Most of the current implementations of QNNs are a adaptation of the traditional algorithms into the higher-dimension case (Fig. 2), which shows there is room for improvement in this aspect. In this section, the technical details of Quaternion Neural Networks are presented.

Neural Networks are implemented similarly to traditional networks, dividing the learning process in *forward-pass* and *forward-pass backward-pass*.

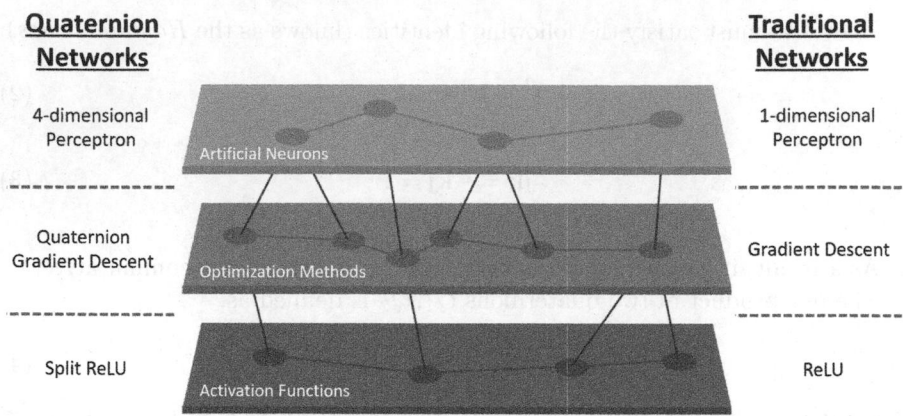

Fig. 2. Main differences of Quaternion Neural Networks and traditional Neural Networks.

3.1 Forward Pass

Firstly, the parameters are defined:

- l is the *number of layers* of the QNN (including input and output layers). It verifies that $0 \leq l \leq L$.
- N_l is the *number of neurons* in layer l.
- ω_{nm}^l is the *weight* of the connection of neuron n (in layer $l-1$) and neuron m (in layer l).
- θ_n^l is the *bias* of neuron n in layer l. It verifies that $1 \leq n \leq N_l$.
- R is the total *number of inputs*.
- Q_r is a normalised *input* data point with $1 \leq r \leq R$.
- t_r is the *label* associated to the input data point Q_r.

The most common activation function in Quaternions Neural Networks (QNN) is the *split ReLU function*, defined as:

$$\beta(Q) = \alpha(a)\mathbf{1} + \alpha(b)\mathbf{i} + \alpha(c)\mathbf{j} + \alpha(d)\mathbf{k} \tag{8}$$

where $\alpha : \mathbb{R} \longrightarrow \mathbb{R}$ is the common ReLU function defined by $\alpha(x) = \max(0, x)$.

The output of neuron n in layer l, denoted as γ_n^l, is defined as [1,27]:

- $\boxed{l = 0}$ $\gamma_n^0 = Q_r^n$ (input layer)
- $\boxed{l = L}$ $\gamma_n^L = t_n$ (output layer)
- $\boxed{1 \leq l < L}$ $\gamma_n^l = \beta(S_n^l)$ (hidden layer l).

$$\text{with } S_n^l = \sum_{m=0}^{N_{l-1}} \omega_{nm}^l \otimes \gamma_m^{l-1} + \theta_n^l \tag{9}$$

3.2   Learning Phase

The error e observed between the expected value t and the actual label is defined depending on the layer:

- $\boxed{l = L}$ $e_n^L = t_n - \gamma_n^l$ (output layer)
- $\boxed{1 \leq l < L}$ $e_n^l = \sum_{n=1}^{N_{l+1}} \omega_{h,n}^{l+1} \otimes \delta_n^{l+1}$ (hidden layers)

where δ is the gradient calculated as:

$$\delta_n^l = e_n^l \otimes \frac{\partial \beta(S_n^l)}{\partial S_n^l}, \text{ with } \frac{\partial \beta(S_n^l)}{\partial S_n^l} = \begin{cases} 1 & \text{if } n \ S_n^l > 0 \\ 0 & \text{otherwise} \end{cases} \qquad (10)$$

3.3   Optimisation Methods

Some optimisation methods typically used in literature [9,34] are:

- Quaternion Gradient Descent Algorithm.
- Quaternion Gauss-Newton Algorithm.
- Quaternion Levenberg-Marquardt Algorithm.
- RMSProp (Root Mean Square Propagation) Algorithm.
- ADAM (Adaptive Moment Estimation) Algorithm.

All the previous algorithms are based on their real equivalent, but other quaternion-based methods have been recently developed [35]. Those methods define new quaternion operations with interesting properties which allow to define the product and the chain rule, as well as gradients and the Hessian matrix.

Instead of decomposing the quaternion problem into complex or value problems, methods such as [22] use advanced properties of Clifford Algebras to optimise convex functions.

Normalisation is commonly used for accelerating the training of the network. In this aspect, Eq. 6 can be used to adapt the range of the data.

4   Use Cases

Quaternion-based Neural Networks is a developing topic which has not been subject to a thorough research yet. However, several breakthroughs have already been achieved. They refer mostly to acoustic and visual problems, which are the cases more suitable for QNNs.

4.1   Visual Cases

The most promising field for quaternion-based networks is image processing as the 4D neurons can naturally adapt to pixels. A great variety of studies have shown the potential quaternion networks have for solving many different types of problems related to colour images. As for grayscale images, QNNs show a equal or lower performance than classical CNNs [36].

The unique properties of quaternions algebras, such as the Hamilton product (Eq. 7), allow to capture statistical dependencies in the data which traditional networks cannot reach. This is due to the relation of two quaternions (which represent rotations) via the Hamilton product, which is associated with a geodesic over a sphere in the $\mathbb{R}^3$. In a natural way, rotations are greatly modelled by QNNs [18].

Xuanyu Zhu *et al.* [36] carried out a comparison of quaternion-based vs real-valued neural networks in which both of them have the same hyper-parameters (number of neurons, number of layers, optimisation method, learning rate, learning rate decay and epochs). As a result, the total number of parameters in the quaternion approach is higher, and the complexity is also higher. Also, it had a better performance than the real-valued CNNs, as its loss function converges more quickly.

Similar approaches have been used in more practical cases such as Prostate Cancer Gleason Grading [10], where its was used in combination with several SVM. Its results and overall accuracy were above any previous work. Also it have been applied to facial expression recognition [31] achieving a good performance, which make QNNs applications feasible for computational facial expression recognition.

Image colour denoising is also covered in [36], showing the natural adaptation of QNNs to colour image processing, resulting in more detailed structures after denoising. Colour information is lost during the encoding phase in real-valued networks, but quaternion-valued networks take full advantage of their 4D structure.

As a specific case of image colour denoising, colour night vision has also been investigated [21]. Better results are obtained for extracting colour information in gloomy images when using quaternion-valued networks.

Common filters in image processing (such as Wavelets, Gaussian Density Estimation, Principal Component Analysis, etc.) have also been generalised and used in the quaternion cases. In [7], a generalisation of the wavelet transform and of the Gauss Density Estimation into the quaternion field is proposed for bank note classification. Colour Image compression has also benefited from QNNs by using the quaternion-adapted version of PCA (Principal Components Analysis) [23]. Quaternion Fourier Transform has also been used as a preprocressing filter for QNNs, and has been applied for recognition of spoken words using images of spatio-temporal representations [3].

4.2   Related Cases

One of the main conclusions of research works in the topic of acoustic applications is the following: QNNs have a faster performance than real-valuated networks, as they require less epochs to train. This is probably related to the fact that the total number of parameters required to train a QNN is lower than to traditional networks. A slightly increased accuracy is often claimed as well, but studies' limitations and the slight improvements found do not show any any trust-worthy proof of such a claims.

Titouan Parcollet et al. [28] proposed a based on quaternions applied to Deep Neural Networks. It is used for spoken language understanding, evaluated using the DECODA[1] dataset. In their experiments, one fact clearly emerged: QNNs have a faster performance that real-valuated networks, as they require less epochs to train.

QNNs also showed a 1–2% increased performance in their experiments, but their experiments had some serious limitations in this aspect. The dataset they used is rather small for deep learning testing (1,242 phone conversations) and they only used hidden layers with 512 neurons in each hidden layer.

In [29], the goad is also spoken language understanding, using the TIMIT[2] dataset. They set the number of hidden layer to four, and vary the number of neurons of all hidden layers. Networks are only trained for 25 epochs. Quaternion-based show a better accuracy in this case, of around 0.5%, which is probably due to their faster performance.

Interestingly, the authors point out that there is a direct correspondence between the number of neurons in both cases (real-valued networks have four times more neurons, as each neuron of the QNN has four dimensions), but the total number of parameters is greatly reduced in the quaternion case (3.8 million parameters vs 9.4 million parameters, quaternion-based and real-based respectively). As a result of the lower complexity of the network, quaternion-based networks can be trained faster. This finding is supported by [32] where a lower number of parameters in required to achieve other NLP milestones.

5   Research Challenges

Although quaternions are believed to be the natural extension of Deep Neural Networks to high-level problems (such as colour image processing, or natural spoken language understanding), the research efforts of the scientific community are still at an early stage. Many existing issues have yet not been fully addressed and many new challenges have been addressed by adapting the existing method for real numbers, ignoring the possibilities of quaternion algebras. In this section, we summarise the topics with the greatest potential and the current research issues.

5.1   Clifford Algebras

Potential applications of Clifford Algebras properties have not yet been used in QNNs. Theoretical progress in this aspect is likely to provide powerful tools to

---

[1] The DECODA dataset is focused on speech mining methods from spontaneous speech recorded in call-centers. Its many application is to measure the robustness and weak-supervision performance of any model [4].

[2] The TIMIT dataset consists of 4,288 sentences, from 536 different speakers. It contains the audio examples and its transcription [8].

many other aspects of these networks, and to result in enhanced performance of algorithms eventually.

Several points for improvement have been identified:

- **Activation functions** are mainly adapted from the real case. Alternative implementations such a split-type activation function have been proposed [14].
- A case which cannot be implemented in the real case is **Moebius transformations**. However, the quaternion implementation already showed good performance on robot vision [25].
- The **hypershepre neuron** has also been proposed in [25] as a promising path for robot vision.
- An approach the topic of quaternion RNNs dynamics from the point of view of **energy functions** have also gained some attention [20].

5.2 Image Processing

One of the keys applications of QNNs is colour image processing. The objective in this case is to use the spatial properties of Clifford algebras (such as rotations) in combination with a quaternion representation of individual pixels components. A typical representation of a pixel is the following:

$$p = 01 + r\mathbf{i} + g\mathbf{j} + b\mathbf{k} \tag{11}$$

where r, g, b represent the *red, green* and *blue* colour of the pixel respectively.

3D image processing is also an interesting topic which has only been researched using traditional CNNs, both for medical applications [17] and shape-related processing tasks [16,30,33]. Quaternions theory for 3D vector spaces has already been researched [5], but there is a surprising lack of any application of this theory.

Furthermore, some authors have suggested that back-propagation of gradients can be represented by reverse rotations of colour vectors with respect to the forward propagation of inputs [36].

5.3 Related Topics

Quaternion Neural networks have many possible applications beyond images. Among the most promising applications stand out: acoustic patter recognition, NLP, design of control systems [6], memorization of music [13], sensors enhancement [24], data processing [19], surface reconstruction [2] and radars [26].

Most of the studies covering these fields are adapted from real-based networks and do not take into account all the possibilities of quaternion algebras. RNNs and CNNs could greatly benefit from well-tailored learning algorithms adapted to hyper-complex numbers (in particular, related to the hyper-complex Hamilton product) and the use of natural operations their numerical field, such as rotations.

Furthermore, a GPU implementation and processing optimisation is much needed for QNNs.

6   Conclusion

Quaternion Neural Networks have recently emerged as a compelling alternative for high-level data processing. The current trend of machine learning is rapidly broadening the possibilities of computers and smart machines, ultimately turning the long-held promise of wide-spread usage of Artificial Intelligence into a reality.

However, despite the recent breakthroughs of CNNs and RNNs, high-level data (such as colour images and natural phone dialogues) has proven that deep learning technologies are not mature enough to tackle such as challenges. The scientific community has paid attention to these research challenges in the last years and now has advanced mathematical tools at their disposal, such as complex and hyper-complex numbers.

In this paper, we have provided an introduction to quaternions and how to develop a neural network with them, as well as surveying the state of the art covering its most important concepts, prominent characteristics and key technologies in all research directions. As the development of Quaternion Neural Networks is still at an early stage, we hope our work will provide a better understanding of the its research challenges, and pave the way for further research in this area.

Acknowledgments. This research has been supported by the project "Intelligent and sustainable mobility supported by multi-agent systems and edge computing (InEDGE-Mobility): Towards Sustainable Intelligent Mobility: Blockchain-based framework for IoT Security", Reference: RTI2018-095390-B-C32, financed by the Spanish Ministry of Science, Innovation and Universities (MCIU), the State Research Agency (AEI) and the European Regional Development Fund (FEDER).

References

1. Arena, P., Fortuna, L., Muscato, G., Xibilia, M.G.: Multilayer perceptrons to approximate quaternion valued functions. Neural Netw. **10**(2), 335–342 (1997)
2. Bayro-Corrochano, E., Scheuermann, G.: Geometric Algebra Computing: in Engineering and Computer Science. Springer Science & Business Media, Berlin (2010)
3. Bayro-Corrochano, E., Trujillo, N., Naranjo, M.: Quaternion fourier descriptors for the preprocessing and recognition of spoken words using images of spatiotemporal representations. J. Math. Imaging Vis. **28**(2), 179–190 (2007)
4. Bechet, F., et al.: Decoda: a call-centre human-human spoken conversation corpus. In: LREC, pp. 1343–1347 (2012)
5. Chou, J.C.: Quaternion kinematic and dynamic differential equations. IEEE Trans. Robot. Autom. **8**(1), 53–64 (1992)
6. Cui, Y., Takahashi, K., Hashimoto, M.: Design of control systems using quaternion neural network and its application to inverse kinematics of robot manipulator. In: Proceedings of the 2013 IEEE/SICE International Symposium on System Integration, pp. 527–532. IEEE (2013)
7. Gai, S., Yang, G., Wan, M.: Employing quaternion wavelet transform for banknote classification. Neurocomputing **118**, 171–178 (2013)

8. Garofolo, J.S., Lamel, L.F., Fisher, W.M., Fiscus, J.G., Pallett, D.S.: DARPA TIMIT acoustic-phonetic continous speech corpus CD-ROM. NIST speech disc 1-1.1. NASA STI/Recon technical report n, **93** (1993)
9. Gaudet, C.J., Maida, A.S.: Deep quaternion networks. In: 2018 International Joint Conference on Neural Networks (IJCNN), pp. 1–8. IEEE (2018)
10. Greenblatt, A., Mosquera-Lopez, C., Agaian, S.: Quaternion neural networks applied to prostate cancer gleason grading. In: 2013 IEEE International Conference on Systems, Man, and Cybernetics, pp. 1144–1149. IEEE (2013)
11. He, K., Gkioxari, G., Dollár, P., Girshick, R.: Mask r-cnn. In: Proceedings of the IEEE international conference on computer vision, pp. 2961–2969 (2017)
12. He, K., Zhang, X., Ren, S., Sun, J.: Deep residual learning for image recognition. In: Proceedings of the IEEE Conference on Computer Vision and Pattern Recognition, pp. 770–778 (2016)
13. Hirose, A.: Complex-Valued Neural Networks: Theories and Applications, vol. 5. World Scientific, Singapore (2003)
14. Hitzer, E., Nitta, T., Kuroe, Y.: Applications of clifford's geometric algebra. Adv. Appl. Clifford Algebras **23**(2), 377–404 (2013)
15. Hu, J., Shen, L., Sun, G.: Squeeze-and-excitation networks. In: Proceedings of the IEEE Conference on Computer Vision and Pattern Recognition, pp. 7132–7141 (2018)
16. Jourabloo, A., Liu, X.: Large-pose face alignment via CNN-based dense 3D model fitting. In: Proceedings of the IEEE Conference on Computer Vision and Pattern Recognition, pp. 4188–4196 (2016)
17. Kamnitsas, K., et al.: Efficient multi-scale 3D CNN with fully connected CRF for accurate brain lesion segmentation. Med. Image Anal. **36**, 61–78 (2017)
18. Kobayashi, M., Muramatsu, J., Yamazaki, H.: Construction of high-dimensional neural networks by linear connections of matrices. Electronics and Communications in Japan (Part III: Fundamental Electronic Science), **86**(11), 38–45 (2003)
19. Kolanowski, K., Świetlicka, A., Kapela, R., Pochmara, J., Rybarczyk, A.: Multisensor data fusion using elman neural networks. Appl. Math. Comput. **319**, 236–244 (2018)
20. Kuroe, Y.: Models of clifford recurrent neural networks and their dynamics. In: The 2011 International Joint Conference on Neural Networks, pp. 1035–1041. IEEE (2011)
21. Kusamichi, H., Isokawa, T., Matsui, N., Ogawa, Y., Maeda, K.: A new scheme for color night vision by quaternion neural network. In: Proceedings of the 2nd International Conference on Autonomous Robots and Agents, vol. 1315. Citeseer (2004)
22. Liu, Y., Zheng, Y., Lu, J., Cao, J., Rutkowski, L.: Constrained quaternion-variable convex optimization: a quaternion-valued recurrent neural network approach. IEEE Trans. Neural Netw. Learn. Syst. **31**(3), 1022–1035 (2019)
23. Luo, L., Feng, H., Ding, L.: Color image compression based on quaternion neural network principal component analysis. In: 2010 International Conference on Multimedia Technology, pp. 1–4. IEEE (2010)
24. Marins, J.L., Yun, X., Bachmann, E.R., McGhee, R.B., Zyda, M.J.: An extended kalman filter for quaternion-based orientation estimation using MARG sensors. In: Proceedings 2001 IEEE/RSJ International Conference on Intelligent Robots and Systems. Expanding the Societal Role of Robotics in the the Next Millennium (Cat. No. 01CH37180), vol. 4, pp. 2003–2011. IEEE (2001)
25. Olver, H., Sommer, G.: Computer Algebra and Geometric Algebra with Applications. Springer, Berlin (2005)

26. Oyama, K., Hirose, A.: Phasor quaternion neural networks for singular point compensation in polarimetric-interferometric synthetic aperture radar. IEEE Trans. Geosci. Remote Sens. **57**(5), 2510–2519 (2018)
27. Parcollet, T., Morchid, M., Bousquet, P.M., Dufour, R., Linarès, G., De Mori, R.: Quaternion neural networks for spoken language understanding. In: 2016 IEEE Spoken Language Technology Workshop (SLT), pp. 362–368. IEEE (2016)
28. Parcollet, T., Morchid, M., Linares, G.: Deep quaternion neural networks for spoken language understanding. In: 2017 IEEE Automatic Speech Recognition and Understanding Workshop (ASRU), pp. 504–511. IEEE (2017)
29. Parcollet, T., et al.: Quaternion recurrent neural networks (2018). arXiv preprint arXiv:1806.04418
30. Su, H., Qi, C.R., Li, Y., Guibas, L.J.: Render for cnn: Viewpoint estimation in images using CNNS trained with rendered 3D model views. In: Proceedings of the IEEE International Conference on Computer Vision, pp. 2686–2694 (2015)
31. Takahashi, K., Takahashi, S., Cui, Y., Hashimoto, M.: Remarks on computational facial expression recognition from hog features using quaternion multi-layer neural network. In: Mladenov, V., Jayne, C., Iliadis, L. (eds.) EANN 2014. CCIS, vol. 459, pp. 15–24. Springer, Cham (2014). https://doi.org/10.1007/978-3-319-11071-4_2
32. Tay, Y., et al.: Lightweight and efficient neural natural language processing with quaternion networks (2019). arXiv preprint arXiv:1906.04393
33. Wang, P.S., Liu, Y., Guo, Y.X., Sun, C.Y., Tong, X.: O-CNN: Octree-based convolutional neural networks for 3D shape analysis. ACM Trans. Graph. (TOG) **36**(4), 1–11 (2017)
34. Xu, D., Zhang, L., Zhang, H.: Learning algorithms in quaternion neural networks using GHR calculus. Neural Netw. World **27**(3), 271 (2017)
35. Xu, D., Xia, Y., Mandic, D.P.: Optimization in quaternion dynamic systems: gradient, hessian, and learning algorithms. IEEE Trans. Neural Netw. Learn. Syst **27**(2), 249–261 (2015)
36. Zhu, X., Xu, Y., Xu, H., Chen, C.: Quaternion convolutional neural networks. In: Proceedings of the European Conference on Computer Vision (ECCV), pp. 631–647 (2018)

A Solar Thermal System Temperature Prediction of a Smart Building for Data Recovery and Security Purposes

José-Luis Casteleiro-Roca[1]([✉])(iD), María Teresa García-Ordás[2](iD),
Esteban Jove[1](iD), Francisco Zayas-Gato[1](iD), Héctor Quintián[1](iD),
Héctor Alaiz-Moretón[2](iD), and José Luis Calvo-Rolle[1](iD)

[1] University of A Coruña, CTC, Department of Industrial Engineering, CITIC,
Avda. 19 de Febrero S/n, 15405 Ferrol, A Coruña, Spain
{jose.luis.casteleiro,esteban.jove,f.zayas.gato,hector.quintian,
jlcalvo}@udc.es
[2] University of León, Department of Electrical and Systems Engineering,
Escuela de Ingenierías, Campus de Vegazana, 24071 León, Spain
{mgaro,hector.moreton}@unileon.es

Abstract. This paper perform a comparison between different clustering algorithms, that their optimal number of clusters has been calculated throw different performance measurements. The comparison takes into account the prediction of the thermal solar panel output temperature to conclude what is the best clustering division. The used dataset is extracted from a Bioclimatic house that belongs to Sotavento Galicia Foundation, and it is composed of the most important variables in the thermal solar energy generation system.

Silhouette, Calinski-Harabasz, and Davies-Bouldin were used to achieve the optimal number of clusters and then, Artificial Neural Networks and Polynomial Regression were trained, with several configurations, to create a hybrid intelligent model for regression. Very good results were obtained with this procedure, that allows to reduce the computational cost of creating a hybrid model without knowing the number of clusters for the dataset.

Keywords: Clustering · Optimal clusters · Regression · Solar thermal panel · Hybrid intelligent model

1 Introduction

Preservation of the environment is an important key to minimize the human impact and reduce the climate change effects. One of the most important efforts is reducing the energy consumption in the cities and, of course, in the houses. In this sense it can be necessary to mention the Smart Cities concept [22].

The Smart Cities definition, despite is continuously under revision, involves the reduction in energy consumption, but it is important to emphasized that

© Springer Nature Switzerland AG 2020
C. Analide et al. (Eds.): IDEAL 2020, LNCS 12490, pp. 468–476, 2020.
https://doi.org/10.1007/978-3-030-62365-4_44

it also means more efficiency in general, increasing the quality of life, promote the sustainable development, etc. To achieve these objectives, Smart Cities use technology and innovation since their conception, and this fact helps making true the main aim: the environment preservation [1,3,7,12,26].

A direct way to increase the efficiency in the Smart Buildings (buildings that have similar *objectives* than Smart Cities) is the reduction of their dependence on the foreign energy. In this sense, it is normal to use models to predict some variables, with the source is renewable. As the renewable energy could be not sufficient all time in a house, the systems must have another auxiliary energy. If a model could predict the renewable energy produced by the house, it would be possible to optimize the consumption of the other source [13,19,20,24].

To increase the performance of the predictions, models with intelligent techniques are usually created, or hybrid ones. This last type of models is based on the division of a system in several clusters with similar characteristics and then one model is created for each cluster [4–6,14,16,17,21]. In this paper, it has been used four clustering techniques to divide the dataset; after that, the performance of the division is measured and compared to decide the optimal number of clusters. Once the clusters are created, two regression algorithms have been used to create the internal (or local) models: Artificial Neural Networks, and Polynomial Regression.

2   Case Study

Sotavento Galicia Foundation was created in 1997 with the aim of studying new renewable energies. As part of this foundation they have a experimental wind farm, to study different types of wind turbines and a *Bioclimatic house*, that is a house built with the aim of reducing the environmental impact. In this house, there are several renewable energies like geothermal, solar or wind; and the architecture of the house was studied to be more effective in energy consumption.

This research is focused only in the thermal solar energy, representing a part of the thermal energy generation of the house (that includes geothermal and biomass energy too). Figure 1 is a schematic of this part, where it shows the most important components: the thermal solar panel, the pump and the inertia tank. The variables used in the developed model in this paper are S1, S2, S3, S4 and flow-meter.

3   Model Approach

Fig. 2 shows the conceptual model approach. This model uses the input temperatures of the two thermal solar panels, the etilenglicol flow throw them, and the solar radiation measure in the *Bioclimatic house* location; to predict the output temperature of the lower panel. As this research is focused on the lower thermal solar panel, the output temperature from the upper panel (S3) is not used.

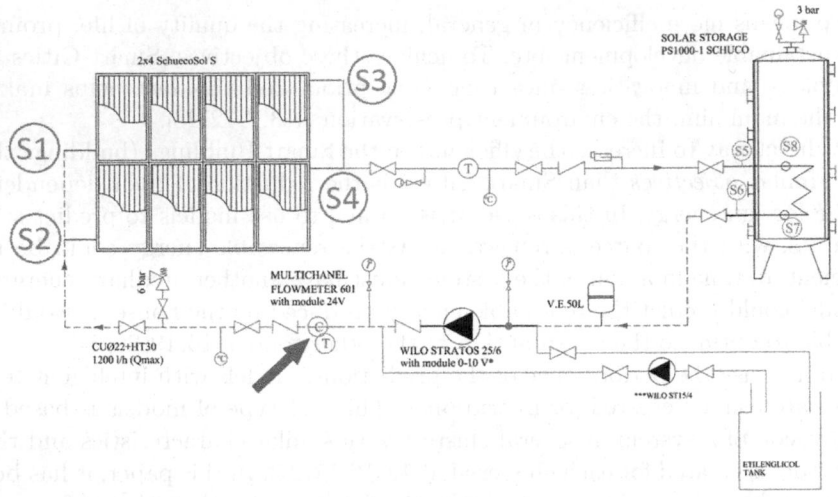

Fig. 1. Solar thermal energy layout

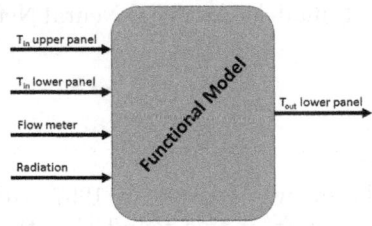

Fig. 2. Conceptual model of the research

The real model used is a hybrid intelligent model, that allows to increase the performance of the prediction. In these type of models, the dataset was divided into different clusters, and an intelligent regression technique is used for each cluster. In this research, it has been used different techniques (for clustering and regression) to compare the obtained results, and to achieve the best model.

The procedure to create the hybrid model is describe in the next steps. The first step is the division of the training dataset using different clustering techniques. In this research, Agglomerative clustering, Gaussian mixture, K-Means and Spectral clustering were used.

For the second step, the regression phase in each cluster, Artificial Neural Networks and Polynomial Regression have been used. Each algorithm has been trained with several configurations, changing internal parameters like the number of neurons, or the polynomial degree. The local models, the models for the different clusters, have been tested using *K-Fold* Cross Validation. The data for each cluster is sub-divided *K* times to create a model with the *K-1* groups of data, and to test this model with the non-used *K* group.

The performance for each regression algorithm in each cluster is calculated after train K models, and it has been used the K testing groups. The *K-Fold* Cross Validation allows to measure a more realistic performance of the tested algorithm as all the data available for training is used in the testing phase. To select the best local models, all the configurations for the different regression techniques have been compare, choosing the ones that achieve less error value.

3.1 Data Processing

The training data used in this paper was pre-processed before start the modelling process. Wrong samples were removed previously and, moreover, the samples during the night are not taken into account. Firstly, the dataset is composed with 52689 samples, but after discarding the wrong and the night samples, it has only 26665 samples.

A validation dataset has been extracted from the dataset, to perform a final test with data that has never included in the training phase. The samples from 5 days have been chosen randomly as validation data; a total of 366 samples are included. Then, the training dataset has 26299 samples.

The training dataset has been normalized to be in the range [0,1]. The Max-Min Scaler has been used for this propose [23,25].

3.2 Clustering Techniques

Agglomerative Clustering. This clustering technique is Hierarchical clustering type, and the idea is to build the clusters by merging or splitting them [8]. Dendrograms are used to represent this type of clustering technique.

Gaussian Mixture. In Gaussian mixture clustering technique [18], each cluster is defined by its centroid, covariance and the cluster's weight. It uses an Expectation-Maximization algorithm [9] to calculate the values of the mean, covariance and weight.

K-Means. This clustering algorithm tries to minimize the inertia by dividing the data in clusters [10]. K-Means algorithm uses only the mean to calculate the centroids.

Spectral Clustering. In Spectral clustering technique, the relation between the samples in each cluster is derive from similarity graphs.

3.3 Regression Algorithms

Artificial Neural Networks. One of the typical topology of the ANN is the Multi Layer Perceptron. This ANN is a basic feed-forward topology, as the signal *flows* always in the same way without any feedback [2,11,15].

Polynomial Regression. The regression algorithm called Polynomial Regression, is one of the eldest technique used for regression. It is based on the use of several basic functions with different degree, and the maximum degree of these functions define the degree of the Polynomial Regression.

4   Results

The results of this research are explained in this section. One of the objectives of this paper is to achieve a good method to find the optimal number of clusters for a dataset. The results section is divided in three different subsections, that includes the clustering results, the local model regression results, and the validation of the hybrid models.

4.1   Clustering Results

The optimal number of clusters have been extracted thanks to Silhouette, Calinski-Harabasz and Davies-Bouldin measurements, searching from two to nine as possible number of groups.

Once the clusters are created, the next step is the training the different regression algorithm for each one.

4.2   Regression Results - Local Models

The ANNs are configure with one hidden (or internal) layer, with a different number of neurons. The ANNs was trained to tune the internal coefficients of the neurons, and the number of neurons in the internal layer was varied from 1 to 15.

For the other regression algorithm, the Polynomial Regression, it has been used two configurations: first and second degree. The training is used to adjust the coefficients of the internal basic functions.

As an example of the local model regression results, Table 1 shows the Mean Squared Error (MSE) and the Mean Absolute Error (MAE) obtained for each configuration of the regression algorithm for the Aggloremative clustering technique.

There are another three tables similar than Table 1, one for each of the other clustering techniques. They are omitted to not extend unnecessary the results section, but in Table 2 there is a resume of the best regression algorithm for each local model of the different clustering technique.

Table 1. Regression errors for Agglomerative clustering

	MSE (Mean Squared Error)				MAE (Mean Absolute Error)			
	Cluster 1	Cluster 2	Cluster 3	Cluster 4	Cluster 1	Cluster 2	Cluster 3	Cluster 4
ANN1	0.0036	5.2942	0.0015	0.0008	0.0476	0.0049	0.0243	0.0224
ANN2	0.0032	5.2732	0.0014	0.0007	0.0451	0.0049	0.0235	0.0220
ANN3	0.0029	0.0000	0.0012	0.0007	0.0431	0.0049	0.0219	0.0213
ANN4	0.0028	5.1609	0.0013	0.0006	0.0422	0.0048	0.0216	0.0209
ANN5	0.0027	5.1184	0.0010	0.0007	0.0408	0.0047	0.0210	0.0215
ANN6	0.0026	0.0000	0.0010	0.0007	0.0399	0.0047	0.0204	0.0208
ANN7	0.0026	0.0000	0.0011	0.0006	0.0400	0.0047	0.0201	0.0204
ANN8	0.0030	4.6543	0.0010	0.0007	0.0400	0.0046	0.0197	0.0215
ANN9	0.0025	4.8737	0.0010	0.0007	0.0394	0.0046	0.0198	0.0212
ANN10	0.0026	8.7407	0.0010	0.0007	0.0397	0.0047	0.0196	0.0218
ANN11	0.0025	4.6378	0.0010	0.0007	0.0391	0.0045	0.0195	0.0209
ANN12	0.0025	0.0513	0.0012	0.0007	0.0392	0.0077	0.0201	0.0212
ANN13	0.0028	4.6712	0.0012	0.0007	0.0391	0.0045	0.0201	0.0214
ANN14	0.0024	4.6956	0.0013	0.0008	0.0386	0.0045	0.0198	0.0213
ANN15	0.0025	0.0001	0.0059	0.0007	0.0388	0.0047	0.0206	0.0210
Poly1	0.0035	5.2769	0.0014	0.0008	0.0472	0.0048	0.0230	0.0227
Poly2	0.0041	5.2414	0.0013	0.0008	0.0439	0.0048	0.0209	0.0224

Table 2. Best regression algorithms for the different clustering techniques

	Cluster 1	Cluster 2	Cluster 3	Cluster 4
Agglomerative clustering	ANN14	ANN11	ANN11	ANN7
Gaussian mixture	ANN10	ANN14	ANN15	ANN12
K-Means	ANN10	ANN14	ANN15	ANN12
Spectral clustering	ANN15	ANN9	ANN14	–

4.3   Validation Results

To validate the four different clustering technique, as it is explained in Sect. 3.1, it is used a different dataset, isolated from the training phase. The samples of five days is used to created this dataset. To predict the temperature using the created model, it is necessary to adjust the inputs to the range [0,1].

Figure 3 shows the results of the prediction obtained with the four hybrid models (one for each clustering technique). The shown values are in real temperature; scaled using the adjusting values calculated with the training dataset.

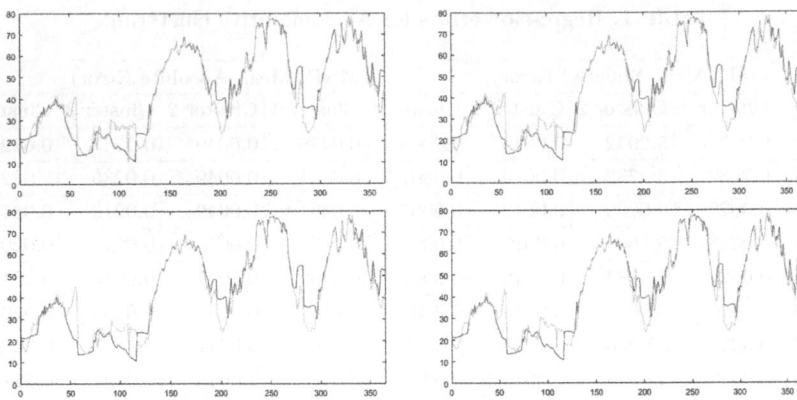

Fig. 3. Validation data for the different clustering technique. Real temperature in blue, and predicted in green (Color figure online)

5   Conclusions

In this work, there has been tested four clustering techniques, and the used of different measurements to achieve the optimal clusters number for a dataset. Agglomerative clustering, Gaussian mixture, K-Means and Spectral clustering has been used to divide the dataset.

To choose the optimal division for the used dataset, it has been used Silhouette, Calinski-Harabasz and Davies-Bouldin and measures for each configuration, obtained different optimal division for some clustering technique.

The results obtained in the validation test shows that the best error values have been achieve with Spectral clustering, dividing the dataset in three clusters. A maximum absolute error of $3.7\,°C$ is obtained for the predicted temperature, that represent less that 15 %. Final hybrid intelligent model includes three ANN.

Acknowledgments. Junta de Castilla y León - Consejería de Educación. Project: LE078G18- UXXI2018/000149. U-220.

References

1. Amin, M.: Smart grid. Public Utilities Fortnightly (2015)
2. del Brío, B., Molina, A.: Redes neuronales y sistemas borrosos. Ra-Ma (2006)
3. Caragliu, A., Del Bo, C., Nijkamp, P.: Smart cities in europe. J. Urban Technol. **18**(2), 65–82 (2011)
4. Casteleiro-Roca, J.L., Barragán, A.J., Segura, F., Calvo-Rolle, J.L.,Andújar, J.M.: Fuel cell output current prediction with a hybrid intelligent system. Complexity **2019** (2019)
5. Casteleiro-Roca, J.L., et al.: Short-term energy demand forecast in hotels using hybrid intelligent modeling. Sensors, **19**(11) (2019). https://doi.org/10.3390/s19112485

6. Casteleiro-Roca, J.L., Javier Barragan, A., Segura, F., Luis Calvo-Rolle, J., Manuel Andujar, J.: Intelligent hybrid system for the prediction of the voltage-current characteristic curve of a hydrogen-based fuel cell. Revista Iberoamericana de Automática e Informática Ind. **16**(4), 492–501 (2019)
7. Dameri, R.P.: Smart city definition, goals and performance. Smart City Implementation. PI, pp. 1–22. Springer, Cham (2017). https://doi.org/10.1007/978-3-319-45766-6_1
8. Defays, D.: An efficient algorithm for a complete link method. Comput. J. **20**(4), 364–366 (1977). https://doi.org/10.1093/comjnl/20.4.364
9. Dempster, A.P., Laird, N.M., Rubin, D.B.: Maximum likelihood from incomplete data via the EM algorithm. J. R. Stat. Soc. Ser. B (Methodol.) **39**(1), 1–22 (1977)
10. Forgy, E.W.: Cluster analysis of multivariate data: efficiency versus interpretability of classifications. Biometrics **21**, 768–769 (1965)
11. Harston, A.M.C., Pap, R.: Handbook of Neural Computing Applications. Elsevier Science, Amsterdam (2014)
12. Hollands, R.G.: Will the real smart city please stand up? intelligent, progressive or entrepreneurial? City **12**(3), 303–320 (2008)
13. Jia, M., Komeily, A., Wang, Y., Srinivasan, R.S.: Adopting internet of things for the development of smart buildings: a review of enabling technologies and applications. Autom. Constr. **101**, 111–126 (2019)
14. Jove, E., Casteleiro-Roca, J., Quintián, H., Méndez-Pérez, J., Calvo-Rolle, J.: Anomaly detection based on intelligent techniques over a bicomponent production plant used on wind generator blades manufacturing. Revista Iberoamericana de Automática e Informática Industrial **17**(1), 84–93 (2020)
15. Jove, E., Alaiz-Moretón, H., García-Rodríguez, I., Benavides-Cuellar, C., Casteleiro-Roca, J.L., Calvo-Rolle, J.L.: PID-ITS: an intelligent tutoring system for PID tuning learning process. In: Pérez García, H., Alfonso-Cendón, J., Sánchez González, L., Quintián, H., Corchado, E. (eds.) SOCO/CISIS/ICEUTE -2017. AISC, vol. 649, pp. 726–735. Springer, Cham (2018). https://doi.org/10.1007/978-3-319-67180-2_71
16. Jove, E., et al.: Attempts Prediction by Missing Data Imputation in Engineering Degree. In: Pérez García, H., Alfonso-Cendón, J., Sánchez González, L., Quintián, H., Corchado, E. (eds.) SOCO/CISIS/ICEUTE -2017. AISC, vol. 649, pp. 167–176. Springer, Cham (2018). https://doi.org/10.1007/978-3-319-67180-2_16
17. Jove, E., Casteleiro-Roca, J.L., Quintián, H., Méndez-Pérez, J.A., Calvo-Rolle, J.L.: A fault detection system based on unsupervised techniques for industrial control loops. Expert Syst. **36**(4), e12395 (2019). https://doi.org/10.1111/exsy.12395
18. McLachlan, G., Peel, D.: Finite Mixture Models. Wiley Series in Probability and Statistics, John Wiley & Sons Inc, Hoboken, NJ, USA September 2000. https://doi.org/10.1002/0471721182,http://doi.wiley.com/10.1002/0471721182
19. Montgomery, L., Magazine, E.H.: Home Automation: A Complete Guide to Buying, Owning and Enjoying a Home Automation System. EH Publishing, Inc., Framingham (2014)
20. Panteli, C., Kylili, A., Fokaides, P.A.: Building information modelling applications in smart buildings: From design to commissioning and beyond a critical review. J. Clean. Product. 121766 (2020)

21. Quintián, H., Casteleiro-Roca, J.-L., Perez-Castelo, F.J., Calvo-Rolle, J.L., Corchado, E.: Hybrid intelligent model for fault detection of a lithium iron phosphate power cell used in electric vehicles. In: Martínez-Álvarez, F., Troncoso, A., Quintián, H., Corchado, E. (eds.) HAIS 2016. LNCS (LNAI), vol. 9648, pp. 751–762. Springer, Cham (2016). https://doi.org/10.1007/978-3-319-32034-2_63
22. Rosenzweig, C., Solecki, W., Hammer, S.A., Mehrotra, S.: Cities lead the way in climate-change action. Nature 467(7318), 909–911 (2010)
23. Scikit-learn: Min max scaler (2018).http://scikit-learn.org/stable/modules/generated/sklearn.preprocessing.MinMaxScaler.html
24. Sinopoli, J.M.: Smart Buildings Systems for Architects, Owners and Builders. Butterworth-Heinemann, UK (2009)
25. Tomás-Rodríguez, M., Santos, M.: Modelling and control of floating offshore wind turbines. Revista Iberoamericana de Automática e Informática Industrial 16(4) (2019)
26. Vega Vega, R.A., et al.: Intrusion detection with unsupervised techniques for network management protocols over smart grids. Appl. Sci. 10(7), 2276 (2020)

A Fault Detection System for Power Cells During Capacity Confirmation Test Through a Global One-Class Classifier

Esteban Jove[✉][iD], José-Luis Casteleiro-Roca[iD], Héctor Quintián[iD],
Francisco Zayas-Gato, and José Luis Calvo-Rolle[iD]

University of A Coruña, CTC, Department of Industrial Engineering, CITIC,
Avda. 19 de Febrero S/n, 15405 Ferrol, A Coruña, Spain
{esteban.jove,jose.luis.casteleiro,hector.quintian,f.zayas.gato,
jlcalvo}@udc.es

Abstract. Power cells have presented an increasing popularity during last decades due to its importance in electric mobility, electronic devices and energy management systems. The international expansion of green policies to promote electric cars and renewable energies, has resulted in the need of ensuring their quality and reliability performance. In this context, detecting any early deviation from the correct operation must be addressed. Hence, this work is focused on the fault detection in a Lithium Iron Phosphate – LiFePO4 (LFP) cell. This is achieved by means of different one-class techniques, whose performance is assessed through artificially generated anomalies. After analysing the behaviour of each tested technique, the chosen classifier presents a successful performance.

Keywords: Power cell · Fault detection · Anomaly detection · One-class

1 Introduction

During last decades, different governments have developed green policies to address the critical problem of greenhouse gasses emission caused by fossil fuels use [2,21]. Among these policies, the promotion of renewable energies are especially relevant [4]. However, the use of renewable resources can result in intermittent energy generation. This problem can be tackled by means of power cells that store energy in cases where the generation exceeds the demand and supply energy when the demand is greater [5,9]. In this context, the implementation of Smart Grids is playing a key role in the system management [6,18].

Besides the importance of energy storage systems for renewable energies, the promotion of electric mobility alternatives has suffered a marked increase [10]. The main weakness of electric vehicles is their range, that can not compete with traditional technologies so far. Furthermore, the use of batteries are generalized in all portable electronic devices, like smartphones, tablets or laptops. All these

© Springer Nature Switzerland AG 2020
C. Analide et al. (Eds.): IDEAL 2020, LNCS 12490, pp. 477–484, 2020.
https://doi.org/10.1007/978-3-030-62365-4_45

applications are advancing toward low weight, great charging speed, and high efficiency in general terms [8].

Batteries present an electrolyte and two different electrodes, a cathode and an anode. A chemical red-ox reaction is generated and produces an electron flow between electrodes. In cases where the battery is charging, a reduction process happens at the anode and the cathode is oxidised. In case of discharge, the anode is oxidised and the cathode reduced [9].

Due to the high applicability of power cells, the scientific community has focused their efforts and interests on improving these devices. Then, ensuring a good performance and fulfilling quality, safety and reliability standards, emphasizes the importance of developing trustworthy tests process prior to commercialization. This means that the appearance of any kind of early deviation of the battery from its normal operation must be detected.

The general goal of this research is to develop a system capable of detecting fault situations during the test process of a Lithium Iron Phosphate – LiFePO4 (LFP). Then, with real data from a capacity confirmation test, registered during normal situations, different semi-supervised one-class algorithms are applied with the aim of modelling the battery behavior. These techniques, whose application led to significant good performance in many fault detection problems [12,15], are tested using a dataset with artificially created faults. The anomalies are generated by generating slight deviation in the State Of Charge (SOC) of the battery during the test.

The present document is divided into five different sections. After this section, a detailed explanation of the case of study and dataset is carried out. The different techniques used to validate the fault detection system are presented in next section. The obtained results are analysed in Sect. 4 and, finally, the conclusions and future works are exposed in last section.

2   Power Cell Capacity Confirmation Test

As stated above, it is necessary to develop a testing process to validate the behaviour of a power cell. Between other features, such as voltage, current or power, the capacity, measured in ampere-hour, is one of the most important in a cell. The capacity confirmation test (CCT) is a common procedure to evaluate the correct performance of a battery. This process, which is shown in Fig. 1, consists of forcing to the battery to a finite number of cycles, each one of them is divided into next steps:

1. First, the battery is charged by delivering to it a constant current from the tester. This implies that the battery voltage increases from V_a up to a value V_b.
2. Then, the current flow is cut and the battery voltage is settled until it reaches the value V_c.
3. After the resting step, a constant current is demanded by the tester until the battery reaches the value V_d.

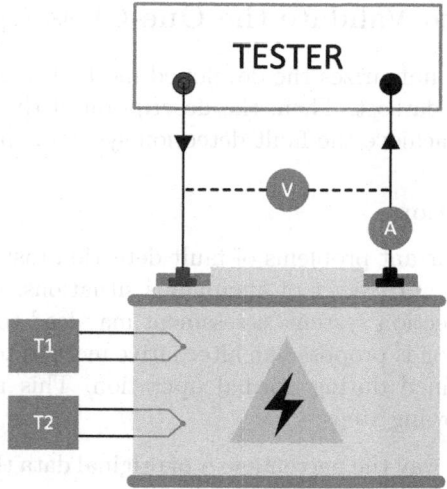

Fig. 1. Scheme of CCT

4. Finally, as it happens after the charging step, the battery is left to rest until it recovers the value V_a. At this moment, a cycle is finished and can be restarted.

This work is focused on a LiFeBATT X-1P [1] battery, composed of a LiFePO4 cell, with a nominal capacity of $8A \cdot h$, and $3.3V$ of nominal voltage. For this device, $V_a = 3\,V$, $V_b = 3.65\,V$, $V_c = 3.3\,V$ and $V_c = 2\,V$. During the process, the voltage, current and SOC are registered. Also, the temperature at two cell points $T1$ and $T2$ are measured. An example of the evolution of these variables during one test cycle is presented in Fig. 2.

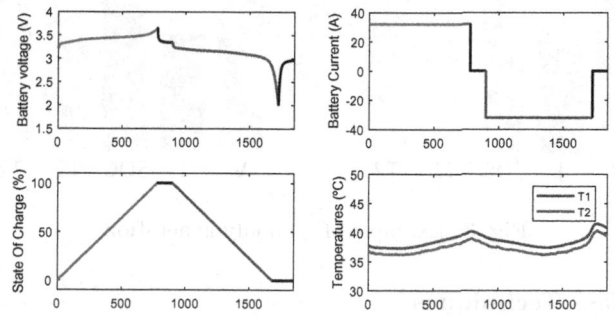

Fig. 2. Voltage and current during one test cycle

The dataset, gathered with one Hertz sample frequency, is comprised of the voltage, current, SOC and temperatures during CCT. At each cycle, approximately 1685 samples are registered. Nine cycles are tested, so the final dataset has 15165 samples.

3   Techniques to Validate the One-Class Approach

The present section summarizes the developed method to generate an anomaly set from the original dataset. Then, the description of the four one-class techniques considered to achieve the fault detection system is presented.

3.1   Fault Generation

One of the most important problems of fault detection tasks is the unfeasibility of obtaining a representative set of anomalous situations. Under these circumstances, the fault detection systems assessment may lead to misinterpretation.

Then, in this section is proposed an alternative method to generate anomalies from a dataset obtained during normal operation. This method is described according to the following steps:

1. Select in a random way the percentage α of original data that will be converted to anomalies.
2. For all the data chosen in step 1, the variables are modified a certain variation $\pm\beta$.

In this work, $\alpha = 25$, so 25 % of the original data was converted to anomalous instances. Then, the anomaly generation consisted of modifying the SOC value, which would represent a fault in the battery capacity. For each instance, the value of β was set at 5, 10, 15, 20 and 25 %. This sweep may help to describe the evolution of each technique when the deviation increases. An example of anomaly generation with $\beta = 15$ is shown in Fig. 3.

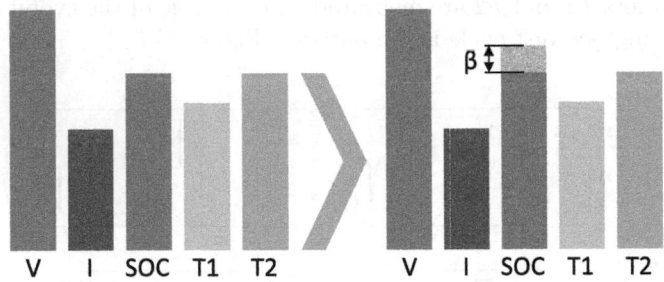

Fig. 3. Example of anomaly generation

3.2   One-Class Techniques

As stated in previous sections, the main goal is to implement a global classifier capable of detecting anomalies during the capacity confirmation test in a LFP battery. Then, the classifier presents the topology shown in Fig. 4. Using information about the current state, it must be able to identify the appearance of anomalous situations.

A brief introduction about the techniques applied to achieve this system are detailed in next subsections.

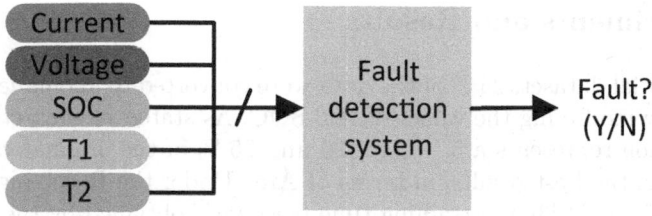

Fig. 4. Global classifier topology

Autoencoder. The use of Artificial Neural Networks (ANN) with Autoencoder topology has the main goal of replicating the input pattern at the output, by means of a nonlinear reduction in the hidden layers. After the dimension reduction, the data is decompressed at the output [19]. As this ANN is trained with data from only normal operation, the network is trained to replicate a certain target pattern. Then, when one test instance is anomalous, the ANN should not be able to replicate the input values, and the reconstruction error, calculated as the difference between input and output should be significantly higher [14].

Gaussian Model. One of the simplest and common way to face the anomaly detection consists of the application of probability density functions to model the training set. The use of Gaussian approximation represents a fast and effective way to determine the appearance of anomalies. The good performance of this technique is directly related to the normal shape of the training set [7]. The criteria to determine the anomalous behavior of a test sample is based on the Gaussian function value, whose threshold is adjusted during the training phase.

Minimum Spanning Tree. The well-known Minimum Spanning Trees (MST) technique aims to obtain a graph with a set of edges of an undirected connected segments with the lowest total weight [17]. This algorithm, initially developed and applied to power grids, can be used in one-class classification tasks [20]. When the graph is computed, the criteria to detect anomalies is based on the distance from the instance to the closest edge. If this value exceeds a certain threshold, it is labeled as anomalous [16].

Principal Component Analysis. Although Principal Component Analysis (PCA) algorithm has been used especially for dimension reduction problems [22,23], it can be applied to detect anomalies [12]. This approach is based on projecting the original data over the principal components, which represent the directions with greater variability. Then, the distance between a test sample and its projection is defined as the reconstruction error.

4   Experiments and Results.

From the initial dataset, 25 % of the data were converted to anomalies. They are generated by modifying the values of real SOC. As stated in subsection 3.1, the SOC deviation represents a 5, 10, 15, 20 and 25 % of the original value. Then, for each case, the best results, in terms of Area Under the Receiving Operating Curve (AUC, in %) [3] and training time (seconds), obtained by the four tested techniques are presented in Table 1.

Table 1. Obtained results for each technique

Technique	Deviation β (%)	AUC (%)	Training time (s)
Autoencoder	5	52.106	4.973
	10	55.077	7.298
	15	63.960	6.634
	20	59.892	7.521
	25	70.252	6.347
Gaussian	5	50.960	0.006
	10	52.737	0.005
	15	53.814	0.013
	20	54.609	0.011
	25	55.412	0.007
MST	5	67.800	36.382
	10	77.061	34.719
	15	81.319	41.458
	20	84.493	36.445
	25	87.120	39.015
PCA	5	82.893	0.041
	10	81.796	0.059
	15	83.026	0.034
	20	85.883	0.069
	25	89.719	0.065

5   Conclusions and Future Works

This work presents a fault detection system to determine the appearance of abnormal operation in a LFP power cell. Furthermore, a detailed analysis of the performance of four different one-class techniques is carried out. The use of Gaussian classifier is highlighted as the fastest technique, since its computation time is really low. However, its performance is not successful, so it must be discarded. Focusing on the classifier behaviour, the best AUC is achieved using

PCA for all SOC deviations. Then, it is the best technique for the case of study. Although MST can compete with PCA, it is significantly slower, which is a critical feature, especially if an edge computing topology is implemented. Despite the Autoencoder offer better results as the SOC deviation increases, it is far from PCA.

From the good performance shown in the results sections, it can be concluded that this method is presented as an interesting tool to determine slight degradation during battery test. This may help manufacturers to develop test procedures and improve battery reliability.

As future works, the possibility of applying clustering techniques to divide the data into different groups, could help to improve the classifier performance. Furthermore, using alternative one-class techniques, such as SVDD or kNN, or projection techniques [13] may lead to interesting results. Finally, a combination of this kind of one-class techniques with imputation techniques [11] could help to recover the missing data in case of sensor errors.

References

1. LiFeBATT X-1P 8Ah 38123 Cell (3 2011). http://www.solarvan.co.uk/Life/LiFeBATT8Ah.pdf
2. Aláiz-Moretón, H., Castejón-Limas, M., Casteleiro-Roca, J.L., Jove, E., Fernández Robles, L., Calvo-Rolle, J.L.: A fault detection system for a geothermal heat exchanger sensor based on intelligent techniques. Sensors 19(12), 2740 (2019)
3. Bradley, A.P.: The use of the area under the ROC curve in the evaluation of machine learning algorithms. Pattern Recog. 30(7), 1145–1159 (1997). https://doi.org/10.1016/S0031-3203(96)00142-2
4. Casteleiro-Roca, J.L., Gómez-González, J.F., Calvo-Rolle, J.L., Jove, E., Quintián, H., Gonzalez Diaz, B., Mendez Perez, J.A.: Short-term energy demand forecast in hotels using hybrid intelligent modeling. Sensors 19(11), 2485 (2019)
5. Casteleiro-Roca, J.L., Javier Barragan, A., Segura, F., Luis Calvo-Rolle, J., Manuel Andujar, J.: Intelligent hybrid system for the prediction of the voltage-current characteristic curve of a hydrogen-based fuel cell. Revista Iberoamericana de Automática e Informática Industrial 16(4), 492–501 (2019)
6. Casteleiro-Roca, Jet al.: Power cellsoc modelling for intelligent virtual sensor implementation. J. Sens. 2017 (2017)
7. Chandola, V., Banerjee, A., Kumar, V.: Anomaly detection: a survey. ACM Comput. Surv. (CSUR) 41(3), 15 (2009)
8. Chaturvedi, N., Klein, R., Christensen, J., Ahmed, J., Kojic, A.: Modeling, estimation, and control challenges for lithium-ion batteries. American Control Conference (ACC), pp. 1997–2002 June 2010
9. Chukwuka, C., Folly, K.: Batteries and super-capacitors. Power Engineering Society Conference and Exposition in Africa (PowerAfrica), pp. 1–6 2012. IEEE July 2012. https://doi.org/10.1109/PowerAfrica.2012.6498634
10. Wagner, I.: Worldwide number of battery electric vehicles in use from 2012 to 2018 (2019). https://www.statista.com/statistics/270603/worldwide-number-of-hybrid-and-electric-vehicles-since-2009/
11. Jove, E., et al.: Missing data imputation over academic records of electrical engineering students. Logic J. IGPL (12 2019). https://doi.org/10.1093/jigpal/jzz056,https://doi.org/10.1093/jigpal/jzz056

12. Jove, E., Casteleiro-Roca, J.L., Quintián, H., Méndez-Pérez, J.A., Calvo-Rolle, J.L.: Anomaly detection based on intelligent techniques over a bicomponent production plant used on wind generator blades manufacturing. Revista Iberoamericana de Automática e Informática industrial (2019)

13. Jove, E., Casteleiro-Roca, J.L., Quintián, H., Méndez-Pérez, J.A., Calvo-Rolle, J.L.: A new method for anomaly detection based on non-convex boundaries with random two-dimensional projections. Inf. Fusion, **65** (2020)

14. Jove, E., Casteleiro-Roca, J.L., Quintián, H., Simić, D., Méndez-Pérez, J.A., Luis Calvo-Rolle, J.: Anomaly detection based on one-class intelligent techniques over a control level plant. Logic J. IGPL (2020)

15. Jove, E., Gonzalez-Cava, J.M., Casteleiro-Roca, J.L., Pérez, J.A.M., Calvo-Rolle, J.L., de Cos Juez, F.J.: An intelligent model to predict ANI in patients undergoing general anesthesia. In: Pérez García, H., Alfonso-Cendón, J., Sánchez González, L., Quintián, H., Corchado, E. (eds.) SOCO/CISIS/ICEUTE -2017. AISC, vol. 649, pp. 492–501. Springer, Cham (2018). https://doi.org/10.1007/978-3-319-67180-2_48

16. Juszczak, P., Tax, D.M., Pe, E., Duin, R.P., et al.: Minimum spanning tree based one-class classifier. Neurocomputing **72**(7–9), 1859–1869 (2009)

17. Pettie, S., Ramachandran, V.: An optimal minimum spanning tree algorithm. J. ACM (JACM) **49**(1), 16–34 (2002)

18. Qian, H., Zhang, J., Lai, J.S.: A grid-tie battery energy storage system. In: Control and Modeling for Power Electronics (COMPEL), 2010 IEEE 12th Workshop on, pp. 1–5 (June 2010). https://doi.org/10.1109/COMPEL.2010.5562425

19. Sakurada, M., Yairi, T.: Anomaly detection using autoencoders with nonlinear dimensionality reduction. In: Proceedings of the MLSDA 2014 2nd Workshop on Machine Learning for Sensory Data Analysis, p. 4. ACM (2014)

20. Tax, D.M.J.: One-class classification: concept-learning in the absence of counter-examples [ph. D. thesis]. Delft University of Technology (2001)

21. Tomás-Rodríguez, M., Santos, M.: Modelling and control of floating offshore wind turbines. Revista Iberoamericana de Automática eInformática Industrial **16**(4) (2019)

22. Vega Vega, R., Quintián, H., Calvo-Rolle, J.L., Herrero, Á., Corchado, E.: Gaining deep knowledge of android malware families through dimensionality reduction techniques. Logic J. IGPL **27**(2), 160–176 (2019)

23. Wu, J., Zhang, X.: A pca classifier and its application in vehicle detection. In: IJCNN 2001 International Joint Conference on Neural Networks. Proceedings (Cat. No. 01CH37222), vol. 1, pp. 600–604. IEEE (2001)

A Deep Metric Neural Network
with Disentangled Representation for Detecting
Smartphone Glass Defects

Gwang-Myong Go[1,2], Seok-Jun Bu[1], and Sung-Bae Cho[1(✉)]

[1] Department of Computer Science, Yonsei University, Seoul 03722, South Korea
{scooler,sjbuhan,sbcho}@yonsei.ac.kr
[2] Samsung Electronics, Co., Ltd., Suwon 16706, South Korea

Abstract. For defect inspection using computer vision, deep learning models have been introduced to improve the conventional rule-based pattern analysis. A lot of data is a prerequisite to the success of them, but the on-the-spot industrial field suffers from lack of data. In this paper, we propose a deep metric neural network to improve the performance even with insufficient data imbalanced in class. The model is verified with the dataset of new products by evaluating the accuracy with 10-fold cross-validation. Our model is based on the data in the smallest category, 1.2 K, which achieves the highest performance of 90.42% using sampled pairs without using all the data for training. High accuracy has been achieved and proven applicability in the industry compared to the conventional machine learning models.

Keywords: Metric few-shot learning · Deep learning · Convolutional neural network · Smartphone glass inspection · Defect detection

1 Introduction

In the smartphone market, quality issues have continuously arisen in terms of production from the fact that glass, which has a fatal impact on surface quality accounts for more than 90 percent of the total products. The visual inspection by the worker is very difficult to meet the strict quality standards in terms of fatigue so it is necessary to introduce the automatic optical inspector (AOI) system inevitably [1]. The conventional approach to inspecting product defects is to examine the surface of the product using a computer vision algorithm [2–5]. However, since various types of defects existed on the surface of the product due to the complexity of the production process and management methods, the rule-based approach has a quantitative limitation and several deep learning models have been proposed to compensate for this.

In this process, sufficient datasets are very important for learning high-performance deep learning models that reflect industry trends, but it is very difficult and limited to collect defect data due to initial process setup and production model changes. In addition, the reliability of the obtained image is also declining due to the classification of data by

© Springer Nature Switzerland AG 2020
C. Analide et al. (Eds.): IDEAL 2020, LNCS 12490, pp. 485–494, 2020.
https://doi.org/10.1007/978-3-030-62365-4_46

the inspection worker from the labeling point of view. An important requirement is to construct a deep learning model using only single class data from normal materials that can be easily constructed. In this respect, designing a highly accurate and applicable model is one of the biggest issues in the field [6–14].

Figure 1 shows the process of acquiring the image data, and Fig. 2 analyzes the images acquired through this process. The image is used for learning and evaluation through a pre-processing process and is increased and refined as the production process progresses. From this point of view, constraints for models applicable to actual industrial sites exist as follows.

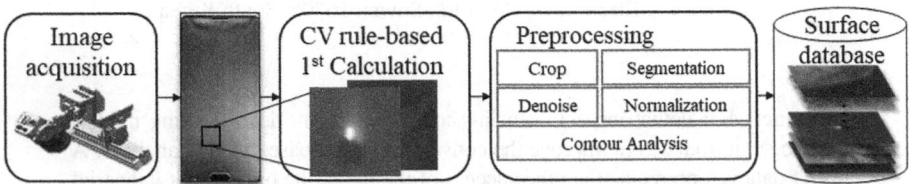

Fig. 1. Acquisition process of smartphone surface defect detection data

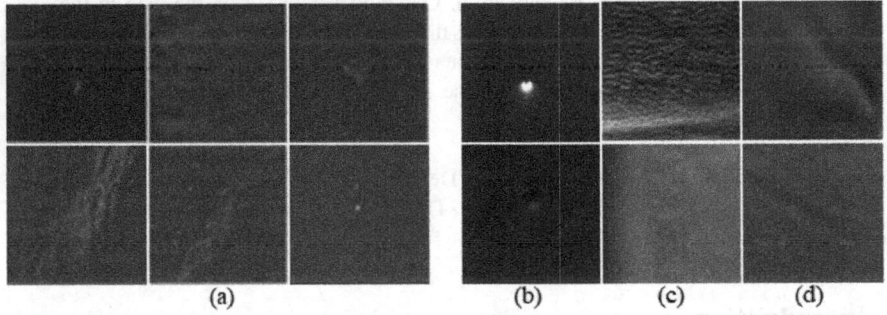

Fig. 2. Acquired data details: (a) OK, (b)–(d) NG

- The dataset for learning and evaluation should be changed periodically (every 6 months) and cannot be reused because the material and shape of the product are changed.
- In the early stage of the production process, an imbalance occurs in the number of data for each class to be classified, and model learning should be conducted with very small number of training data.
- There is a case where classification cannot be performed normally because the texture is similar.
- The time for model learning and reasoning should be small (Because it must be inferred using the desktop in the manufacturing facility).

From the viewpoint of satisfying the above constraints and applicable to industrial sites, this paper proposes a deep metric learning method. As suggested in several previous studies, a more suitable model is selected through comparison of Siamese network and Triplet neural network, which are the most robust models with the highest performance among them and are modified and applied to our domain. In Fig. 3, comparative analysis is performed on the two models that are reviewed (Table 1).

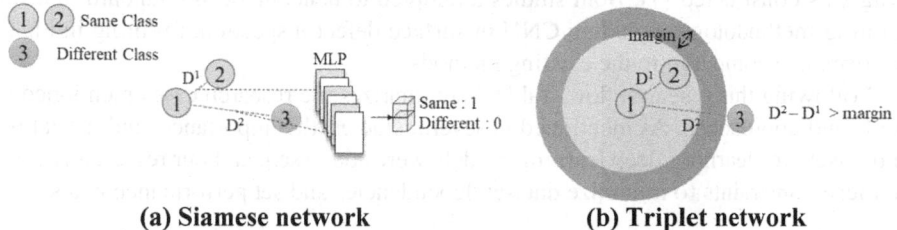

(a) **Siamese network**   (b) **Triplet network**

Fig. 3. Comparison of two models that extract input features and map them to the latent space based on distance: Siamese networks make up the same or different classes through MLP based on distance, but if the triplet network using margin parameters is used as the number of classes increases, accuracy according to the characteristics of input data can be improved.

Table 1. Characteristics of acquired data and specification of primary method

Type	Characteristics	Detection Method
(b) Dent	Imprinted shape	Contour analysis
(c) Mold mark	Imprinted mold	Blob shape
(d) Distortion	Warping shape	Pixel shift analysis

To overcome these constraints, this paper proposes a deep metric neural network with a hierarchical structure that allows the neural network to directly learn the expression of separation for each class based on the dataset most recently acquired at the production site and achieves high accuracy for similar inputs. The proposed method using the dataset of about 1 k per class for the corresponding model achieves higher classification accuracy than the conventional deep learning algorithms, and the model is qualitatively analyzed through the analysis of misclassification data clustering in the visualized feature space.

2 Related Works

The recent deep learning-based image defect detection field is divided into two main research fields: pre-processing and feature extraction [6, 8, 9, 11, 12] and modeling based on traditional vision algorithms [2–5]. The defect detection model using the deep learning approach has attracted a lot of attention due to its high accuracy, efficiency, and automated functions, and has a great influence on the quantitative rule-base detection system utilizing the computer vision.

Borwankar et al. presented a data-driven approach to the field of surface defect inspection by combining traditional image processing algorithm and machine learning-based classification algorithm [5]. Natarajan et al. proposed the ensemble of multiple convolutional neural networks (CNNs) for practical complement in terms of accuracy, and an end-to-end defect classification network combining feature extraction and modeling was constructed [7]. Both studies attempted to benchmark the standard machine learning methodology including CNN in surface defect inspection, resulting in higher performance compared to the existing methods.

Following this research flow, Table 2 summarizes the research cases mentioned by fields and approaches. As mentioned in several studies, the importance, and limitations of datasets for learning deep learning models were addressed, and our research is based on these constraints to minimize dataset dependencies and set performance goals.

Table 2. Related works on defect detection using machine learning algorithms

Authors	Image preprocessing	Spatial feature extraction, modeling
Zhou [14]	–	CNN, Class activation map
Borwankar [5]	Rotated wavelet transform	k Nearest neighbor
Natarajan [7]	–	Transfer learning, CNN ensemble
Dong [9]	CNN-based segmentation	U-Net, Random Forest
Fu [10]	Non-uniform illumination, Camera noise, Motion blur	Pre-trained squeeze-net (CNN)
Sun [13]	–	Transfer learning, CNN, ImageNet
Yang [11]	CNN	Convolutional autoencoder
Staar [6]	Flip, shift augmentation	CNN triplet learning

3   The Proposed Method

Figure 4 shows an overall model of smartphone glass defect detection in which a separate neural network is arranged after learning to extract the characteristics of class based on the product image collected through the tester optical system in the production process, to generate a learning pair for the triplet neural network among deep metric neural networks, and to arrange the relative distance between each of the feature vectors through in the latent space based on the classification list [15–17].

The image obtained by the optical system is first determined using the traditional computer vision pattern recognition rule, and then goes through crop, denoise, contour analysis, segmentation, and normalize processes. It does not utilize the pre-processed data as a whole pair, but only a part of it is used for learning and evaluation through batch sampling.

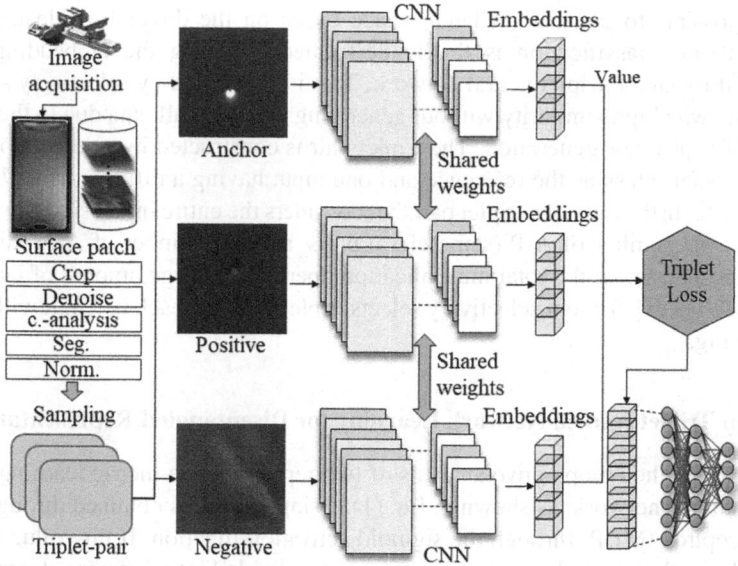

Fig. 4. Overview of the entire glass inspection detection process: The image collected through the optical system is pre-processed, and a triplet pair is generated through a sampling process.

3.1 Input/Output Characteristics

As can be seen through the misclassification analysis in the previous study, CNN ensemble, if input textures are similar, classification cannot be performed normally, so the reliability of the entire model is reduced [1]. Based on this, if the characteristics of the new dataset are analyzed in accordance with the previous research results and the constraints of the study, it can be classified into inter-class similarity and intra-class difference (in Fig. 5). This is the input with a high probability of misclassification. For this reason, input analysis was preceded.

The proposed model, aiming to achieve high accuracy for inputs with high similarity, allows the neural network to directly learn the disentangled representation by extracting the characteristics of the similarity of the corresponding inputs.

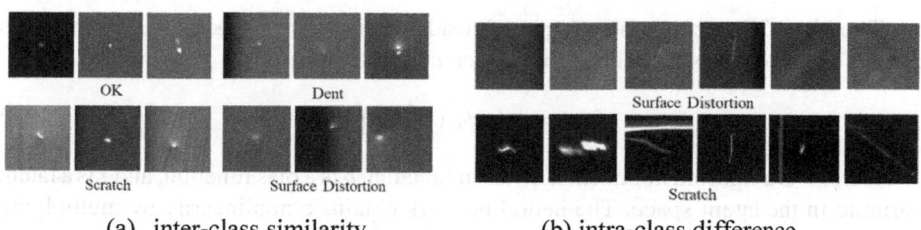

    (a) inter-class similarity   (b) intra-class difference

Fig. 5. Analysis of input variation: Analysis of the cases where the intra-class difference is larger than the model to learn and classify for class classification.

It is possible to construct a latent space based on the direct Euclidean distance between them. Classification is finally performed by using the embedding vector extracted through the triplet neural network. This is performed by selectively sampling only inputs with high similarity without generating a pair for all data due to the characteristics of triplet pair generation. The triplet pair is constructed by selecting one input having the same class as the reference and one input having a different class based on one reference. In the case of a triplet pair that considers the entire input, since the number of classes, the number of A-P (same class) pairs, and the number of A-N (difference class) pairs increase to the total multiplication operation, a large amount of learning is required, so our experiment selectively selects triplet pairs for each batch. It will be used for screening.

3.2 Deep Triplet Neural Network Learning for Disentangled Representation

Figure 3 shows the comparative analysis of two representative metric learnings. In the case of Siamese network, as shown in Eq. (1), a single value is obtained through multilayer perceptron (MLP) through the sigmoid activation function. If this value is one, it is divided into the same class, and if it is zero, it is divided into different classes.

$$\sigma\left(\sum_j \alpha j \left| h_{1,D1-1}^{(j)} - h_{2,D2-1}^{(j)} \right|\right) \tag{1}$$

In this case, the same or different classes are configured through the MLP based on the distance, but if the triplet network using the margin parameter is used as the number of classes increases, accuracy according to characteristics of input data can be improved.

Conversion operation ϕ_c which is designed to model spatial correlation by learning useful filters based on data and pooling operation ϕ_p, which extracts the representative value of the input, can be represented by Eq. (2) for the i-row j-column node output x_{ij}^l in the lth layer. At this time, the pooling distance τ for the $(m \times m)$-sized convolution filter w_f and the $(k \times k)$-sized pooling area is used.

$$\phi_c^l(\bar{x}) = \sum_{a=0}^{m-1} \sum_{b=0}^{m-1} w_{ab} x_{(i+a)(j+b)}^{l-1}$$
$$\phi_p^l(\bar{x}) = \max x_{ij \times \tau}^{l-1} \tag{2}$$

The given 2D vector x is compressed inside the neural network by the computation of Eq. (3) and outputs the reconstruction feature vector.

$$f(x) = \sigma(w\phi_C(x) + b) = z \tag{3}$$

Here, σ is a sigmoid nonlinear activation function, b is a bias function, and z is a latent variable in the latent space. The neural network obtains a nonlinearity by multiplying the input x by a weight and applying an activation function by adding a bias constant to finally obtain a vector representation in the latent space. This can be interpreted as intuitively modeling the characteristics of the input.

We define the triplet loss using Euclidean distance in the expression space of each feature vector by grouping three neural networks that share each weight (4).

$$L(A, P, N) = max\Big(\|f(A) - f(P)\|^2 - \|f(A) - f(N)\|^2 + \alpha, 0\Big) \qquad (4)$$

Here, A and P are selected from the same classification class, and N is selected from different classes from A and P and used for learning the neural network. The A, P, and N training pairs perform sampling without using the entire data. Sampling is performed in batch-random strategy. In the experiment, 450 A-N pairs were used based on 450 A-P pairs. In the case of a triplet pair considering the entire input, a large amount of learning is required because the number of classes, the number of A-P pairs, and the number of A-N pairs increase to the overall multiplication operation. Therefore, in our experiment, triplet pairs are selectively selected and used for each batch.

4   Experimental Results

4.1   Dataset

The dataset for experiments was collected from the actual production process, consisting of the product images of the most recent production model: a total of 25 k gray scale images in four categories. Table 3 shows the specification of the dataset. Because the imbalance occurs for each classification class, a model is required to overcome this. Finally, they are normalized from -1 to 1 with a pixel resolution of 112×112 and used for training and evaluation of the models.

Table 3. Input Data Characteristics

	Dent	Mold	Fold	Control
Instances	4,636	1,390	1,210	18,574
Original size	112×112	Flexible	Flexible	112×112
Re-size	112×112			
Pre-processing	crop $\rightarrow$ contour analysis $\rightarrow$ segmentation $\rightarrow$ normalize			

4.2   Performance Evaluation

Figure 6 shows the results of 10-fold cross-validation of classification accuracy with machine learning algorithms including the proposed model. In the experiment, sampling was performed with all 5 K data considering the time complexity. In the case of the proposed triplet neural network, the classification accuracy increases to 90.42%. The SVM only achieves the classification accuracy of 71.90%. The validity of the proposed model is proved as a result of learning a deep metric neural network to solve the problem of classification of similarity of inputs for analysis of misclassification in previous studies.

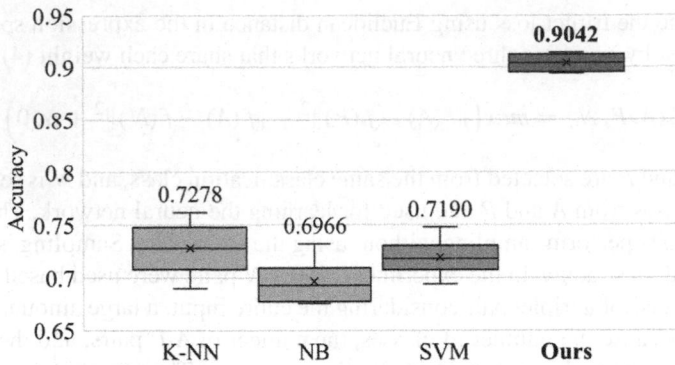

Fig. 6. 10-fold cross validation result: Since data with a small number of imbalances was used, the accuracy was improved in the traditional machine learning methods. (K-NN: K-Nearest Neighbor, NB: Naïve Bayes, SVM: Support Vector Machine)

To perform an analysis of the confusion matrix misclassification cases in Table 4, Fig. 7 visualizes the activation values of the layer just before the output with the t-SNE algorithm and tracks the misclassification data. Table 5 analyzes the precision and recall of the proposed model. Since the deep metric neural network is used, it can be confirmed that the recall is high.

Table 4. Confusion matrix analysis

		Predicted			
		Dent	Mold	Fold	Control
Actual	Dent	326	6	14	29
	Mold	4	347	2	14
	Fold	8	8	337	12
	Control	24	11	17	293

4.3 Defect Analysis

Figure 7 shows the comparison of the proposed model by analyzing t-SNE visualization by analyzing the results of the dataset that is misclassified in the previous study [1]. As an improvement in the previous work, first, in case of input with similar texture among different class data, the misclassification cases decrease, and second, the degree of clustering of the same class increases to make the model more robust in classification. Finally, the model training is conducted using a small amount of dataset with imbalance than the previous work.

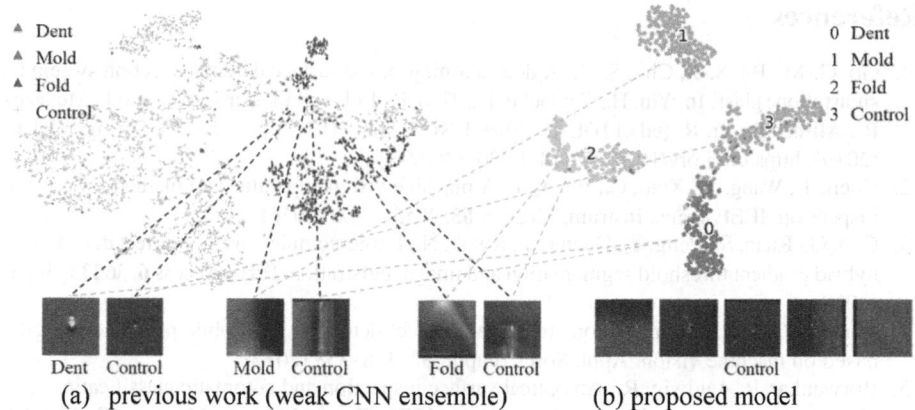

(a) previous work (weak CNN ensemble)   (b) proposed model

Fig. 7. Comparison of prior work with t-SNE clustering: As a result of tracking the input that was misclassified in the previous study, normal classification was performed, and clustering was normally performed for each class to be classified.

Table 5. Precision and recall for defect types

Defect type	Dent	Mold	Fold	Control	Average	**Accuracy**	F1-score
Precision	0.9006	0.9328	0.9108	0.8419	0.8965	**0.8974**	0.8967
Recall	0.8693	0.9455	0.9233	0.8493	0.8969		

5 Conclusions

In this paper, we proposed a hierarchical deep metric model that can improve the misclassification due to the similarity of input of imbalanced learning data between classes and when the defect data acquisition is limited for the glass defect detection and classification model. The contribution of this paper is that a dataset in the actual process and a model that satisfies the aforementioned constraints are proposed. The method is based on the deep ensemble model of the previous work through the metric neural network that introduced the batch sampling method, and we proved the robustness of the proposed model through quantitative analysis of precision and recall.

In the future, additional misclassification will be analyzed to track the misclassification to improve the model constituting the deep metric neural network, improve the method for constructing data for curriculum learning, and modify the loss function based on n-dimensional Euclidean distance.

Acknowledgments. This work was partly supported by Institute of Information & Communica tions Technology Planning & Evaluation (IITP) grant funded by the Korean government (MSIT) (No. 2020-0-01361, Artificial Intelligence Graduate School Program (Yonsei University)) and Samsung Electronics Co., Ltd.

References

1. Go, G.-M., Bu, S.-J., Cho, S.-B.: A deep learning-based surface defect inspection system for smartphone glass. In: Yin, H., Camacho, D., Tino, P., Tallón-Ballesteros, Antonio J., Menezes, R., Allmendinger, R. (eds.) IDEAL 2019. LNCS, vol. 11871, pp. 375–385. Springer, Cham (2019). https://doi.org/10.1007/978-3-030-33607-3_41

2. Chen, T., Wang, Y., Xiao, C., Wu, Q.J.: A machine vision apparatus and method for can-end inspection. IEEE Trans. Instrum. Measur. **65**, 2055–2066 (2016)

3. Cao, G., Ruan, S., Peng, Y., Huang, S., Kwok, N.: Large-complex-surface defect detection by hybrid gradient threshold segmentation and image registration. IEEE Access **6**, 36235–36246 (2018)

4. Jian, C., Gao, J., Ao, Y.: Automatic surface defect detection for mobile phone screen glass based on machine vision. Appl. Soft Comput. **52**, 348–358 (2017)

5. Borwankar, R., Ludwig, R.: An optical surface inspection and automatic classification technique using the rotated wavelet transform. IEEE Trans. Instrum. Measur. **67**, 690–697 (2018)

6. Staar, B., Lutjen, M., Freitag, M.: Anomaly detection with convolutional neural networks for industrial surface inspection. Procedia CIRP **79**, 484–489 (2019)

7. Natarajan, V., Hung, T.Y., Vaikundam, S., Chia, L.T.: Convolutional networks for voting-based anomaly classification in metal surface inspection. In: IEEE International Conference on Industrial Technology, pp. 986–991 (2017)

8. Gupta, E., Kushwah, R.S.: Combination of global and local features using DWT with SVM for CBIR. In: International Conference on Reliability, Infocom Technologies and Optimization, pp. 1–6 (2015)

9. Dong, X., Taylor, Chris J., Cootes, Tim F.: Small defect detection using convolutional neural network features and random forests. In: Leal-Taixé, L., Roth, S. (eds.) ECCV 2018. LNCS, vol. 11132, pp. 398–412. Springer, Cham (2019). https://doi.org/10.1007/978-3-030-11018-5_35

10. Fu, G., Sun, P., Zhu, W., Yang, J., Cao, Y., Yang, M.Y., Cao, Y.: A Deep-learning-based approach for fast and robust steel surface defects classification. Opt. Lasers Eng. **121**, 397–405 (2019)

11. Yang, H., Chen, Y., Song, K., Yin, Z.: Multiscale feature-clustering-based fully convolutional autoencoder for fast accurate visual inspection of texture surface defects. IEEE Trans. Autom. Sci. Eng. **99**, 1–18 (2019)

12. Krizhevsky, A., Sutskever, I., Hinton, G.E.: Imagenet classification with deep convolutional neural networks. In: Advances in Neural Information Processing Systems, pp. 1097–1105 (2012)

13. Sun, J., Wang, P., Luo, Y.K., Li, W.: Surface defects detection based on adaptive multiscale image collection and convolutional neural networks. IEEE Trans. Instrum. Measur. **68**(12), 1–11 (2019)

14. Zhou, B., Khosla, A., Lapedriza, A., Olivia, A., Torralba, A.: Learning deep features for discriminative localization. In: IEEE Conference on Computer Vision and Pattern Recognition, pp. 2921–2929 (2016)

15. Schroff, F., Kalenichenko, D., Philbin, J.: Facenet: a unified embedding for face recognition and clustering. In: IEEE Conference on Computer Vision and Pattern Recognition, pp. 815–823 (2015)

16. Hoffer, E., Ailon, N.: Deep metric learning using triplet network. In: Feragen, A., Pelillo, M., Loog, M. (eds.) SIMBAD 2015. LNCS, vol. 9370, pp. 84–92. Springer, Cham (2015). https://doi.org/10.1007/978-3-319-24261-3_7

17. Wu, S., Wu, Y., Cao, D., Zheng, C.: A fast button surface defect detection method based on siamese network with imbalanced samples. Multimedia Tools Appl. **78**, 34627–34648 (2019)

Improving Performance
of Recommendation System Architecture

Gil Cunha[1] ⓘ, Hugo Peixoto[1,2] ⓘ, and José Machado[1,2(✉)] ⓘ

[1] University of Minho, Campus Gualtar, 4710 Braga, Portugal
`a77249@alunos.uminho.pt`
[2] Centro Algoritmi, University of Minho, Campus Gualtar, 4710 Braga, Portugal
`{hpeixoto,jmac}@di.uminho.pt`

Abstract. The exponential appearance of online stores has implied higher market competitiveness and, consequently, companies need to adopt certain strategies to obtain greater prominence and gain clientele. This paper explores an architectural approach to incorporate a recommendation system in online stores, in order to offer a solution to achieve those goals. Developing the recommendation system infrastructure with NodeJS, based on a REST API, and according to microservices architecture concepts, has proven to be very efficient when it comes to managing great volumes of requests and data, and be capable to serve multiple tenants within a short response time. Clustering techniques were also implemented to increase the system's performance and capability of handling requests.

Keywords: E-commerce · Recommendation system · Software architecture

1 Introduction

Over the years, electronic commerce (*e-commerce* [10]) has been growing with the advancement of technology and the existence of easier ways for people to do business online. Since consumers are able to search and compare different products from diverse sales platforms, it is crucial for companies to captivate and keep their clients interested. In order to maintain customers' loyalty, it is necessary to provide them with a good experience during the purchasing process in the store. Offering a personalized service suggests to users that they are important as a client to the store and gives them confidence in the company and its products.

Such a personalized experience can be achieved with the implementation of a recommendation system on the sales platform. A recommendation system (RS) is able to present products to shoppers according to their expectations and needs, based on various factors about consumers and market information.

The implementation of a recommendation system aims to increase the number of clients, sales and, consequently, the profit margin, by solving some of the

© Springer Nature Switzerland AG 2020
C. Analide et al. (Eds.): IDEAL 2020, LNCS 12490, pp. 495–506, 2020.
https://doi.org/10.1007/978-3-030-62365-4_47

problems mentioned above and providing platforms with an advantage in the market compared to competitors.

The subject of this paper is, therefore, part of the development of a recommendation system and its implementation in e-commerce platforms, focusing more specifically on its *architectural component*. This component reflects the structure of the entire system, contemplating the development of an infrastructure capable of extracting, storing and analyzing relevant data from e-commerce platforms, for the integration of a *recommendation engine*. This infrastructure is responsible for managing data and communication necessary for the proper functioning of the system and presenting recommendations in online stores, thus functioning as a bridge between e-commerce platforms and the system's recommendation engine. The quality of data used for recommendations has an impact on the accuracy of results and the system's performance. In addition, its potential can be extended to explore various others applications, mainly in the area of *Business Intelligence* [11], to support in decision making and adopt better business strategies.

The project was developed under the context of the *Beevo* [3] company. Beevo provides e-commerce *Business to Client (B2C)*, *Business to Business (B2B)* and *Business to Employee (B2E)* solutions, for mid-market and large companies, offering a digital platform for their *e-business*. The company builds and maintains other companies' e-commerce platforms, including online stores and respective business logic. The recommendation system architecture was built to serve all platforms of Beevo's domain, in order to increase its arsenal of services available to their customers, improve their online stores quality and expand the *Business Intelligence* (BI) sector.

In this document, however, Machine Learning (ML) algorithms and techniques applied in the *recommendation engine* to generate recommendations are not be exposed.

2   State of the Art

A recommendation system [13], in the e-commerce area, consists of a mechanism that applies one or more ML techniques to data from online sales platforms, with the objective of suggesting products or services to its clients. In most cases, recommendations for a certain client are represented by sets of items, sorted by the probability of that client buying them.

There are several types of approaches to produce recommendations, the most popular being *Collaborative Filtering* [14] and *Content-Based Filtering* [9]. With this in mind, these are the main types of recommendation addressed in this project:

– *Collaborative Filtering:* The premise in this type of approach is the search for similarities between customers, according to their actions and preferences. *User-based* recommendations take into account the similarity between customers' profile, i.e., products purchased by a certain client will be recommended to another client who has similar tastes and behaviors. On the other hand, *item-based* recommendations are supported in products' characteristics. For example, user A has a similar buying pattern as the user B. Consequently, products with similar attributes to those that user A has purchased in the past, may be suggested to user B.

- *Content-based Filtering:* In this type of approach, the user's shopping history is important. The characteristics of products from previous purchases made by the client are analyzed and compared with the remaining candidate products. Products that have more in common with those that the user has purchased are recommended. For example, in a certain online music store, a user bought some albums in the Jazz category. According to the user's shopping history, the system may recommend other similar albums, i.e. of the same category (Jazz), to that user in the next visit to the online store.

These and other approaches have their benefits and drawbacks in terms of recommendations accuracy, depending on the context in which they are inserted. In order to rectify the disadvantages that each method presents and to make the system more robust, a *hybrid approach* is generally adopted. The objective of this approach is to combine several techniques so to fix the limitations of some with the advantages of others. In this case, by combining collaborative and content-based filtering. This way, both concepts are used to form a more efficient recommendation system. Following the previous example, with a hybrid approach, the system would recommend not only other Jazz albums to the client (content-based filtering), but would also take into account the album's rating given by the other users (collaborative filtering).

Developing an efficient recommendation system may require a large economic investment and time, and this can prove to be an obstacle for companies with less economic power. In today's market, *SaaS (Software as a Service)* and *PaaS (Platform as a Service)* Recommendation Systems [1] offer several solutions to this situation. The system is hosted in third-party servers, which are responsible for the system's updates and availability, whereas companies only have to adapt their platforms to communicate with the system to obtain recommendations.

There are several SaaS/PaaS recommendation systems on the market, such as Yusp and Amazon Personalize systems. **Yusp** [8] is a personalization engine, developed by Gravity R&D company, the same team that competed in the Netflix Prize [5]. This service offers customization features for e-commerce platforms, having several case studies from large companies in which their revenues have increased significantly thanks to these solutions. It provides control dashboards where the client can customize and adapt the recommendation engine to the needs of his platform and his customers. It is also possible to obtain detailed analysis reports, thus giving several insights about the business to help make better decisions for the future.

Developed by Amazon, **Amazon Personalize** [2] is a PaaS that allows its users to create and implement a personalized recommendation system on their online stores. To start the process of integrating this service, the user provides information about the platform and its customers, and inventory of items they wish to recommend. Then, Amazon Personalize consumes this information and, in background, processes and analyzes it, by selecting the most relevant data and applying the ML algorithms that best suit that information. Lastly, the system trains and adjusts recommendation models to the context of online platforms. To access recommendations, the user simply needs to request them through an API or JavaScript library provided by the service. Users may integrate widgets on their platforms, such as showcases or recommendation banners, which connect to the service recommendation models through its API. When it is necessary to show recommendations for a certain client, widgets send a request to the service, which in turn returns a list of product IDs recommended for that specific client.

In general, the recommendation process workflow is illustrated in the following diagram (Fig. 1):

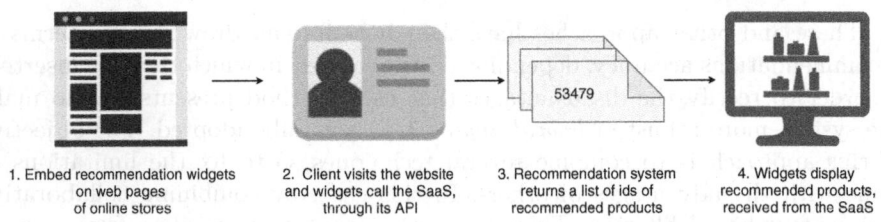

| 1. Embed recommendation widgets on web pages of online stores | 2. Client visits the website and widgets call the SaaS, through its API | 3. Recommendation system returns a list of ids of recommended products | 4. Widgets display recommended products, received from the SaaS |

Fig. 1. SaaS recommendation system workflow

3   Architecture

The architecture of the recommendation system encompasses not only the communication between the *recommendation engine* and the e-commerce platforms but also between components within the system. The system's workflow can be divided into the following steps:

1. Extract data from e-commerce platforms, necessary for the engine to train recommendation models. This data is related to clients of online stores, shopping activities and item inventory.
2. Store the extracted data in a database capable of adapting to different contexts of distinct e-commerce platforms. The system presents tools capable of processing and analyzing the data collected, in order to provide various *Business Intelligence* features to the company, such as future perspectives on its business and support in decision making.
3. After recommendations being generated by the engine, the system stores and makes them available at platforms' demand, whenever necessary.

Nevertheless, while developing an architecture that corresponded to the imposed objectives, several **challenges** were raised:

- *Custom data sets:* E-commerce platforms from different contexts will be considered and the architecture must be able to store different types of data.

- *Data analysis:* The data analysis tools used must be versatile and capable of processing variables of various types to create informative and useful dashboards.
- *Cold Start problem:* When entities (users or items) are recently registered in the platform, the recommendation system has limited information about them to be able to produce accurate recommendations. Nevertheless, new customers should get relevant results and new products should be included in recommendations.
- *Scalability:* The RS must deal with a large number of items and active users simultaneously, maintaining a short response time and good performance.
- *Multi-tenancy:* The problem of multitenancy refers to a software architecture in which an instance of the application is hosted on a server and serves multiple users. The system must be generic and capable of handling several different platforms and users.
- *Security:* Data used to produce recommendations must be protected, as they are related to confidential information, such as personal client information, products and order history. Therefore, privacy must be protected.

Taking the previous points into consideration, it was defined that the **approach** to build this architecture would be based on an Application Programming Interface (API), in JavaScript, supported on the concepts of Platform as a Service (PaaS), Representational State Transfer (REST) and the State of the Art referred in this document.

In the architecture's design plan, it was determined that the recommendation system would be hosted on a server maintained by the company itself, to serve the e-commerce platforms of its domain. The communication process between RS and platforms is accomplished through an API, developed based on REST principles, due to the advantages and increasing usage of this architectural style. To achieve this, *Node.js* was used as the server engine with great performance, scalability and lightweight, building the API in JavaScript. As it can be observed on the **diagram** (Fig. 2), each e-commerce platform communicates with the RS through an application, designated as *Business Intelligence Application*. This application was developed to operate as a *gateway*, connecting platforms to the system through its *API*, which in turn acts as an intermediary between e-commerce platforms and the recommendation engine. Thus, both components form a communication channel, between online stores and the RS, handling platforms' requests and delivering recommendation results, produced by the engine.

The reason to choose *Elasticsearch* is the unstructured data expected in the recommendation process and the need for a generic architecture and adaptable database. Elasticsearch allows the use of JSON documents, meaning that fields can vary from document to document and data structure can be changed over time. Therefore, it has advantages both in storage structure, performance and flexibility. It can handle entities with distinct attributes, e.g., different products may have different characteristics. Data from POST requests is stored in the Elasticsearch database. Built to work along with Elasticsearch, *Kibana* is a powerful data

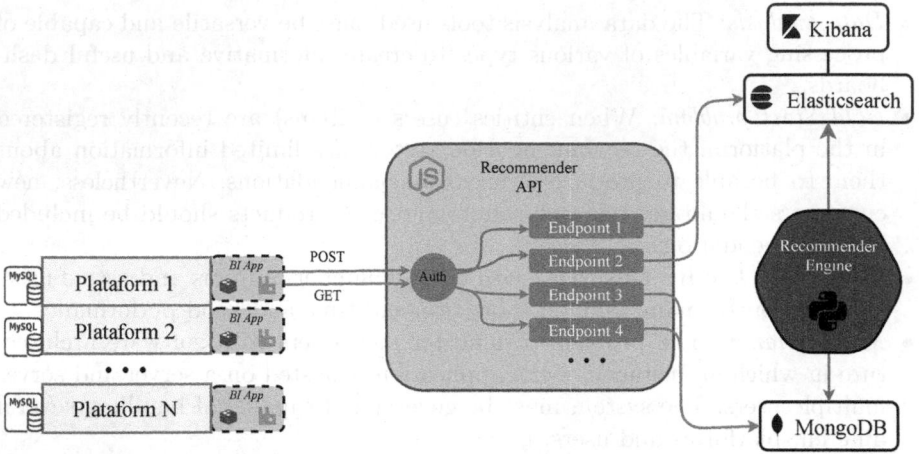

Fig. 2. Recommendation System architecture

analytics and visualization tool, that enables users to create dashboards on top of the content stored on Elasticsearch. Thus, it complements Elasticsearch, increasing its potential to explore e-commerce business intelligence opportunities.

The *Recommendation Engine* is the core of the system. It is responsible to produce recommendations from various types, based on platform e-commerce data. The architecture component assures any type of data necessary to this component. When it comes to producing recommendations, the engine connects to the Elasticsearch database so it can consume the data previously stored and apply ML techniques. This process is done in *offline mode*, where results are refreshed once every day, by default.

After this process, it stores recommendation results in a MongoDB database. *MongoDB* stores data in the form of JSON style documents, which keeps the consistency with data format used in HTTP requests and Elasticsearch. This database has high query performance and is easy to scale, making it a good choice to preserve recommendation results. These results must be always ready on platform demand, being possible with MongoDB's replication, high availability and fast access to data.

For security concerns, each request is filtered by an *authentication* and *authorization* mechanism, before reaching an endpoint. This security measure validates and ensures that the author of the request has the right permissions to make use of the API, protecting it from potential threats. It was implemented authentication, with JWT (*JSON Web Tokens* [6]), and authorization mechanisms with an *Access Control List* (ACL).

Since there are several different contexts within the business of an online store, there was a necessity of having a flexible **recommendation structure** to present different types of recommendations, depending on which fits better on the context of the web page being displayed to the user. Recommendations are based on data extracted from the platforms, which is related to the three unique **entities** considered by the recommendation system:

- *Clients:* represents customers of e-commerce platforms. These can be regular users (B2C) or companies (B2B) that make purchases in online stores.
- *Products:* represents products available for purchase in online stores, being the logic of all inventory and stock of items in the stores.
- *Order-Items:* represents the relation between an item and a client. An order is a set of one or more order-items, where each order-item corresponds to the link between the order and each different product (item) from that order. For example, if a client purchases 3 different products in a single order, then that order will originate 3 order-items. This entity was considered, instead of the entire order, as it is possible to represent information in more detail, facilitating the training of recommendation models by the system engine and, consequently, increase the accuracy of recommendations.

E-commerce platforms send business information about clients, products and orders' history to the RS, which preserves this data for later analysis and processing by the recommendation engine. In turn, the engine produces and stores recommendations of different types and formats. However, results are sent in the same format to all platforms: a *list of recommended product IDs*, regardless of the type and context of the recommendation. Distinct web pages of an online store may present different scenarios, hence *four* **types of recommendations** have been developed to cover various perspectives and functionalities:

Popularity: In this type of recommendation, a list of IDs related to "hot products" is returned. These are the most popular products at the moment and with the highest likelihood of being purchased by clients. The list of IDs is ordered by a *score*, which varies from 0 to 1 (0%–100%), and represents the probability of a product being purchase by a customer.

Hybrid: These recommendations are oriented to each client of the platform. Given the client ID, the RS returns a list of recommended product IDs computed specifically for that client. As one may conclude by the type's name, these recommendations result from a combination of ML algorithms, following the Collaborative and Content-based filtering approaches. When stored at MongoDB, each recommended product has associated a *list of categories* to which it belongs. This list of categories will be used by the RS filters that will be explored further in this document.

Similar Products: Recommendations are oriented to each product of the platform. Given a certain product ID, the RS returns a list of product IDs that are similar to that specific product, sorted by similarity score.

Cart Suggestions: These recommendations serve as suggestions for completing clients' current shopping cart. Given the product IDs found in a shopping cart, the RS returns a list of product IDs that are best suited to complement a client's current cart, completing the order. Each suggestion produced by the recommendation engine is stored with a *support* [4] value associated, that determines the probability of certain products being purchase together.

Each type of recommendation aims to serve a specific store feature (*recommendation widgets*), explored further in this paper.

Following the development of the communication process in the RS, it needed to be complemented at the company's side. Hence, a **BI App** (*Business Intelligence Application*) was developed to allow communication from e-commerce platforms' endpoints. Each platform is capable of integrating the *BI App*, earning the ability to interact with the recommendation system and, consequently, obtain product recommendations.

BI App is responsible for extracting and filter data of Client, Product and Order-Item entities, from the platform's database, and send it to RS. At a first stage, the application must do an initial population of the RS's database, by fetc.hing current data related to the entities mentioned, filtering it by selecting only the relevant fields for recommendations and send it to the RS's server API, which in turn will handle and store this information.

Further events' generated by consequent activities in e-commerce platforms, such as the creation or updates of entities are communicated to the RS, so it can be constantly up-to-date. This is possible, due to triggers fired when a client registers, or changes their profile (*Client*); a product information is created or its attributes updated (*Product*); a client makes an order or the order's status is changed (*Order-Item*).

Whenever such events occur, BI App sends updated information about the respective entity to RS, through HTTP requests to its API. This way, it is possible to combat the *cold start problem*, always keeping the RS database consistent and synchronized according to the platform's. It guarantees that all entities registered on the platforms are inserted in recommendations, ensuring that there are no gaps of information to provide recommendations to recent clients and include new products in results.

BI App uses *RabbitMQ* software as a *message broker*, to enable *asynchronous communication* of information sent by the mentioned triggers. This way, *bottleneck* problems are less prone to occur when there's intense activity on platforms. *RabbitMQ* receives messages from triggers and distributes them across several queues, balancing the load. In turn, these queues process messages asynchronously and sends the information almost in real-time to the RS server. On the other hand, *Redis* is used to cache results received from the RS. By saving recommendations, for a certain time, the application no longer needs to send repeated requests for the same recommendations, therefore reducing the number of calls to the server and, consequently, the load on the communication flow, increasing overall system performance.

Recommendation widgets request recommendations according to the type that they were configured, displaying the recommended products that were received through vitrines and caching the results to load web pages faster next time. In this project, *four* different features were developed, which use different types of recommendations, and were implemented in several widgets, strategically inserted in the platforms.

In the *Homepage*, a showcase of *"Recommended Products"* presents a set of recommended products specific to the user, based on *hybrid* or *popularity* recommendations. *Product Details* page also has a showcase of recommended products, visually similar to the one used in Homepage. However, recommendations are from *similar-products* type, meaning that products that are recommended have similarities to the selected product.

As for suggestions to complete the shopping cart, these are available on a *Side Cart*, accessible at any point during the purchase process on the store. Whenever a client adds or removes a product to their cart, this showcase is updated displaying the appropriate suggestions. The platform informs the RS of the current products in the cart, which aggregates all lists of suggestions for each of these products, orders them by support value and limits the result, returning only to the number of suggested products required by the widget.

Lastly, a new ordering option was added to *Product Listing*: *"Order by Recommended"*. With this option, products from Product Listing are ordered according to recommendations given by the RS. Here, once again, recommendations are from types *hybrid* or *popularity*. In this case, two scenarios had to be considered: a *generic* product listing, when the store page is displayed showing all available products, and a *specific* listing, when the user filters the product by category. In this last situation, it was necessary to ensure that all recommended products belonged to the category selected by the client. Since each recommended product of hybrid and popularity types has its category associated, and taking advantage of MongoDB's capacity to be a very efficient search engine, it was possible to filter the recommendations and send only those related to the category of the page. In order for this process to occur, the platform sends a request to the RS API, through the BI Application, indicating in the query parameters the number of recommended products it expects to receive and the categories they should belong to. If no category is indicated, the RS returns the recommendations without any filtering, this being the case of the first scenario mentioned.

4   Results

To assess the performance of the proposed architecture, *load tests* were developed collecting some evaluation metrics from the server. The workload consisted of requests generated synthetically and randomly, either in time and information, intending to simulate real users and prevent the software used from caching, after receiving the same requests in the same order for some time. About 1500 users were simulated in total, at a rate of 2 users created per second, of which 375 (25%) represent *logged users* and 1125 (75%) correspond to *anonymous users*, who are not authenticated on the platform. Logged users sent requests between 5 to 10 s, whereas anonymous users send at a rate of 7 to 14 s, simulating limited knowledge about the platform and a longer search for products.

NodeJS is *single-threaded* by default, using a single core of a machine while others remain idle. However, it is possible to implement a *clustering* module, using *Process Manager 2 (PM2)* [15], which enables an automatic usage of *Node's Cluster API* and a built-in load balancer, giving the application the ability to run in multiple processes. With this module, the parent process can be forked into several child processes, all sharing server ports and handling a large volume of requests concurrently.

Table 1. System's server load test performance results

	Throughput (requests/second)	Reliability (fail-ures/second)	Median response time (ms)	Average response time (ms)	95 percentile response time[3] (ms)
Single-core	205.8	5.8	90	230	280
Multi-core (8× CPU)	239.4	6.5	69	107	170

95% of the requests are served before this time

Examining the values in Table 1, one can verify that the response time is shorter when clustering is implemented: the *average response time* is influenced by momentary peaks due to some connection errors, caused by the huge load made by test users (the limit of connections to the server is sometimes exceeded). Hence the most reliable value is the *median response time*, presenting a value below 100 ms, which does not produce a significant impact on the loading time of web pages nor in the user experience on the platform. In addition to some connection errors, failures include responses with HTTP 404 code, regarding requests for recommendations that the system does not have, due to the recommender engine not having produced for some products (e.g. *cart suggestions*).

Furthermore, a clustering approach increases the system capacity to scale, allowing it to serve a greater number of requests while maintaining a reduced response time, making it more reliable.

When carrying out load tests it was noticed that most of the simulated requests addressed to popularity recommendations. Thus, and since these recommendations are the same for all users, it was proposed to cache them in the server, using Redis, in an attempt to improve the performance of the system. However, when implementing the Redis component on the server-side, there were no changes in response time. On the other hand, the load that is submitted to MongoDB is reduced, allowing it to have more capacity to respond to the remaining requests. This can be a solution, if there is a large number of requests that require multiple and distinct recommendations and compromise MongoDB's performance. Otherwise, the implementation of a Redis server-side component does not pay off.

5   Discussion

A *microservice* [7] approach was assumed, rather than a *monolithic* architecture, because working as a single application may represent data congestion problems (*bottleneck*) when faced with a scenario where platforms are hit by numerous users, generating a great amount of events/requests. A microservice architecture allowed to build a robust and flexible system, due to the independence of each component, capable of handling requests and managing resources efficiently. This way, the service can attend multiple tenants - *multi-tenancy* - instead of creating an instance to serve each e-commerce platform - *single-tenancy*.

There are several suggestions to improve the overall performance of the architecture, such as converting the protocol used in requests, from HTTP/1 to HTTP/2 [12]: HTTP/2 is more efficient and faster than the first due to the ability of multiplexing, header compression and binary format. As for scalability, we opted for *horizontal scaling* by adding more machines into the pool of resources, instead of adding more power (CPU, RAM, ...) to the existing machine (*vertical scaling*). This way it is possible to better manage resources of each component and implement an *Nginx* element to serve as a load balancer and security mechanism, in the future.

6   Conclusions

Given the increasing affluence of online information, recommendation systems appear as an effective strategy to combat multiple decisions and divergent options available to users on online platform. The RS works as a PaaS, containing all the necessary infrastructure and computation for the production of recommendations. In turn, e-commerce platforms that use the RS only have to worry about communicating with the system to receive such recommendations, through their respective applications.

The trigger mechanism implemented in the architecture allows the employment of *online computing*, giving the recommendation engine the possibility to generate recommendations on-the-fly, if it has the capacity to do so, and consequently enable platforms to present result in real-time, in response to users activities or other events.

Furthermore, it is possible to analyse and obtain statistics from data collected from the platforms in real-time, through Kibana dashboards, allowing greater control over data management by users. Hence, the data stored by the system is not only useful for producing recommendations, but also helping to form new perspectives about the market and support in business decisions - *Business Intelligence*.

Acknowledgments. This work has been supported by FCT – Fundação para a Ciência e Tecnologia within the RD Units Project Scope: UIDB/00319/2020.

References

1. Afify, Y., Moawad, I., Badr, N., Tolba, M.: A personalized recommender system for SaaS services. Concurr. Comput. Pract. Exp. **29** (2016). https://doi.org/10.1002/cpe.3877

2. Amazon Web Services, I.: Amazon personalize (2019). https://aws.amazon.com/personalize/. Accessed 12 Apr 2020
3. Beevo: Beevo - business ecommerce evolution (2015). https://www.beevo.com/. The Business eCommerce Evolution for medium-sized and large companies
4. Gupta, S., Mamtora, R.: A survey on association rule mining in market basket analysis. Int. J. Inf. Comput. Technol. 4(4), 409–414 (2014)
5. Hafner, K.: Netflix prize still awaits a movie seer, June 2007. https://www.nytimes.com/2007/06/04/technology/04netflix.html. Accessed 12 Apr 2020
6. Jones, M., Bradley, J., Sakimura, N.: JSON Web Token (JWT). RFC 7519, May 2015. https://doi.org/10.17487/RFC7519. https://rfc-editor.org/rfc/rfc7519.txt
7. Kharenko, A.: Monolithic vs. microservices architecture, September 2015. http://www.antonkharenko.com/2015/09/monolithic-vs-microservices-architecture.html. Accessed 12 Apr 2020
8. Ltd, G.R.D.: Yusp - personalization engine. (2017). https://www.yusp.com/. Accessed 12 Apr 2020
9. Meteren, R.: Using content-based filtering for recommendation, June 2000
10. Nanehkaran, Y.A.: An introduction to electronic commerce. Int. J. Sci. Technol. Res. 2, 190–193, April 2013
11. Negash, S., Gray, P., Burstein, F., Holsapple, C.: Business Intelligence, pp. 175–193, January 2008. https://doi.org/10.1007/978-3-540-48716-6_9
12. Ramadan, N., Abdelwahab, I.: Impact of implementing http/2 in web services. Int. J. Comput. Appl. 147, 27–32 (2016). https://doi.org/10.5120/ijca2016911182
13. Ricci, F., Rokach, L., Shapira, B.: Introduction to Recommender Systems Handbook, pp. 1–35. Springer, Boston (2011). https://doi.org/10.1007/978-0-387-85820-3_1
14. Schafer, B., J, B., Frankowski, D., Dan, Herlocker, Jon, Shilad, Sen, S.: Collaborative filtering recommender systems, January 2007
15. Strzelewicz, A.: Process manager 2 (2013). https://pm2.keymetrics.io/. Accessed 05 June 2020

Automated Learning of In-vehicle Noise Representation with Triplet-Loss Embedded Convolutional Beamforming Network

Seok-Jun Bu and Sung-Bae Cho[✉]

Department of Computer Science, Yonsei University, Seoul 03722, South Korea
{sjbuhan,sbcho}@yonsei.ac.kr

Abstract. In spite of various deep learning models devised, it is still a challenging task to classify in-vehicle noise because of the reverberation and the variance in the low-frequency band generated from the narrow interior space. Considering the impulsive characteristics of the vehicle noise and the multi-channel sampling environment at the same time, it is essential to automatically learn the disentangled noise representation as well as parameterize the conventional beamforming operation. We propose a method to overcome the above two major hurdles by parameterizing a beamforming operation based on convolutional neural network. Moreover, we improve the structure of the beamforming network by explicitly learning of the distance between vehicle noises within the triplet network framework. Experiments with the dataset consisting of a total 241,958,848 time-series collected by a global motor company show that the proposed model improves the classification accuracy by 5% compared to the latest deep acoustic models. The detailed analysis shows that the proposed method can potentially compensate for the disjoint issues between the learning and validation vehicle types.

Keywords: Multichannel sensor array · Deep metric learning · Learnable beamformer · In-vehicle noise

1 Introduction

In consideration of market trends that emphasize a quieter driving experience, various deep learning-based models for absorption or attenuation of in-vehicle noise based on the analysis of body vibration have been developed [1, 2]. Among the noises generated while driving, in-vehicle noise from interior finishes is fatal for both manufacturers and users, accounting for 50% of the total vehicle repair cases [3]. Due to the painstaking process of disassembling the car and the subjective evaluation of experts, the deep learning model appeared to suggest a new paradigm in automated classification process of in-vehicle noise classification. However, the deep learning-based in-vehicle noise classification models encountered two major hurdles: One is the reverberation and refraction caused by the interior materials and the narrow space, and the other is the transient and impulsive nature of the noise mainly occurring in the low frequency band.

© Springer Nature Switzerland AG 2020
C. Analide et al. (Eds.): IDEAL 2020, LNCS 12490, pp. 507–515, 2020.
https://doi.org/10.1007/978-3-030-62365-4_48

Although the multi-channel sampling environment is proposed to overcome the difficulties mentioned as depicted in Fig. 1(a), the performance bottleneck of the existing direction-of-arrival (DOA) beamformer from the limited environment caused the performance degradation of the deep learning-based classification model as shown in Fig. 1(b). Moreover, because the in-vehicle noise is mainly expressed in the low frequency band and being absorbed mostly by internal finishes, it is difficult to construct the in-vehicle noise classification model with conventional convolutional neural network (CNN) and recurrent neural network (RNN).

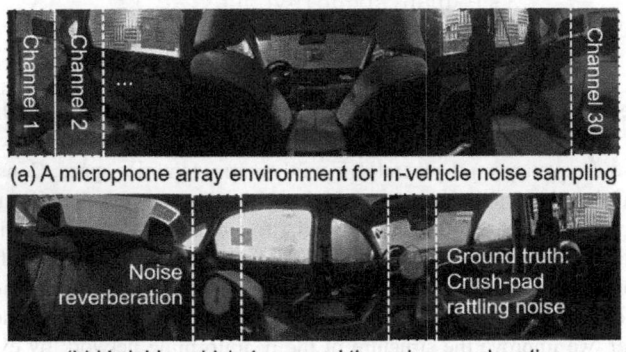

(a) A microphone array environment for in-vehicle noise sampling

(b) Variable vehicle types and the noise reverberation

Fig. 1. (a) The multichannel sampling environment to consider the impulsive and transient nature of in-vehicle noise and reflective and refractive characteristics of in-vehicle environment. (b) Performance degradation of conventional beamformers due to the reverberation.

Taken together, instead of the existing deep learning model including the standard CNN-RNN approach, a novel mechanism of beamforming and modeling is needed in the in-vehicle noise classification domain at the same time. We propose the triplet-loss embedded convolutional beamformer that can explicitly learn the difference metrics of in-vehicle noise and model the feature space and the DOA-weights.

The key idea is to parameterize the weights of the channels via the convolutional filters and learns the disentangled features from triplets in an end-to-end manner. The proposed method outperforms the conventional deep learning models including standard triplet networks with the accuracy of 0.9823 that is verified by 10-fold cross validation and chi-squared test. In addition, we conduct the grid-wise parameter exploration experiments to provide the guidelines for multi-channel speech recognition or other manufacturing domains.

2 Related Works

Based on the similarities of the fields and the techniques used, we refer to recent metric learning based acoustic modeling studies. Table 1 summarizes the significant speech recognition studies in terms of feature extraction and metric learning in the last two years.

Table 1. Related works on acoustic modeling with respect of feature extraction and metric learning

Author	Feature extraction	Metric learning
Zhang [4]	STFT spectrogram, CNN-RNN	Triplet, thresholding
Bredin [5]	MFCC, Bi-LSTM	Triplet, thresholding
Yang [6]	Fisher encoding after extracting prosodic descriptor	Triplet, thresholding
Novoselov [7]	MFCC, Time-delay DNN	Modified triplet loss (Cosine score)
Wang [8]	MFCC, Bi-LSTM	Modified triplet loss (Centroid-based score)
Turpault [9]	MFCC, CNN-BiGRU	Semi-supervised triplet pair selection
Zhao [10]	STFT spectrogram, CNN-LSTM autoencoder	Modified embedding method (with de-noising loss)

In the speech recognition field, the triplet-loss concept has been introduced to explicitly model the difference metrics between vocal characteristics [11]. Zhang and Koishida introduced the triplet loss into speaker verification field to encode the short utterances and achieved better performance over the conventional i-vector system [4].

Bredin used Mel-frequency cepstral coefficients (MFCC) and recurrent neural networks to verify the superiority of metric learning in speaker verification field [5]. Based on proven excellence, Yang et al. proposed an application of estimating blood pressure from prosodic features from patients [6]. Novoselov et al. and Wang et al. proposed the use of cosine and centroid-based metrics instead of Euclidean distance metric [7, 8]. Although it has not been verified yet that the modified triplet loss can be used in general domains, it is at least effective in the modeling of phonetic features. Turpault et al. introduced a semi-supervised method inspired by data augmentation [9]. Zhao et al. proposed a combination of de-noising autoencoder and triplet network that shows the robustness in signal level [10].

The work presented here has focused on the modification of the loss function or embedding method of metric learning concepts, which takes advantage of optimization for signal level feature space generation from single-channel spectrums. While the present study is related to recent approaches in deep beamforming [12], it capitalizes on a multi-channel environment, which was not considered in these earlier studies.

3 The Proposed Method

3.1 Convolutional Neural-Based Beamforming Network

We present the overall architecture of the triplet-loss embedded convolutional neural beamformer that parameterizes the weights of beamformer with spatial filters in CNN and generate the feature space of in-vehicle noise as depicted in Fig. 2. The major modification from standard triplet networks is that the data-driven beamformer is implemented with CNN in an end-to-end manner to consider the multichannel environment. The term of beamforming refers to the design of a spatio-temporal filter which operates on the outputs of the multichannel signal array [13].

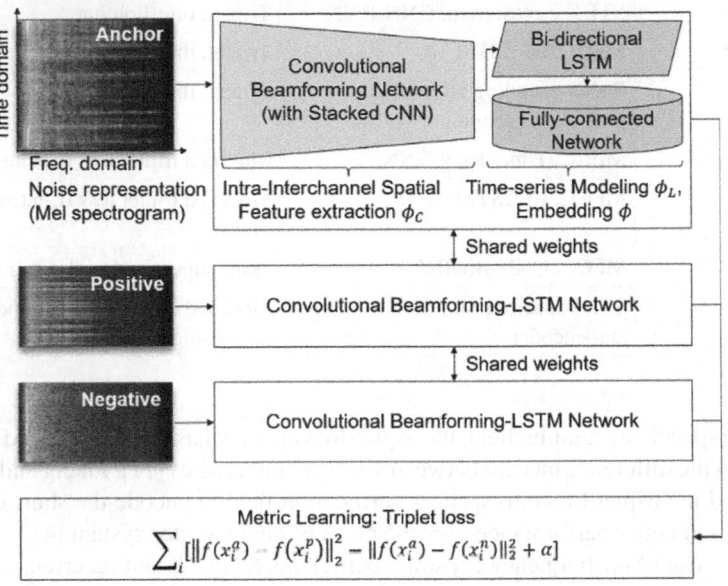

Fig. 2. The overall architecture: A spatial feature extractor ϕ_c that models the hidden correlations from the intrachannel (global) and interchannel (local), respectively, and temporal feature extractor ϕ_L with embedding network ϕ that models time-series information selectively using a gate operation.

In the field of classifying in-vehicle noise where reflections and refractions occur in a narrow space, it is impossible to improve both intelligibility and quality when using a selective single channel. In order to improve the classification accuracy of existing deep models, it is necessary to adopt the concept of beamforming:

$$y(t) = \sum_{i=1}^{J} \sum_{p=0}^{t-1} w_{i,p}^* x_i(t - p) \tag{1}$$

where $w_{i,p}^*$ represents the weight of global spatial filter from the combination of the J sensors at time t.

The conventional beamformer, however, is a non-parametric model that stores time-invariant coefficients calculated as a constant, which has practical limitations in the field of classifying in-vehicle noise: we still need to parameterize and learn the dynamic properties from reverberations. Specifically, two of the most important factors that should be considered in classifying the in-vehicle noise are DOA and robustness to reverberant. Neural networks are the best candidates to be adopted into beamforming to consider dynamic environmental factors in consideration of its complex mapping capability. The convolutional neural-based beamformer proposed in this paper learns the translational-invariant filter as well as the time-invariant filter which consider localized features of in-vehicle noise and spatial features in time-frequency in a data-driven manner:

$$\phi_c^l\left(x_{i_f}(t_\omega)\right) = \sum_{a=0}^{c_m-1} \sum_{b=0}^{c_n-1} w_{ab} x_{(i+a)(f+b)}^{l-1}(t_\omega) \tag{2}$$

where the output ϕ_c^l from the i th node of the l th convolutional layer performs the convolution operation on x^{l-1} using $c_m \times c_n$ sized filter f_w [14, 15]. From the conventional beamformer viewpoint, it can be assumed that DOA and adaptive beamforming are simultaneously learned in a data-driven manner. From the standard CNN viewpoint, the inter-channel is modeled as distributed over the time axis before LSTM-DNN model and embeds the time-series features.

3.2 Triplet-Loss Embedded Metric Learning

In terms of traditional delay-and-sum beamformers, the most important function is to multiply sensor data input from multiple channels by an optimal weight and output it as a time-series features [16]. To introduce the adaptive and learning capacity to the delay-and-sum beamformer, several studies proposed the data-driven beamformers based on machine learning techniques such as SVMs [17, 18] or beamformers that combine prior knowledge in terms of Bayesian approach [19].

We use the LSTM network to combine and model the multi-channel in-vehicle noise to fully exploit the non-linear mapping capacity of the neural network. A conventional LSTM network computes a mapping from an input sensor sequence $x = (x(t), \ldots, x(t - \omega))$ to an output temporal feature map $y(t)$ by calculating the network unit activations.

After the time-series features from the spatial features are extracted by LSTMs, the typical DNN is used to complete the embedding function $\phi(\cdot)$:

$$\phi^l(x_i(t_\omega)) = \phi_L^{l-1}\left(\phi_C^{l-2}(x_i(t_\omega))\right) \tag{3}$$

The weights of the convolutional neural-based beamformer and LSTM-DNN are optimized using the backpropagation algorithm based on gradient descent optimization, by minimizing the L2 distance-based triplet loss function L_T:

$$L_T = \sum_i \left[\left\| \phi(x_i^a) - \phi(x_i^p) \right\|_2^2 - \left\| \phi(x_i^a) - \phi(x_i^n) \right\|_2^2 + \alpha \right] \tag{4}$$

where α is an empirically defined margin that is enforced between positive p and negative n pairs and anchor a [6].

4 Experimental Results

4.1 In-vehicle Noise Dataset and Preprocessing

Table 2 summarizes the in-vehicle rattling noise dataset collected for training and test. 30 sensors are arranged in a spherical shape. We have converted each rattling noise file to a STFT spectrogram, with the purpose to compare the proposed method with the conventional approaches. As the result, the rattling noise window dataset forms $(n, 7, 156, 30)$ sized matrix where each column represents the number of window instances, size of window, intensity per frequency band, and the number of channels.

Table 2. Type and length per in-vehicle noise sampled in controlled environment

Class index	Noise type	Length [s]	Windows Sampled	Class index	Noise type	Length [s]	Windows sampled
0	Shaker noise	6.8	332	5	Seat rail 2	12.0	596
1	Center console armrest 1	2.4	110	6	Seat backboard	23.4	1162
2	Center console armrest 2	11.0	544	7	Sun visor	20.8	1036
3	Center console armrest 3	11.0	546	8	Crash pad	12.3	610
4	Seat rail 1	12.2	602	9	Passenger seat armrest	7.9	390

4.2 Classification and Embedding Performance

In Fig. 3, 10-fold cross-validation is conducted by approaching the in-vehicle noise classification task as a pattern recognition problem that classifies $(\omega, 156, 30)$ vectors. CNN-LSTM [20] proved to be difficult to classify in-vehicle noise by achieving 0.9329 test accuracy. Triplet loss approach [4] achieves 0.3% of performance improvement over CNN-LSTM despite the loss of stability and its heavy computational cost. On the other hand, the proposed triplet-loss embedded convolutional beamformer achieves the accuracy of 0.9823, which improves the triplet network by 5%. Table 3 quantitatively analyzes the confusion matrix and shows the precision and recall per noise type. As expected, the precision for seat rail rattling noise is the lowest as 0.89 and the recall between the center-console armrest and seat backboard rattling noise is relatively low.

The feature space generated [21] by the optimized triplet network is depicted in Fig. 4. At the bottom, each noise window is mapped in 2-dimensions based on t-SNE embedding algorithm. While the distinction between center-console armrest noise, sun visor rattling noise and passenger seat armrest noise in the original distribution is ambiguous, the similarities and differences between the noises are explicitly learned and distinct in the generated feature space.

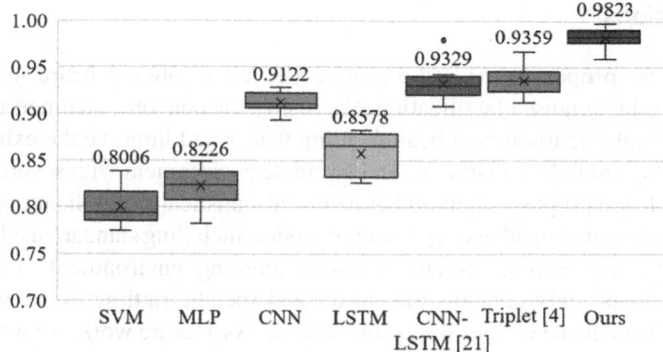

Fig. 3. Comparison of 10-fold cross-validation accuracy with other machine learning algorithms, including standard CNN-LSTM and triplet network

Table 3. Classification performance with respect to noise type

Noise type	0	1	2	3	4	5	6	7	8	9	Avg.
Precision	0.99	1.00	0.97	1.00	1.00	0.89	1.00	1.00	1.00	0.98	0.98
Recall	1.00	0.95	0.98	0.98	0.99	1.00	0.95	1.00	0.99	1.00	0.98

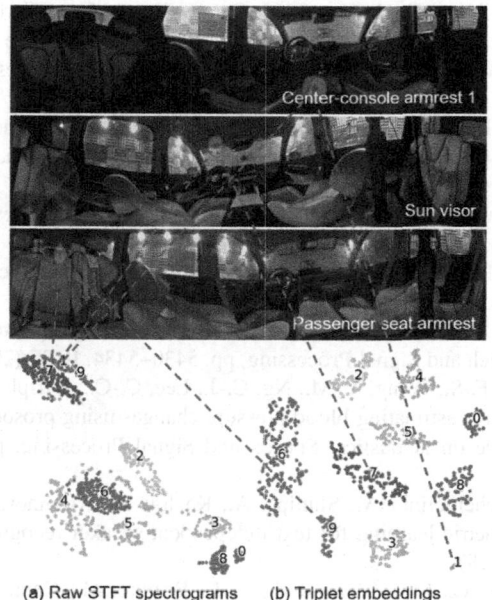

(a) Raw STFT spectrograms   (b) Triplet embeddings

Fig. 4. Comparison of the distribution of original noise data with the generated feature space. Ambiguous interior noise is identified after embedding.

5 Conclusions

In this paper we propose a triplet-loss embedded convolutional neural-based beam-former for in-vehicle noise classification. We justify the convolutional neural networks in accordance with the traditional beamforming theory and improve the existing triplet network architecture to be suitable for an in-vehicle environment with severe reverbera-tion. As a result, the proposed method has achieved statistically significant performance improvement compared to other deep learning models including standard triplet network.

Considering the current in-vehicle noise sampling environment is the sound-absorbing room and only contains the shaker and specific rattling noise, whereas the environment includes the various background noise. As a future work, we will introduce a recent Bayesian deep learning approach to cope with the uncertainty of noisy environ-ments. It is also necessary to verify whether the proposed method can be extended to general acoustic modeling fields with multi-channel environment.

Acknowledgments. This work was partly supported by Institute of Information & Communica-tions Technology Planning & Evaluation (IITP) grant funded by the Korean government (MSIT) (No. 2020-0-01361, Artificial Intelligence Graduate School Program (Yonsei University)) and Hyundai Motors, Inc.

References

1. Bu, S.-J., Park, N., Nam, G.-H., Seo, J.-Y., Cho, S.-B.: A Monte Carlo search-based triplet sampling method for learning disentangled representation of impulsive noise on steering gear. In: International Conference on Acoustics, Speech and Signal Processing, pp. 3057–3061. IEEE (2020)
2. Bu, S.-J., Cho, S.-B.: Classifying in-vehicle noise from multi-channel sound spectrum by deep beamforming networks. In: International Conference on Big Data, pp. 3545–3552. IEEE (2019)
3. Cerrato, G.: Automotive sound quality-accessories, BSR, and brakes. Sound Vibr. **43**, 10 (2009)
4. Zhang, C., Koishida, K.: End-to-end text-independent speaker verification with triplet loss on short utterances. In: Interspeech, pp. 1487–1491 (2017)
5. Bredin, H.: TristouNet: triplet loss for speaker turn embedding. In: International Conference on Acoustics, Speech and Signal Processing, pp. 5430–5434. IEEE (2017)
6. Yang, H.-C., Tsai, F.-S., Weng, Y.-M., Ng, C.-J., Lee, C.-C.: A triplet-loss embedded deep regressor network for estimating blood pressure changes using prosodic features. In: Inter-national Conference on Acoustics, Speech and Signal Processing, pp. 6019–6023. IEEE (2018)
7. Novoselov, S., Shchemelinin, V., Shulipa, A., Kozlov, A., Kremnev, I.: Triplet loss based cosine similarity metric learning for text-independent speaker recognition. In: Interspeech, pp. 2242–2246 (2018)
8. Wang, J., Wang, K.-C., Law, M.T., Rudzicz, F., Brudno, M.: Centroid-based deep metric learning for speaker recognition. In: International Conference on Acoustics, Speech and Signal Processing, pp. 3652–3656. IEEE (2019)
9. Turpault, N., Serizel, R., Vincent, E.: Semi-supervised triplet loss based learning of ambient audio embeddings. In: International Conference on Acoustics, Speech and Signal Processing, pp. 760–764. IEEE (2019)

10. Zhao, F., Li, H., Zhang, X.: A robust text-independent speaker verification method based on speech separation and deep speaker. In: International Conference on Acoustics, Speech and Signal Processing, pp. 6101–6105. IEEE (2019)
11. Mingote, V., et al.: Language recognition using triplet neural networks. In: Interspeech, pp. 4025–4029 (2019)
12. Xiao, X., et al.: Deep beamforming networks for multi-channel speech recognition. In: International Conference on Acoustics, Speech and Signal Processing, pp. 5745–5749. IEEE (2016)
13. Markovich, S., Gannot, S., Cohen, I.: Multichannel eigenspace beamforming in a reverberant noisy environment with multiple interfering speech signals. IEEE Trans. Audio Speech Lang. Process. **17**, 1071–1086 (2009)
14. Sainath, T., Parada, C.: Convolutional neural networks for small-footprint keyword spotting. In: Interspeech, pp. 1478–1482 (2015)
15. Kim, T.Y., Cho, S.B.: Predicting residential energy consumption using CNN-LSTM neural networks. Energy **182**, 72–81 (2019)
16. Ribeiro, L.N., de Almeida, A.L., Mota, J.C.: Tensor beamforming for multilinear translation invariant arrays. In: International Conference on Acoustics, Speech and Signal Processing, pp. 2966–2970. IEEE (2016)
17. Ramón, M.M., Xu, N., Christodoulou, C.G.: Beamforming using support vector machines. IEEE Antennas Wirel. Propag. Lett. **4**, 439–442 (2005)
18. Salvati, D., Drioli, C., Foresti, G.L.: A weighted MVDR beamformer based on SVM learning for sound source localization. Pattern Recognit. Lett. **84**, 15–21 (2016)
19. Bell, K.L., Ephraim, Y., Van Trees, H.L.: A Bayesian approach to robust adaptive beamforming. IEEE Trans. Signal Process. **48**, 386–398 (2000)
20. Donahue, J., et al.: Long-term Recurrent Convolutional Networks for Visual Recognition and Description. In: IEEE Conference on Computer Vision and Pattern Recognition, pp. 2625–2634 (2015)
21. Kim, J.Y., Cho, S.B.: Electric energy consumption prediction by deep learning with state explainable autoencoder. Energies **12**, 739 (2019)

Sequence Mining for Automatic Generation of Software Tests from GUI Event Traces

Alberto Oliveira[1(✉)], Ricardo Freitas[1], Alípio Jorge[1], Vítor Amorim[2], Nuno Moniz[1], Ana C. R. Paiva[3], and Paulo J. Azevedo[4]

[1] LIAAD-INESC TEC, FCUP-University of Porto, Porto, Portugal
{alberto.p.oliveira,alipio.jorge}@inesctec.pt
[2] RandTech Computing, R&D, Porto, Portugal
vitor.amorim@rtcom.pt
[3] INESC TEC, FEUP-University of Porto, Porto, Portugal
[4] INESC TEC, University of Minho, Braga, Portugal

Abstract. In today's software industry, systems are constantly changing. To maintain their quality and to prevent failures at controlled costs is a challenge. One way to foster quality is through thorough and systematic testing. Therefore, the definition of adequate tests is crucial for saving time, cost and effort. This paper presents a framework that generates software test cases automatically based on user interaction data. We propose a data-driven software test generation solution that combines the use of frequent sequence mining and Markov chain modeling. We assess the quality of the generated test cases by empirically evaluating their coverage with respect to observed user interactions and code. We also measure the plausibility of the distribution of the events in the generated test sets using the Kullback-Leibler divergence.

Keywords: Software testing · Frequent pattern mining · Markov chains · Data mining

1 Introduction

Software development is a complex and continuous process that requires frequent changes in the code [1]. Each change can introduce errors that affect the ability of the software maker to timely deliver a quality product [2]. Errors in software can cause distrust in software users but can also lead to substantial economic losses [3] and even the sacrifice of human lives [4]. Taking into account that software development is becoming increasingly more agile [5], with systems undergoing constant changes, the moments for introducing errors are multiplying.

The software industry typically relies on test cases that are executed before each release [6]. Although the automation of test checking is a common practice [7], the set of tests is bounded to the ones previously defined and planned. Moreover, the design of test cases is mostly based on human expertise [8]. However,

© Springer Nature Switzerland AG 2020
C. Analide et al. (Eds.): IDEAL 2020, LNCS 12490, pp. 516–523, 2020.
https://doi.org/10.1007/978-3-030-62365-4_49

manually devising software tests demands much time, costs and effort of human software testers [8]. Correctly selecting the tests and evaluating their outputs is crucial in order to efficiently improve the quality of software [8].

In this paper, we propose an adaptable framework for learning software test generators from user interaction data. It has been developed in the context of a software company that produces the web-based application Anywhere+, a platform for managing insurance products. Nevertheless, our proposed approach can be used with any web GUI-based software. In our pipeline, the first step is to store the user interaction logs. A browser plugin captures this data as users work normally. From this data, we discover sequential micro-patterns using sequence mining. The third step is to chain the discovered patterns into a global Markov chain model. Finally, this model is used to generate test cases based on these patterns automatically. The approach is tested on real data in terms of coverage and plausibility of the generated patterns. As a result, we have obtained stable growth rates in terms of coverage – adding more generated tests increases our coverage metric value, even reaching full coverage for one of the cases – and very low values of the Kullback-Leibler's divergence between the distribution of actions in user sessions and in artificially generated tests.

This paper is organized as follows. We first discuss related work. Then we give an overview of the software's deployment pipeline. We describe how data is collected, how frequent sequences are found and how we use Markov chains to produce our test generator. We wrap up with evaluation and conclusions.

2   Related Work

Given the importance of the software development process and the tremendous possibilities that AI can bring to it [9], this is a fertile ground for AI research. Many works can be found in the last two decades with contributions to different phases of the process and in particular to test generation.

Isabella and Retna [10] present a general overview of test case generation for GUI based testing. This includes generation of test cases, repairing infeasible test suites and multiple GUI testing tools over various types of software, as well as its usage advantages and disadvantages. Conroy et al. [11] proposed a generic method for generating tests for testing web services from their reference legacy GUI applications. This work mainly relies on the concept that GUI elements are programming objects whose values can be set and retrieved and whose methods are associated with actions that users perform on these elements, which is very similar to our solution's plugin purpose.

The closest work to ours in spirit and method is the one accomplished by Zhou et al. [12]. It first builds a Markov usage model based on improved state transition matrix (STM), which is a table-based modeling language. It then generates a software reliability test method, including test case generation and test adequacy determination using the previously created Markov usage model. An improved Kullback discriminant was chosen as the judgment criteria of convergence from the test chain to the usage chain in order to measure if the testing process is sufficient.

Lastly, the approach of Last et al. [8] aims to automate the input-output analysis of execution data based on a machine learning methodology. This methodology relies on the info-fuzzy network (IFN), which has a tree-like structure. The network is used to predict output values given test-cases.

3   The Software Deployment Framework

The main contribution of this paper is the software test generation process that relies on captured GUI events data. This is part of a broader software deployment framework for an insurance ERP called Anywhere+ developed by RandTech Computing (https://rtcom.pt). It has been extended to incorporate this Artificial Intelligence component, from data collection to test generation. This framework automates the complex software update process, including code, databases and tests, and enables safer, faster and more frequent updates. It is flexible enough to be easily adapted to other deployment flows. The framework follows a modular structure (Fig. 1).

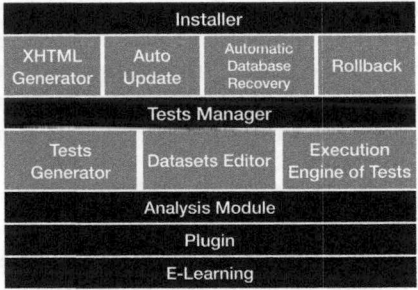

Fig. 1. Software deployment framework

The Plugin component collects interaction data from the browser as the application is used. This is high-level data that represents business events. In the Analysis Module this data is used to induce a model for generating tests. The Tests Generator uses the model to generate software tests. Since some actions of the tests require specific values (for example, filling in the name of a client), this is provided by a specific dataset of attribute-value pairs created using the Datasets Editor. When the software tests are automatically executed, the E-learning component captures the sequence of screenshots that can later be used for user training. The Installer is the component that deploys new versions of the software. The XHTML generator automatically transforms the XHTML files which compose the application's UI. This generator is capable of assigning graphical widgets to high-level functional categories corresponding to embedded business concepts. This is important to give semantics to the events to be logged by the Plugin. The AutoUpdate component warns users of new updates. The Automatic Database Recovery changes the structure of the database if needed, and Rollback is there to recover the previous database if anything goes wrong.

4   Data Acquisition

As the application is used, GUI events are continuously recorded by the Plugin component. This provides a memory of the real sessions that will drive the building of test case scenarios.

In order to define and execute the software tests, it is fundamental that the framework can recognize the various business concepts. For that, we have defined a syntax for the XHTML generator which recognizes a set of patterns used on the UI design of the Anywhere+ application. The XHTML files which compose the Anywhere+ application are transformed by inserting the reference to the business concepts. Despite its complexity, this transformation occurs transparently and automatically, both for programmers as for users and is triggered at each build of the application.

4.1   Data Format

User activities are recorded on a text file, following a simple and optimized structure for the analysis task. The interactions' format firstly contain the *timestamp*, *session id*, *tab id* and the *business concept* separated by a comma. Secondly, there are three components separated by semicolon: action/command, target and value.

The action field is mandatory, but target and value parameters may be void. The captured user sessions are the input for the Analysis Module that builds output for the Test Generator. This is a Markov model whose states are sequences of user actions. The Test Generator pre-processes the interaction events logged and looks for frequent sequences with a given maximum length. The resulting sequences are chained into a single Markov model. This model is built by identifying all the initial states from the sequences and then, for each initial state, it explores the next states. The transition probabilities are estimated using the number of transitions from a determined current state to a next state. The Markov chain model is then output in JSON format.

5   Frequent Sequence Identification

To generate the frequent sequence patterns, we considered various frequent sequence mining algorithms. These algorithms can be categorized by their search approach as breadth-first search or depth-first search. Depth-first search algorithms need less database scans in order to obtain all frequent sequences so they are more computationally efficient with larger databases.

Fournier-Viger et al. [13] proposed the data structure CMAP (Co-occurrences Map, CMAP) capable of keeping a co-occurrences map of items extracted from a single database scan and also a new approach to the sequence pruning stage based on this data structure and on the co-occurrences' properties. The CM-SPAM algorithm is an optimized version of SPAM [14] for the frequent pattern mining task. The SPAM algorithm first constructs a vertical sequence representation of the sequences database and obtains the set of frequent items, according to the given minimum support parameter. It then searches for candidate patterns based on the set of frequent items. SPAM uses bitmaps for faster pattern joining operations. The algorithm outputs sequences and their respective frequency.

This library uses the IBMGenerator format for sequence databases, which is represented by a binary file of integers ordered by little-endian, where positive values represent events, -1 represents the separation between events and -2 represents the end of a sequence.

6   Software Test Generator

A Markov model provides us with flexible, easily understandable representations of the operational profiles of given programs or software systems [15]. The Markov property says that the probability distribution of future states of a process relies only upon the current state. Therefore, a Markov model captures the time-independent probability of being in state s_1 at time $t + 1$ knowing that the state at time t was s_0. The relative frequency of event transitions during program executions provides a probability estimate for each possible immediate state.

For the deployment framework, the Markov model is represented as a dictionary structure, $<key, values>$, where key is the current initial state (associated with the previous actions) and $values$ is a list of future actions associated to their probabilities. The states in our application are the actions performed by the user. This model is stored in JSON format. When used to reproduce sequences, Markov models can lead to infinite cycles of alternating states. To avoid that, we have adopted an end-of-sequence token which, when generated, terminates the sequence.

Given the Markov model, tests are generated in multiple ways using a proportional sampling criterion. This approach, proposed by Zhou et al. [12], divides the choice interval space, making it between $[0, 1]$ and splitting by the occurrence probabilities of each action (for example, if we have A = 0.3, B = 0.2 and C = 0.5, our interval will be split into intervals of $[0, 0.3]$, $[0.3, 0.5]$ and $[0.5, 1]$). A random number between 0 and 1 is generated and the action is chosen according to where the generated value fits (for example if the generated value is 0.4, the chosen action will be B).

The generated tests correspond to interaction paths that could be followed by the platform's users. The test generation algorithm takes three parameters, N, L and $markov$. N represents the order of the Markov model, i.e., the number of actions to be taken into account for the next action of the Markov chain. L is the number of sequences we want to generate, and $markov$ is the Markov model generated by the Analysis module.

7   Evaluation

The current evaluation of our approach focuses on code coverage and plausibility. In our experiments, we generate large numbers of tests and observe how these two dimensions evolve. We aim at assessing the quality of the generated tests, as well as determining the minimum number of tests that must be generated to ensure quality.

7.1 Metrics

The metrics used are Proportion of Actions Covered (PAC) and the Kullback-Leibler's (KL) divergence [16]. PAC is the ratio between the number of distinct actions in real sessions and the number of distinct actions in sessions built by the Analysis Module. This metric does not measure the code coverage directly since it is based on the *de facto* users' actions. If a part of the code is never involved in real sessions, it is not tested. However, the more PAC grows, the more code is tested. The Kullback-Leibler's divergence [16] is a comparison measure between two probabilities' distributions. Using it, we can compare the distribution of events in the real session with the generated ones. The expression to calculate this divergence is presented below.

$$D_{\mathrm{KL}}(P||Q) = \sum_i P(i) log \frac{P(i)}{Q(i)} \tag{1}$$

7.2 Results

Code Coverage. We performed the test for the code coverage using a growing number of N (from 1 to 6) and a maximum value for L (600). For each value of N, we generated 0 to 600 tests and cumulatively measured the coverage of the tests. In Fig. 2, we observe how PAC grows with the number of generated tests, reaching a value of 1 to $N = 1$ and nearly 1 for the remaining N. Although there is no clear relation between N and the measured of PAC, higher values of N do not seem to pay off in terms of coverage. As we will see, this is also the case with plausibility. In any case, a number of 600 test cases already offers excellent coverage. The parameter N (size of the frequent sequences) does not seem to have a clear influence. This indicates that the number of tests has to be relatively large to assure high coverage.

Kullback-Leibler's Divergence. We have executed an experiment similar to the previous one for measuring the plausibility of the generated tests. Now we measure KL of the produced distribution of events given the observed one. We obtained the results shown in Fig. 3. We see that KL tends to zero for all values of N (a KL value close to 0 indicates that the generated sequences are a good representation of real sequences). With $L > 2500$ test cases, we already obtain plausible distributions for all N. This shows that we can find various safe pairs, L and N. Combining both evaluation dimensions, and taking into account that lower values of N and L are preferable for computational reasons, right combinations would be $N \in \{1, 2, 3\}, L \geq 2500$.

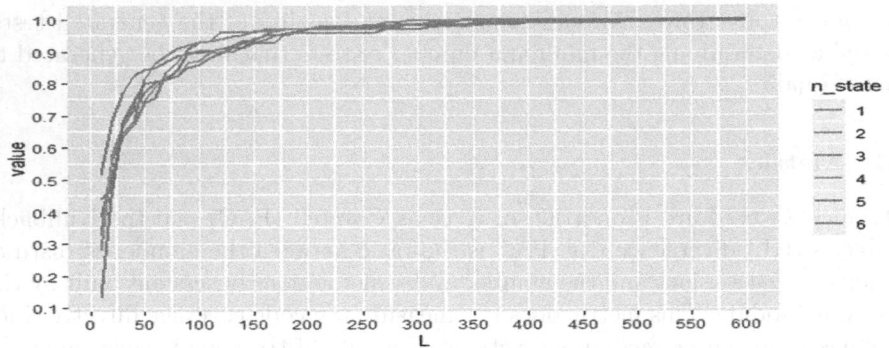

Fig. 2. Evaluation results for code coverage (PAC)

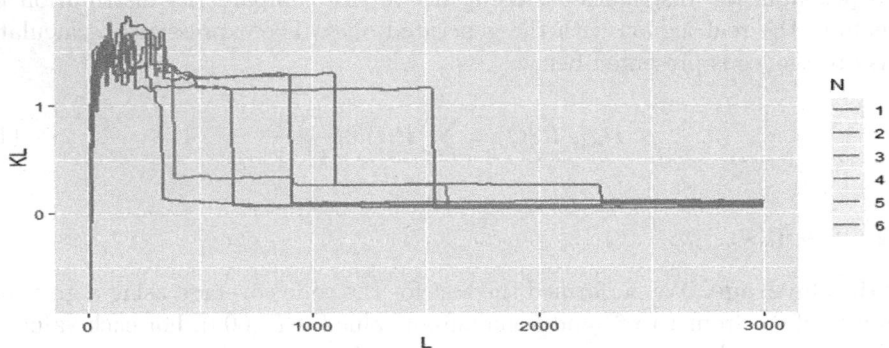

Fig. 3. Evaluation results for the Kullback-Leibler divergence

8   Conclusions

With this work, we have managed to implement a tool that automatically generates software tests based on GUI event logs. This proposed solution has a high degree of adaptability for easy adoption by other systems. However, to implement this methodology, it is necessary for some degree of permanent system users', in order to obtain useful results. For further research/improvements, we will deepen the presented empirical study and consider other dimensions. Currently, the Kullback-Leibler's divergence does not compare distributions of micro sequences but only of individual items. Despite the merits of the PAC metric, which give more importance to more frequent user actions, we should also measure the plain coverage of code. These metrics are continuously and automatically obtained throughout the software development and deployment process. It is, therefore, important to provide developers with dashboard tools for easy access to these performance indicators. In another line of evaluation, we are designing an A/B test methodology that enables the direct comparison of the performance of automatic software testing with manual test design.

Acknowledgment. This work is financed by the Northern Regional Operational Program, Portugal 2020 and the European Union, through the European Regional Development Fund (https://www.rtcom.pt/wordpress/rute-randtech-update-and-test-environment/). Also, this work is financed by National Funds through the Portuguese funding agency, FCT - Fundação para a Ciência e a Tecnologia, within project UIDB/50014/2020.

References

1. Ajouli, A., Henchiri, K.: MODEM: an UML profile for MODEling and Predicting software Maintenance before implementation (2019). https://doi.org/10.1109/ICCISci.2019.8716421
2. Choetkiertikul, M., Dam, H.K., Tran, T., Ghose, A.: Predicting delays in software projects using networked classification (T). In: 2015 30th IEEE/ACM International Conference on Automated Software Engineering (ASE), pp. 353–364 (2015). https://doi.org/10.1109/ASE.2015.55
3. Grossman, L.: Metric math mistake muffed mars meteorology mission (2010). https://www.wired.com/2010/11/1110mars-climate-observer-report/
4. Kelion, L.: Fatal a400m crash linked to data-wipe mistake. BBC (2015). https://www.bbc.com/news/technology-33078767
5. Babar, M.A., Brown, A.W., Mistrik, I.: Agile Software Architecture: Aligning Agile Processes and Software Architectures, 1st edn. Morgan Kaufmann Publishers Inc., San Francisco (2013)
6. Florea, R., Stray, V.: The skills that employers look for in software testers. Softw. Qual. J. **27**, 1449–1479 (2019)
7. Anderson, B.: Best automation testing tools for 2019 (top10 reviews), October 2017. https://medium.com
8. Last, M., Friedman, M., Kandel, A.: Using data mining for automated software testing. Int. J. Software Eng. Knowl. Eng. **14**(4), 369–393 (2004). https://doi.org/10.1142/S0218194004001737
9. Giudice, D.L.: How AI Will Change Software Development And Applications Key takeaways (2016)
10. Isabella, A., Retna, E.: Study paper on test case generation for GUI based testing. CoRR, vol. abs/1202.4527 (2012). http://arxiv.org/abs/1202.4527
11. Conroy, K., Grechanik, M., Hellige, M., Liongosari, E., Xie, Q.: Automatic test generation from GUI applications for testing web services, pp. 345–354, October 2007
12. Zhou, K., Wang, X., Hou, G., Wang, J., Ai, S.: Software reliability test based on Markov usage model. JSW **7**(9), 2061–2068 (2012). https://doi.org/10.4304/jsw.7.9.2061-2068
13. Fournier-Viger, P., Gomariz, A., Campos, M., Thomas, R.: Fast vertical mining of sequential patterns using co-occurrence information, May 2014
14. Ayres, J., Flannick, J., Gehrke, J., Yiu, T.: Sequential pattern mining using a bitmap representation, July 2002
15. Gutjahr, W.J.: Software dependability evaluation based on Markov usage models. Perform. Eval. **40**(4), 199–222 (2000). https://doi.org/10.1016/S0166-5316(99)00052-8
16. Kullback, S.: Information Theory and Statistics. Wiley, New York (1959)

Deep Learning Based Algorithms for Welding Edge Points Detection

Ander Muniategui(✉) ⓘ, Jon Ander del Barrio ⓘ, Xabier Zurutuza ⓘ,
Xabier Angulo ⓘ, Iñaki Silanes ⓘ, Uxue Irastorza ⓘ, Aitor García de la Yedra ⓘ,
and Ramón Moreno ⓘ

LORTEK-BRTA, Arranomendi kalea 4A, 20240 Ordizia, Spain
{amuniategui,jadelbarrio,xzurutuza,xangulo,isilanes,uirastorza,
agarciadelayedra,rmoreno}@lortek.es

Abstract. Low defective rates and high variability in industrial processes, make difficult to develop accurate, reliable and fast automatic quality control systems. This paper presents two novel methods for the Quality Control of welded parts based on computer vision. The first method, **BG**, uses a modification of the bisection method, and a tuned version of the GoogleNet. The second method, **BGN**, is an ensemble of GoogleNet CNNs and a Convolutional Auto-Encoder to absorbe observed data variability. The CAE is used to select the best CNN from the ensemble to use for each input image. Both methods have been tested on more than 10^5 images and run in less than 0.2 s in a standard i7 CPU with mae values around 0.04–3.63 pixels, standard deviation of absolute errors around 1.15–7.48 pixels, and a percentage of correct predictions between 95–99.97%.

Keywords: Deep learning · Convolutional neural networks · Auto-encoder · Edge detection · Weld · Quality control

1 Introduction

Current technological advances are enabling an industrial manufacturing scenario with an expected defective rate measured in parts per million (ppm) [1, 2]. Production chains are continuously being improved and getting closer to perfection. In this sense, Zero Defect Manufacturing (ZDM) paradigm, fostered by Industry 4.0 for quality control, is being applied to reduce scrap and money expenditures [3]. Among its benefits are an increase in sales and competitiveness, customer loyalty and image consolidation [1]. However, a lot of effort is needed to integrate new technology and to make use of the huge amounts of data required to apply machine learning techniques [1, 4].

Defects can occur at any stage of the production chain. In case defects arise at early stages, they can be amplified and diffused in next steps leading to malfunctions and issues. Some of the defects can be avoided by adjusting manufacturing operations applying any of the four strategies: detection, repair, prediction and prevention [5]. However, some defects cannot be avoided and must be detected and removed from the productive chain. The development of accurate prediction models is complex and hence,

© Springer Nature Switzerland AG 2020
C. Analide et al. (Eds.): IDEAL 2020, LNCS 12490, pp. 524–532, 2020.
https://doi.org/10.1007/978-3-030-62365-4_50

most of the strategies are directed to detect and repair. Studies have shown that ZDM has been mainly applied to Additive Manufacturing, Casting and Welding processes [3] with a clear trend to prediction helped by recent evolution of hardware and software.

In the last years an increasing flexibility of production chains to accommodate to the heterogeneity, more demanding customers, and faster evolving technology has been observed. The trend is to reduce batch sizes and to increase variety requiring tool changes and setups of productive chains. Furthermore, the increase in decentralized production systems, entails a complex and challenging scenario for quality control applications [1, 6]: it demands higher precision and flexibility of used methods.

This is the case of production of safety components for capital goods, where several quality check points are added to the productive line to ensure that all defective parts are detected. When welding processes are present in the manufacturing process, it is important to realize that defects may appear even if the welding process is fully optimized. In fact, process variability causes unexpected defects that must be correctly identified. In the last years, defect detection addressed by means of visual inspection strategies is being replaced by computer-vision techniques.

Deep learning (DL) is transforming industry quality control [4, 7–9]. These methods cover a variety of applications – image, video, text and speech – and techniques – segmentation, classification, time series, clustering and so on – [10, 11]. Techniques such as transfer-learning, fine-tuning, data augmentation and networks of networks allow to accommodate existing nets to new sets of data in an easy way [12].

This paper presents a method for Quality Control of welded parts based on the Bisection Method [13], the GoogleNet [14] CNN network and a Convolutional Auto-Encoder (CAE) [11, 15]. This method has been implemented for real-time performance (<0.5 s in CPU) for quality control. It has been trained and validated with a manually labeled dataset with up to 10^5 images. This algorithm is simple, easy to train, fast to deploy in the productive line and it does not need much previous knowledge about DL.

The paper is structured as follows: Sect. 2 explains the methods. Section 3 shows the experimental results, and, in Sect. 4 conclusions and further work are indicated.

2 Methods

In this section the **BG** and **BGN** classification methods are described. **BG** is based on the Bisection method that uses GoogleNet as the evaluator function. **BGN** extends **BG** with an ensemble of GoogleNet CNNs and a CAE. The CAE accounts for the color variability of input images and it is used to select the best GoogleNet of the ensemble.

2.1 BG Method

The parts to be inspected consist of a cylindrical base with a weld on top of it. The aim here is to identify the edge that separates both parts (see Sect. 3.1 setup). Not all the edge needs to be detected to ensure the quality of the welded part: the detection of the outermost left and right points of the edge suffice. These points will be referred as *Cut Points* (CP) and they indicate the transition from base to the weld.

BG algorithm is based on the Bisection Method and GoogleNet. The precision and accuracy of the algorithm relays on those of the trained classifier. The Bisection Method is only used to guide the algorithm through the boundary edge to identify the CP. The algorithm consists of two steps. First, the edges of the part within the images are determined. Boundary is then divided into left- and right-hand sides. Second, the algorithm finds the CP separately in each side using the bisection method and GoogleNet network as function evaluator. The algorithm is illustrated in Fig. 1 (a).

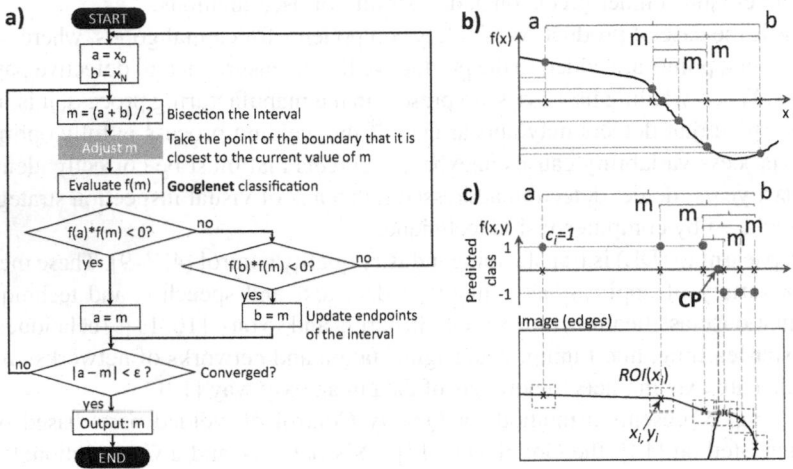

Fig. 1. a) Diagram flow b) Graphical representation and c) adaptation of the bisection method.

The Bisection Method: An Overview

The bisection method is a mathematical tool used to find the roots of continuous functions based on the occurrence of a change of sign within the function around the root. The method consists on iteratively bisecting an input interval known to contain the root of interest. In each iteration the subinterval at which the endpoints change the sign of the evaluated function are kept. The algorithm stops once the length of the interval is lower than a given tolerance value. See Fig. 1 (a–b).

Let f(x) be a continuous function. Let a and b be the endpoints of an interval [a, b] that satisfies $a < b$ and $f(a) \cdot f(b) < 0$. Then, the algorithm iteratively repeats the following steps. First, find the midpoint $m = (a + b)/2$. Second, check if this point corresponds to the root: $f(m) = 0$. In case is not, substitute b by m if $f(a) \cdot f(m) < 0$ holds or substitute a by m if $f(m) \cdot f(b) < 0$ holds. Repeat these steps with the new interval until $|a - b| < \varepsilon$.

Tuned Bisection Algorithm

For the purposes of this paper, the bisection method has been slightly modified (see Fig. 1(c)). In this case, a two variable function $f(x, y)$ is used. The cut points to be found are the roots of $f(x, y)$. Here x and y correspond to the coordinates of the image. A binary classifier is used as the function f to be evaluated that returns a 1 or a -1 if the point (x, y) corresponds to the base or to the weld and $f(x^{CP}, y^{CP}) = 0$ for the CP.

Let (x_i, y_i), with i in 0 to N, be any point of the set of points that correspond to one of the sides of the part's boundary. Boundary points are sorted, i.e.: $i = 0$ corresponds to the inferior point of the boundary and $i = N$ to the superior point. For simplicity of notation, intervals will be referred by index i instead of by the whole notation of the point (x_i, y_i). Thus, endpoints will accordingly be referred as $a = (x_0, y_0)$ and $b = (x_N, y_N)$.

Let $\textbf{ROI}(x_i, y_i)$ be a Region Of Interest centered at point (x_i, y_i) and L x L x 3 in size (here $L = 50px$, among proven values, 25, 20, 50, 75 and 100, this value gave best results). Following index notation, $\textbf{ROI}(x_i, y_i)$ will be referred as $\textbf{ROI}_i$. Let $c_i = \{-1, 1\}$ be the class assigned to $\textbf{ROI}_i$, where $c_i = 1$ if $\textbf{ROI}_i$ belongs to the base region while $c_i = -1$ if $\textbf{ROI}_i$ belongs to the welding region. Note that $c_0 = 1$ and $c_N = -1$.

The bisection method is applied in the following way. First, a is set to (x_0, y_0) and b to (x_N, y_N); inferior and superior points of the boundary. Then, the mid-point is determined, by first calculating $(x_m, y_m) = ((x_0 + x_N)/2, (y_0 + y_N)/2)$ and then finding the corresponding point of the boundary that is closest to this point. Second, the classifier is run to determine the class of $\textbf{ROI}_m$ and checked if $f(a) \cdot f(m) < 0$ or $f(m) \cdot f(b) < 0$. In the first case, m is assigned to a, otherwise, to b. This is iteratively repeated until the length of the interval is smaller than a threshold ε, i.e.: distance between centers of the ROI of the new interval is lower than ε. A pre-trained GoogleNet is used to classify the ROI.

Tuned GoogleNet for ROI Classification
The size of input images for standard GoogleNet is $224 \times 224 \times 3$. In order to speed up the execution time of the BG method, the size of the ROI images used in the classification has been reduced to $50 \times 50 \times 3$. To accommodate GoogleNet to these image input sizes, the last pool layer was removed see below figure Fig. 2(a).

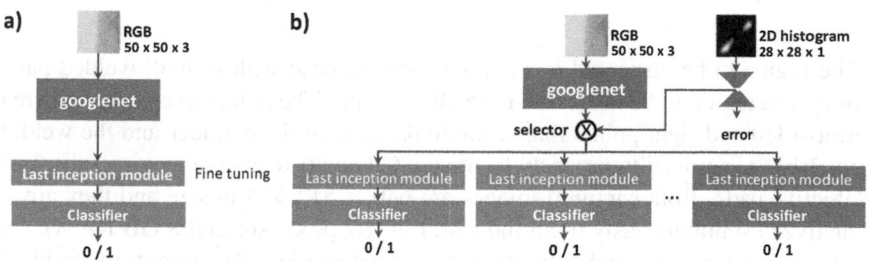

Fig. 2. a) GoogleNet classifier used and b) Alternative classifier with the CAE.

2.2 BGN Method

The major variability observed within these datasets concerns the color of the weld. To deal with it, the classifier has been extended to a network of classifiers in a similar way to that used in Expert Gate [12] (see Fig. 2(b)). Color variations can be identified training a CAE with 2D histograms obtained from Hue and Saturation values of the HSV color space [16]. First, a 2D histogram is created from HS values of the top part. Afterwards, this histogram is binned to 28×28 in size. This binned map is a chromatic representation and hence it is robust to variations on illumination sources. Then, a CAE is trained using these 2D histograms to obtain a classifier able to cluster chromatic groups.

Input images are grouped by color and GoogleNet is fine-tuned separately for each group obtaining a set of classifiers one for each color group. Finally, an ensemble of classifiers is built by merging their last inception layers. Note that during training, all previous layers of GoogleNet have been kept frozen. Therefore, these previous layers coincide for all the trained classifiers. The output of the color classifier is used as a gate to select the last inception layer to be used during classification.

3   Experimental Results

This section is divided as follows. First, the data used to train and validate the algorithms is introduced. Second, the performance of **BG** and **BGN** methods are shown.

3.1   Setup

The setup is based on the use of two horizontally placed color cameras that simultaneously capture images on either side of the part, see Fig. 3(a). To ensure homogeneous illumination and to avoid glare and shadows, a dome-type illumination system is used.

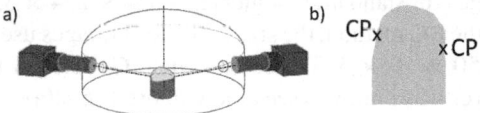

Fig. 3.   Description of the a) Inspection system. b) Example of a captured image, CP = Cut Point.

The region to be inspected is a cylinder-shaped head with a small welded part on top of it. The region to be inspected is small (~3 mm). The points to be detected are the outermost left and right points that separate the base of the cylinder and the weld, CP in Fig. 3(b). It is crucial to properly locate the CP points to ensure a correct discarding of defective parts. Both captured images are $640 \times 512 \times 3$ in size and they are sent and analyzed simultaneously to an industrial PC (i7 processor and 8 GB RAM). Their outputs are combined to send a unique decision output (defective or not) to the PLC.

3.2   Description of the Data

To check the effectiveness of the algorithm, more than 120,000 images were visually inspected, and their CP annotated. Four datasets were created, A1, A2, B1 and B2 (see Table 1). Datasets A1 and A2 show small variations in chroma values, while, datasets B1 and B2 are highly variable.

In both, **BG** and **BGN** methods, the accuracy of the results relies on the performance of the classifier and therefore, the only part of the algorithm to be trained is the classifier. For the **BG** method, this means GoogleNet. For the **BGN** method, it means the CAE and the ensemble of GoogleNet networks. The CAE is used to classify data by color values and the ensemble to adjust the classifier to differences in color.

Table 1. Datasets

Dataset	Welding machine	# images	Observations
A1	A	24,662	Small chroma variations
A2	A	26,123	Small chroma variations
B1	B	52,015	Highly variable chroma
B2	B	20,106	Highly variable chroma

3.3 Results for BG Method

GoogleNet has been trained for each dataset separately using Digits[1] with pre-trained weights[2]. In order to use GoogleNet as a classifier for the **BG** method, it must be trained with reference ROI images $50 \times 50 \times 3$ in size. First, in every image the boundaries of the parts are determined. Then, 15 points of the boundary are selected at random and the **ROI$_i$** centered on each of the selected i^{th} point is cropped. Note that the class of each point is known. Every cropped image is labeled with its c_i value (-1 or 1). A balanced dataset is created by selecting half of the ROI images of class -1 and half of class 1. Finally, GoogleNet is trained using 25% of cropped images for validation and 10% for testing. Obtained accuracy and loss values were of 0.9996 and 0.00172.

Trained GoogleNet model was added to the bisection method and **BG** algorithm was run for the four datasets. The tolerance of **BG**, ε, was set to 10px. Predicted (x, y) coordinates of the CP were compared with the annotated ones. As error measurement, the distance between predicted and annotated coordinates is used. Mean Absolute Error (mae) and the standard deviation of the absolute error (std) are shown in Table 2. The percentage of images with both CP correctly predicted had been also added to this table. These values have been calculated for predicted absolute errors above 5, 10, 20 and 30px. Observe that as expected, due to the reduced variability of datasets A, the algorithm correctly predicts the 99.85% and 99.97% of the CP. On the contrary, these values are lower for dataset B (90.16% and 93.66%) due to color variabilities.

3.4 Results for BGN Method

Images on dataset B are highly variable in color values. GoogleNet for **BG** had been trained by selecting images from B1 and B2 datasets at random, thereby obtaining an imbalanced training dataset from color point of view.

To deal with color variability, a CAE was trained with 2D histograms obtained from Hue and Saturation values. The dataset used for training consisted of three groups of chromatic sets, denoted as 0, 1, and 2 with 1600, 1200 and 1100 images, respectively.

---

[1] https://developer.nvidia.com/digits.

[2] http://dl.caffe.berkeleyvision.org/.

Table 2. Results for BG model: Mean Absolute Errors (mae), Standard Deviation of the Error (std) and Percentages of Correct Predictions (PCP) for errors lower than of 5, 10, 20 and 30 px.

DataSet	mae	std	5px	10px	20px	30px
A1	0.04	1.29	99.85%	99.94%	99.94%	99.96%
A2	0.06	1.15	99.48%	99.92%	99.97%	99.97%
B1	3.67	15.65	93.66%	93.84%	94.64%	94.93%
B2	7.10	22.83	90.16%	90.54%	90.85%	90.91%

The last trained layer of the encoder was used to train a color classifier using 25% of the images for validation and 10% for testing. The confusion matrix of the color classifier is shown in Table 3. The top-1 accuracy of the model was of 0.95.

Table 3. Confusion matrix for CC.

	0	1	2	Per-class accuracy
0	417	6	0	0.9858
1	6	202	48	0.7891
2	0	1	278	0.9964

Predicted classes were used to divide datasets by color groups and used to train GoogleNet (training accuracies > 0.995 and losses < 0.002). Trained models were merged to create the **BGN** model that was run for images of datasets B1 and B2. Results are shown in Table 4. This allowed to increase the percentage of correctly identified cut points (PCP) from 90.85% and 94.64% to 97.09% and 97.73% for an error > 20px.

Table 4. Mae, std and PCP for the BGN model.

DataSet	Mae	std	5 px	10 px	20 px	30 px
B1	2.14	6.38	84.34%	91.72%	97.09%	99.03%
B2	3.87	7.48	64.61%	88.15%	97.73%	98.83%

Table 5 shows the results separated for each chromatic group. The mae and std values increases considerably for classes 0 and 1, and the CP correctly predicted reduces considerably for errors of 5 and 10px. This is mainly because groups 0 and 1 concentrate the highest variability in the data. However, their values are reasonably low for errors > 20px. Note from the confusion matrix that the groups 0 and 1 are not easy to distinguish: 48 out of 256 histograms of group 1 have been classified as group 2.

Table 5. Mae, std and PCP values for the BGN model divided by predicted color group.

DatasSet	Color group	# img	mae	std	5 px	10 px	20 px	30 px
B1	0	4,994	14.53	9.41	5.37%	37.83%	74.62%	92.39%
B1	1	3,218	7.88	4.54	10.85%	78.03%	98.94%	99.72%
B1	2	43,802	0.31	3.66	98.74%	98.87%	99.51%	99.73%
B2	0	2,217	12.52	9.56	9.65%	46.5%	85.84%	94.9%
B2	1	5,562	7.81	4.67	10.65%	80.62%	99.21%	99.49%
B2	2	12,326	0.53	5.77	98.86%	99.03%	99.2%	99.24%

3.5 Execution Time and Convergence

Both algorithms were run in a CPU i7, 8 GB, 2 core industrial computer. The time required by GoogleNet to classify $50 \times 50 \times 3$ images was of 0.01 s. The algorithms were stopped if an error lower than a prefixed tolerance was reached or if the maximum number of iterations of 10 was reached. The overall time for the algorithms was of 0.2 s.

4 Conclusions

In this paper an image analysis algorithm based on the Bisection Method and GoogleNet has been introduced. It runs in less than 0.25 s for two images in parallel in a CPU. It has been tested with more than 10^5 manually annotated images and collected from different productive lines. The algorithm has been forced to stop in case the expected error was lower than 10 pixels or in case the number of iterations reached 10. Obtained mae relay between 0.04 and 3.63 pixels, a standard deviation of absolute errors around 1.15 and 7.48 pixels, and a percentage of correct predictions between 95 and 99.97%.

To accommodate to process variability, observed as chromatic changes in images, a Convolutional Auto-Encoder (CAE) has been added. By adding the CAE, process alterations can be directly identified within the image and thus, quality control could be directed towards process quality control. Ongoing work will consist of simplifying the classifier: to speed up the algorithm and to reduce required memory usage in the CPU.

Acknowledgments. This work has been founded by the project KK-2019/00095 (Departamento de Desarrollo Economico e Infraestructuras del Govierno Vasco. Programa ELKARTEK 2019

References

1. Godina, R., Matias, J.C.O.: Quality control in the context of industry 4.0. In: Reis, J., Pinelas, S., Melão, N. (eds.) IJCIEOM 2018. SPMS, vol. 281, pp. 177–187. Springer, Cham (2019). https://doi.org/10.1007/978-3-030-14973-4_17
2. Alcácer, V., Cruz-Machado, V.: Scanning the industry 4.0: a literature review on technologies for manufacturing systems. Eng. Sci. Technol. Int. J. **22**(3) (2019). https://doi.org/10.1016/j.jestch.2019.01.006

3. Psarommatis, F., May, G., Dreyfus, P.A., Kiritsis, D.: Zero defect manufacturing: state-of-the-art review, shortcomings and future directions in research. Int. J. Prod. Res. 1–17 (2019). https://doi.org/10.1080/00207543.2019.1605228

4. Bajic, B., Cosic, I., Lazarevic, M., Sremčev, N., Rikalovic, A.: Machine Learning Techniques for Smart Manufacturing: Applications and Challenges in Industry 4.0 (2018)

5. Psarommatis, F., Kiritsis, D.: A scheduling tool for achieving zero defect manufacturing (ZDM): a conceptual framework. In: Moon, I., Lee, G.M., Park, J., Kiritsis, D., von Cieminski, G. (eds.) APMS 2018. IAICT, vol. 536, pp. 271–278. Springer, Cham (2018). https://doi.org/10.1007/978-3-319-99707-0_34

6. Van den Broeke, M., Boute, R., Van Mieghem, J.: Platform flexibility strategies: R&D investment versus production customization tradeoff. Eur. J. Oper. Res. (2018). https://doi.org/10.1016/j.ejor.2018.03.032

7. Muniasamy, A., Alasiry, A.: Deep learning: the impact on future eLearning. Int. J. Emerg. Technol. Learn. (iJET) 15, 188 (2020). https://doi.org/10.3991/ijet.v15i01.11435

8. Tabernik, D., Šela, S., Skvarč, J., Skočaj, D.: Segmentation-based deep-learning approach for surface-defect detection. J. Intell. Manuf. 31(3), 759–776 (2019). https://doi.org/10.1007/s10845-019-01476-x

9. Villalba-Diez, J., Schmidt, D., Gevers, R., Ordieres-Meré, J., Buchwitz, M., Wellbrock, W.: Deep learning for industrial computer vision quality control in the printing industry 4.0. Sensors 19, 3987 (2019). https://doi.org/10.3390/s19183987

10. Sengupta, S., et al.: A review of deep learning with special emphasis on architectures, applications and recent trends. Knowl.-Based Syst. 194 (2020). https://doi.org/10.1016/j.knosys.2020.105596

11. Käding, C., Rodner, E., Freytag, A., Denzler, J.: Fine-tuning deep neural networks in continuous learning scenarios. In: Chen, C.-S., Lu, J., Ma, K.-K. (eds.) ACCV 2016. LNCS, vol. 10118, pp. 588–605. Springer, Cham (2017). https://doi.org/10.1007/978-3-319-54526-4_43

12. Aljundi, R., Chakravarty, P., Tuytelaars, T.: Expert Gate: Lifelong Learning with a Network of Experts, pp. 7120–7129 (2017). https://doi.org/10.1109/cvpr.2017.753

13. Ehiwario, J.C.: Comparative study of bisection, Newton-Raphson and secant methods of root-finding problems. IOSR J. Eng. 4, 01–07 (2014). https://doi.org/10.9790/3021-04410107

14. Szegedy, C., et al.: Going deeper with convolutions. In: The IEEE Conference on Computer Vision and Pattern Recognition (CVPR), pp. 1–9 (2015). https://doi.org/10.1109/cvpr.2015.7298594

15. Guo, X., Liu, X., Zhu, E., Yin, J.: Deep Clustering with Convolutional Autoencoders, pp. 373–382 (2017). https://doi.org/10.1007/978-3-319-70096-0_39

16. Baek, N., Park, S.M., Kim, K.J., Park, S.B.: Vehicle color classification based on the support vector machine method. Proc. Commun. Comput. Inf. Sci. 2, 1133–1139 (2007). https://doi.org/10.1007/978-3-540-74282-1_127

Detecting Performance Anomalies in the Multi-component Software a Collaborative Robot

Héctor Quintián[1], Esteban Jove[1]([✉]), José Luis Calvo-Rolle[1],
Nuño Basurto[2], Carlos Cambra[2], Álvaro Herrero[2],
and Emilio Corchado[3]

[1] CTC, Department of Industrial Engineering, CITIC, University of A Coruña,
Avda. 19 de febrero s/n, 15405 Ferrol, A Coruña, Spain
{hector.quintian,esteban.jove,jlcalvo}@udc.es
[2] Grupo de Inteligencia Computacional Aplicada (GICAP), Departamento de
Ingeniería Informática, Escuela Politécnica Superior, Universidad de Burgos,
Av. Cantabria s/n, 09006 Burgos, Spain
{nbasurto,ccbaseca,ahcosio}@ubu.es
[3] Edificio Departamental, Campus Unamuno, University of Salamanca,
37007 Salamanca, Spain
escorchado@usal.es

Abstract. The detection of anomalies (affecting hardware or software) is an open challenge for cyber-physical systems in general and robots in particular. Physical anomalies related to the hardware components of such systems have been widely researched. However, scant attention has been devoted so far to study the anomalies affecting the software components. In order to bridge this gap, the present paper proposes the application of different classifiers to a robot performance dataset for the first time. The applied supervised models are targeted at detecting synthetically-induced software anomalies, having a detrimental impact on the performance of a collaborative robot. Obtained results demonstrate that the applied Machine Learning models can successfully address the target problem, with acceptable detection rates.

Keywords: Anomaly detection · Performance · Software · Robot · Pattern classification

1 Introduction

As it is widely acknowledged, robotics is a driving force in many different fields varying from the Industry 4.0 to elderly care. More precisely, collaborative robots are considered as a disrupting technology that is greatly contributing to the progress in such fields. At the same time, there is an increasing demand of reliability and robustness of this kind of robots, as they cooperate and interact with human beings. In order to meet this demand, anomaly/failure detection can been applied to robots.

© Springer Nature Switzerland AG 2020
C. Analide et al. (Eds.): IDEAL 2020, LNCS 12490, pp. 533–540, 2020.
https://doi.org/10.1007/978-3-030-62365-4_51

Up to now, many Machine Learning (ML) models have been previously applied to different robot issues such as control [20,21] and communications [1]. Since the seminal studies in this field [13], most anomaly-detection previous work has been focused on the identification of hardware anomalies [7], while the software components of robots have remained unexplored for a long time. To go one step further on this interesting topic, this work deals with the detection of performance anomalies in the software system of a robot, composed of different modules. In order to do that, different ML models based on supervised learning are applied to a dataset [15] that is publicly available [16]. This is a pioneering dataset containing tagged anomalies generated in the robot while carrying out different tasks. Further details are explained in Sect. 3.

Initially, authors of the dataset [17] applied some methods inspired by *Support Vector Machines* (SVM), such as One-Class SVM, in order to detect the induced anomalies. In a sequel paper by Wienke et al. [18], an analysis of each one of the components of the robot was performed. Each one of them is compared with the other ones to know the potential changes that can take place in the use of resources. That is, from the values that they collect from one component, authors try to predict the behavior of another component regarding the use of resources. [18] includes the analysis of the individual components of the robot to discover the potential changes that can take place in the use of resources. Such an analysis makes enables the prediction of how changes in the operation of one component may affect the other ones. More recently, [2] presents results from One-Class SVM that have been enhanced by applying data balancing algorithms to this dataset.

In addition to SVMs, some other ML models have been previously applied for anomaly/failure detection in robots. That is the case of [8], that proposed the application of Artificial Neural Networks (ANN) to detect minor failures in the sensors and control surfaces of an autonomous underwater vehicle. In [3] Logistic Regression (LR), Regularized Logistic Regression (RLR), Random Forest (RF) and Extreme Learning Machines were applied to detect faults related to the degradation of one specific component in a robotic arm. Quite recently, [12] has proposed the application of LR and SVMs, together with RF and Ensemble Stacking to identify gearbox failures in industrial robots. Furthermore, results are improved by different data-preprocessing alternatives.

Differentiating from the above-mentioned previous work, the present paper proposes the novel application of different one-class classifiers for the detection of performance anomalies in robots rather to identifying hardware failures. This is the first time that this problem is addressed by applying the ML models proposed in the present work.

The rest of this paper is organized as follows: the applied ML models are introduced in Sect. 2 while the dataset under analysis in described in Sect. 3. The obtained results are presented and discussed in Sect. 4 while the conclusions and proposals for future work are stated in Sect. 5.

2   Applied Classifiers

A brief introduction of the one-class classifiers applied to carry out the anomaly detection task are presented in the next subsections.

2.1   Autoencoder Artificial Neural Network

The use of ANN with the Autoencoder topology has the main goal of replicating the input pattern at the network output, by means of a nonlinear reduction of the dimensionality happening in the hidden layers. After the dimension reduction, the data is decompressed at the output [6]. As this ANN is trained with data from only normal operation, the network is trained to replicate a certain target pattern.

2.2   k-Nearest Neighbor

The k-Nearest Neighbor (kNN) algorithm performs the anomaly detection based on the distances between objects. This means that, for a given test point x, its local density is calculated by implementing an hyper-sphere that contains the kth nearest neighbors [10]. Then, if this density is high, it means that the data belongs to the target set. Otherwise, if the density is far below a specific threshold (calculated during the training time), the anomaly is detected [11].

2.3   Principal Component Analysis

Although the Principal Component Analysis (PCA) algorithm has been used especially for dimension reduction problems [19], it can be also applied to detect anomalies [4]. This approach is based on projecting the original data over the principal components, which represent the directions with greatest variance. Then, the distance between a test sample and its projection is defined as the reconstruction error.

2.4   Support Vector Data Description

The one-class Support Vector Data Description (SVDD) represents a widely used boundary method [11]. It is based in the well-known SVM, whose main target is to project the data over a feature space with higher dimension and, then, implement an hyper-plane that maximizes the distance between the origin and the target set. From this idea, in [11], it is proposed the concept of implementing an hyper-sphere containing training samples instead of an hyper-plane. Once this hyper-sphere is calculated, its boundaries are used to determine the anomalous nature of a test instance.

3 Robot Performance Dataset

The present research is carried out with a dataset obtained from a component-based robotic system, in which faults have been induced by software. The induction of fault by software implies that the objective is to make a variation in the system's counters, thus presenting different records in the collected data. The analysed dataset has been released by researchers at the University of Bielefeld (Germany) and is publicly available at [16]. Further details about the data are described in [15]. The data were collected from the robot during its participation in a competition called RoboCup@Home, in which the robots are required to carry out a set of tasks in which the different components of the robot interact. The data were gathered during 71 trials in which the same actions were taken; the robot had to perform various actions always in the same order. The induced errors are distributed among the different trials.

As previously mentioned, the robot is composed of different modules, that can be from different manufacturers, which are linked through a middleware intercommunicating all of them. The middleware used in this robot is event-based and is called Robotics Service Bus (RSB) Middleware [14]. Together with the RSB middleware, there was a tool (rsbag) that collects the information that flows through the middleware. For the transmission of the information, there are events coded as notifications. Two participants deal with the information exchange between the bus and the usercode: the *informer* which is in charge of supervising that the information goes from the usercode to the bus and the *listener* which does it in the opposite direction. Additionally, the robotic system includes a framework called *BonSAI*, that is in charge of coordinating the sensors and the actuator components. It is done by creating RSB participants when necessary for a state or transition. There is also a Finite State Machine which is in charge of controlling the execution flows and representing them. All of the described components and flows can be seen in greater detail in the Fig. 1.

A 1-s sampling of the data taken from the robot (a data instance) was performed. This data from each component was obtained from different indicators, such as the current size of virtual memory occupied by a task, number of written bytes, amount of threads a process has in execution, etc.

In the present work, it is studied the *statemachine* component that centralizes the control of the state of the whole system, being crucial for the performance of the robot. Its operation is based on the model elaborated by Siepmann and Wachsmuth [9]. The *statemachine* component has been selected as it is the one suffering from the highest number of anomalies (3), that are:

- *btlAngleAlgo*: When computing the location of people in front of the robot, a mathematical error is made.
- *bonsaiParticipantLeak*: Unused RSB participants are not deleted.
- *bonsaiTalkTimeout*: The wrong RSB is used, resulting in a timeout until the restoration.

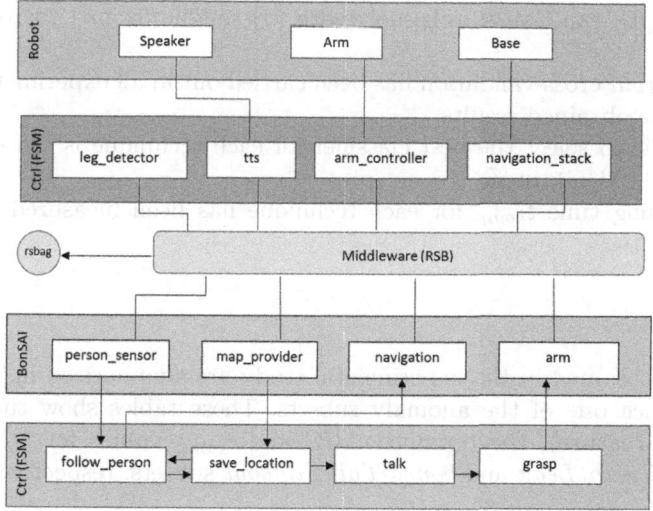

Fig. 1. Overview of the robot system architecture. Adapted from [15].

A subset of the whole dataset has been generated for each one of the anomalies. From the 71 trials, only 33 are included in the analysed subsets as not all of the trials are valid according to the criteria established by the dataset authors [15].

A total of 21885 instances and 186 features for each one of the three subsets are analysed. The initial data had a large amount of missing values in some columns, so all features containing any of them were eliminated.

4   Experiments and Results

This section presents the experimental setup and obtained results when applying the above-introduced classifiers to the robot performance dataset.

4.1   Experimental Setup

To check the classifier performance over the three subsets, the next experimental setup was considered for each one of them:

- Data conditioning was tested with three configurations: normalization 0 to 1, normalization using Z-Score and using raw data (without normalization).
- Several different combinations of parameter values were tested for each classifier:
    - Autoencoder: the number of neurons in the hidden layer (n_a) ranging from 1 to $(n-1)$, being n the dimension of the dataset.
    - kNN: the number of neighbors (k) ranging from 1 to 20.
    - PCA: the number of components (n_c) ranging from 1 to $(n-1)$, being n the dimension of the dataset.

- SVDD: the Gaussian kernel width (σ) ranging from 1 to 10 with a 0.5 step.
- A 10 $k-fold$ cross-validation has been carried out in all experiments in order to validate obtained results.
- The metric to select the best classifier for each technique is the Area Under the Curve (AUC), in %.
- The training time t_{comp} for each technique has been measured and is also shown.

4.2   Results

The results obtained in the experimental study are summarized in Tables 1, 2, and 3 for each one of the anomaly subsets. These tables show the best configurations (Feature), the obtained AUC and t_{comp} values for $btlAngleAlgo$, $bonsaiParticipantLeak$ and $bonsaiTalkTimeout$ subsets, respectively.

Table 1. Configuration and classification results for the $btlAngleAlgo$ subset

Classifier	Preprocessing	Feature	AUC (%)	t_{comp}(s)
Autoencoder	–	$n_a = 1$	83.17	9.48
kNN	–	$k = 10$	80.35	0.01
PCA	Zscore	$n_c = 148$	**86.02**	0.03
SVDD	0–1	$\sigma = 2.5$	74.43	0.11

For all the three subsets, the best classification results, according to the AUC metric, have been obtained with PCA and a high number of components (148, 183, and 186 respectively). Additionally, in all cases the training time of this method is the second smallest (0.03), quite close to the smallest one (0.01).

From a general perspective, it can be mentioned that the results significantly vary from one classifier to the other ones when applied to the three subsets.

Table 2. Configuration and classification results for the $bonsaiParticipantLeak$ subset

Classifier	Preprocessing	Feature	AUC (%)	t_{comp}(s)
Autoencoder	–	$n_a = 11$	85.96	34.90
kNN	0–1	$k = 1$	86.29	0.01
PCA	0–1	$n_c = 183$	**86.30**	0.03
SVDD	0–1	$\sigma = 3$	82.98	0.10

Table 3. Configuration and classification results for the *bonsaiTalkTimeout* subset

Classifier	Preprocessing	Feature	AUC (%)	$t_{comp}(s)$
Autoencoder	–	$n_a = 141$	72.83	85.03
kNN	0–1	$k = 1$	86.23	0.01
PCA	0–1	$n_c = 186$	**86.40**	0.03
SVDD	0–1	$\sigma = 3$	78.72	0.11

5 Conclusions and Future Work

The present work address the problem of anomaly detection in the software component of a collaborative robot. Three different anomalies linked with the *statemachine* component are induced, obtaining a successful classification performance in general terms. This achievement may represent an interesting breakthrough to improve robot performance, making the process more cost-efficient.

Four different one-class classifiers have been applied. Among them, PCA has attained the bests classification rates, in the three subsets. However, although it presents the best performance and a significantly low computation time, kNN is competitive in both fields, especially when applied to the *bonsaiParticipantLeak* and *bonsaiTalkTimeout* subsets. It is worth highlighting that, even though Autoencoder exhibited good classification results, its training times are significantly higher. This is an important disadvantage, especially when an online training stage is sought.

As future work, the possibility of applying alternative one-class methods is to be considered [5]. Furthermore, detecting other robot anomalies, not only related to the software can represent an interesting path to advance the present proposal.

References

1. Alsamhi, S.H., Ma, O., Ansari, M.S.: Survey on artificial intelligence based techniques for emerging robotic communication. Telecommun. Syst. **72**(3), 483–503 (2019). https://doi.org/10.1007/s11235-019-00561-z
2. Basurto, N., Cambra, C., Álvaro Herrero: improving the detection of robot anomalies by handling data irregularities. Neurocomputing (2020, in press)
3. Costa, M.A., Wullt, B., Norrlöf, M., Gunnarsson, S.: Failure detection in robotic arms using statistical modeling, machine learning and hybrid gradient boosting. Measurement **146**, 425–436 (2019). https://doi.org/10.1016/j.measurement.2019.06.039
4. Jove, E., Casteleiro-Roca, J.L., Quintián, H., Méndez-Pérez, J.A., Calvo-Rolle, J.L.: Anomaly detection based on intelligent techniques over a bicomponent production plant used on wind generator blades manufacturing. Revista Iberoamericana de Automática e Informática industrial (2019)
5. Jove, E., Casteleiro-Roca, J.L., Quintián, H., Méndez-Pérez, J.A., Calvo-Rolle, J.L.: A new method for anomaly detection based on non-convex boundaries with random two-dimensional projections. Inf. Fus. **65**, 50–57 (2020)

6. Jove, E., Casteleiro-Roca, J.L., Quintián, H., Simić, D., Méndez-Pérez, J.A., Luis Calvo-Rolle, J.: Anomaly detection based on one-class intelligent techniques over a control level plant. Logic J. IGPL **28**(4), 502–518 (2020)
7. Lu, H., Li, Y., Mu, S., Wang, D., Kim, H., Serikawa, S.: Motor anomaly detection for unmanned aerial vehicles using reinforcement learning. IEEE Internet of Things J. **5**(4), 2315–2322 (2018). https://doi.org/10.1109/JIOT.2017.2737479
8. Ranganathan, N., Patel, M.I., Sathyamurthy, R.: An intelligent system for failure detection and control in an autonomous underwater vehicle. IEEE Trans. Syst. Man Cybern. Part A Syst. Hum. **31**(6), 762–767 (2001)
9. Siepmann, F., Wachsmuth, S.: A modeling framework for reusable social behavior. In: De Silva, R., Reidsma, D. (eds.) Work in progress workshop proceedings ICSR, pp. 93–96 (2011)
10. Sukchotrat, T.: Data mining-driven approaches for process monitoring and diagnosis. Ind. Manuf. Eng. (2009)
11. Tax, D.M.J.: One-class classification: concept-learning in the absence of counterexamples Ph.D. thesis. Delft University of Technology (2001)
12. Vallachira, S., Orkisz, M., Norrlöf, M., Butail, S.: Data-driven gearbox failure detection in industrial robots. IEEE Trans. Industr. Inf. **16**(1), 193–201 (2020)
13. Visinsky, M., Cavallaro, J., Walker, I.: Robotic fault detection and fault tolerance: a survey. Reliab. Eng. Syst. Saf. **46**(2), 139–158 (1994). https://doi.org/10.1016/0951-8320(94)90132-5. http://www.sciencedirect.com/science/article/pii/0951832094901325
14. Wienke, J., Wrede, S.: A middleware for collaborative research in experimental robotics. In: 2011 IEEE/SICE International Symposium on System Integration (SII), pp. 1183–1190, December 2011. https://doi.org/10.1109/SII.2011.6147617
15. Wienke, J., Meyer zu Borgsen, S., Wrede, S.: A data set for fault detection research on component-based robotic systems. In: Alboul, L., Damian, D., Aitken, J.M.M. (eds.) TAROS 2016. LNCS (LNAI), vol. 9716, pp. 339–350. Springer, Cham (2016). https://doi.org/10.1007/978-3-319-40379-3_35
16. Wienke, J., Wrede, S.: A Fault Detection Data Set for Performance Bugs in Component-Based Robotic Systems. https://doi.org/10.4119/unibi/2900911
17. Wienke, J., Wrede, S.: Autonomous fault detection for performance bugs in component-based robotic systems. In: 2016 IEEE/RSJ International Conference on Intelligent Robots and Systems (IROS), pp. 3291–3297. IEEE (2016)
18. Wienke, J., Wrede, S.: Continuous regression testing for component resource utilization. In: IEEE International Conference on Simulation, Modeling, and Programming for Autonomous Robots (SIMPAR), pp. 273–280. IEEE (2016)
19. Wu, J., Zhang, X.: A PCA classifier and its application in vehicle detection. In: IJCNN 2001. International Joint Conference on Neural Networks. Proceedings (Cat. No. 01CH37222), vol. 1, pp. 600–604. IEEE (2001)
20. Xiao, B., Yin, S.: Exponential tracking control of robotic manipulators with uncertain dynamics and kinematics. IEEE Trans. Industr. Inf. **15**(2), 689–698 (2019). https://doi.org/10.1109/TII.2018.2809514
21. Zhao, D., Ni, W., Zhu, Q.: A framework of neural networks based consensus control for multiple robotic manipulators. Neurocomputing **140**, 8–18 (2014). https://doi.org/10.1016/j.neucom.2014.03.041

Prediction of Small-Wind Turbine Performance from Time Series Modelling Using Intelligent Techniques

Santiago Porras[1], Esteban Jove[2]($\boxtimes$), Bruno Baruque[3],
and José Luis Calvo-Rolle[2]

[1] Departamento de Economía Aplicada, University of Burgos,
Plaza Infanta Doña Elena, s/n, 09001 Burgos, Burgos, Spain
sporras@ubu.es
[2] Departamento de Ingeniería Industrial, University of A Coruña,
Avda. 19 de febrero s/n, 15495 Ferrol, A Coruña, Spain
{esteban.jove,jlcalvo}@udc.es
[3] Departmento de Ingeniería Informática, University of Burgos,
Avd. de Cantabria, s/n, 09006 Burgos, Burgos, Spain
bbaruque@ubu.es

Abstract. The present research work deals the model creation obtaining for power generation prediction of a small-wind turbine, based on the atmospheric variables of its location. For testing purposes, a real dataset has been obtained of a bio-climate house located in Sotavento Experimental Wind Farm in the north of Spain. A deep study of the system and atmospheric variables has been performed. Then, some different regression techniques have been tested for accomplishing prediction, obtaining excellent results.

Keywords: Small-wind turbine · Time series · Regression tree · K-nearest neighbours · Gradient boosted ensembles · Auto-regressive neural networks

1 Introduction

Over past decades, the problem of climate change derived from the use of fossil fuels, has enhanced the promotion of alternative clean energies. To achieve this goal, both national and international government institutions have developed and applied different strategies to the greenhouse effect problem. One fundamental strategy consisted of rising the price of CO2 emission rights that, in countries like Spain, represented a 300% of increase [4]. In this context, the promotion of sustainable an energy generation system has been presented as a key factor [2,7].

However, this new trend towards the renewable technologies is still in a development phase. In 2007, 15% of the energy production around the world was obtained from renewable energies. Although the hydroelectric power has

© Springer Nature Switzerland AG 2020
C. Analide et al. (Eds.): IDEAL 2020, LNCS 12490, pp. 541–548, 2020.
https://doi.org/10.1007/978-3-030-62365-4_52

been traditionally the most exploited renewable source, there has been a significant development of other technologies, especially the wind power. In [9], the worldwide wind power installed suffered a remarkable increment, from 17 GW (gigawatts) in 2000 to 514 GW in 2017. Focusing on works related to the future of this energy, it is estimated that the wind power will represent almost the 23% of the global energy production in 2030 [1,5].

This situation emphasizes the importance of developing and improving wind generation systems to make them more efficient, since they may play a key role in a near future. A possible way to face the problem of improving wind technology is presented by the use of intelligent techniques to predict the power generated by a system. An accurate estimation of the performance of a wind facility can present a key step to energy management. This can be achieved by monitoring atmospheric variables, such as pressure or wind speed, which can vary from year to year, with the season, on a daily basis, or even in seconds [5], the use of intelligent techniques can present a good approximator of the behavior of the system.

The present work deals with the power generation prediction of a wind turbine placed in a bioclimatic house. The prediction is carried from an original dataset of 50834 samples registered during one year.

2   Case of Study

This work faces the problem of predicting the energy generated in a wind turbine placed at the Sotavento bioclimatic house. This is a sustainable building funded by the Sotavento Galicia Foundation, whose main goal is to make use of different renewable energy sources, and contribute to disseminate their benefits. This facility is placed in the borders of the provinces of A Lugo and Coruña, in the autonomous region of Galicia.

The electric power can be obtained from three different sources. First, a small-wind turbine takes advantage of wind speed to generate 1.5 kW. Furthermore, a photovoltaic field comprised of twenty two solar panels is sized to produce 2,7 kW. In cases where wind and solar production are not enough for cover the electricity demand, the power grid is also available. The Domestic Hot Water (DHW) system is supplied through three different technologies. First, eight solar panels are located over the roof to transfer the solar radiation to a thermal accumulator. Then, a biomass boiler can be configured to deliver from 7 to 20 kW, with a yield of pellets of 90%. Finally, a geothermal system comprised of an horizontal collector can also absorb the heat from the ground.

2.1   Small-Wind Turbine

As stated above, the present research deals with the estimation of the electric power generated by an small-wind turbine, whose main features are described in this section.

The small-wind turbine is a BORNAY INCLIN 1.500 model, which has two carbon fiber and fiberglass blades. The rotatory movement is converted in Alternating Current (AC) by means of a three phase synchronous generator with neodymium permanent magnets. An AC to DC (Direct current) stages, combined with an electronic inverter, ensures the fixed frequency and level voltage (230 Vrms 50 Hz).

2.2 Dataset Description

The dataset used during this case study was collected from April 2017 to March 2018. Therefore, we have all the observations of the power generated by the turbine for a whole year. Which have been taken at 10 min intervals, making 52560 observations. After carrying out the corresponding filtering of erroneous data and outliers, the observations are reduced to 50836. The values of the series are representing power watts transmitted from the wind turbine system to the home installation. Its minimum value is 0.0, its 75 percentile: 62.426350 and the maximum value registered 2009.58. The values have also been normalized to the range [0,1] for easier handling by the algorithms. Figure 1 shows the series for the whole year, after normalization.

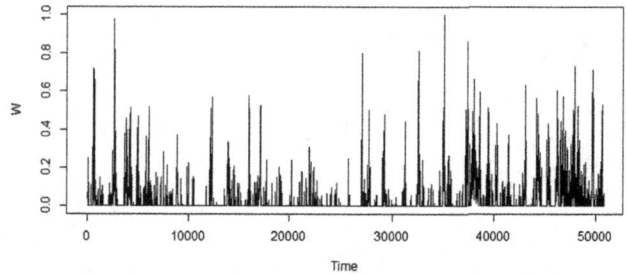

Fig. 1. Representation of the power (W) generated over a whole year by the system. Data has been normalized on the interval [0,1].

3   Power Prediction Study

The study presented in this work is centred around the measurements performed using mini wind turbine system that is attached to the house. The objective of our study is to see if the short-term prediction of the power that will be generated by this installation can be achieved in a precise way by analysing the power readings previous obtained. This analysis could be potentially be used by a more complex system in order to achieve an optimization of the bio-climatic home resources at a given moment.

3.1   Lag Selection Study

One of the problems faced in this study is choosing how many previous observations to use to predict the power values in the following instants of time. As this is a real problem, we must make a prediction using recent observations over time. There is little point in using distant observations over time, as they have no influence on the current state. For this, the Box-count method is applied:

Box-Count Method. Originally, this method was designed to estimate the fractal dimension of a geometry and is widely used in image recognition or signal analysis for identify self-similarity. Applied to time series as explained in [6] it divides the series in boxes and in a recursive way creates small boxes until the observations in each box has the same similarity. This measure give us the number of observations, which are correlated, and we can choose this value as the lag value. In this case of study Fig. 2 shows the result of Box-count with a value of 0.883. Analyzing the graphs obtained and the value thrown by the Box-count method, we can conclude that the optimal value of the lags is between eight and nine, following a restrictive criterion, we will select the lower limit, therefore we set 8 steps backwards to include that information in the methods.

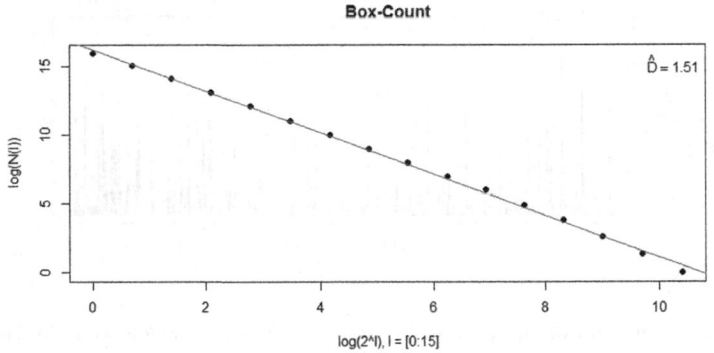

Fig. 2. Representation of the Box-Count function over the power data series analysed.

4   Applied Techniques

This section provides a brief description of the techniques employed for the prediction of short-term power generation problem. As can be seen, a variety of models have been used, ranging from classical statistical methods, such the ARIMA to more recent approaches such as neural networks.

ARIMA. Or Autoregressive Integrated Moving Average and their variants have been widely used in wind power turbine forecast problems like in [8]. The idea in this method is use the past observation to predict the future ones. ARIMA model is defined by three parameters: p, q and d. p: corresponds with the AR or autoregressive part, also knows as the lag order; q: is the number of moving average terms also called the moving average window and d: is the difference used to convert the time series in stationary also know as the degree of differencing.

Nearest Neighbours Regressor. This regression algorithm is based on the popular K-Neighbours classification algorithm [10]. When it is requested to obtain a prediction, it finds the K closest samples within the training data and returns a prediction that is a mean of the corresponding outputs of the selected neighbours. In this study a modification over the initial algorithm has been used: the calculation of the average of the outputs is a weighted average that takes into consideration the distance of that sample with the query one.

Gradient Boosted Decision Trees. The decision tree regressor is an approach to the regression problem of this kind of models used to generate a set of quite simple decision rules, derived from the input data available. In this particular problem instance, the Classification and Regression Tree (CART) algorithm has been used [11]. The tree is constructed by observing one of the variables of the data instance and calculating a threshold that when applied to the data available for analysis, would better split the dataset in order to deduce the final predicted value. The gradient boosting approach is what is called an ensemble model.

Autoregressive Neural Networks. Autoregressive neural networks, also known as, NNAR networks, this technique applied in time series studies predict the future values based in the previous ones. Consists in feed forward neural networks with a single hidden layer and lagged inputs. The network is fitted with lagged values of the time series as inputs and a single hidden layer a number of nodes equal to a half of the input node plus one. The network is computed several times according to the random starting weights. The network is applied iteratively for forecasting the steps ahead required. In turbine power problems this networks has been applied in [3].

5   Experiments and Results Discussion

5.1   Experimental Setup

As it has been explained in Sect. 3.1, the preliminary analysis of the time series data obtained from the installation yielded the confirmation of a previous intuition: the long term prediction of the power generation would be very difficult. The system depends almost completely of wind energy, that is by its own nature, very difficult to predict. The wind conditions vary enormously from one time instant to another and the system does not include any loop or feedback mechanism that feeds back, such as geothermal installations.

As the initial step, a baseline prediction has been calculated to compare different prediction techniques against this one. We have used some the most widespread quality measures in literature for regression tasks: the mean absolute error (MAE), the root mean squared error (RMSE) and the coefficient of determination (R^2). The base predictor is the value of the series at a given time instant as the prediction of two different time horizons: 60 minutes and 90 minutes. The experiments also involve calculating if the prediction is also affected by the season in which the prediction is made.

Table 1 shows the errors yielded if we predict each of the horizons using the series.

Table 1. Errors obtained with the baseline predictor

Time horizon predicted	Measure	Spring	Summer	Autumn	Winter
60 min.	MAE	0.0163	0.0097	0.0107	0.0231
	RMSE	0.0401	0.0265	0.0352	0.0502
	R^2	0.4067	0.6724	0.7401	0.7226
90 min.	MAE	0.0186	0.0108	0.0122	0.0257
	RMSE	0.0441	0.0290	0.0383	0.0542
	R^2	0.2825	0.6084	0.6928	0.6768

5.2   Experimental Results

To carry out the experiments, the first step, is to divide the time series into training and test sets. For this, the following division was established: March and April (Spring), June and July (Summer), September and October (Autumn) and December and January (Winter) as training sets. May (Spring), August (Summer), November (Autumn) and February (Winter) are the test sets. The instances are adjusted to the temporal reality of the stations. All training sets exceeds 8000 instances with an average of 8524 and the test sets have an average of 4171.

The second step is to calculate the lags of each subset. The number of lags is 8. For this purpose, a dataset has been conform as follow: For each sample, the value of the power at time t, the eight previous values (lags) from $t - 1$ to $t - 8$, and the values at times $t + 6$ and $t + 18$ that correspond to the two time horizons to be predicted.

Then, methods detailed in Sect. 4 are trained. For MLP and the NNAR, a cross validation has been performed. The trained models have been applied to the test sets, calculating both time horizons $t + 6$ and $t + 9$. Finally, the results obtained have been compared with the actual values and the error measures have been calculated, with the results shown in Tables 2 and 3. Analysing these

Table 2. Results of the prediction in the 60 min. time horizon for each of the regression models

Model	Measure	Spring	Summer	Autumn	Winter
ARIMA	MAE	0.0556	0.0466	0.0622	0.0963
	RMSE	0.0866	0.0829	0.1221	0.2003
	R^2	**0.4237**	**0.6806**	**0.7498**	**0.7338**
K-nearest neighbours	MAE	0.0172	0.0099	0.0123	0.0237
	RMSE	**0.0384**	**0.0259**	0.0430	**0.0486**
	R^2	**0.4545**	**0.6870**	0.6141	**0.7403**
Gradient boosted decision trees	MAE	0.0200	0.0110	0.0129	0.0246
	RMSE	**0.0375**	**0.0251**	0.0391	**0.0462**
	R^2	**0.4800**	**0.7053**	0.6809	**0.7656**
Autoregressive neural networks	MAE	0.0265	0.3796	0.0219	0.0502
	RMSE	0.0535	0.3871	0.0712	0.1069
	R^2	**0.4113**	**0.6877**	0.7256	**0.7277**

Table 3. Results of the prediction in the 90 min. time horizon for each of the regression models

Model	Measure	Spring	Summer	Autumn	Winter
ARIMA	MAE	0.0580	0.0526	0.0557	0.0870
	RMSE	0.0901	0.0936	0.1098	0.1813
	R^2	**0.3072**	**0.6170**	**0.7070**	**0.6820**
K-nearest neighbours	MAE	0.0198	0.0110	0.0139	0.0261
	RMSE	**0.0424**	**0.0288**	0.0468	**0.0536**
	R^2	**0.3359**	**0.6147**	0.5413	**0.6847**
Gradient boosted decision trees	MAE	0.0233	0.0124	0.0148	0.0276
	RMSE	0.0510	**0.0278**	0.0435	**0.0510**
	R^2	**0.7145**	**0.6391**	0.60423	**0.0714**
Autoregressive neural networks	MAE	0.0300	0.3880	0.1129	0.0509
	RMSE	0.0523	0.3961	0.1944	0.1058
	R^2	**0.2878**	**0.6278**	**0.7056**	0.6695

results, it can be concluded in general that all models work better with the 60-minute forecast horizon. This result was somewhat predictable because the more distanced in time is the time horizon, the more precision is lost.

The model with the best results is K-Nearest Neighbours predictor. Even if the difference is not very high, it is able to outperform the baseline predictor in three of the seasons. It is worth noting that error results are all calculated with normalized data and the higher value in the series is 2009W, so an error measure difference of 10^{-2} or 10^{-3} is worth to be taken into account. Also, is interesting to note that, Autumn seems to be very hard to predict, probably because is a period of with very unstable weather and sudden climatic changes.

6   Conclusions and Future Work

In this work, a wind turbine installation in a bioclimatic house has been presented. As a research station, it is constantly being monitored during its operation conditions and power generation outputs. The present study has confirmed that long term predictions of the power output of the system, from previous outputs, is difficult; due to external weather dependency. Nevertheless, the results displayed hint at the possibility of using automated learning algorithms for predicting the power output that the system will generate in the short term.

Future work includes both using more complex models, able to better capture the dynamics of the system, and also including additional information such as the weather conditions at the moment or even future predictions.

References

1. Aláiz-Moretón, H., Castejón-Limas, M., Casteleiro-Roca, J.L., Jove, E., Fernández Robles, L., Calvo-Rolle, J.L.: A fault detection system for a geothermal heat exchanger sensor based on intelligent techniques. Sensors 19(12), 2740 (2019)
2. Casteleiro-Roca, J.L., et al.: Short-term energy demand forecast in hotels using hybrid intelligent modeling. Sensors 19(11), 2485 (2019)
3. Do Nascimento Camelo, H., Lucio, P., Junior, J., De Carvalho, P., Santos, D.: Innovative hybrid models for forecasting time series applied in wind generation based on the combination of time series models with artificial neural networks. Energy 151, 347–357 (2018)
4. Government of Spain: Real decreto-ley 15/2018, de 5 de octubre, de medidas urgentes para la transición energética y la protección de los consumidores (2018). bOE-A-2018-13593
5. Infield, D., Freris, L.: Renewable Energy in Power Systems. Wiley, Hoboken (2020)
6. Olsson, J., Niemczynowicz, J., Berndtsson, R., Larson, M.: An analysis of the rainfall time structure by box counting-some practical implications. J. Hydrol. 137(1–4), 261–277 (1992)
7. Owusu, P.A., Asumadu-Sarkodie, S.: A review of renewable energy sources, sustainability issues and climate change mitigation. Cogent Eng. 3(1), 1167990 (2016)
8. Shi, J., Qu, X., Zeng, S.: Short-term wind power generation forecasting: Direct versus indirect arima-based approaches. Int. J. Green Energy 8(1), 100–112 (2011)
9. Sorknæs, P., Djørup, S.R., Lund, H., Thellufsen, J.Z.: Quantifying the influence of wind power and photovoltaic on future electricity market prices. Energy Convers. Manag. 180, 312–324 (2019)
10. Wahid, F., Ghazali, R., Fayaz, M., Shah, A.S.: A simple and easy approach for home appliances energy consumption prediction in residential buildings using machine learning techniques. J. Appl. Environ. Biol. Sci. 7, 108–119 (2017)
11. Wang, R., Lu, S., Feng, W.: A novel improved model for building energy consumption prediction based on model integration. Appl. Energy 262, 114561 (2020)

Review of Trends in Automatic Human Activity Recognition Using Synthetic Audio-Visual Data

Tiago Jesus[1,2], Júlio Duarte[1,2], Diana Ferreira[1,2], Dalila Durães[1,2],
Francisco Marcondes[1,2], Flávio Santos[1,2], Marco Gomes[1,2],
Paulo Novais[1,2], Filipe Gonçalves[2], Joaquim Fonseca[2],
Nicolas Lori[1,2(✉)], António Abelha[1,2], and José Machado[1,2]

[1] Centre Algoritmi, University of Minho, 4710-057 Braga, Portugal
nicolas.lori@algoritmi.uminho.pt,
{abelha,jmac}@di.uminho.pt
[2] Bosch Car Multimedia, 4705-820 Braga, Portugal

Abstract. An in-depth study of knowledge and technologies was made related to the various scientific, technical, and industrial domains necessary for the acquisition of skills and capabilities for the design and development of a multisensory fusion system for vehicle cockpits. After an extensive literature review, it was possible to determine the baselines of the solution to be developed and obtain a pipeline prototype.

Keywords: Autonomous car · Multisensory fusion · Audio-visual synthetic data

1 Introduction

For a primary statement, there are very few researchers focusing on in-vehicle action recognition, presumably due to privacy issues/concerns. Therefore, several elements for this approach will need to be built, at least in part, from scratch. This is the key-risk for this publication and the results will reflect this reality.

1.1 Internet of Things (IoT)

In this novel paradigm, embedded sensors and Internet connectivity are installed in smart objects, serving as facilitators to interactions, communication, and integration with the surrounding environment to provide intelligent and useful services [10].

Focusing on the automotive area, a lot of works have been produced on vehicular communications specifically for inter-vehicular, intra-vehicular and vehicle to infrastructures WSNs (Wireless Sensor Networks). The increasing demand for driver safety and assistance in a modern vehicle, has brought attention to the intra-vehicle WSNs.

© Springer Nature Switzerland AG 2020
C. Analide et al. (Eds.): IDEAL 2020, LNCS 12490, pp. 549–560, 2020.
https://doi.org/10.1007/978-3-030-62365-4_53

As more automotive designers implement IoT into their designs, more and smarter cars are manufactured. A recent trend in the automotive industry are smart cars enabled with IoT. Many researchers [11,12,18,27] identify some of the important implications of IoT in transportation:

i. It allows the use of cloud-based intelligent monitoring control system for tracking the location of vehicles in real time;
ii. It allows the communication between equipment through devices attached to vehicles (inter-equipment connection), allowing drivers to avoid delays or accidents;
iii. It can improve passenger comfort and convenience by alerting them about delays via their mobile devices;
iv. It allows a predictive maintenance by providing to vehicles the capacity of transmitting defect-indication data directly to engineers. Predictive maintenance can identify components in need of repair/replacement.

In this sense, important data can be captured from various IoT connections and devices. For example, sensors installed in vehicles offer the ability to track maintenance needs, driver safety, fuel usage and other related metrics in real time. The data collected from these sensors can help companies to optimize performance and can lead to profitable outcomes for themselves through better user experiences. These data can be used to improve the way that they design, upgrade and maintain devices in the field [23].

The ever-increasing number of connected devices will enable a complex atmosphere with billions of sensors and devices connected to the internet, to ultimately gather, analyze and transmit data in real time. Thus, without the data, IoT would not be able to hold the features and functionalities that have brought them incredible benefits, powerful emphasis and world-wide magnitude.

1.2 Big Data

As cities around the world are increasingly digitized, we are on the verge of an era in which IoT will comprise massive amounts of devices capable of sensing, computing, capturing and operating in the real world [25]. Every day, these devices will generate continuous streams of real-time data on critical infrastructure components and services. The magnitude of the daily explosion of high volumes of data has led to the emerging Big Data paradigm (e.g. [1,2,14–16]).

In a data-driven utopia, data would be highly valued and used in an ethical and effective way. In reality, however, data must travel a long way before it reaches its highest purpose, gaining incremental value as it goes. The essence of the data value chain is to provide a framework through which the data lifecycle can be perceived, transforming low-value inputs (raw data) into high-value outputs (actionable information).

As EU Commissioner Kroes said, "Big Data is the new oil", it is a technology and innovation driver that creates value not only for companies but also for citizens and society [13].

Given the continuous technological advances and the focus towards autonomous driving in recent years, autonomous vehicles are expected to be a major source of Big Data. Over the last few years, the degree of car automation has gradually increased to the point of fully autonomous driving vehicles being a reality in the near future [25]. Using vehicles to form ad-hoc vehicle networks (VANETs) or the Internet of Vehicles (IoVs) has now become a reality. The IoV Big Data is a major enabling technology for conceiving revolutionary self-driving vehicles. As a matter of fact, several researchers [5,6,8,24,28] have already dedicated their efforts towards big sensor data systems using vehicles as sensing elements in a large networked system for intelligent transportation and were able to outline the following insight that to make the self-driving come to life, a convergence of Big Data is required, including the data from on-board sensors, e.g., cameras, radar, Lidar, GPS, and information shared from other connected vehicles, e.g.. road condition and traffic information.

Accordingly, Big Data is a powerful technology that requires attention and investigation to make exciting new mobility features possible and to enable unprecedent vehicular phenomena and experiences while offering efficiency benefits and improving safety.

1.3   Audio-Visual Violence Detection

Figure 1 represents "the basic modes of detection of violence". This figure shows a model for automated detection of human-computer behavior, through selective identification of a person. The method is based on a combination of data from audio, video, and other sensors. The video data is a collection of images of a location. The video data is further processed to extract human and human hierarchical resources [9]. People's actions are detected and processed to detect selectable behavior. A configurable behavior rule can be used to select people's behavior and resources. The hybrid human detector algorithm has been used for human detection, which may include one of several machine learning algorithms.

It's possible to extract six segment-level audio features, which can be used at a next stage by a classifier:

- **Time-Domain Features** express abrupt changes during time of the audio signal. This feature can be calculated by further dividing the frames into N sub-windows of fixed duration;
- **Frequency-Domain Features** expresses abrupt changes in amplitude of the audio signal;
- **Cepstrum** it is an approximately continuous signal, owing to a large part, to the smoothing effect of windowing;
- **Time-frequency** changes in frequency along the time;
- **Energy** admissible poses of the microphones considering the sound energy level and the ILD as acoustic features;
- **Biologically or Perceptually Driven** spectral features based on Gamma-tone Filter Bank.

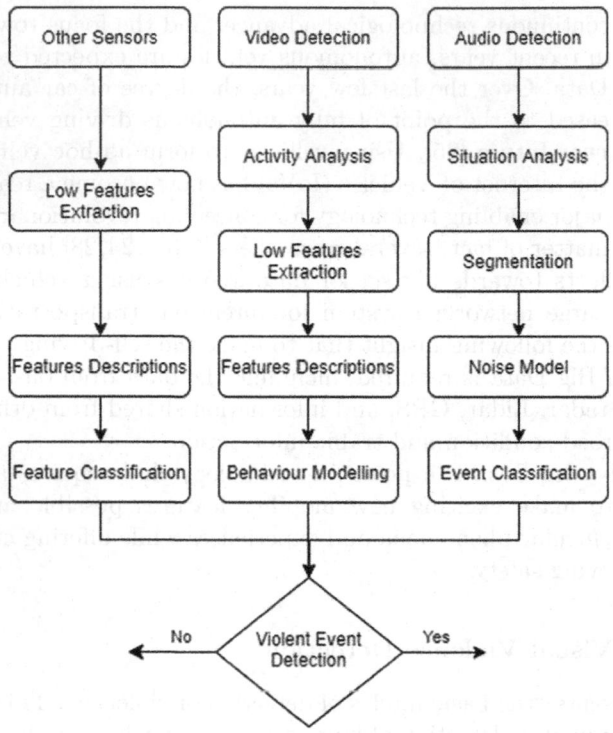

Fig. 1. Basic ways of violence detection.

From the computer vision community point of view, visual tracking is the process of locating, identifying, and determining the dynamic configuration of one or many moving (possibly deformable) objects (or parts of objects) in each frame of one or several cameras. What makes a good visual tracking algorithm is its capacity to handle all the variability in a video sequence caused by the tracked object, the scene and the camera acquiring the scene. Such variability can be caused, for instance, by pose and illumination variations, occlusions, varying and erratic motions [7]. Visual tracking in video sequences raises many problems that can cause a loss of track on the object. Table 1 presents a list of some of the most challenging ones [7].

2   Methods

This section is dedicated to a selection of the datasets publicly available on the web for human activity recognition. There are several datasets, however, we can divide them in several types: visual tracking, motion recognition and action recognition using video and audio.

Table 1. List of some of the most challenging problems while visual tracking objects in video [7].

Problem	Description
Illumination effects	The light scene environment may change due to external conditions (weather, time of day) or internal conditions (lights on or off). Furthermore, depending on the incidence of light, the variations of light can cause object colors to change over time, which can confuse the visual tracking algorithm
Scene clutter	Can happen due to a very textured background or other moving objects in the scene, often similar to the tracked object. This feature can cause some deviations from the visual tracker, resulting in the loss of tracking of the object
Changes in object appearance	Happens because of the projection of 3D movements onto a 2D plane (frames from sequences), the tracked object can have geometric deformations
Abrupt changes in motion	The object velocity can vary with time. This can make the object very hard to track because its movement can become unpredictable and, therefore, the object can be lost
Occlusions	Difficult to handle because some parts of the object can disappear from the scene
Similar appearances	When different objects have similar appearances in the video sequence it may be difficult for the visual tracker to discriminate between these objects

Several public datasets for visual tracking exist such as the datasets: BEHAVE, BoBoT, CAVIAR, Ross, and SPOT. Table 2 describes some of these datasets.

Regarding motion recognition, some examples of dataset are: HDM05, MSR-Daily Activity3D, UTKinect, ActivityNet, MSR-Action3D, RGBD-HuDaAct, CAD-60, MSRC-12, YouTube 8M, and Hollywood 3D. Some of the datasets were analyzed and the information was is summarized in Table 3.

Most of the developed audio scene datasets are not publicly accessible. Table 4 summarizes some of the ones publicly available which are commonly used in audio scene recognition.

There are several dataset repositories for human action. However, it was not possible to find any specific in-vehicle human action dataset. Therefore, a

Table 2. Publicly available datasets used in visual tracking.

Dataset	Description
BoBoT	It presents twelve video sequences in .avi format. All frames have a size of 320×240 pixels and their numbers vary from 305 to 1308. Ground truth commentaries are also given for each sequence. They match the coordinates of the target object's bounding box and its size. This dataset was used in several recent works [7,26]
CAVIAR	The CAVIAR (Context Aware Vision using Image based Active Recognition) project, from MIA Labs was dedicated to the development of algorithms to richly describe and understand video scenes. It contains a lot of information, such as rectangular bounding boxes' locations and sizes, head and feet positions, body direction, etc. The CAVIAR dataset is very popular and used by a lot of computer vision research teams
SPOT	It proposes six very challenging video sequences that were collected from Youtube. It's dedicated to track simultaneously multiple objects, sometimes with similar appearances. However, the movements of the objects in a same sequence are related to each other

Table 3. Publicly available datasets used in motion recognition.

Datasets	References	Classes
MSR-Action3D	[26]	Contains 20 actions: high arm wave, draw x, hammer, hand catch, horizontal arm wave, forward punch, high throw, draw circle, bend, forward kick, side kick, jogging, draw tick, two hand wave, hand clap, tennis serve, pick up, golf swing, throw and side-boxing
UTKinect dataset	[26]	Actions include sit down, stand up, wave, clap hands, walk, throw, push, pick up, carry and pull
ActivityNet	[21]	The dataset divided into three subgroups by application domain such as untrimmed videos classification, trimmed videos classification, and Activity detection on all the untrimmed videos
YouTube 8M	[21]	Provides 7 million videos 12 billion/audio/visual features, 4716 classes, and 3.4 average labels per videos. Each video sequence has 120–500 s in length and more than one thousand views per video
Hollywood 3D	[21]	It consists of 14 activities classes such as no actions, runs, punches, kicks, shots, eat, drive, uses phone, kiss, hug, stands up, sit downs, swims, and dances. There are 650 manually labelled videos in the dataset approximately

dataset must be created from scratch and may require the usage of Synthetic Sensor Data Generation for enabling the proof of concept. A possibility to be explored is to find datasets that produce an approximate result considering the in-vehicle dataset. Nevertheless, this must be explored after a primary dataset is created for comparison.

As pointed out by [22], popular datasets that contain human action have become increasingly detailed over those used for early works. Since then, there have been many advances in how datasets are created. Improvement of quality datasets allows for more complex models, hence the use of challenging datasets allow for the evaluation of the robustness and generalizability of these models. It is crucial, however, to consider that many approaches depend upon inputs that are absent in common/popular datasets, which then require the creation of domain-specific datasets. Meanwhile, some essential characteristics of these popular datasets (that can be found in the literature) must be highlighted. Namely, the main and most relevant can be expressed according to Table 5 [21]:

For the preparation of a human activity behavior dataset from scratch, it is necessary to define a basic structure that respects the most popular characteristics in similar datasets. To set a baseline and what are the proper criteria and workflow for generating a new dataset, some guidelines and recommended best-practices must be followed.

Table 4. Datasets used in Environmental Audio Scene Recognition (EASR) Task [4].

Datasets	Environmental audio scene classes
DARES-G1 (10 contexts)	Basketball game, beach, inside an office facility, inside a bus, grocery shop, inside a car, street, hallways, restaurant, and stadium with track and field events
LITIS-Rouen audio scene dataset (19 contexts)	Busy street, bus, cafe, car, train station hall kid game hall, market, metro-Paris, metro-Rouen, high-speed train, billiard pool hall, quiet street, plane, student hall, restaurant, pedestrian street, shop, train, and tube station
TUT-CASR (18 contexts)	Six high-level classes are: (1) vehicles, (2) outdoors, (3) public/social, (4) offices/meetings/quiet, (5) home, and (6) reverberant places
IEEE/AASP DCASE 2013 Scene dataset (10 contexts)	Bus, busy street, restaurant, office, open-air market, park, quiet street, supermarket, tube (subway train), and tubestation (subway station)
TUT-2016 dataset (15 contexts)	Beach, bus, cafe, car, city center, forest path, grocery store, home, library, metro station, office, residential area, train, tram, and urban park
TUT-2017 dataset (15 contexts)	Bus, cafe/restaurant, car, city center, forest path, grocery store, home, lakeside beach, library, metro station, office, residential area, train (traveling, vehicle), tram (traveling, vehicle), and urban park

3   Results

The state of the art was properly studied, and we concluded that there are five articles that constitute the state-of-the-art and that we now analyse in detail, first the pipeline analysis, and then its data structure.

In [17], a two-stage pipeline was developed to handle pose estimation. The first stage of the proposed cascade model is based in the Faster-RCNN method applied on top of a ResNet-101 CNN to predict the location and scale of boxes likely to contain people. The boxes containing people are then used in the second stage, where the locations of each keypoint for each of the boxes are predicted with a fully convolutional ResNet. A novel keypoint-based Non-Maximum- Suppression (NMS) mechanism is used to avoid duplicate pose detections. Hence building directly on the object keypoint similarity (OKS) metric (called OKS-NMS), instead of a cruder box-level approach called IOU NMS. Finally, a novel keypoint-based confidence score estimator is proposed, which leads to greatly improved average precision compared to using the Faster-RCNN box scores for ranking the final pose proposals.

In [19], two algorithms of the 2016 COCO Keypoints Challenge were evaluated - [17] and [3] - and made four contributions. The first was the taxonomization of the types of error that are typical of the multi-instance pose estimation frameworks. The second contribution was sensitivity analysis of those errors with respect to measures of image complexity. The third was side-by-side comparison of two leading human pose estimation algorithms highlighting key differences in behaviour that are hidden in the average performance numbers. Finally, the last contribution was the assessment of which types of datasets and benchmarks would be most productive in guiding future research.

In [20], a comprehensive survey was presented, as a review paper, of both handcrafted and learning-based action representations, offering comparison, analysis, and discussions on these approaches.

In [29], a weakly-supervised transfer learning method was proposed. It used mixed 2D and 3D labels in a unified deep neural network that presents two-stage cascated structure. The network augmented a 2D pose estimation sub-network by use of a 3D depth regression sub-network.

Table 5. Popular dataset characteristics.

Characteristic	Definition
Classes	The choice of action classes greatly affects the diversity and coverage of the dataset. Although certain datasets construct a rich hierarchy of action classes, in most datasets the following groupings are usually present in some form: Person-Object, Person only and Person-Person
Focus	A majority of the datasets either consider activities performed during daily life or do not have a very specific domain focus

(*continued*)

Table 5. (*continued*)

Characteristic	Definition
Modality	Human action datasets usually preserve the temporal dimension, unlike simpler image classification tasks, although there exist large datasets for image-based activity classification. Singh (Singh, 2019) highlights that the natural representation of datasets is in the form of clips of 2D images, and most datasets use this format extensively. However, the introduction of low-cost 3D sensors such as Microsoft Kinect, has brought great interest in using depth information. A non-visual sensor is used to record an entirely different class of datasets
Data source	The source from which dataset is acquired determines to a large degree how well an algorithm trained on it will perform on unseen data. The datasets consisting mainly of video clips containing one or a pair of actors performing an activity in an indoor lab setting and are called, respectively, recorded or scripted datasets
Annotation method	For generated, recorded, and crowdsourced datasets, the video label is already known at the start of video creation. On the other hand, for annotated datasets, accurate and precise annotation and subsequent verification of labels are essential for supervised learning schemes
Annotation type	The type of annotation determines the degree to which an action is localized in time and space. When temporal localization is not an important concern, as in activity classification problems, the entire sequence or video clip is directly labeled with its corresponding class, this is called a sequence level annotation and is the case for the majority of datasets due to constraints of more complex annotations. For datasets focusing on detecting activities specifically, frame range or action segment annotation specifies an interval related to a particular class
Evaluation	In general, a human action algorithm can have two basic tasks with different evaluation protocols. In action classification task (trimmed activity recognition), the action being performed in a particular clip is identified and thus becomes a multiclass classification problem (with/without null class). In the cases where the sequence level ground truth is available, most datasets use multiclass accuracy as the metric. Although realistic datasets are unbalanced and long-tailed, some alternative metrics such as recall, precision and F score are often used instead

4   Conclusion and Discussion

This work started by introducing the subjects of IoT and Big Data. These are major subjects that need to be addressed when trying to create a pipeline for human activity recognition either inside or outside a vehicle. It then continues by introducing one of the main topics: violence detection. To perform the detection of violence we need to apply human activity recognition algorithms capable of analysing the faintest signs of violence in a certain situation. Thus, a study was made about the state of the art so as to obtain a better understanding of action recognition.

After studying the literature, we concluded that, although not perfect, a typical pipeline exists to handle automatic human activity recognition. Furthermore, we found several datasets, publicly available whilst searching the literature. To accurately perform action recognition, a two-stage model should be applied as it achieved better results according to the literature.

Finally, by knowing this typical pipeline exists, and having several datasets at hand, we can adapt the pipeline to our specific needs in the future and develop a pipeline prototype to detect situations where violence occurs.

Acknowledgments. This work has been supported by FCT – Fundação para a Ciência e Tecnologia within the R&D Units Project Scope: UIDB/00319/2020. Human and material resources have also been supported by the European Structural and Investment Funds in the FEDER component, through the Operational Competitiveness and Internationalization Programme (COMPETE 2020) [Project number 039334; Funding Reference: POCI-01-0247-FEDER-039334].

References

1. Analide, C., Novais, P., Machado, J., Neves, J.: Quality of knowledge in virtual entities. In: Encyclopedia of Communities of Practice in Information and Knowledge Management, pp. 436–442. IGI Global (2006)
2. Brandão, A., et al.: A benchmarking analysis of open-source business intelligence tools in healthcare environments. Information **7**(4), 57 (2016)
3. Cao, Z., Simon, T., Wei, S.E., Sheikh, Y.: Realtime multi-person 2D pose estimation using part affinity fields. In: Proceedings of the IEEE Conference on Computer Vision and Pattern Recognition, pp. 7291–7299 (2017)
4. Chandrakala, S., Jayalakshmi, S.: Environmental audio scene and sound event recognition for autonomous surveillance: a survey and comparative studies. ACM Comput. Surv. (CSUR) **52**(3), 1–34 (2019)
5. Chaqfeh, M., Lakas, A., Jawhar, I.: A survey on data dissemination in vehicular ad hoc networks. Veh. Commun. **1**(4), 214–225 (2014)
6. Dikaiakos, M.D., Iqbal, S., Nadeem, T., Iftode, L.: VITP: an information transfer protocol for vehicular computing. In: Proceedings of the 2nd ACM International Workshop on Vehicular Ad Hoc Networks, pp. 30–39 (2005)
7. Dubuisson, S., Gonzales, C.: A survey of datasets for visual tracking. Mach. Vis. Appl. **27**(1), 23–52 (2015). https://doi.org/10.1007/s00138-015-0713-y
8. Gerla, M.: Vehicular cloud computing. In: 2012 The 11th Annual Mediterranean Ad hoc Networking Workshop (Med-Hoc-Net), pp. 152–155. IEEE (2012)
9. Gilbert, A., Illingworth, J., Bowden, R.: Action recognition using mined hierarchical compound features. IEEE Trans. Pattern Anal. Mach. Intell. **33**(5), 883–897 (2010)
10. Kim, K.J.: Interacting socially with the internet of things (IoT): effects of source attribution and specialization in human-IoT interaction. J. Comput. Med. Commun. **21**(6), 420–435 (2016)

11. Leng, Y., Zhao, L.: Novel design of intelligent internet-of-vehicles management system based on cloud-computing and internet-of-things. In: Proceedings of 2011 International Conference on Electronic & Mechanical Engineering and Information Technology, vol. 6, pp. 3190–3193. IEEE (2011)

12. Lumpkins, W.: The internet of things meets cloud computing [standards corner]. IEEE Consum. Electron. Mag. **2**(2), 47–51 (2013)

13. María Cavanillas, J., Curry, E., Wahlster, W.: New horizons for a data-driven economy: a roadmap for usage and exploitation of big data in Europe. Springer Nature (2016)

14. Neto, C., Brito, M., Lopes, V., Peixoto, H., Abelha, A., Machado, J.: Application of data mining for the prediction of mortality and occurrence of complications for gastric cancer patients. Entropy **21**(12), 1163 (2019)

15. Neves, J., Martins, M.R., Vilhena, J., Neves, J., Gomes, S., Abelha, A., Machado, J., Vicente, H.: A soft computing approach to kidney diseases evaluation. J. Med. Syst. **39**(10), 131 (2015)

16. Neves, J., Vicente, H., Esteves, M., Ferraz, F., Abelha, A., Machado, J., Machado, J., Neves, J., Ribeiro, J., Sampaio, L.: A deep-big data approach to health care in the AI age. Mob. Netw. Appl. **23**(4), 1123–1128 (2018)

17. Papandreou, G., et al.: Towards accurate multi-person pose estimation in the wild. In: Proceedings of the IEEE Conference on Computer Vision and Pattern Recognition, pp. 4903–4911 (2017)

18. Qin, E., Long, Y., Zhang, C., Huang, L.: Cloud computing and the internet of things: technology innovation in automobile service. In: Yamamoto, S. (ed.) HIMI 2013. LNCS, vol. 8017, pp. 173–180. Springer, Heidelberg (2013). https://doi.org/10.1007/978-3-642-39215-3_21

19. Ruggero Ronchi, M., Perona, P.: Benchmarking and error diagnosis in multi-instance pose estimation. In: Proceedings of the IEEE International Conference on Computer Vision, pp. 369–378 (2017)

20. Sargano, A.B., Angelov, P., Habib, Z.: A comprehensive review on handcrafted and learning-based action representation approaches for human activity recognition. Appl. Sci. **7**(1), 110 (2017)

21. Singh, R., Sonawane, A., Srivastava, R.: Recent evolution of modern datasets for human activity recognition: a deep survey. Multimed. Syst. 1–24 (2019)

22. Singh, T., Vishwakarma, D.K.: Video benchmarks of human action datasets: a review. Artif. Intell. Rev. **52**(2), 1107–1154 (2018). https://doi.org/10.1007/s10462-018-9651-1

23. Uden, L., He, W.: How the internet of things can help knowledge management: a case study from the automotive domain. J. Knowl. Manag. **21**, 57–70 (2017)

24. Xu, W., et al.: Internet of vehicles in big data era. IEEE/CAA J. Automatica Sinica **5**(1), 19–35 (2017)

25. Zaslavsky, A., Perera, C., Georgakopoulos, D.: Sensing as a service and big data. arXiv preprint arXiv:1301.0159 (2013)

26. Zhang, J., Li, W., Ogunbona, P.O., Wang, P., Tang, C.: RGB-D-based action recognition datasets: a survey. Pattern Recogn. **60**, 86–105 (2016)

27. Zhang, Y., Chen, B., Lu, X.: Intelligent monitoring system on refrigerator trucks based on the internet of things. In: Sénac, P., Ott, M., Seneviratne, A. (eds.) ICWCA 2011. LNICST, vol. 72, pp. 201–206. Springer, Heidelberg (2012). https://doi.org/10.1007/978-3-642-29157-9_19

28. Zhou, H., et al.: Chaincluster: engineering a cooperative content distribution framework for highway vehicular communications. IEEE Trans. Intell. Transp. Syst. **15**(6), 2644–2657 (2014)
29. Zhou, X., Huang, Q., Sun, X., Xue, X., Wei, Y.: Towards 3D human pose estimation in the wild: a weakly-supervised approach. In: Proceedings of the IEEE International Conference on Computer Vision, pp. 398–407 (2017)

Atmospheric Tomography Using Convolutional Neural Networks

C. González-Gutiérrez[1,2]([✉]) [ID], O. Beltramo-Martin[3] [ID], J. Osborn[4],
José Luís Calvo-Rolle[5], and F. J. de Cos Juez[1,6]

[1] Instituto Universitario de Ciencias y Tecnologías Espaciales de Asturias,
Oviedo, Spain
gonzalezgcarlos@uniovi.es, fjcos@uniovi.es
[2] Department of Computer Science, University of Oviedo, Oviedo, Spain
[3] Aix Marseille Univ, CNRS, CNES, LAM, Marseille, France
olivier.beltramo-martin@lam.fr
[4] Department of Physics, Durham University, South Road, Durham DH1 3LE, UK
james.osborn@durham.ac.uk
[5] Department of Industrial Engineering, University of A Coruña,
Ferrol, A Coruña, Spain
jose.rolle@udc.es
[6] Department of Exploitation and Exploration of Mines, University of Oviedo,
Oviedo, Spain

Abstract. We present an application of Convolutional Neural Networks
(CNN) to atmospheric tomography that is required for compensating
optical aberrations introduced by the atmospheric turbulence using dedicated tomographic Adaptive Optics (AO) systems. We compare the state
of the art Minimum Mean Square Error (MMSE) reconstructor with a
Multi-Layer Perceptron (MLP) and a CNN architecture and show that
the CNN performs up to 15%–20% better than the MMSE and is more
robust to atmospheric profile variations up to 10% compared to the MLP.
Such results pave the way to implement CNN architectures to revisit
atmospheric tomography for astronomical telescopes equipped with AO.

Keywords: Adaptive optics · Tomography · Convolutional neural
networks

1 Introduction

Adaptive Optics (AO) has become a game changer for achieving high angular resolution and high sensitivity astronomical observations from the ground
by compensating in real-time the optical aberrations introduced by the atmospheric turbulence [13]. The correction is ensured thanks to a closed-loop scheme
involving a Wave Front Sensor (WFS) that measures the incoming wavefront
from a star and a Deformable Mirror (DM) that corrects for it. However, the
corrected field of view is limited to few arc-seconds (arcsec) only. In order to

© Springer Nature Switzerland AG 2020
C. Analide et al. (Eds.): IDEAL 2020, LNCS 12490, pp. 561–569, 2020.
https://doi.org/10.1007/978-3-030-62365-4_54

increase the corrected field of view (fov), advanced systems, so-called tomographic systems, have emerged and relied on multiple WFSs measurements and linear tomographic reconstruction processes [12]. However, atmospheric tomography necessitates the knowledge of the atmospheric profile [13] that may vary significantly in few minutes [3]. Besides, in an attempt to mitigate the numerical complexity of the tomographic reconstruction, one may rely on a simplified expression of the WFS operator, i.e. the relation between the measurement and the incoming wavefront before combining multiple sources of measurement in the whole fov.

In this context and in line with previous efforts on this issue [10,14], it is presented an analysis that compares the state-of-the-art tomographic reconstructor with Artificial Neural Networks (ANN) solutions. The use of ANN is particularly relevant to learn both the WFS operator and the atmospheric profile from the data and without increasing the complexity during real-time operations. In Sect. 2 the complete methodology is presented including, (i) the end-to-end simulations tool that served to generate the training and test data in Sect. 2.1, (ii) a description of the two architectures implemented in Sect. 2.2, a Multi-Layer Perceptron (MLP) and a Convolutional Neural Network (CNN), and (iii) a discussion that details the training methodology. A performance analysis of those two architectures compared to the MMSE reconstructor is presented in Sect. 3.

2   Methodology

2.1   End-to-End Simulations

Object–Oriented Matlab Adaptive Optics (OOMAO) is a Matlab toolbox dedicated to AO systems simulation, data processing and real-time control [2]. It is a modular and efficient tool that allows the generation of end-to-end simulations of AO systems. We used OOMAO to generate synchronously a phase, ϕ, map, Shack-Hartmann (SH) WFS detector pixels and WFS measurements s. In order to compare our results with the literature, we have simulated the CANARY AO system, an adaptive optics test facility on the 4.2 m William Herschel Telescope, La Palma. Our configuration includes three Natural Guide Stars (NGSs) distributed over a fov of 1' [7]. Therefore, the goal of our experiment is to estimate ϕ in the fov center from the three sets of WFS measurements. Besides, so as to perform a supervised training of our ANNs, we have also simulated the corresponding ground-truth ϕ and the corresponding WFS measurements s we would have from a WFS pointing at a star in the fov center. Furthermore, for each configuration we have tested, we have simulated two sets of training data, one including WFS noise (detector and shot noise) and another one with no measurement noise. The simulation parameters that served to generate training data are reported in Table 1. In order to generate training data, we have followed the same strategy proposed by [11] that trained the ANN with a 2-layers atmosphere having a single altitude layer that produces spatial decorrelation of the wavefront. The altitude of this layer was tuned from 10 m to 15.5 km and for each setting, we have simulated 1000 temporally-independent frames of control loop data, i.e., on/off-axis WFSs slopes and pixel intensity. Also, we have generated noisy data including detector read-out noise and shot noise that are scaled with respect to σ_{ron} and R-mag in Table 1.

Table 1. Simulation settings to generate training data. The Fried parameter, r_0, defines the integrated turbulence strength and L_0 defines the turbulence outer scale.

Telescope		Off-axis NGSs	
Telescope diameter	4.2 m	λ	640 nm
Central obscuration	28.5%	Position	40.6", 53", 47.9"
Temporal sampling	150 Hz	R-mag	10.2, 8.7, 9.9
Atmosphere		**WFSs**	
# Layer	2	# Lenslets	7
r_0	16 cm	# Pixels/subaperture	16
L_0	25 m	WFS pixel scale	220 mas
Weights	50, 50	Transmission	10%
Altitude	10 m to 15.5 km	σ_{ron}	0.7 e-

2.2 MMSE Reconstructor

Assuming that s is the vector that concatenates WFSs measurements and ϕ the optical aberration that must be corrected for, one of the common approach consists in estimating the Minimum Mean Square Error (MMSE) reconstructor that minimizes the following criterion

$$\varepsilon^2 = ||\phi - \boldsymbol{R}_{\mathrm{MMSE}}\boldsymbol{s}||^2_{\mathcal{L}_2},\tag{1}$$

where $||x||_{\mathcal{L}_2}$ is the norm $\mathcal{L}_2$ of vector x and R_{MMSE} is the MMSE reconstructor. This latter is derived from the following

$$R_{\mathrm{MMSE}} = \mathcal{C}_{\phi s}\mathcal{C}_{\mathrm{ss}}^{-\dagger},\tag{2}$$

where $\mathcal{C}_{\phi s}$ is the covariance matrix of ϕ with the measurements s, while $\mathcal{C}_{ss}$ is the covariance matrix of WFS measurements. From a given atmosphere configurations, one may straightforwardly calculate the covariance matrices [1]. Contrary to an ANN, the MMSE reconstructor is optimized for a single atmospheric profile.

2.3 Multi-layer Perceptron: CARMEN

Multi-layer Perceptrons are one of the most common neural network options and their ability for reconstruction have already been proven in the AO field

[10]. Based on those experiments, a MLP has been created using the SH-WFS centroids as inputs and the on-axis star centroids as outputs. The neural network topology is composed by an input layer with 216 neurons, a single hidden layer with also 216 neurons and an output layer with 72 using all of them an hyperbolic tangent as activation function.

In the training process, a dataset with the characteristics described in Sect. 2.1 is used. The altitude of the second layer is moved 155 times, being the changes in height equally distributed between 10 m and 15.5 km and generating 1000 samples per situation. With this dataset, the MLP is trained using the Nadam algorithm [5] with a starting learning rate of 0.001, along with the mean squared error function and a batch size of 256 samples.

2.4 Convolutional Neural Networks

Convolutional neural networks have been widely used for different challenges during the last years, such as pattern recognition, language processing or image classification [9]. In the present work, our CNN uses the complete image of all the SH-WFS as inputs, instead of using the centroids of each sub-aperture as CARMEN does [4]. For this configuration, three different guide stars are used, so the input to the neural network will be the three images of 112×112 pixels provided by each sensor. On the other hand, the output of the neural network will be the 72 values of the slopes of an on-axis SH-WFS, as it was done in the on-sky experiment with CARMEN. To create the topology of the neural network, different approaches were used. The first experiments were conducted by adapting typical architectures such as AlexNet [8] and GoogLeNet [15] to the presented problem. These topologies provided interesting results, since they were able to improve MMSE reconstructor performance, but not CARMEN's. Since increasing the number of layers did not help to improve the ability to recover the ideal slopes, a completely different approach was taken. Instead of adding a high amount of layers to the neural network, a really small solution was proposed, as can be seen in Fig. 1. The first layer is composed by 48 filters of 4×4 pixels with a stride of 4 pixels, which creates 48 images of 28×28 pixels at its output. The second one repeats the process and uses again 48 filters of 4×4 pixels with a stride of 4, generating 48 images of 7×7 pixels. The last convolutional layer increase the amount of filters and their size, having 96 filters of 7×7 pixels and applying a stride of 7 pixels. All the convolutional layers use a ReLU activation function. Once the convolutional stage has finished, 96 images of only one pixel are generated. These images are directly connected to the output through a fully-connected layer which has 6912 weights and no activation function on its output. Different changes for the current neural network were tested, such as replacing strides for max pooling, changing the activation function for ReLU variations (Leaky ReLU, PReLU) or increasing the amount of hidden layers in the fully connected stage, but none of them were able to improve the obtained performance.

The same dataset as the MLP is used for training, with a total of 155,000 samples. Nadam algorithm was used for backpropagation using an initial learning

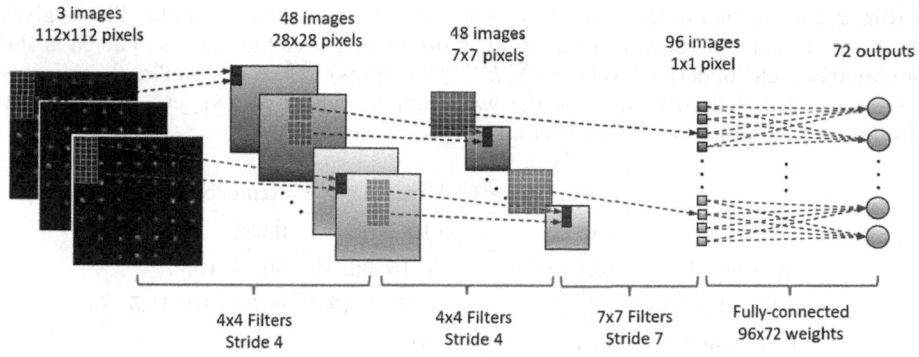

Fig. 1. Illustration of the CNN architecture

rate of 0.001, with mean squared error function and a batch size of 16. In this case, some of the usual solutions for increasing the training quality for CNNs were tested, such as batch normalization, dropout or l2 regularization, but none of them provided any improvement either in reducing the error or increasing the speed of the training process.

3   Results

To compare the previously presented reconstructors, different scenarios have been created with the OOMAO simulator. In these tests, telescope, NGS and WFSs parameters are the ones described in Table 1, but to get a better performance evaluation of the different reconstructors, more complex atmospheres are needed. To achieve this, three different test cases are proposed as in [11], being good, median and bad seeing atmospheric scenarios. Each profile has four turbulent layers at different heights and different r_0 for each case. The parameters used for generating these tests are summarized in Table 2. In order to perform a fair comparison of ultimate performance between all the reconstructors, the MMSE algorithm is adjusted with the turbulence parameters before each prediction, while both neural networks are trained only once and can be tested in any scenario.

We present in Table 3 different metrics obtained with the three reconstructors with and without the noise. We have considered four metrics: the Strehl-ratio (100% gives the diffraction-limit), the Full Width at Half Maximum (FWHM) of the Point Spread Function (PSF), the 50% Encircled Energy that is the angular size that contains 50% of the intensity of the PSF and the Wave-Front Error (WFE) that measures the distance of the tomographic reconstruction to the ground truth. Results highlight clearly that the CNN is the best reconstructor in all situations with an improvement of 15% up to 20% on each metric compared to the state of the art MMSE. Furthermore, the CNN reaches a better tomographic reconstruction than the MLP and especially in the presence of measurement noise with an improvement almost up to 10%.

Table 2. Description of the three atmospheres used to generate test data. The r_0 gives the total strength of the turbulence that is discretized over four layers. The θ_0 is the isoplanatic angle [6] derived as $\theta_0 = (\sum h^{-5/3} \times \text{weights})^{-3/5}/r_0$. A smaller θ_0 indicates a stronger spatial decorrelation of the wavefront for which we expect a less efficient absolute tomographic reconstruction error.

	Atmosphere 1	Atmosphere 2	Atmosphere 3
r_0 (m)	0.16	0.12	0.085
Weights (%)	65,15,10,10	45, 15, 30, 10	80, 5, 10, 5
Altitudes (km)	0,4,10,15.5	0, 2.5, 4, 13.5	0, 6.5, 10, 15.5
θ_0 (arcsec)	3.74	3.31	2.43

Table 3. Table of metrics obtained with each reconstructor applied in three different scenario including or not the WFS noise. Systematically, the utilization of neural networks enhance the performance compared to the MMSE and the CNN performs slightly better than the MLP, especially on noisy data.

Test name	Reconstructor	Metrics			
		Strehl ratio (%)	FWHM (mas)	E50 (mas)	WFE (nm)
Atm 1	MMSE	30.9	125.6	290.5	290.5
No noise	MLP	36.7	115.1	264.6	264.6
	CNN	36.8	114.6	264.2	264.2
Atm 1	MMSE	29.3	127.5	486.1	297.8
Noise	MLP	33.0	118.2	464.6	280.0
	CNN	34.1	117.1	453.9	275.7
Atm 2	MMSE	27.4	129.6	529.7	307.3
No noise	MLP	29.0	125.5	512.5	300.0
	CNN	30.1	124.0	500.7	294.7
Atm 2	MMSE	25.9	131.5	539.4	314.7
Noise	MLP	26.4	128.6	533.4	313.0
	CNN	28.0	126.7	517.0	304.8
Atm 3	MMSE	21.0	154.5	560.9	346.0
No noise	MLP	24.5	139.2	525.4	324.9
	CNN	26.0	135.1	515.3	316.4
Atm 3	MMSE	20.0	157.4	568.5	352.4
Noise	MLP	22.4	144.0	545.1	337.5
	CNN	24.2	138.7	529.7	326.6

In an attempt to assess the robustness of tomographic reconstructors, we have generated test data from a 2-layers atmosphere with a variable altitude layer from 1 km to 15 km. We have calculated WFEs using the MLP, the CNN and two MMSE reconstructors optimized for a 2-layers atmosphere but with an altitude layer set to either 7 km or 12 km. We present in Fig. 2 the WFE with respect to the altitude for those four reconstructors, as well as the theoretical

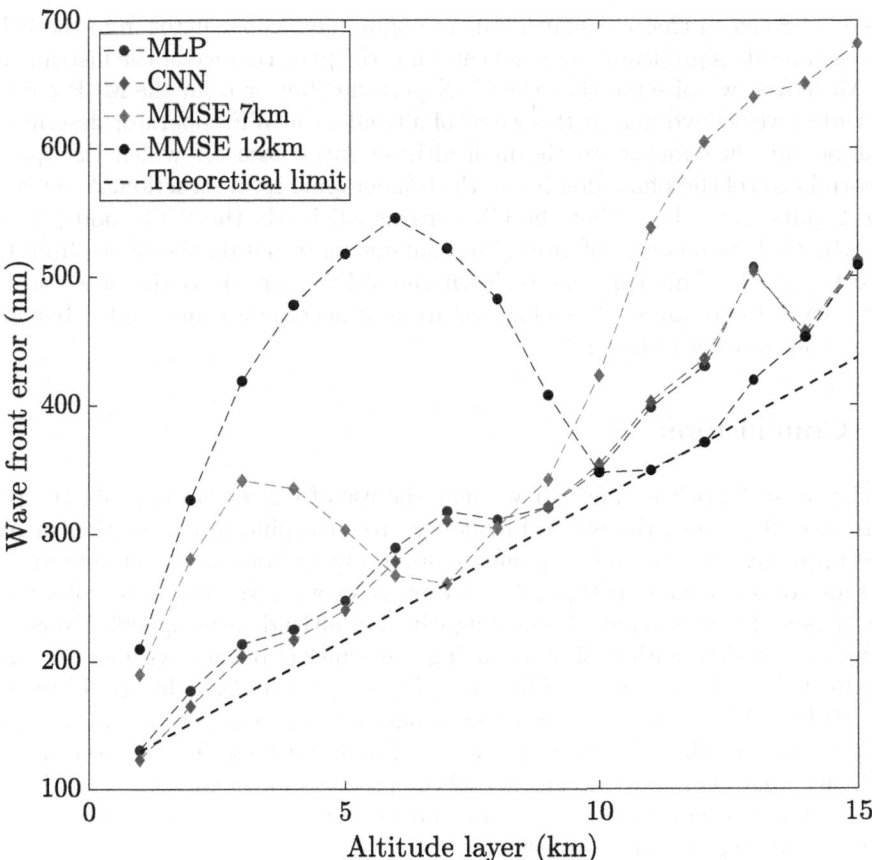

Fig. 2. WFE with respect to the layer altitude obtained on noise-free simulations. Each MMSE reconstructor is optimized to perform atmospheric tomography assuming a fixed altitude layer (7 or 12 km). Overall, the MMSE reconstructor is not robust to altitude variations while ANNs allow to maintain the wfe close to the theoretical limit. There is no noise in the simulation.

limit. As a result, both CNN and MLP stay close to the theoretical limit which degrades with increasing altitude due to the diminishing of the isoplanatic angle. On the contrary, the MMSE reconstructor performs optimally when applied to a set of data for which it is optimized, but degrades quickly when the altitude layer deviates by up to 1 km from the optimal altitude. This effect is expected from the MMSE reconstructor and the Fig. 2 illustrates how robust the ANN tomographic reconstruction is. Therefore, one of the strong advantages of ANN for atmospheric tomography is that they do not need to be updated during the observations, contrary to the MMSE reconstructor that is usually updated every 10 ms and even at shorter time scales. This update rate is extremely problematic for the next generation of Extremely Large Telescope instrumentation, where the large WFSs mean that calculating the tomographic reconstructor from the WFS slopes can take significant time and can limit the performance of the instrument.

Moreover, we observe that the CNN performs better than the MLP for low altitudes; we believe that at this range of altitude, the WFS operator description is important in contrast to the high altitude layer case for which the spatial decorrelation of the phase dominates the tomographic reconstruction. As a result, Fig. 2 puts into a light that the CNN learns efficiently the WFS model where the MLP relies on center of gravity measurements to obtain the slopes from the detector pixels. This capacity to learn the WFS operator is the other strong motivation to deploy CNN techniques from atmospheric tomography, but also wave-front sensing in general.

4   Conclusions

In the present work we have shown how the use of CNNs can improve the performance of state of the art techniques for tomographic reconstruction in AO. The improvement obtained for harder observing conditions (lower r0s) and in presence of sensor noise is specially relevant. Also, we have proven the robustness of CNNs and their strength for dealing with low altitude atmospheric layers and being able to deal with rapidly changing atmospheric profiles without the need of adjust the neural network. The possibility of using a CNN that could replace the MLP or MMSE as a tomographic reconstructor is also very promising since it opens the possibility of remove the center of gravity algorithm present in AO real-time controllers and integrating all the process in a single algorithm. Therefore, testing this new reconstructor in an optic bench and in a real telescope will be the next steps for this project.

Acknowledgments. The authors acknowledge Spanish ministry projects MINE CO AYA2017-89121- P, and support from the European Union's Horizon 2020 research and innovation program under the H2020-INFRAIA-2018-2020 grant agreement No 210489629. This work has been partially funded by the French National Research Agency (ANR) program APPLY - ANR-19-CE31-0011. This work also benefited from the support of the WOLF project ANR-18-CE31-0018 of the and the OPTICON H2020 (2017–2020) Work Package 1. James Osborn acknowledges support from the UKRI Future Leaders Fellowship (UK) (MR/S035338/1).

References

1. Noll, R., et al.: Zernike polynomials and atmospheric turbulence. JOSA **6**, 207–211 (1976)
2. Correia, C., et al.: Object-oriented matlab adaptive optics. In: Sixth International Conference of AO4ELT (2019)
3. Costille, A., et al.: Impact of CN2 profile on tomographic reconstruction performance: application to E-ELT wide field AO systems. In: Adaptive Optics Systems III. Proceedings of the SPIE International Conference, vol. 8447 (2012)

4. deCosJuez, F.J., et al.: An ANN-based smart tomographic reconstructor in a dynamic environment. Sensors **12**(7), 8895–911 (2012)
5. Dozat, T.: Incorporating nesterov momentum into adam. Technical report (2015)
6. Fried, D.L.: Limiting resolution looking down through the atmosphere. JOSAA (1917–1983) **56**, 1380 (1966)
7. Gendron, E., et al.: Status update of the CANARY on-sky MOAO demonstrator. In: Adaptive Optics Systems II. Proceedings of the SPIE International Conference, vol. 7736 (2010)
8. Krizhevsky, A., et al.: ImageNet classification with deep convolutional neural networks. In: Pereira, F., Burges, C.J.C., Bottou, L., Weinberger, K.Q. (eds.) Advances in Neural Information Processing Systems 25, pp. 1097–1105 (2012)
9. LeCun, Y., et al.: Deep learning. Nature **521**(7553), 436–444 (2015)
10. Osborn, J.: Open-loop tomography with artificial neural networks on CANARY: on-sky results. MNRAS **441**, 2508–2514 (2014)
11. Osborn, J., et al.: Using artificial neural networks for open-loop tomography. Opt. Express **20**, 2420–2434 (2012)
12. Rigaut, F., Neichel, B.: Multiconjugate adaptive optics for astronomy. Ann. Rev. Astron. Astrophys. **56**, 277–314 (2018)
13. Roddier, F.: The effects of atmospheric turbulence in optical astronomy. Progress in optics, vol. 19. North-Holland Publishing Co., Amsterdam (1981), 281–376 (1981). Roddier, F.: Adaptive Optics in Astronomy. Cambridge University Press, Cambridge (1999)
14. Suàrez-Gómez, S.L.: Experience with artificial neural networks applied in multi-object adaptive optics. Publ. Astron. Soc. Pac. **131**(1004), 108012 (2019)
15. Szegedy, C.: Going deeper with convolutions. In: Computer Vision and Pattern Recognition (CVPR) (2015)

Workshop on Machine Learning in Smart Mobility

Workshop on Machine Learning in Smart Mobility

Sara Ferreira[2,3], Henrique Lopes Cardoso[1,3(✉)],
and Rosaldo J. F. Rossetti[1,3]

[1] Artificial Intelligence and Computer Science Laboratory (LIACC), Porto, Portugal
[2] Research Centre for Territory, Transports and Environment (CITTA),
Porto, Portugal
[3] Faculty of Engineering, University of Porto, Porto, Portugal
{sara,hlc,rossetti}@fe.up.pt

The workshop on **Machine Learning in Smart Mobility (MLSM)** was co-located with the 21st International Conference on Intelligent Data Engineering and Automated Learning—IDEAL 2020, held online on November 4–6.

The MLSM workshop gathered both the machine learning (ML) community and transportation practitioners to discuss how cutting-edge ML technologies can be effectively applied to improve the performance of transportation and mobility systems on a sustainable basis, according to three important dimensions: economic, environmental, and social. The meeting aimed to serve as an appropriate platform to germinate new ideas towards building innovative applications of machine learning into smarter, greener, and safer mobility systems, stimulating contributions that emphasize on how theory and practice are effectively coupled to solve real-life problems in contemporary transportation, naturally including all sorts of mobility modes and their intrinsic interactions. Indeed, contemporary transportation is evolving rapidly on a more intelligent basis, and the concept of Intelligent Transportation Systems (ITS) has become already a reality among us, supporting the infrastructure leading to the emergence of the so-called Smart Mobility, and to a whole bunch of Mobility-as-a-Service (MaaS) options as we witness today. Also, when placed within the framework of Smart Cities, smart mobility gains more complexity and brings about new performance measures such as equity and social impact, privacy and security, ethical and legal compliance, explainable decision-support, while environmental sustainability is strongly emphasized.

Discussions focused mainly on application-oriented, integrative, and multidisciplinary perspectives of ML in smart mobility. The MLSM Workshop contributed to the IDEAL Conference with an appropriate forum to foster constructive debates on emerging and challenging topics in intelligent data analysis, data mining and their associated learning systems, and new paradigms in the very dynamic and evolving domain of urban mobility. It intended to leverage the cross-fertilization between ML and Smart Mobility, offering the support for a more effective and improved decision-making platform underlying urban mobility planning and management tasks, to which data is paramount.

C. Analide et al. (Eds.): IDEAL 2020, LNCS 12490, pp. 573–574, 2020.
https://doi.org/10.1007/978-3-030-62365-4_55

A total of 16 manuscripts were submitted for consideration, which were reviewed by at least three members of our International Program Committee. As a result of the rigorous review process, 5 papers were selected to shape a very interesting and diverse workshop program. Additionally to the presentations of contributed papers, an important part of the MLSM Workshop technical program featured a session entirely dedicated to the subject "New Training Modules to Increase Usage of 'Soft' Modes of Transport," hosted by the H2020 SIMUSAFE Project. Members of the SIMUSAFE Project, practitioners, the scientific community, and guests in general debated about ways of promoting a safe use of soft modes and their seamless integration with other actors on urban road networks. One important goal of such a debate particularly envisioned how to raise awareness and educate citizens about the benefits of new and softer transit modes in urban settings, accomplishing thus one major and significant contribution of the SIMUSAFE Project to the community.

This first edition of the MLSM Workshop constitutes an integral part of the SIMUSAFE (Simulation of Behavioral Aspects for Safer Transport) Project, funded by the European Union's Horizon 2020 research and innovation programme under grant agreement No 723386, whose support is greatly appreciated and recognized. This event was also promoted as an initiative of the IEEE ITS Society's Technical Activities Sub-committee on Artificial Transportation Systems and Simulation. The organizers are equally indebted to the members of the International Program Committee and additional reviewers, whose expertise, comments, and constructive critics were fundamental to the success and quality of the MLSM Workshop. The support and encouragement of the organizers of the IDEAL 2020 conference are greatly appreciated as well.

August 30, 2020   Sara Ferreira
Porto   Henrique Lopes Cardoso
                                                                    Rosaldo Rossetti

Driver Monitoring System Based on CNN Models: An Approach for Attention Level Detection

Myriam E. Vaca-Recalde[1]([⊠]), Joshué Pérez[1]([⊠]), and Javier Echanobe[2]([⊠])

[1] Tecnalia, Basque Research and Technology Alliance (BRTA), Edif 700, Derio, Spain
{myriam.vaca,joshue.perez}@tecnalia.com
[2] Department of Electricity and Electronics, University of the Basque Country,
Leioa, Spain
franciscojavier.echanove@ehu.eus

Abstract. Drivers provide a wide range of focus characteristics that can evaluate their attention level and analyze their behavioral states while driving. This information is critical for the development of new automated driving functionalities that support and assist the driver according to his/her state, ensuring safety for them and other users on the road. In this sense, this paper proposes a Driver Monitoring System (DMS) based on image processing and Convolutional Neural Networks (CNN), that analyzes two important driver distraction aspects: inattention of the road and drowsiness. Our approach makes use of CNN models for detecting the gaze and the head direction, which involves training datasets with different pre-defined labels. Additionally, the system is complemented with the drowsiness level measurement, using face features to detect the time that the eyes are closed or opened, and the blinking rate. Crossing the inference results of these models, the system can provide an accurate estimation of driver attention level. The different parts of the presented DMS have been trained in a Hardware-in-the-loop driving simulator with an eye fish camera. It has been tested as a real-time application recording driver with different characteristics.

Keywords: Driver Monitoring System · Convolution Neural Network · Artificial intelligence · Advanced Driver Assistance System (ADAS).

1 Introduction

Drowsiness and driver distractions are among the principal causes of fatal accidents on the road [1,2]. Hence, the design and validation of real-time advanced Driver Monitoring Systems (DMS) that can indicate to the system to take control of the vehicle, whatever the distraction reason [3], have become an important research and application field [4]. This information is critical for new Advanced Driver Assistance Systems (ADAS) that support and assist the driver according to his/her state.

Different sensor strategies are used to monitor the driver status inside the vehicle. Two examples are car-mounted cameras (in the dashboard or next to the

C. Analide et al. (Eds.): IDEAL 2020, LNCS 12490, pp. 575–583, 2020.
https://doi.org/10.1007/978-3-030-62365-4_56

driver [5,6]), and measuring the vital signs using intrusive devices [7]. Concerning the algorithms, different approaches have been studied in the last decade, using mathematical models, machine learning methods, and deep neural networks.

A Hidden Markov Model (HMM) is used in [8] where he driver's eye-blinking and head nodding are separately modeled. The system combines both behavioral states to decide if the driver is drowsy or not, but it does not take into account a possible inattention of the driver. In [9] and [10] it is applied a Viola-Jones method to detect the face and a linear Support Vector Machines (SVM) classifier to detect fatigue and distraction. The work [9] compares the SVM with mean intensity and Scale-Invariant Feature Transform (SIFT) techniques. Authors in [11] develop a weighted ensemble of classifiers, using a genetic algorithm and Convolutional Neural Networks (CNN), to identify distracted driving postures. CNN is also used in [12] to estimate the driver gaze zone.

Although the driver state has been analyzed by different approaches, most of them do not analyze distraction and drowsiness simultaneously, nor do they usually combine different variables of the eyes and head at the same time. The proposed approach evaluates and combines both estates using eye-tracking [13] [12] and head position [6] for defining different states of driver distraction.

Our contribution relies on creating a system capable of offering real-time information about the state of the driver to the rest of the automated vehicle systems in a short amount of time, with low computational resources and small databases. We divide the problem into two sub-systems: one for gaze (blinking rate and eyes direction), and the other one to detect head movements, and train both systems separately. The definition of driver state relies on the value of those systems simultaneously, i.e., if the head position system gives "center" output, and the eyes system gives "left", it is considered a distraction since the driver does not have full attention on the road. CNN was used to determine each of these methods. Two datasets were created using the HW-in-the-loop simulator and considering real-world conditions. These datasets are the combination of a non-labeled public dataset [14], and own recorded images to improve it.

This article is divided as follows. Section 2 contains the description of the overall system, together with the algorithms used to provide the blinking rate, the gaze, and head tracking information. The experimental tests and results are explained in Sect. 3. Section 4 closes with the conclusions of this work.

2   System Description

The proposed driver distraction monitoring system is divided in three main modules as shown in Fig. 1: head detection method, gaze method, and decision system.

2.1   Head Position Detection Method

Head Dataset. A synthetic dataset had been created in a car simulator (see Fig. 2) with a total of twenty different persons in different positions gaining

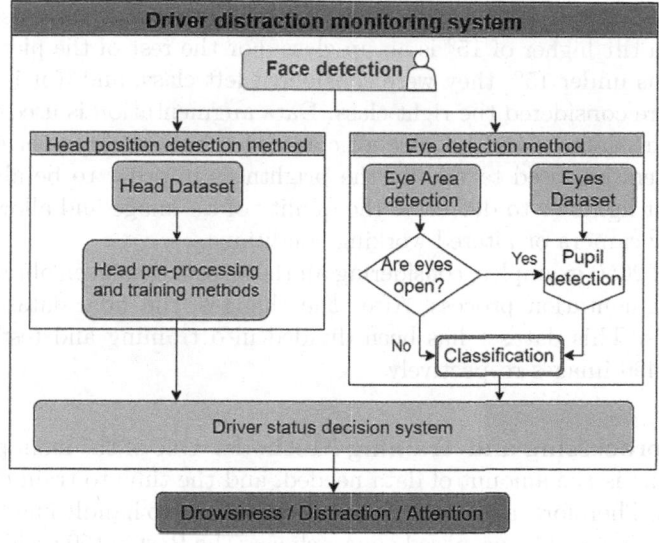

Fig. 1. Driver distraction monitoring system scheme.

a total of 180 photos per class. Videos were recorded manually to get tagged and classify photos in extreme and intermediate positions. The dataset was merged with another dataset from internet published in [14][1] to get diversity. This dataset provides two non-labeled series of 93 images of 15 people with variations of pan and tilt angles ([−90; 90]°) following the markers showed in Fig. 3.

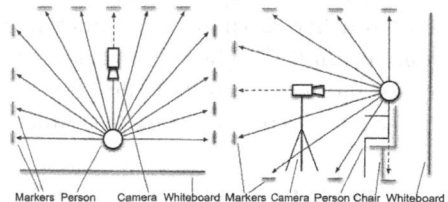

Fig. 2. Car simulator environment   **Fig. 3.** Tilt and pan angles scheme [14]

The head position estimator uses five classes for the head (up, left, center, right, and down). Tilt and pan angles divide the dataset according to the mentioned classes.

---

[1] http://www-prima.inrialpes.fr/perso/Gourier/Faces/HPDatabase.html.

Each photo of the dataset, where tilt was inferior to $-15°$, is considered as a down class; a tilt higher of $15°$ is an up class. For the rest of the pictures, if the pan angle was under $15°$, they were considered left class, and if it is was above $15°$, they were considered the right class. Data augmentation is used to increase the size of the dataset and balance all classes to the same number of samples, i.e. some filters are used to modify the brightness in order to be able to work with different lights or to decreases the quality of an image and allows simulate a low-quality camera or altered working conditions.

A total of 2650 examples, considering all the classes, has been obtained before the data augmentation process. After the changes, the final dataset contains 13462 images. This dataset has been divided into training and test sets, with 10966 and 2496 images respectively.

Head Pre-processing and Training Methods. One of the main problems of Deep Learning is the amount of data needed, and the time to train the network from scratch. Therefore, a pre-trained network is a way to handle fine-tuning with previous knowledge. The proposed approach uses the RestNet50 architecture [15] and only the last four layers were trained for the proposed system. In addition to these last layers, some more were added: an average polling 2D layer, a flatter layer, and two fully connected dense layers (respectively 256 and 5 matching the number of classes, that were defined above).

The ResNet50 is a powerful network that overfits rapidly in case of a low dataset. To prevent it, a dropout was introduced between the two last layers. The training was performed in two phases. In the first phase, the training uses a Stochastic Gradient Descent (SGD) optimizer with a learning rate of 10^{-3} and a momentum of 0.9. These parameters are used to find the direction of the best accuracy.

When the model started overfitting, the training is stopped, and the second phase started. The weights of the first phase are loaded, and the training is relaunched with new parameters in the optimizer (learning rate 10^{-5} and momentum 0.7). These parameters slow down the learning process so, it is necessary more time to find the best weights. These training phases perform with a categorical cross-entropy loss function on 20 epochs and a batch size of 24.

2.2   Eyes Detection Method

The methods used to extract the eyes areas and pre-process the data. It also concerns the dataset creation, the network used to predict the gaze, and the way it was trained.

Pupil Detection. The position and size of the eyes are extracted from the landmarks vector obtained by the DLIB library [16]. Once the eyes are detected (left eye landmarks: [37–42], right eye landmarks: [43–48]), its opening is analyzed. The system warns of drowsiness when the eyes are closed more than 3 frames (73 ms *approximately*) in a row or if there is a lack or excess of the average blinking rate [17]. If the eyes are open, the landmarks are used to mask the image and extract a cropped image of the eyes and the code handles two pictures corresponding to the subject's eyes.

The dominance of the white color in the eye can be used to detect the pupil position, which is a well-known and used parameter in the literature for the determination of the gaze driver [18–20]. In this work, a binary threshold filter has been applied to the eyes area to get the iris in black and the rest of the eye in white (see Fig. 4). The changes of light and reflections can affect the detection of the pupil, making the correct distinction of both parts difficult. The threshold has been set to 60 following a trial and error procedure (the values over the threshold will be set to 255 (white pixels), otherwise, they will be set to 0 (black pixel)).

Once the pupil is detected, each eye is divided into two parts with a tolerance to approximate the gaze direction (see Fig. 5). For these two parts, the numbers of white pixels will be counted. In the case of a side gaze, the number of white pixels will be minor in the side where the pupil is located. If both sides have approximately the same number of white pixels, the gaze is centered.

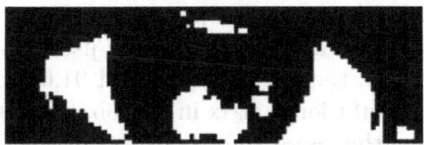

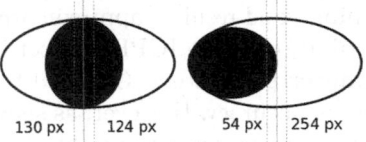

Fig. 4. Example of threshold filter in an eye   Fig. 5. Eye split method

Eyes Dataset. A dataset is created from the recorded videos for gaze detection. The dataset is pre-processed to get a .csv file containing all the white pixels counted of each gaze picture. After that, the images are classified into three classes: left, center, and right. A total of 770 images were obtained and classified into the mentioned classes. The number of images was divided into 80% for training and 20% for testing, i.e. 621 and 149 images respectively.

Classification. The CNN used to estimate the gaze is created using Keras framework. Due to the small dataset, the network contains three fully connected layers (respectively 180, 60, 3 neurons) and two dropout layers used to prevent overfitting. It is trained with a categorical cross-entropy loss function and an Adam optimizer [21] during 120 epoch and the batch size is set to 5. The network provides a [3, 1] vector with the score of each class.

2.3   Driver Status Decision System

The third module is the decision system. It combines the results obtained from the previous head positioning and gaze detection systems. According to the defined set of rules, this part identifies whether the driver is distracted, asleep, or attentive to the road. The gaze indicates whether or not the driver looks at the road. However, when complementing it with the head position, the state of the driver is defined correctly. For example, even if the driver looks at the road, if the head is down or turned right or left, the system indicates distraction since it may mean that s/he will be eventually distracted.

3 Experimental Results

This section shows the results achieved from the experiments carried out to evaluate the systems. Both head and eye systems are first separately evaluated to demonstrate their effectiveness. The final result is explained afterward. The results are divided into the network training and testing results and, on the other hand, real-time results in videos recorded in the simulator.

3.1 CNN Training and Testing Results

Head Position Estimation Results. Experiments showed the network performing good results (approximately 80–85%) depending on parameters like the distribution of the ICPR dataset in 5 classes, the data augmentation used, and optimizer parameters. On 1600 testing images, the network provided 91.6% of average accuracy. However, as shown in the confusion matrix in Fig. 6b, the network struggles with the right class samples, that is, it only reaches 72% while the rest of the classes reaches or exceeds 85%. This implies that a better study of the image classification for the right-label images is needed. Also, this could be for the position of the camera between the ICPR dataset and our images.

Gaze Estimations Results. The results for estimating the gaze are shown in the confusion matrix in Fig. 6.a. It has been obtained more than 90% in each class of the test dataset. Moreover, it has shown that the network can predict correctly the gaze in more than 95% of cases. Errors are often due to light problems were the threshold filter became not enough aggressive to separate the iris from the white of the eye. An adaptive threshold filter, that could be dynamically adapted to the light to pre-process samples, in a better way, could improve the results. On the other hand, training this network with more samples would provide also better performance.

3.2 Real Conditions Tests

Real conditions tests were carried out in the car simulator used to create the dataset (see Fig. 2) and a Basler camera acA1920-40uc, that delivers 41 frames per second at 2.3 MP resolution, with a 12 mm focal length lens was set up on board behind the steering wheel.

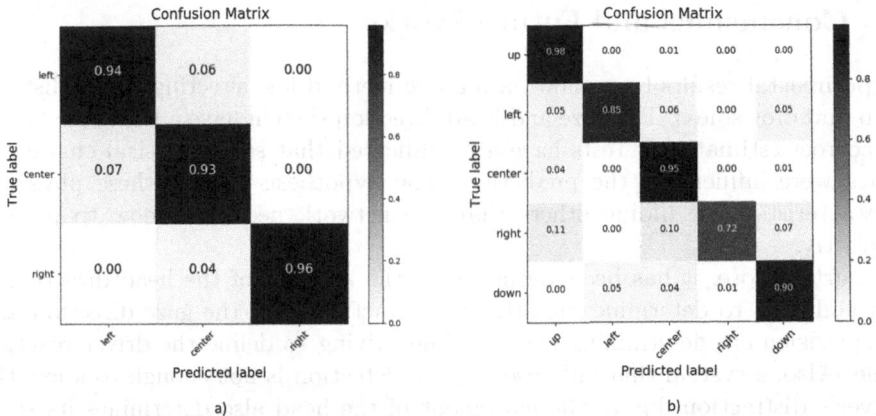

Fig. 6. a) Gaze estimation network results. b) Head position estimation network results.

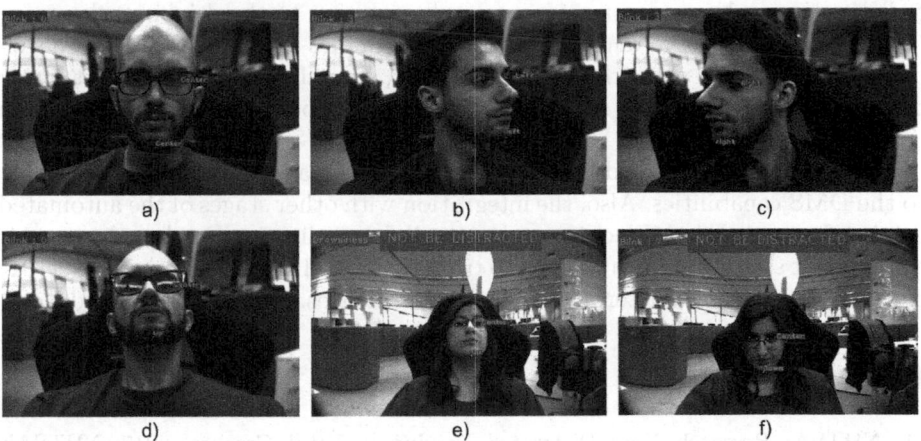

Fig. 7. Examples frames from real-time application, a) Frontal identification of gaze and head direction. b) Detection of complete left distraction, c) Detection of complete right distraction. d) Fail detection up head and gaze. e) Detection of sleeping person with head up. f) Detection of distraction despite looking at the road but the head is down.

The system has been tested on people with different characteristics, i.e.: beard, glasses, long hair, slanted eyes, big eyes, different zooms, etc. It performs in real-time with good results. Some examples are shown in Fig. 7. The gaze is correctly estimated in almost all cases, but errors are produced when the subject adopts a full lateral head position or in the presence of daily glasses due to the environment light and its reflection (see Fig. 7d). Also, in the case of long hair the prediction could be false due to the hair movement or possible occlusions of the eyes (see Fig. 7f).

4   Conclusions and Future Works

Experimental results have shown a reliable method for detecting driver distraction and drowsiness. The gaze and head direction system have a high percentage of correct estimations. Tests have also indicated that some physical characteristics were influencing the prediction. The hypothesis is that these physicals characteristics are hiding others that the network needs to know to predict correctly.

Furthermore, it has been verified that the analysis of the head direction is not sufficient to determine the driver's distraction since the gaze direction and range vision are determining factors while driving to define the driver reaction time. Also, a system that only uses a gaze detection is not enough to know the driver's distraction due to the movement of the head also determines its state and even where it will look in the future.

The analysis of the blinking rate allows increasing the knowledge on driver distraction, for example, if the driver does not blink enough, it is interpreted as a distraction as he may not paying enough attention to the driving task.

Having an extensive database with many samples is important for the development of a neural network system. However, a small database does not imply that a system with good results cannot be developed. It is one of the main contributions of this work.

Future works include adding other variables such as hands position detection to the DMS capabilities. Also, the integration with other stages of the automated driving framework (e.g., decision and control) for applications in shared control is a clear next step. Other frameworks to train the neural networks (e.g. ResNet101) or changes of its parameters (e.g. epochs, batch sizes...) will be evaluated.

References

1. NHTSA: Research Note Distracted Driving in Fatal Crashes, 2017. NHTSA's National Center for Statistics and Analysis, DOT HS, vol. 812, no. April, pp. 1–8 (2019)
2. Fountas, G., et al.: FINAL REPORT factors affecting perceived and observed aggressive driving behavior: an empirical analysis of driver fatigue, and distracted driving. Technical report (2019)
3. Prat, F., Gras, M.E., Planes, M., Font-Mayolas, S., Sullman, M.J.M.: Driving distractions: an insight gained from roadside interviews on their prevalence and factors associated with driver distraction. Transp. Res. Part F: Traff. Psychol. Behav. **45**, 194–207 (2017)
4. Rolison, J.J., Regev, S., Moutari, S., Feeney, A.: What are the factors that contribute to road accidents? An assessment of law enforcement views, ordinary drivers' opinions, and road accident records. Accident Anal. Prevent. **115**, 11–24 (2018)
5. Magdalena Nowara, E., Marks, T.K., Mansour, H., Veeraraghavany, A.: SparsePPG: towards driver monitoring using camera-based vital signs estimation in near-infrared. In: IEEE Computer Society Conference on Computer Vision and Pattern Recognition Workshops, vol. 2018-June, pp. 1353–1362 (2018)

6. Raman, K.J., et al.: Fatigue monitoring based on Yawning and head movement. In: 2018 6th International Conference on Information and Communication Technology, ICoICT 2018, pp. 343–347. IEEE, May 2018

7. Zhang, C., Wang, W., Chen, C., Zeng, C., Anderson, D.E., Cheng, B.: Determination of optimal electroencephalography recording locations for detecting drowsy driving. IET Intel. Transp. Syst. **12**(5), 345–350 (2018)

8. Choi, I.H., Jeong, C.H., Kim, Y.G.: Tracking a driver's face against extreme head poses and inference of drowsiness using a hidden Markov model. Appl. Sci. **6**(5), 137 (2016)

9. Naz, S., Ziauddin, S., Shahid, A.: Driver fatigue detection using mean intensity, SVM, and SIFT. Int. J. Interact. Multimed. Artif. Intell. **5**(4), 86 (2019)

10. Lupu, D., Necoara, I.: Primal and dual first order methods for SVM: applications to driver monitoring. In: 2018 22nd International Conference on System Theory, Control and Computing, ICSTCC 2018 - Proceedings, pp. 565–570. Institute of Electrical and Electronics Engineers Inc., November 2018

11. Eraqi, H.M., Abouelnaga, Y., Saad, M.H., Moustafa, M.N.: Driver distraction identification with an ensemble of convolutional neural networks. J. Adv. Transp. **2019** (2019)

12. Vora, S., Rangesh, A., Trivedi, M.M.: Driver gaze zone estimation using convolutional neural networks: a general framework and ablative analysis. IEEE Trans. Intell. Veh. **3**(3), 254–265 (2018)

13. Desmet, C., Diependaele, K.: An eye-tracking study on the road examining the effects of handsfree phoning on visual attention. Transp. Res. Part F: Traff. Psychol. Behav. **60**, 549–559 (2019)

14. Gourier, N., Hall, D., Crowley, J.L.: Estimating face orientation from robust detection of salient facial structures. In: FG Net Workshop on Visual Observation of Deictic Gestures (POINTING), pp. 17–25 (2004)

15. Chollet, F., et al.: Applications - Keras documentation (2015)

16. Dlib: D-lib C++ library (2016)

17. Haq, Z.A., Hasan, Z.: Eye-blink rate detection for fatigue determination. In: India International Conference on Information Processing, IICIP 2016 - Proceedings. Institute of Electrical and Electronics Engineers Inc., July 2017

18. Naqvi, R.A., Arsalan, M., Batchuluun, G., Yoon, H.S., Park, K.R.: Deep learning-based gaze detection system for automobile drivers using a NIR camera sensor. Sensors (Switzerland) **18**(2), 456 (2018)

19. Panicker, A.D., Nair, M.S.: Open-eye detection using iris-sclera pattern analysis for driver drowsiness detection. Sadhana - Acad. Proc. Eng. Sci. **42**(11), 1835–1849 (2017)

20. Vicente, F., Huang, Z., Xiong, X., De La Torre, F., Zhang, W., Levi, D.: Driver gaze tracking and eyes off the road detection system. IEEE Trans. Intell. Transp. Syst. **16**(4), 2014–2027 (2015)

21. Kingma, D.P., Ba, J.L.: Adam: a method for stochastic optimization. In: 3rd International Conference on Learning Representations, ICLR 2015 - Conference Track Proceedings. International Conference on Learning Representations, ICLR (2015)

Road Patterns Identification and Risk Analysis Based on Machine Learning Framework: Powered Two-Wheelers Case

Milad Leyli-abadi$^{(\boxtimes)}$ ⓘ, Abderrahmane Boubezoul ⓘ, and Stéphane Espié ⓘ

TS2-Simu&Moto, Gustave Eiffel University, IFSTTAR, Champs-sur-Marne, France
{milad.leyli-abadi,abderrahmane.boubezoul,stephane.espie}@univ-eiffel.fr

Abstract. Analysis of motorcyclists' behaviour and the risk that this mode of transport incurs for the mobility as well as their safety have gained more attention in recent years. In the context of the European project SimuSafe, various experimentations have been done using instrumented motorbikes in a naturalistic riding study and the riders' behaviours are subjected to self-confrontation interviews with traffic psychologists. This paper aims at the identification of different riding patterns using machine learning techniques and allows for a deeper understanding of pattern-specific risk exposure from multi-source data (video footage and interviews). More specifically, we focus on the roundabout pattern analysis as it is the most important source of collisions and a set of rules are designed using a decision tree to analyse their related risks. The generated rules may also fuel a multi-agent simulator to reflect the riders' real-world behaviours.

Keywords: Riding patterns identification · Hidden Markov model · Event specific risk analysis

1 Introduction

The H2020 SimuSafe[1] project aims noticeably at the design of a multi-driver Driving Simulator, and at the refinement of Multi-Agent Simulators (MAS) providing realistic interactions between the users, simulated or real, of a virtual road network. The ambition is to design the digital twin of a potentially complex road network where simulators controlled by Humans and simulated avatars of pedestrians, cyclists, motorcyclists and car drivers act and interact in a realistic manner. Such a digital twin can be used to study the behaviour of road users in various safe or unsafe situations.

To simulate realistic Human behaviours, it is necessary to design a simulation platform where each individual agent acts and reacts with its current and anticipated context. As an example, a simulation model for unsignalized crosswalks is proposed in [6] which considers the collisions between vehicles and

---

[1] http://simusafe.eu/.

© Springer Nature Switzerland AG 2020
C. Analide et al. (Eds.): IDEAL 2020, LNCS 12490, pp. 584–593, 2020.
https://doi.org/10.1007/978-3-030-62365-4_57

pedestrians. To account for collisions, it has been assumed that a given portion of drivers and pedestrians is distracted and do not pay attention to the road. It has been shown that subjects' distraction has a linear relation with the collisions frequency. Furthermore, the pedestrian fatalities were influenced by the collision speed.

Naturalistic studies have been used since decades to observe the car drivers' behaviour. More rarely and thanks to the video recording of the driver's journeys, self-confrontation interviews conducted by traffic psychologists have been used to identify the motivation of each decision undertaken by the drivers. Furthermore, the context dependent elements are also taken into account for the development of their decision [5]. Nowadays, data mining techniques can be used to help conducting self-confrontation interviews, and to identify patterns used by drivers in specific road situations. Thanks to such techniques, one can detect some specific events (abrupt breaking/accelerating for instance) in the collected data and put markers in the video to help conducting the interviews [1].

The Motorbike riding is a very complex task compared to that of a four wheeled vehicles. This complexity comes from the fact that the rider is strongly involved in the riding task with the aim to maintain the dynamic stability of his/her vehicle. This task is even more complex during emergency (near miss) events, such as in the case of harsh braking.

Little work has been undertaken so far concerning the modelling and simulation of motorcyclists' behaviour. Within SimuSafe, the Simu&Moto research team at Gustave Eiffel University is trying to refine the behavioural model for motorcyclists that it has been developing for several years within the Archisim traffic simulation model [8]. The Archisim simulation model aims at simulating the traffic following the emergence philosophy used in multi-agents systems: the traffic phenomenon comes during the simulation thanks to the various actions and interactions occurring among the various actors of the road situation [11] [4]. The difficulty for elaborating the motorcyclists' behaviours being to identify the rules they used in various encountered contexts.

To tackle this difficulty, we propose a methodology based upon a machine-learning framework which is used for riding patterns recognition. In this article, the riding patterns are referred to different road (infrastructure) segments which should not be confused with traffic patterns or states that are studied in various research works [3,9,13]. The problem is formulated as a classification task to identify the class of riding patterns using data collected from GPS sensor during a naturalistic study. Furthermore, machine learning techniques are used to extract context specific rules from the annotations of the video recordings. These annotations concern rider's journeys and are determined by traffic psychologists. Finally, we propose to use verbal data corresponding to the riders' decisions in various encountered situations. These data are gathered thanks to the self-confrontation interviews conducted by psychologists with riders and help to better understand their decisions. The work presented in this paper focuses on the identification of roundabout situations and on the analysis of their related risk for the riders.

2   Data Description

The data analysed in this work are issued from naturalistic riding study (NRS) within SimuSafe project, and comprise 48 trajectories performed by 6 riders. The study is done from August 2019 to October 2018 in Burgos (Spain). We note by $\mathbf{S} = \{S_1, \ldots, S_6\}$ the set of subjects where $S_i = \{Tr_1, \ldots, Tr_J\}$ with J indicating the variable number of trajectories Tr performed by each rider. Each trajectory $Tr_i = \{(x_1, y_1), \ldots, (x_T, y_T)\}$ is a time series represented by a set of manoeuvres where the pair (x_t, y_t) designates the longitude and latitude of the GPS coordinates and T is its length. The GPS tracks of the riders during each trajectory is captured by a mobile phone which is the main source for the feature extraction and behaviour analysis in the following sections. In addition to the sensory data, some descriptive information concerning environment characteristics, event characteristics, riders' status, riders' behaviour and risky riding are also available. To conduct the behavioural analysis, a total of 270 events were extracted from the videos of the road and of the rider. These data will be used to characterise and to estimate the risk related to roundabouts.

2.1   Preprocessing

As experiment took place in urban area, the GPS signal was interrupted from time to time due to the masking effects (urban corridors). The preprocessing step aims at identifying the missing segments and to interpolate the data by ensuring the sampling frequency at 1 Hz. To this end, the Kalman filter is used for which the transition (A), observation (H) and covariance (Q) matrices are initialised as follows:

$$
A = \begin{pmatrix} 1 & 0 & \delta t & 0 \\ 0 & 1 & 0 & \delta t \\ 0 & 0 & 1 & 0 \\ 0 & 0 & 0 & 1 \end{pmatrix}, \quad H = \begin{pmatrix} 1 & 0 & 0 & 0 \\ 0 & 1 & 0 & 0 \end{pmatrix}, \quad Q = \begin{pmatrix} 1e\text{-}8 & 0 \\ 0 & 1e\text{-}8 \end{pmatrix}
$$

where δt is the the time elapsed since the previous iteration which is set to 1 second. The state vector is defined by $\mathbf{x} = (x_t, y_t, \dot{x}_t, \dot{y}_t)^T$ where the parameters are respectively the position on the x and y axes and the velocity along these axes. The technical details of the Kalman filters are beyond the scope of this article and the reader may refer to [7]. Furthermore, the Ramer–Douglas–Peucker algorithm [12] is applied on the output of the Kalman filter to reduce the number of data points and to keep one data point per meter.

2.2   Feature Extraction

To identify the different riding patterns and to analyse the risky behaviours, some features should be extracted from GPS signals. These features are classified into two groups: features related to rider's dynamic behaviours and those related to the infrastructure. The first group of features involves the rider's specific behaviour when interacting with various road patterns, e.g., velocity, acceleration and jerk. The second group is related directly to the infrastructure without considering the rider's behaviour, e.g., bearing, curvature, centripetal acceleration and direction. In this way, each trajectory is composed of set of features $F = \{\mathbf{F}_1, \ldots, \mathbf{F}_T\}$ that are observed over time and $\mathbf{F}_t \in \mathbb{R}^p$ where p determines the dimensionality of the feature space.

3 Riding Patterns Identification Methodology

As it was previously mentioned, the correct identification of all the road patterns is a difficult task because of the dynamic behaviour of motorcyclists and similarity between these patterns. The riding patterns we aim to recognize are straight segments (SL), left/right segments (LT and RT) and roundabouts (RA). To cope with this problem, we adopt a two-stage methodology. The following sections describe each stage in greater details.

3.1 First Stage: Identification of Road Patterns

To identify the different riding patterns in an unsupervised fashion while taking into account the temporal dependency of the manoeuvres, a Gaussian hidden Markov model (GHMM) [10] is used. For a set of observations $y = \{y_1, \ldots, y_T\}$ with hidden discrete states $\mathbf{z} = \{z_1, \ldots, z_T\}$ where $z_t \in \{\text{LT,RT,SL}\}$, the model is defined by:

$$P(z_{1:T}, y_{1:T}) = \prod_{t=1}^{T} f(y_t|z_t) = P(z_1)P(y_1|z_1) \prod_{t=2}^{T} P(z_t|z_{t-1})P(y_t|z_t), \quad (1)$$

where $f(y|z)$ is the probability density function of the Gaussian random variable Y_t given the state z_t denoted by $y_t^{z_t}$ and $y_t^{z_t} \sim \mathcal{N}(\mu_{z_t}; \sigma_{z_t}^2)$. In our case, the observations are the multivariate real valued extracted features introduced previously. The graphical representation of this model is shown in Fig. 1.

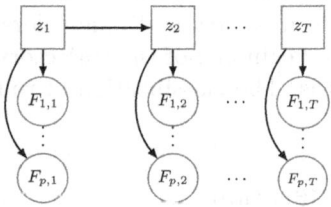

Fig. 1. Graphical representation of a Hidden Markov model. $z_t \in \{\text{LT,RT,SL}\}$ is the state at time t and $F_{j,t}$ is the j-th feature observed at time t

The model is completely determined by the set of parameters $\phi = (\pi, A, \mu_1, \ldots, \mu_m, \sigma_1^2, \ldots, \sigma_m^2)$, where π is the vector of initial probabilities, A is the transition matrix, μ and σ are the mean and variance of the Gaussian distribution which should be estimated, and m corresponds to the number of states. The parameters are estimated using an iterative Expectation-Maximization (EM) algorithm [2], alternatively called the Baum-Welch algorithm.

3.2   Second Stage: Classification of Roundabouts

Once the three of four riding patterns are identified in the previous stage, the roundabouts that are misidentified as left/right turns are manually labelled for six trajectories. This constitutes the learning set for a classification algorithm which aims at identifying the roundabout patterns for the remaining 42 trajectories (test set). The previously introduced extracted features are used for the classification of roundabouts.

4   Results

In the following, we present separately the results obtained at each stage of the proposed methodology. Afterwards, by incorporating the descriptive data introduced in the data section, we focus on the analysis of risky events.

4.1   Trajectory Segmentation

This section presents the results obtained by GHMM for segmentation of the road patterns into four major clusters, i.e., right/left turns, straight lines and other patterns. It should be noticed that the clustering is performed completely in a non-supervised fashion and no ground-truth labels were available for the performance evaluation. Figure 2 shows a raw trajectory (in left) and the associated clustering result (in right) which is performed using the proposed method. The emplacement of road patterns are annotated manually in Fig. 2a to facilitate the reading. In Fig. 2b, the left turns, right turns and straight lines are shown respectively with green, yellow and blue colors. It can be observed that the two roundabouts in the beginning of the trajectory are misidentified respectively as left and right turns. The classification of roundabouts is addressed in next section.

4.2   Roundabout Classification

This section presents the results associated to the classification of roundabouts. Firstly, the roundabout patterns that are misidentified as left/right turns in the previous step, are manually labelled for six trajectory. It constitutes the learning set for a classification approach.

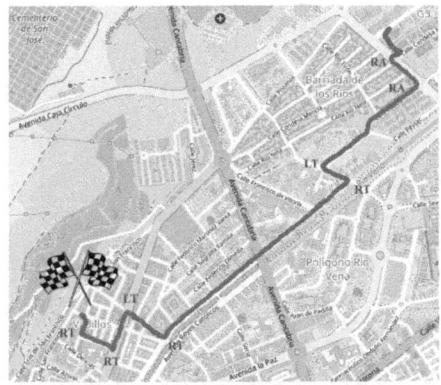

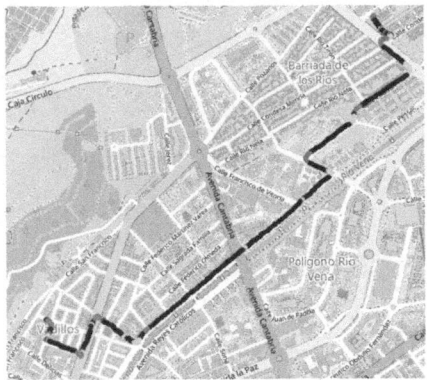

(a) A trajectory performed by a subject, RT: right turn, LT: left turn, RA: roundabout

(b) Segmented trajectory by GHMM

Fig. 2. The trajectory performed by the driver and the segmentation results obtained by using GHMM

Table 1 summarizes the evaluation results for classification of roundabouts concerning the remaining 42 trajectories. The evaluated methods are Long Short-Term Memory (LSTM) network, Random Forest (RF), Gradient Boosting (GB), Support Vector Machines (SVM), K-nearest neighbors (KNN) and Logistic Regression (LR). The evaluation metrics are accuracy, F-score, precision and recall. The best results are highlighted in bold. It can be observed that the best results are obtained using LSTM network in terms of all the metrics. This model takes into account the temporal dependency among the riding patterns.

Table 2 shows also the confusion matrix obtained by the LSTM network. Among 97 roundabout patterns, 79 cases are correctly identified. Figure 3 shows an example of roundabout classification using LSTM network (original trajectory in left and classified trajectory in right). The rider crosses two times the roundabout situated at the middle and crosses once the roundabout located on the right part of the figure. Three crossings are correctly identified (red portions in Fig. 3b).

Table 1. Comparison table for the classification of roundabouts

Classifier	Accuracy	F-score	Precision	Recall
LSTM	**0.97**	**0.85**	**0.89**	**0.81**
RF	0.95	0.67	0.79	0.58
GB	0.94	0.63	0.74	0.55
SVM	0.70	0.21	0.14	0.46
KNN	0.90	0.52	0.47	0.59
LR	0.93	0.61	0.69	0.54

Table 2. Confusion matrix using LSTM network

		Predicted class	
		Roundabout	Other
True class	Roundabout	79	18
	Other	9	1003

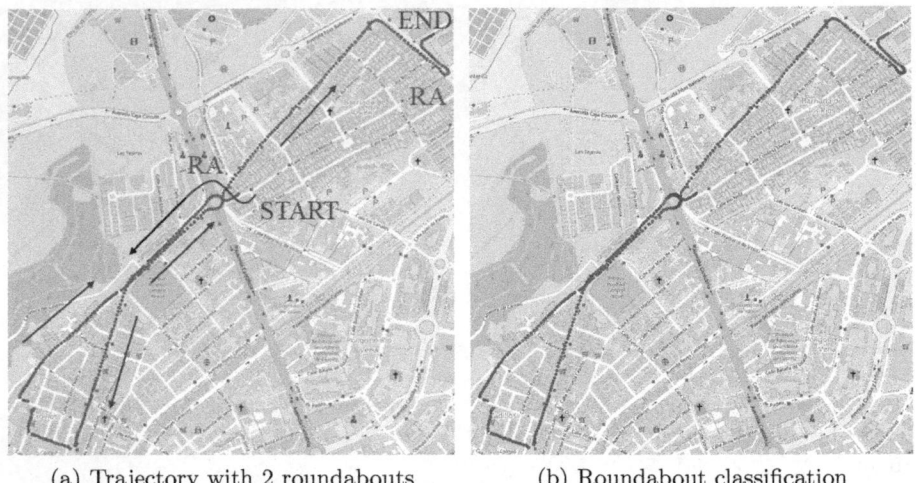

(a) Trajectory with 2 roundabouts   (b) Roundabout classification

Fig. 3. A trajectory presenting 2 roundabout patterns and 3 roundabout crossings in (a), and roundabout classification (shown in red) using LSTM network in (b) (Color figure online)

4.3   Risky Events Analysis

In parallel with the collected sensory data, traffic psychologists have also analyzed the subjects' behaviors. This analysis has been done using the video data recording the rider's facial expression and the infrastructure. The facial expression provides information such as distraction and attention. On the other hand, the videos of infrastructure shed light on the rider's decision when interacting with different road segments and other road users (pedestrians, cars,etc.). The annotations resulting from this analysis comprise the environment characteristics, events characteristics, and riders' behaviour, and the ranking of the risk level. For each rider and the corresponding events, 47 features have been made available, with one feature indicating the related event's risk level.

In this work, the aim is to analyse the risk linked to the roundabout events. A decision tree model based on video annotations has been designed (see Fig. 4) to achieve this purpose. More specifically, this decision tree is designed based on different local learners issued from the random forest approach introduced in the previous section. The decision tree model is preferred because it is human interpretable. The characteristics and events are shown using the rectangles on this tree, and two levels of risk are depicted using green (no risk) and red (First level risk) colours. The leaves of the tree involve the number of samples, which are the situations in the corresponding risk class.

The decision tree model associates the road situations and the riders' behaviour to a risk level determined by the psychologists. This model is established from the psychologists' point of view on the road situations and corresponding risk levels.

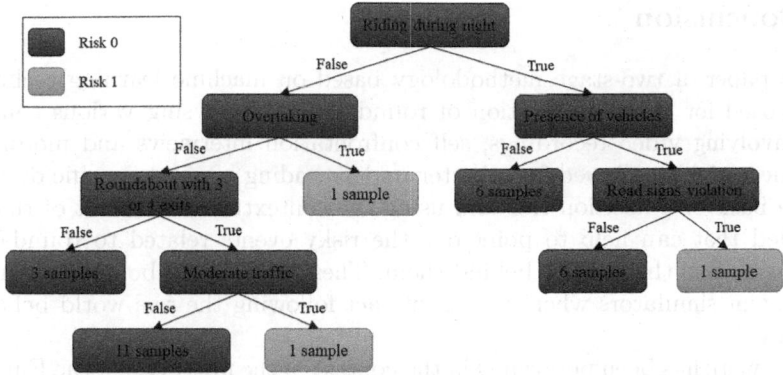

Fig. 4. Decision tree to identify the risky behaviours during roundabout situations

A set of rules can be deduced from this decision tree concerning roundabout situations:

- If *riding at night*, in the *presence of vehicles* and *violation of road signs*, the risk increases to the first level;
- If *riding during day* and *overtaking* in roundabout, the risk increases to the first level;
- If *riding during day*, no *overtaking*, *roundabout involves 3 or 4 exits* and there is a *moderate traffic*, the risk increases to the first level;
- If none of the three above mentioned sequence of events and characteristics visited, no risk is announced for the roundabout pattern.

In addition to these descriptive data, a self confrontation interview with the rider is also available. The following provides two examples extracted of such interviews:

- *"If I come to a roundabout and there is no cars, I enter the roundabout without stopping"*;
- *"In a roundabout, if I'm riding on the left but I want to leave the roundabout, I slow down waiting for the cars on the right to pass by"*.

The rider's verbal comments provide the rider's own standpoint concerning the situation. Following the conducted set of rules, we can associate the first behaviour to a no risk event (with no *presence of vehicles*) and the second behaviour may be associated to a first level risk, because of a *moderate traffic* in the roundabout. It can be observed that the generated decision tree is in accordance with the rider's decisions. The potential discrepancies may be explained by the following facts: a different judgement concerning the same situation (risky/not risky), the different levels of risk acceptance, the ambiguities in the phrasing (what means a "dense traffic"?), etc. It raises the question of the need to build up a common culture between the experimenter and the rider to limit ambiguities.

5   Conclusion

In this paper, a two-stage methodology based on machine learning techniques is proposed for the identification of roundabouts. Analysing various source of data involving video recordings, self confrontation interviews and motorbikes' dynamic variables allowed for a better understanding of event specific decisions. On the basis of a decision tree and using the contextual data, a set of rules are identified that can help to point out the risky events related to roundabouts and to examine the reasons behind them. These rules could be used to validate multi-agent simulators where the agents act following the real-world behaviour of riders.

This work has been performed in the context of the first cycle of the European project SimuSafe. In the course of the following cycles of the project, more data will be made available both from sensors and from traffic psychologists' analysis. The complementary data will be used to refine the deduced rules in this article, which in turn will conduct to the development of a more precise simulation platform. As a perspective, this work will be extended to involve more actors (cars, motorbikes, pedestrians) in the analysis and to study the interactions between them.

References

1. Attal, F., Boubezoul, A., Samé, A., Oukhellou, L., Espié, S.: Powered two-wheelers critical events detection and recognition using data-driven approaches. IEEE Trans. Intell. Transp. Syst. **19**(12), 4011–4022 (2018)
2. Bilmes, J.A., et al.: A gentle tutorial of the EM algorithm and its application to parameter estimation for Gaussian mixture and hidden Markov models. Int. Comput. Sci. Inst. **4**(510), 126 (1998)
3. Celikoglu, H.B., Silgu, M.A.: Extension of traffic flow pattern dynamic classification by a macroscopic model using multivariate clustering. Transp. Sci. **50**(3), 966–981 (2016)
4. Espié, S.: Archisim, multi-actor parallel architecture for traffic simulation. In: Proceedings of the Second World Congress on Intelligent Transport Systems, vol. 4 (1995)
5. Espié, S., Boubezoul, A., Aupetit, S., Bouaziz, S.: Data collection and processing tools for naturalistic study of powered two-wheelers users' behaviours. Accident Anal. Prevent. **58**, 330–339 (2013)
6. Feliciani, C., Gorrini, A., Crociani, L., Vizzari, G., Nishinari, K., Bandini, S.: Calibration and validation of a simulation model for predicting pedestrian fatalities at unsignalized crosswalks by means of statistical traffic data. J. Traff. Transp. Eng. (English Edn.) **7**(1), 1–18 (2020)
7. Krakiwsky, E.J., Harris, C.B., Wong, R.V.: A Kalman filter for integrating dead reckoning, map matching and gps positioning. In: IEEE PLANS 1988, Position Location and Navigation Symposium, Record. 'Navigation into the 21st Century', pp. 39–46. IEEE (1988)
8. Ksontini, F., Mandiau, R., Guessoum, Z., Espié, S.: Affordance-based agent model for road traffic simulation. Auton. Agent. Multi-Agent Syst. **29**(5), 821–849 (2014). https://doi.org/10.1007/s10458-014-9269-x

9. Mihaylova, L., Boel, R., Hegyi, A.: Freeway traffic estimation within particle filtering framework. Automatica **43**(2), 290–300 (2007)
10. Rabiner, L.R.: A tutorial on hidden Markov models and selected applications in speech recognition. Proc. IEEE **77**(2), 257–286 (1989)
11. Saad, F., Espié, S., Djemame, N., Schnetzler, B., Bourlier, F.: Microscopic traffic simulation and driver behaviour modelling: the archisim project. Road Safety In Europe and Strategic Highway Research Program (SHRP), Lille, France, 26–28 September 1994 (VTI Konferenz), vol. 2A: 3 (1995)
12. Saalfeld, A.: Topologically consistent line simplification with the Douglas-Peucker algorithm. Cartogr. Geogr. Inf. Sci. **26**(1), 7–18 (1999)
13. Wang, Y., Papageorgiou, M., Messmer, A.: Real-time freeway traffic state estimation based on extended Kalman filter: a case study. Transp. Sci. **41**(2), 167–181 (2007)

Towards Predicting Pedestrian Paths: Identifying Surroundings from Monocular Video

José Aleixo Cruz[1,2](✉) [ID], Thiago R. P. M. Rúbio[1,2] [ID], João Jacob[1,2] [ID], Daniel Garrido[1,2], Henrique Lopes Cardoso[1,2] [ID], Daniel Silva[1,2] [ID], and Rui Rodrigues[1,2] [ID]

[1] Faculty of Engineering, University of Porto, Porto, Portugal
{up201403526,reis.thiago,joao.jacob,up201403060,hlc,dcs,
rui.rodrigues}@fe.up.pt
[2] Artificial Intelligence and Computer Science Laboratory (LIACC), Porto, Portugal

Abstract. Pedestrian behavior is an essential subject of study when developing or enhancing urban infrastructure. However, most behavior elicitation techniques are inherently bound to be biased by either the observer, the subject, or the environment. The SIMUSAFE project aims at collecting road users' behavioral data in naturalistic and realistic scenarios to produce more accurate decision-making models. Using video captured from a monocular camera worn by a pedestrian, we employ machine learning and computer vision techniques to identify areas of interest surrounding a pedestrian. Namely, we use object detection and depth estimation to generate a map of obstacles that may influence the pedestrian's actions. Our methods have shown to be successful in detecting free and occupied areas from monocular video.

Keywords: Pedestrian behavior · Machine learning · Computer vision

1 Introduction

Understanding pedestrian behavior is vital to study and develop transportation and mobility systems. The actions each pedestrian performs have an impact on the actions of other pedestrians and drivers. Therefore, researchers and engineers must account for the causes and consequences of these actions to build improved infrastructure.

Generally, researchers examine human response by putting subjects in a controlled environment, observing their behavior, or allowing the subjects to describe their behavior in unconfined environments. However, the data retrieved is likely to be biased. On the one hand, the observer's point of view will never be the same as the subject's. On the other hand, it is hard for a subject to clearly explain what he or she has experienced during the experiment.

© Springer Nature Switzerland AG 2020
C. Analide et al. (Eds.): IDEAL 2020, LNCS 12490, pp. 594–601, 2020.
https://doi.org/10.1007/978-3-030-62365-4_58

The SIMUSAFE[1] project collects video and position data from human pedestrians and drivers acting in real-life scenarios, intending to improve driving and traffic simulators' technology by enhancing how simulators assess risk perception and decision making of road users. Our global aim is to use the collected data, which is independent of the observer's or subject's bias, to understand how what a pedestrian observes influences its locomotion and body movement.

In this paper, we focus on the particular issue of extracting relevant environment information from the gathered data. More specifically, we present a methodology for identifying areas of interest surrounding pedestrians, which may affect their decision-making, from video captured by a monocular smart-glass camera worn by pedestrians in their daily commutes.

Our approach is innovative in two ways: we study data gathered in a naturalistic setting, where the pedestrians are performing their daily routines, and we generate an environment representation from the pedestrian's point of view, which can provide more insight than an outside perspective.

The rest of this paper is organized as follows: Sect. 2 contains an overview of the related work and gap analysis; Sect. 3 details the method we applied for identifying a pedestrian's surroundings and shows the results of applying it to our data; Sect. 4 exposes the conclusions as well as future work.

2   Related Work

The taxonomy of pedestrian behavior prediction studies introduced by Ridel *et al.* [9] categorizes studies according to input data. One category encompasses studies that focus on predicting pedestrian behavior by evaluating how pedestrians' surroundings influence their actions. We consider our work to belong to this category and review related works that fit the same class.

From video data of a real crosswalk, Suh *et al.* [10] observed that in that particular crosswalk, most pedestrians had a gap-seeking behavior, crossing when a gap between vehicles is available, regardless of the pedestrian crossing signal indication. The researchers argue that models that do not take this behavior into account may be overestimating the average pedestrian waiting time in a crosswalk. Using statistics from the gathered data, Suh *et al.* have generated a model that more closely resembles the waiting times observed in the studied situation. However, it did not consider scenarios where crosswalks do not have pedestrian signals.

Rasouli *et al.* [7] studied pedestrian crossing behavior from the perspective of an autonomous vehicle. They developed an annotated dataset from real video data that contains information about the environment (e.g., street width, weather) and the actions taken by the pedestrian (e.g., walking, looking, gestures), to improve behavior prediction in crossings with and without pedestrian signals.

Another work by Bock *et al.* [1] involves a system that continuously improves its prediction of pedestrian trajectories at an intersection. The system uses global

---

[1] http://simusafe.eu/.

navigation satellite systems to track pedestrians' positions and an array of other devices, including cameras mounted on vehicles, to gather information about the environment.

From our literature review, we found a lack of studies that consider the environment's characteristics from the pedestrian's point of view. We intend to contribute to the understanding of pedestrian behavior by describing the environment as a pedestrian observes it. Specifically, we wish to extract relevant parts of the pedestrians' surroundings that might affect their decision making from video captured with the monocular camera of smart-glasses that pedestrians wore on their daily commutes.

3   Identifying Surroundings from Monocular Video

Through the use of smart glasses to record pedestrian behavior, we have access to data that reliably describes part of the pedestrians' actual behavior in their regular commutes, while avoiding the cost of setting up a simulation environment. Plus, we have access to much of the information sensed by the pedestrian. While the camera does not record sound nor captures the same field of view as the human eye, it shows most of the surroundings the pedestrian observes at a given point in time.

Considering the pedestrian as an agent in a multi-agent system, the set of obstacles, vehicles, and other nearby pedestrians represents the *environment* at a given point in time, and the pedestrian's body movement represents its *action*. The smart glasses possess a monocular camera that captures images at fixed frame-rate of 30 Hz with 1920 pixels of width and 1080 pixels of height (full-HD resolution). Therefore, we can extract information about the environment 30 times in a second. To do so, we try to map relevant parts of the environment using two techniques: object detection and depth estimation.

3.1   Object Detection

To detect objects in an image, we trained a neural network using YOLO [8] on the KITTI dataset [2]. The network can detect entities such as cars, people, and traffic signs. Because the head of a pedestrian is continually moving, the recorded video is very unsteady. To facilitate object detection, we stabilize the video without re-scaling. The object detection network determines the bounding box of a detected object and labels it with the most likely class. For each frame, we store information about all the detected objects. Let V represent a video, which is a set of frames. We represent a frame as shown in Eq. 1, where *frame_number* is the index of the frame in the video (the first frame of a video will have *frame_number* = 1), t is the frame's timestamp in seconds and *Context* is the set of detected objects.

$$frame = \{frame\_number, t, Context\} \tag{1}$$

We represent each *object* ∈ *Context* as shown in Eq. 2, where *class* is the class to which the object belongs (which can be "car", "person" or "traffic sign"), (x_{min}, y_{min}) is the lower-left corner and (x_{max}, y_{max}) is the upper-right corner of the object's bounding box in pixels.

$$object = \{class, x_{min}, y_{min}, x_{max}, y_{max}\} \tag{2}$$

3.2   Depth Estimation

While the object detection neural network achieves satisfactory results, it does not detect all objects in the frame. Being unable to identify objects nearby, which can be obstacles in the pedestrian's path, is very troublesome. To solve this issue, we also perform a depth estimation for each pixel in a frame to detect other obstacles. Because the video is captured from a monocular camera, we cannot apply techniques available in stereo setups to calculate depth. Therefore, we trained a neural network using the MonoDepth [5] algorithm on the Karlsruhe dataset [3,4], which contains stereo video recorded from a moving vehicle. For each left frame in the stereo camera video, the neural network estimates its disparity map, i.e., how much each pixel of the frame must shift to achieve the image from the right camera. While the neural network requires the left and right frame for training, it only requires one frame as input for inference. Hence, we feed our monocular video frames to the neural network and obtain disparity values for each pixel. Let b be the baseline in meters of a stereo camera setup, f the focal length of both cameras in pixels, z the depth of an object in meters, and d the disparity in pixels of the object. The relation between these values is given by Eq. 3.

$$\frac{d}{b} = \frac{f}{z} \tag{3}$$

While we know the focal length of the camera, there is no baseline measure in a monocular video. Therefore, we cannot have an accurate notion of scale in a video captured by a monocular camera. Hence, for the rest of this paper, we assume that $b = 1$. Using the MonoDepth network's disparity output, we calculate z for each pixel of the image.

To verify whether a pixel represents an obstacle or not, we perform ground estimation following the method described in [6], assuming fixed pitch and roll angles. By having an estimate of the depth of the ground plane z_g, we can assume that pixels with depth $z < z_g + \theta$ are above the ground and should be considered obstacles. θ is a threshold we use to make sure we do not consider most of the noise in the ground depth to be an obstacle.

To determine whether an area in the frame is free or obstructed, we merge the results of both obstacle detection and depth estimation. Let $c_{object} \in [0, 1]$ represent the cost of a pixel as a function of nearby objects detected by our neural network, and let $c_{obstacle} \in [0, 1]$ represent the cost of a pixel as a function of obstacles detected using ground and depth estimation. We generate a cost map, where each pixel of a frame has an associated cost c, with values $c \in [0, 1]$, which is given by $c = \max(c_{object}, c_{obstacle})$. Lower cost values indicate a higher chance of the area represented by the pixel being unoccupied, while higher values indicate that the area is likely to represent an obstruction. To do this, we localize which pixels in the image are inside the bounding box of an object detected with the object detection neural network and attribute $c_{object} = 1$ to pixels inside that bounding box. Pixels surrounding the bounding box represent areas near the obstacle, which the pedestrian will avoid, so we attribute a value $c_{object} \neq 0$, which decreases linearly with the distance to the object. All other pixels have value $c_{object} = 0$. Similarly, if we consider that a pixel does not belong to the ground plane, we attribute a cost $c_{obstacle} = 1$ to it.

We show an example of the cost calculation for a given frame in Fig. 1. Using the original frame as input, we get the bounding boxes of objects detected by our object detection network (drawn in red in the figure), including most of the cars in the original frame and one pedestrian. We use this information to calculate the object cost c_{object} for each pixel of the frame. Using the disparity estimate, we calculate the obstacle cost $c_{obstacle}$ to discriminate obstacles that might not have been detected by the object detection network. In this particular example, we were able to identify the lamp post. However, some areas were considered to be obstacles, even though they are not. We believe this derives from the noise of the disparity estimate and from not considering the camera's pitch and roll angles in our ground plane estimation method. After these intermediate steps, we calculate the total cost c of each pixel, which reveals what areas of the frame are plausibly free for the pedestrian to occupy and which the pedestrian will probably avoid.

3.3   3D Mapping

Using the depth map obtained from the disparity information, we can retrieve the estimated 3D position (x, y, z) in the camera reference frame of a location represented by a pixel. We calculate x and y according to Eq. 4, where (c_x, c_y) is the camera's principal point offset and f_x and f_y are the horizontal and vertical focal lengths, respectively, which we consider to be the same and equal to f.

$$x = \frac{(u - c_x) * z}{f_x}, \quad y = \frac{(v - c_y) * z}{f_y} \tag{4}$$

In an image with w pixels of width and h pixels of height, we obtain a point cloud with $w * h$ points, essentially mapping every pixel in the 2D image to a 3D point in the camera's reference frame, as shown in Fig. 2.

Now that we can map each pixel to a 3D position, we can associate each pixel's cost in our cost map to a 3D position in the camera's frame of reference. Because working with a point cloud is difficult, we average the cost over the vertical axis of the camera reference plane to obtain a 2D image of the costs, from a "top-down" perspective, as shown in Fig. 3.

The 2D cost image provides essential information regarding the pedestrian's surroundings. Namely, which areas are free for pedestrians to move and what

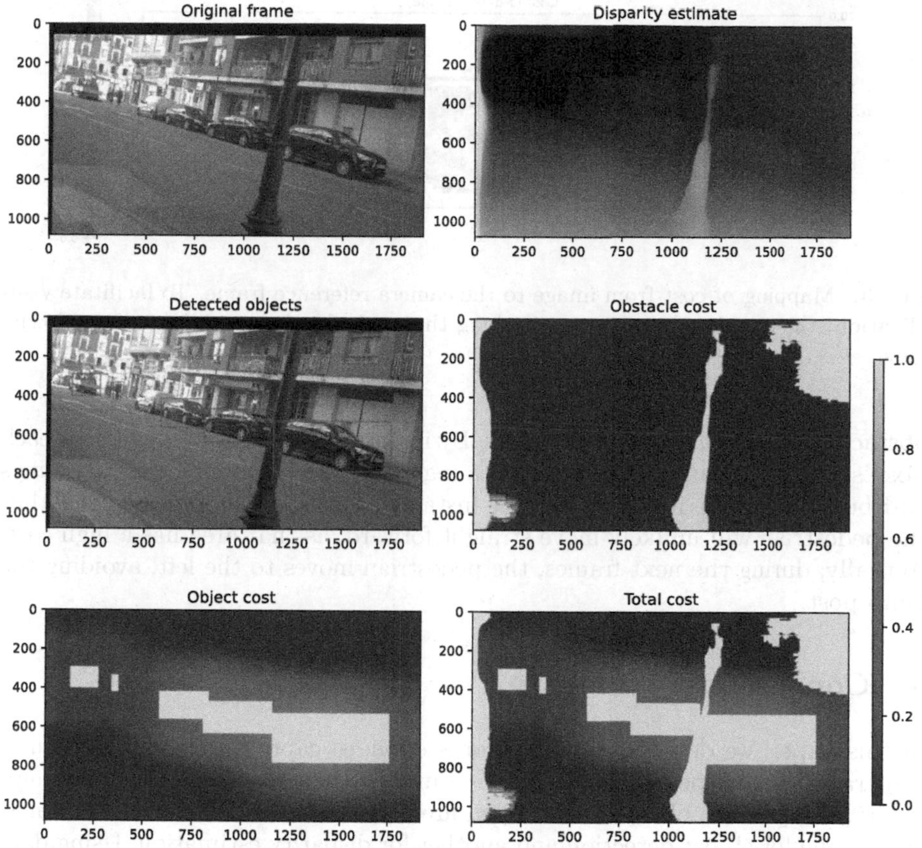

Original frame

Disparity estimate

Detected objects

Obstacle cost

Object cost

Total cost

Fig. 1. Calculation of pixel cost through object and obstacle detection.

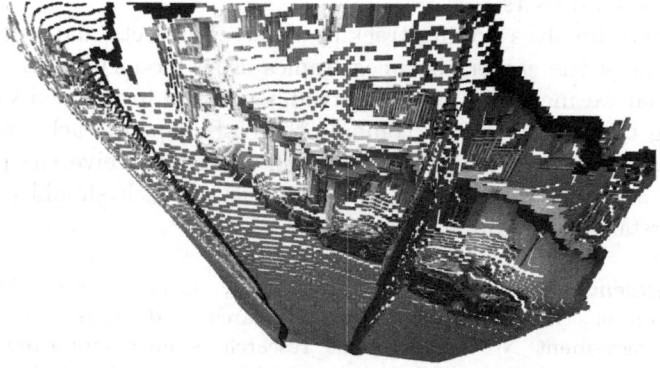

Fig. 2. Point cloud generated mapping pixels to 3D points using the estimated disparity information.

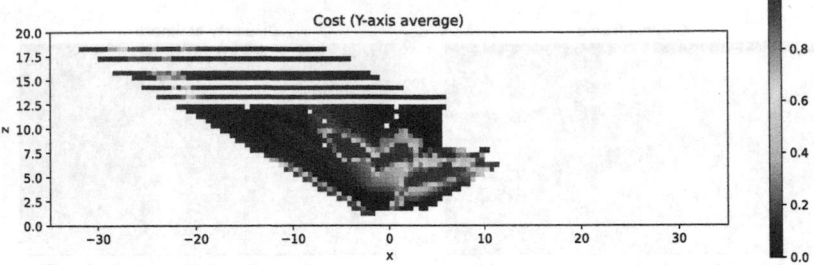

Fig. 3. Mapping of cost from image to the camera reference frame. To facilitate visualization, the cost has been averaged along the vertical axis, providing a "top-down" view.

obstacles are visible to the pedestrians. By integrating our knowledge about each pixel's depth, we can extract more data regarding the pedestrian's surroundings and better estimate its behavior. For instance, from Fig. 3 we can predict that the pedestrian will unlikely move straight forward, as that area has a high cost. Actually, during the next frames, the pedestrian moves to the left, avoiding the lamp post.

4   Conclusions

In this paper, we describe how we processed video captured from a monocular camera worn by a pedestrian to create a map of the pedestrian's surroundings, identifying possible obstacles and open areas. We trained and utilized two networks: one for object detection and another for disparity estimation. Using data from both, we are able to associate a cost and a 3D position to each pixel in video frames. Using this map, we take a step towards understanding a pedestrian's environment, which is essential to comprehend what influences the actions taken by these road users.

Future work involves adding tracking detected objects and estimating the change of pose of the camera between frames using monocular odometry techniques, so that we may infer the pedestrian's position in the real world. Using the resulting trajectory, we can combine the cost maps of each frame to generate a unique cost map for each video, and we can perceive the pedestrian's actions and associate them with the environment, which should enable us to better understand behavior.

Acknowledgments. This work is supported by project SIMUSAFE, funded by the European Union's Horizon 2020 research and innovation programme under grant agreement No 723386. This research is also supported by LIACC (FCT/UID/CEC/0027/2020). Authors are grateful to the SIMUSAFE Consortium's members for their valuable comments and fruitful discussions throughout the SIMUSAFE Project (http://simusafe.eu/), within which this work has been developed.

References

1. Bock, J., Beemelmanns, T., Klösges, M., Kotte, J.: Self-learning trajectory prediction with recurrent neural networks at intelligent intersections. In: VEHITS 2017 - Proceedings of the 3rd International Conference on Vehicle Technology and Intelligent Transport Systems, pp. 346–351 (2017). https://doi.org/10.5220/0006374003460351

2. Geiger, A., Lenz, P., Urtasun, R.: Are we ready for autonomous driving? The KITTI vision benchmark suite. In: Conference on Computer Vision and Pattern Recognition (CVPR) (2012)

3. Geiger, A., Roser, M., Urtasun, R.: Efficient large-scale stereo matching. In: Kimmel, R., Klette, R., Sugimoto, A. (eds.) ACCV 2010. LNCS, vol. 6492, pp. 25–38. Springer, Heidelberg (2011). https://doi.org/10.1007/978-3-642-19315-6_3

4. Geiger, A., Ziegler, J., Stiller, C.: StereoScan: dense 3D reconstruction in real-time. In: Intelligent Vehicles Symposium (IV) (2011)

5. Godard, C., Mac Aodha, O., Brostow, G.J.: Unsupervised monocular depth estimation with left-right consistency. In: The IEEE Conference on Computer Vision and Pattern Recognition (CVPR), July 2017

6. Kırcalı, D., Tek, F.B.: Ground plane detection using an RGB-D sensor. In: Czachórski, T., Gelenbe, E., Lent, R. (eds.) Information Sciences and Systems 2014, pp. 69–77. Springer, Cham (2014). https://doi.org/10.1007/978-3-319-09465-6_8

7. Rasouli, A., Kotseruba, I., Tsotsos, J.K.: Are they going to cross? A benchmark dataset and baseline for pedestrian crosswalk behavior. In: Proceedings - 2017 IEEE International Conference on Computer Vision Workshops, ICCVW 2017, vol. 2018-Janua, pp. 206–213. IEEE, October 2017. https://doi.org/10.1109/ICCVW.2017.33. http://ieeexplore.ieee.org/document/8265243/

8. Redmon, J., Divvala, S., Girshick, R., Farhadi, A.: You only look once: unified, real-time object detection. In: Proceedings of the IEEE Computer Society Conference on Computer Vision and Pattern Recognition, vol. 2016-Decem, pp. 779–788. IEEE, June 2016. https://doi.org/10.1109/CVPR.2016.91. http://ieeexplore.ieee.org/document/7780460/

9. Ridel, D., Rehder, E., Lauer, M., Stiller, C., Wolf, D.: A literature review on the prediction of pedestrian behavior in urban scenarios. In: IEEE Conference on Intelligent Transportation Systems, Proceedings, ITSC, vol. 2018-Novem, pp. 3105–3112. Institute of Electrical and Electronics Engineers Inc., December 2018. https://doi.org/10.1109/ITSC.2018.8569415

10. Suh, W., et al.: Modeling pedestrian crossing activities in an urban environment using microscopic traffic simulation. Simulation **89**(2), 213–224 (2013). https://doi.org/10.1177/0037549712469843

A Semi-automatic Object Identification Technique Combining Computer Vision and Deep Learning for the Crosswalk Detection Problem

Thiago R. P. M. Rúbio[1,2]([✉]) [iD], José Aleixo Cruz[1,2][iD], João Jacob[1,2][iD], Daniel Garrido[1,2], Henrique Lopes Cardoso[1,2][iD], Daniel Silva[1,2][iD], and Rui Rodrigues[1,2][iD]

[1] Faculty of Engineering, University of Porto, Porto, Portugal
{reis.thiago,up201403526,joao.jacob,up201403060,hlc,dcs,
rui.rodrigues}@fe.up.pt
[2] Artificial Intelligence and Computer Science Laboratory (LIACC), Porto, Portugal

Abstract. Object detection in the traffic domain has faced growing relevance through the years in developing autonomous driving mechanisms. As with vehicles, pedestrians face a very dynamic context, and identifying relevant objects from a pedestrian perspective presents many challenges. Improving the detection of some objects, such as crosswalks, is very relevant in this regard. This paper presents a technique that applies a computer vision approach to automatically generate datasets for training YOLO-based deep learning algorithms. An initial precision of 0.82 achieved with the generated dataset, which is increased to 0.84 after manually removing incorrect annotations. Results show that our approach leverages the dataset building process by reducing the manual workload needed. The approach could be used for training other object detection models used in traffic scenarios.

Keywords: Crosswalk identification · Autonomous pedestrian · Computer vision · Deep learning · Traffic

1 Introduction

Object detection and classification is currently a trend in traffic-related applications, mainly in the context of aiding safe driving conditions. We must also consider the appearance of other autonomous entities that coexist with humans and human-operated vehicles in daily traffic scenarios. One of the most relevant is the figure of the *Autonomous Pedestrian*. The term was first introduced by Shao and Terzopoulos [9] to explain the need for realistic pedestrian behaviors in software models. In this paper, we consider autonomous pedestrians the entities that, while not necessarily being a human, must assume typical pedestrian behaviors in a traffic situation, such as walking on a sidewalk, maneuvering between

© Springer Nature Switzerland AG 2020
C. Analide et al. (Eds.): IDEAL 2020, LNCS 12490, pp. 602–609, 2020.
https://doi.org/10.1007/978-3-030-62365-4_59

other pedestrians, among others. This opens new challenges and issues where navigation has fewer norms than for road-users. Pedestrians have to face damaged sidewalks and sometimes need to invade the streets. The crossing points are risky for pedestrians and vehicles, creating a much more dynamic environment – to avoid obstacles, sometimes a risky behavior needs to be assumed. Research for pedestrian scenarios could benefit from the existing basis for autonomous vehicles and could help leverage the generalization for any traffic context, which would benefit all the traffic-related entities.

Object identification becomes an essential task towards such automation and consists of analyzing traffic videos. A camera captures the traffic scene, and according to the detected objects, actions, or recommendations should be issued accordingly. Traditionally, computer vision (CV) approaches are the most used, despite being highly sensitive to dynamic situations such as changing lighting characteristics and uncontrolled scenarios. On the other hand, relying on annotated images for training, Deep Learning (DL) algorithms, such as Faster R-CNN [3] or YOLO [6], have shown good near real-time performance [8]. The higher the number of annotations, the higher the probability the algorithm will learn to identify the objects correctly. However, creating and curating a useful dataset for traffic objects is a repetitive and time-consuming manual task.

We want to address the Crosswalk Detection Problem to contextualize object detection difficulty from a pedestrian's point of view. Although the crosswalk (or zebra-cross) pattern is easily recognizable for humans, it is hard to identify using traditional CV approaches, mainly because the crosswalks can appear in diverse angles and shapes and might be partially occluded by other objects or scene context. Figure 1 shows an example of a pedestrian perspective, where a bounding box identifies the crosswalk.

Fig. 1. A crosswalk annotation with the corresponding bounding box

Some pure CV algorithms have been proposed for crosswalk detection [4,7], but their approaches are not generic enough to be applied in any video, and the possibilities to use DL algorithms are still open depending on the dataset used. To the best of our knowledge, there is no open crosswalk dataset publicly available. Manual annotations are too laborious to create a dataset that is big enough to train a robust crosswalk detection system. It is thus motivating to create and explore the possibilities of new techniques for crosswalk detection.

In this context, the outcomes of the European H2020 project SIMUSAFE (SIMUlator of Behavioural Aspects for SAFEr transportation) are beneficial. One of the goals of this project is to develop realistic, multi-agent behavioral models in traffic environments. Human subjects were given non-intrusive equipment to record daily live activities in traffic (naturalistic observation). The actors comprised drivers (car, motorcycle), pedestrians, and cyclists. The project produced a large dataset of traffic videos for each actor, and after a data fusion approach, we could explore the produced videos.

This work aims to evaluate crosswalk detection using different methodologies, with a few manual annotations as possible. We believe a simple CV algorithm can be used as a first step into classification to create a training dataset for DL. We intend to measure two things: 1) the number of images where a crosswalk is detected; 2) the precision of the model. We formulate the following hypothesis:

Using a simple algorithm to create a training dataset for a DL framework can reduce manual effort with few added errors. A trained detection model can perform better when compared to the simple algorithm.

The methodology of this work, further detailed in Sect. 3, concerns developing a simple custom algorithm for crosswalk detection and manually evaluating the error in the dataset automatically generated by this algorithm. As an outcome of this evaluation process, the correctly identified images compose a curated dataset, to be used for training a refined model. We expect to have a useful model able to detect crosswalks from a pedestrian point of view and help leverage the autonomous pedestrian concept within SIMUSAFE.

2   Related Work

It is relevant to understand the distinction between object instance recognition and object class recognition. Zhang et al. [11] describe object instance recognition as the ability to identify a specific instance of an object from stored examples of the same object, effectively being a matching process where imaging conditions change. However, object class recognition focuses on the ability to identify new instances of a predefined category. Yan et al. [10] enumerate Gabo, Haar, and HOG as the most common identification features to rely on, and the most common classifiers to be SVM, AdaBoost, and Neural Nets. Their work shows that applying Haar-like features and AdaBoost to the problem of vehicle identification in both urban roads and highways can yield a high detection rate.

Mittal et al. [5] present commonly used DL-based methods for the object classification problem. They conclude that while R-CNN can be applied to a large dataset, YOLO v1 is preferred when the number of objects to detect is large, outperforming R-CNN. YOLO has been successfully applied in the classification and counting of different road user types with reasonable accuracy in some particular situations [2].

The automatic detection of crosswalks is not a novel problem [7], and its solution can benefit multiple stakeholders. The detection of crosswalks can be used to help blind pedestrians safely navigate in streets, either by applying CV techniques from the perspective of the pedestrian [4,7] or by analyzing geo-referenced satellite images [1]. Additionally, the detection of crosswalks and ramps can empower mobility-impaired people, particularly those in wheelchairs. This is still an ongoing relevant problem, particularly when applying object instance recognition techniques as these are susceptible to occlusion issues.

While there are still several issues in the CV area to be solved (depth estimation, occlusion), current works have capitalized on machine learning techniques to enable pattern detection under dynamic circumstances. Relatively new algorithms such as YOLO and its revisions have proven useful for traffic object detection. Indeed, they still require high-quality curated datasets, a challenging task considering most of them are manually labeled. In the context of crosswalk detection, more automatic labeling mechanisms are still needed.

3   A Hybrid Approach for Crosswalk Detection

We propose the following methodology: 1) build a simple algorithm to detect crosswalks automatically; 2) use this algorithm to generate a training dataset used as input in the DL approach; 3) compare the results in terms of precision and the number of identified crosswalks. Considering the annotation process, we want to generate annotations for the detected objects (positives) but disregard the images that do not contain crosswalks (negatives). An incorrect annotation is considered a False Positive and helps to increase the error on training the DL algorithm. Thus, we aim to create an algorithm that can identify the higher number of correct crosswalks (True Positives).

Algorithm 1 represents a simple way to identify crosswalks by considering their pattern: a set of (usually) white rectangles that have the same size and placed with the same distance between them. Thus, we transform every video frame into a gray-scale image and detect all the contours in it using a Canny edge detector (Fig. 2, left). With a contour approximation method, we want to combine all rectangles with similar centers and sizes (between thresholds), as they might represent a pattern. If their colors match with a crosswalk pattern and there are at least three matching rectangles, we consider a hit and mark the bounding box for the union set (Fig. 2, right). The algorithm has a simple implementation and complexity of $\mathcal{O}(n \cdot m^2)$, where n is the number of frames on a video (usually 30 frames per second), and m is the number of rectangles detected in each image.

The information stored for each image is the class to which the object belongs (crosswalk), (x_{min}, y_{min}) is the lower-left corner and (x_{max}, y_{max}) is the upper-right corner of the object's bounding box, in percentage, so as to be useful with any scale transformation in the image. The result is a *Dataset*, comprehending a set of images and the corresponding annotations.

The automatically annotated dataset can then be used as input to train a DL-based recognition system, as illustrated in Fig. 3. Such a dataset is called *dirty* because it may contain false positives. The lesser the number of false

Algorithm 1. A simple crosswalk detector

```
1: foreach image ∈ videoFrames do
2:    contours ← CANNY(gray(image))
3:    foreach c ∈ contours do
4:       CrosswalkBoxes ← ∅
5:       foreach candidate ∈ (contours − c) do
6:          if SIMILARCENTERANDSIZE(c, candidate) then
7:             if MATCHCROSSWALKPATTERN(candidate) then
8:                CrosswalkBoxes ← CrosswalkBoxes ⋃ {candidate}
9:       if |CrosswalkBoxes| ≥ 3 then
10:         minX, maxX, minY, maxY ← COORDINATES(CrosswalkBoxes)
11:         ANNOTATECROSSWALK(image, minX, maxY, minY, maxY)
```

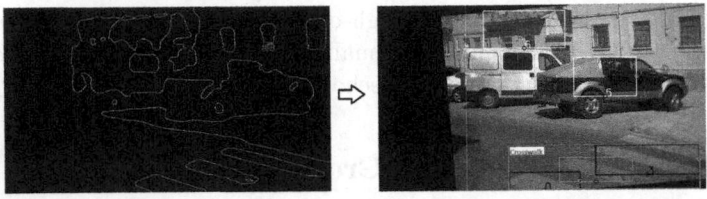

Fig. 2. Snapshot of the simple algorithm detecting a crosswalk

positives in the dataset, the better (fewer objects wrongly identified as crosswalks). The impact of this error is analyzed in Sect. 4. The dashed part of Fig. 3 corresponds to manually cleaning the dataset by allowing only true positives. This process requires significantly less effort than manually producing the whole annotated dataset.

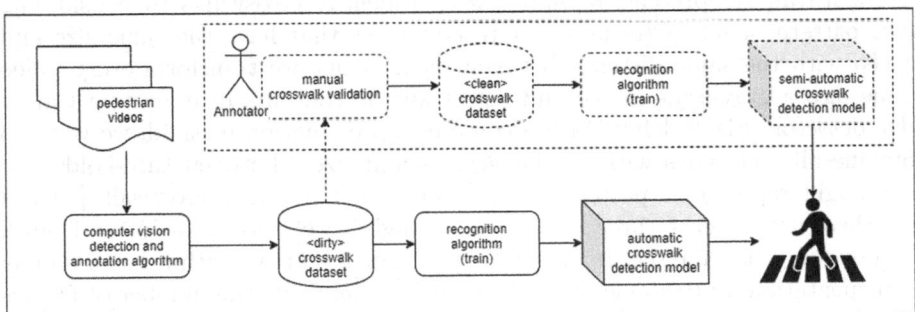

Fig. 3. A hybrid mechanism for automatic dataset annotation and learning

4   Results

From the SIMUSAFE project, we had access to 93 videos (almost 200 GB) of different pedestrian subjects walking on Burgos, Spain. The total duration of the videos was 39,132 s (almost 11 h of video), with a rate of 30 frames per second (fps), meaning we had in total 1,173,960 images. After running the algorithm, a (dirty) dataset of 7,218 images was generated, including true and false positives. Figure 4 shows an example of the results.

(a) True positive   (b) False positive

Fig. 4. Results using the simple algorithm

Table 1 shows the results of the simple CV. As we expected, it could only detect a few crosswalks (7218 in more than 1 million images). Since we did not manually verify all the original video frames, we could not state our model's accuracy. Instead, we proceeded with a manual validation of the automatically generated dataset and discovered the algorithm had correctly identified 3,255 crosswalks, although, in multiple images with more than one crosswalk, the algorithm could not detect them all. This process took about five hours of a single human validator; nevertheless, deciding if the annotation was correct was easier and faster than doing manual annotations. We measured a 45% precision value, as per Eq. 1. As evident in Eq. 2, we could not use the recall metric because it requires the number of False Negatives as a parameter.

$$precision = \frac{True\ Positive}{True\ Positive\ +\ False\ Positive} \tag{1}$$

$$recall = \frac{True\ Positive}{True\ Positive\ +\ False\ Negative} \tag{2}$$

As for the DL algorithm, we selected YOLO [6], a state of the art framework and could process our datasets without any modifications. After 27 hours of training in a Tesla K20 GPU, the algorithm converged. Using the clean dataset (with an 80%-20% split for training and testing, respectively), it achieved the best precision of 0.84, as shown in Table 2. For comparative purposes, we have also trained a YOLO model using the whole dirty dataset and testing in the same clean test set, obtaining a 0.82 precision model.

Precision was lower when training with the dirty dataset because of the noise caused by the false positives. The difference of 2% in precision is minimal, considering that training with the dirty dataset included incorrect annotations. The DL model seemed to achieve a fair generalization, in any case. Applying the DL model to the whole dataset also resulted in a considerable increase in the number of detected crosswalks compared to the simple CV algorithm results. The hybrid algorithm was able to identify 28,489 crosswalks (some of which may be false positives), nearly four times the number of crosswalks identified with the simple algorithm.

Table 1. Results with simple CV

Images analyzed	1,173,960
Annotations generated	7,218
True positives	3,255
False positives	3,963 (55%)

Table 2. Results with the hybrid mode

	Dirty	Clean
Training set	7,218	2,604 (80%)
Test set	651 (20%)	
Precision	0.82	0.84

5   Conclusions and Future Work

This paper presented a methodology to train object identification algorithms in a semi-automated way using simple object detection models as the first step towards dataset generation. Automatic annotation is a crucial process considering that manual annotation is a burdensome process, and simple CV algorithms are usually not precise enough when detecting objects in highly dynamic contexts such as the crosswalks from the pedestrian perspective.

We have validated our proposal on two datasets, one dirty (automatic) and other clean (manually filtering out false positives) to train a DL object classifier. We have used the YOLO algorithm to train a crosswalk recognition model and a pedestrian video dataset from the SIMUSAFE project to show that precision is not profoundly affected by the error present in the dirty dataset, and the overall number of detections is also higher than using the simple version. We have presented two main contributions: 1) a new dataset with pedestrian-perspective crosswalk annotations to be used in other applications and 2) a working model for crosswalk detection with 0.84 precision in the best case. To increase the model's precision, we expect to improve our model by training with a higher number of annotations. This work's applications lie not only on autonomous pedestrians but also on all driving assistance used to navigate congested traffic cities relying on the same infrastructures pedestrians already use.

Finally, we claim that the followed methodology could be useful to train different object classifiers in multiple scenarios.

Acknowledgments. This work is supported by project SIMUSAFE, funded by the European Union's Horizon 2020 research and innovation program under grant agreement No 723386. This research is also supported by LIACC (FCT/UID/CEC/0027/2020). The authors are grateful to the SIMUSAFE Consortium's members for their valuable comments and fruitful discussions throughout the SIMUSAFE Project.

References

1. Ahmetovic, D., Coughlan, J.M., Manduchi, R., Mascetti, S.: Zebra crossing spotter: automatic population of spatial databases for increased safety of blind travelers. In: ASSETS 2015 - Proceedings of the 17th International ACM SIGACCESS Conference on Computers and Accessibility, pp. 251–258 (2015). https://doi.org/10.1145/2700648.2809847

2. Chauhan, M.S., Singh, A., Khemka, M., Prateek, A., Sen, R.: Embedded CNN based vehicle classification and counting in non-laned road traffic. In: ACM International Conference Proceeding Series (2019). https://doi.org/10.1145/3287098.3287118

3. Girshick, R.: Fast R-CNN. In: Proceedings of the IEEE International Conference on Computer Vision, pp. 1440–1448 (2015)

4. Ivanchenko, V., Coughlan, J., Shen, H.: Crosswatch: a camera phone system for orienting visually impaired pedestrians at traffic intersections. In: Miesenberger, K., Klaus, J., Zagler, W., Karshmer, A. (eds.) ICCHP 2008. LNCS, vol. 5105, pp. 1122–1128. Springer, Heidelberg (2008). https://doi.org/10.1007/978-3-540-70540-6_168

5. Mittal, U., Srivastava, S., Chawla, P.: Review of different techniques for object detection using deep learning. In: ACM International Conference Proceeding Series (2019). https://doi.org/10.1145/3339311.3339357

6. Redmon, J., Divvala, S., Girshick, R., Farhadi, A.: You only look once: unified, real-time object detection. In: Proceedings of the IEEE Computer Society Conference on Computer Vision and Pattern Recognition, vol. 2016-Decem, pp. 779–788. IEEE, June 2016. https://doi.org/10.1109/CVPR.2016.91

7. Se, S.: Zebra-crossing detection for the partially sighted. In: Proceedings IEEE Conference on Computer Vision and Pattern Recognition, CVPR 2000 (Cat. No.PR00662), vol. 2, pp. 211–217. IEEE Computing Society (2000). https://doi.org/10.1109/CVPR.2000.854787

8. Shanahan, J.G., Dai, L.: Realtime object detection via deep learning-based pipelines. In: Proceedings of International Conference on Information and Knowledge Management, pp. 2977–2978 (2019). https://doi.org/10.1145/3357384.3360320

9. Shao, W., Terzopoulos, D.: Autonomous pedestrians. Graph. Models **69**(5–6), 246–274 (2007)

10. Yan, Z., Deming, Y., Jun, Z.: Research on vehicle identification method based on computer vision. In: ACM International Conference Proceeding Series, pp. 140–145 (2019). https://doi.org/10.1145/3335656.3335700

11. Zhang, X., Yang, Y.H., Han, Z., Wang, H., Gao, C.: Object class detection: a survey. ACM Comput. Surv. **46**(1), 1–53 (2013). https://doi.org/10.1145/2522968.2522978

Using Deep Learning to Construct a Real-Time Road Safety Model; Modelling the Personal Attributes for Cyclist

Faheem Ahmed Malik$^{(\boxtimes)}$, Laurent Dala, and Krishna Busawon

Faculty of Engineering and Environment, Northumbria University,
Newcastle upon Tyne NE1 8ST, UK
faheem.malik@northumbria.ac.uk

Abstract. This paper is concerned with the modelling of cyclist road traffic crashes by considering the personal attributes, i.e. gender and age of the cyclists. There are 21 different types of variables considered for each crash, which broadly fall into spatial, infrastructure, and environment categories. The study area of Tyne and Wear county in the north-east of England is selected for investigation. Six deep learning-based safety models are constructed using historic crash data. The effectiveness of deep learning methodology for road safety analysis is demonstrated, and it is found that spatial, infrastructural, and environmental conditions affect the safety interactions of a particular cyclist. These variables can be used for determining/predicting safety for a rider at a location. The model can predict age and gender of the rider, which is likely to be the most unsafe based upon the specific input variables. The significant accuracy is obtained for the constructed models with an overall accuracy of 84%. It is hoped that the proposed models can help in better designing of cyclist network, design, and planning, which will contribute to a sustainable transportation system.

Keywords: Cyclist safety · Deep learning · Road safety model · Gender · Age

1 Introduction

The promotion of cycling as a mode of travel has social, economic and environmental benefits [1]. The bicycle has gained a more prominent role in the transportation policy because of its pivotal role in providing the sustainable mode of travel [2]. To achieve a sustainable transport system, cycling mode share has to increase by many folds [3]. However, there are concerns regarding road safety [4]. The critical identified variables from literature affecting the cyclist's safe interaction are infrastructure parameters, trip, and trip maker's attributes. However, there is insufficient evidence to understand the relationship between these variables and safety for cyclist [5]

One of the reported cycling hazards is the personal attribute of the rider, i.e. age, and gender [6]. Women are less likely to make faster journeys, and the spatial and temporal structure of their journey is significantly different from men. They generally cycle at lower speeds and have a more strong liking for low traffic streets [7]. A study in the

© Springer Nature Switzerland AG 2020
C. Analide et al. (Eds.): IDEAL 2020, LNCS 12490, pp. 610–619, 2020.
https://doi.org/10.1007/978-3-030-62365-4_60

Czech Republic found that the males account for around 69% of the crashes, and are more likely to be involved in a fatal crash (80%) than women [8]. Similar a study in the USA [9], found that males are at a higher risk than females (around 5 times more likely than females for the same distance traversed). It is common speculation that men use infrastructure-less safely and more recklessly than women [8]. Another reported personal attribute is the age of the rider. The study in England for the assessment of road safety [10] led them to conclude that the risks for road users are highest in their youth. Their risk falls with age. However, when comparing the different modes of transport, cycling has been reported safer compared to the motorized mode for the younger generation, especially the male population. The study in Sweden [11] to understand the cyclist's injury by age and gender, concluded that the females show a lower incidence than males, however, the elder women are more likely to be involved in a serious crash than younger women. The same results have been reported for males with even more difference between the young and the elderly population. They found that females sustain more work trip injuries than men [11].

Although age and gender have been reported as significant variables, however, there are very few works which have undertaken modelling of this variable or mathematically validate the effect of this variable. This has led to the omission of this variable in the present road safety models [12, 13]. Therefore, the research aims to develop a road safety model for a cyclist using the personal attribute of the user (gender and age), for which the following objectives are defined:

1. To develop an understanding of how safety varies for the cyclist (gender).
2. Test the Hypothesis that unsafeness of the interaction between the user and infrastructure is dependent upon the gender of the user.
3. To develop a safety model with gender and age as an output variable.
4. Identify the most important variable affecting the unsafeness of an identified group of rider.

2   Methodology

The aim requires the development of the model, which can only be achieved through a case study. The study area of the Northeast of England (Tyne and Wear county) is selected for the investigation. It encompasses an area of 210 sq. miles and a population of 1.13 million. The database for road crashes is housed by the Department for Transport (DfT). For each crash, i) Accident Statistics, ii) Contributory Factors, iii) Casualty Record, and iv) Vehicle Record details, are investigated and stored on an online platform. This access was provided by Gateshead city council for the study.

A base input file is constructed for all the cyclist crashes in the study area between 2005–2018. This input file contains the following variables for each crash (Table 1).

There are six different models constructed, using a) spatial, b) infrastructural, and c) environmental variables each for male and female for different age groups using deep learning neural nets. It is a powerful data-driven flexible computational tool which can approximate a wide range of statistical models, without the requirement to hypothesize in advance the relationships between the dependent and independent variable [14]. This

Table 1. Input variable for the proposed model.

No.	Input variable	Values
1.	Spatial	
a)	Month of Journey	Jan-Dec
b)	Journey Day	Monday, Tuesday, Wednesday, Thursday, Friday, Saturday, Sunday
c)	Journey Weekday/Weekend	Weekday. Weekend
d)	Journey Hour	0–23
e)	Number of vehicles	1–5
f)	Journey Purpose	Commuting, work trip, School Journey by Pupil, taking pupil to school, other, Unknown
2.	Environmental	
a).	Lighting conditions	Daylight/Darkness- No Street Lighting, Street Lighting Unknown, Street Lights present and lit, Street Lights present but unlit,
b).	Meteorological conditions	Fine/Rain/Snow-with high winds, without high winds, fog or Mist Hazard, Other
c).	Road surface condition	Dry, Frost/ice, Wet/damp, Snow
3.	Infrastructure	
a)	Road type	Dual Carriageway, One-way street, Roundabout, single carriageway, slip road,
b)	Speed limit	20–70
c)	1st road class	A, B, C, E, U
d)	Road hierarchy level	0-4
e)	Road Hierarchy level and direction	−4 to 4
f).	Junction detail	Crossroad, Mini Roundabout, Multiple Junction, Straight Road, Roundabout, Slip Road, T or Staggered, Private Drive
g).	Junction control	No Control, Traffic Signal, Give way or uncontrolled, Stop sign
h)	2nd road class	A, B, C, E, U
i)	Vehicle maneuver	Changing lanes, Going ahead, Moving off, Overtaking, Parked, Reversing, Slowing/stopping, Turning, U-turn, Waiting to go ahead, waiting to turn
j)	Vehicle junction location	Approaching junction or waiting/parked at junction exit, cleared junction or waiting/parked at junction exit, Entering, Leaving, Mid Junction, Straight Road (Not at or within 20 meters of the junction)

(continued)

Table 1. (*continued*)

No.	Input variable	Values
k)	Road location of vehicle	Bus Lane, Busway, Cycle lane, cycleway, footpath, on layby or hard shoulder, main carriageway, tram/light rail track
l)	Skidding and overturning	No skidding or overturning or jack-knifing, overturned, skidded, overturned and skidded
	Output variable	Risk gender and Age Group

method can be used for simulating different processes and has proved to be a suitable empirical tool to characterize, model, and predict the non-linear process with high accuracy [15]. The main motivation for using deep learning for modelling safety is that the crashes are highly non-linear, and the modeller has no guidance from either theory or even dimensional analysis for modelling. Firstly a learning algorithm is developed, to divide the data set randomly into training (65%), testing (30%), and holdout (5%). This division ensures enough dataset for learning and assessment of the trained model, and ensure that the constructed model is relevant to untrained scenarios [16]. The following network structure is used to construct the model (Table 2).

Step 1: Random weights: Random weights are assigned to each input variable (between the input and hidden, first and second hidden, and between the hidden and output layer).

In the hidden layers, activation function 'Hyperbolic tangent' is used given by:

$$O_a = \tanh(S_a) = \frac{e^{S_a} - e^{-S_a}}{e^{S_a} + e^{-S_a}} \tag{1}$$

O_a is the activation of the *ath* output neuron

In the output layer, activation function 'Softmax' is used given by:

$$O_a = \sigma(S_a) = \frac{e^{S_a}}{\sum_{k=1}^{m} e^{S_k}} \tag{2}$$

m is the number of output neurons

These functions take real numbers as arguments and return real values $[-1, +1]$.

Step 2: Error calculation: The error between the desired output (target) and output obtained, is calculated using the cross-entropy error function.

$$E = -\sum_{a=1}^{m} t_a \ln O_a \tag{3}$$

O_a is the actual output value of the output node a,

t_a is the largest value a, and m is the number of output nodes

Table 2. The network structure of the deep learning model

Network topology	Number of hidden layers	2
	Elements in each layer	350
	Activation function between the hidden layers	Hyperbolic tangent
	Activation function between hidden and output layer	Softmax
	Error function	Cross-entropy
Training	Type	Batch
	Optimization	Scaled conjugate gradient
	Initial lambda	0.000000001
	Initial sigma	0.000000001
	Initial centre	0
	Initial offset	±0.000000001
Stopping and memory criterion	Steps (max) without a change in the error	999,999
	Training (max) time	999,999
	Training (max) epochs	999,999
	Relative change in the training error (min)	0.000001
	Relative change in the training error ratio (min)	0.000001
	Cases to store in the memory (max)	999,999
Hidden layers	Total no. of hidden layers	2
	Total no. of units in the hidden layers	700 (350in each layer)

Step 3: Updating synaptic weights: Based upon the error (step 2), the initial weights are updated. In each epoch, the backpropagation algorithm calculates the gradient of the training error as

i) nodes between the input and hidden layer

$$\frac{\partial E}{\partial w_{ha}} = \sum_{a=1}^{m} (O_a - t_a) x_h w_{ha} (1 - x_h) x_b \tag{4}$$

ii) nodes between the output and hidden and layer

$$\frac{\partial E}{\partial w_{hj}} = (O_a - t_a) x_h \tag{5}$$

In each of the training case (epoch), the weight w_{ih} is updated by adding it

$$\Delta w_{bh} = -\gamma \frac{\partial E}{\partial w_{ha}} \tag{6}$$

$$\Delta w_{bh+1} = w_{bh} + \Delta w_{bh} \tag{7}$$

x is the input variable, and γ is the learning rate.

Step 4: Iteration (scaled conjugate gradient): The updating of weights is iterated until either the minimum change in the training error or the maximum number of these iterations (epochs) is achieved.

To evaluate the performance of the constructed models, Area Under the Curve (AUC) of the Receiver Operating Characteristics (ROC) is used. It is an effective measure of the accuracy of a constructed network [17]. After establishing the credibility and predictive power of the constructed model, the research also aims to develop an understanding of the relationship between the input variables and safety. Therefore, the importance of each of the variable in the prediction model is determined, and the normalized importance concerning the most critical variable.

3   Results and Discussion

The following ROC curves for the six constructed models are presented below, along with the AUC values in Table 3 (Figs. 1 and 2).

From AUC values, it is quite evident that significantly accurate models are constructed, especially through spatial and infrastructural variables. The accuracy is plausible compared with the available models present in the literature (see [13]). For all the output variables, ROC curves are close to the top left corner except environmental variables, depicting the prediction capability of the constructed model. The results lead to infer that the spatial and infrastructural variables directly impact the safe usage of the infrastructure depending upon the personal attribute. However, environmental variables

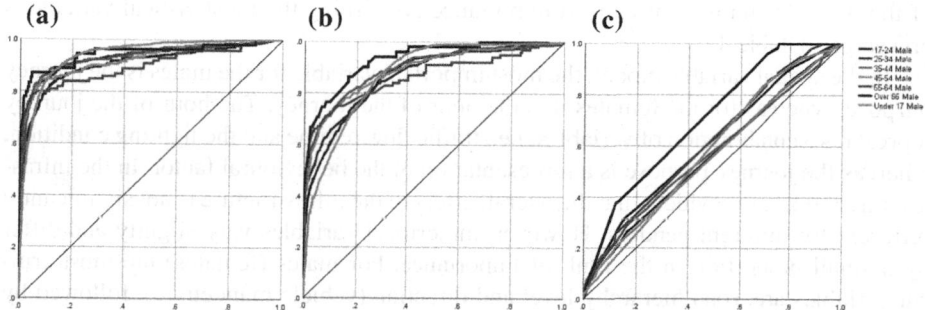

Fig. 1. Receiver operating characteristics curve (sensitivity vs 1-specificity) for male model spatial variable a) Spatial, b) Infrastructure, and c) Environmental variables

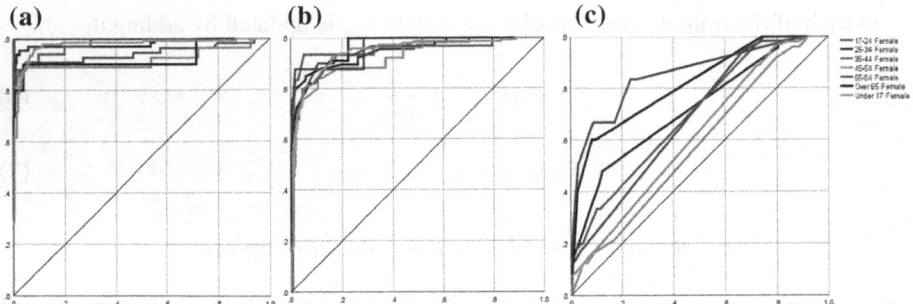

Fig. 2. Receiver operating characteristics curve (sensitivity vs 1-specificity) for female model spatial variable a) Spatial, b) Infrastructure and c) Environmental variables

Table 3. The area under the curve for the three constructed deep learning model

	Spatial		Environment		Infrastructure	
	Male	Females	Male	Females	Male	Females
Under 17	0.91	0.94	0.61	0.68	0.87	0.94
17–24	0.93	0.98	0.56	0.72	0.87	0.95
25–34	0.93	0.98	0.57	0.70	0.90	0.96
35–44	0.92	0.92	0.56	0.58	0.89	0.93
45–54	0.90	1.00	0.62	0.85	0.86	0.96
55–64	0.95	0.92	0.71	0.82	0.93	0.96
Over 65	0.94	0.97	0.58	0.61	0.87	0.94
Total	6.47 (92%)	6.71(96%)	4.21(60%)	4.96 (70%)	6.19 (88%)	6.64 (95%)

impact safety in combination with other variables. This is somewhat an expected result as adverse environmental conditions will badly affect the safety irrespective/across gender and age group, whereas the spatial and infrastructure parameters pose different risks to the rider based upon their personal attribute. In the next step, the importance of each of the variable and the normalized importance concerning the most critical variable is presented in Table 4.

In the spatial variable model, the most important variable for the males is the journey purpose, whereas for the females it is the hour of the journey. The hour of the journey represents a combination of variables, i.e., traffic flow regime and the lighting condition, whereas the journey purpose is a representation of the behavioural factor. In the infrastructure model, it is evident that the overall effect of the infrastructure is not significantly different for different genders. However, the critical variables vary slightly and differ by a small proportion in the rank of importance. For males (females) the most critical variables are; road hierarchy level and direction (vehicle manoeuvre), followed by vehicle manoeuvre (road hierarchy level and direction), junction location of the vehicle (junction details), junction detail (road location of the vehicle), and road location of the

Table 4. Importance of various variables in the three constructed models.

		Male Importance	Male Normalized importance	Female Importance	Female Normalized importance
Spatial	Month	0.170	73.8%	0.192	90.3%
	Day	0.143	62.3%	0.165	77.6%
	Weekday or weekend	0.101	43.9%	0.111	52.4%
	Hour	0.201	87.6%	0.212	100.0%
	Number of vehicles	0.156	67.7%	0.172	81.1%
	Journey purpose of rider	0.230	100.0%	0.149	70.0%
Infrastructure	Road type	0.081	83.3%	0.071	81.7%
	Speed limit	0.076	77.6%	0.069	79.1%
	1st road class	0.074	75.4%	0.067	77.2%
	Road hierarchy level	0.075	77.1%	0.065	74.4%
	Road hierarchy level direction	0.098	100.0%	0.083	95.2%
	Junction detail	0.092	94.5%	0.080	91.8%
	Junction control	0.079	80.7%	0.066	76.1%
	2nd road class	0.082	84.5%	0.070	80.9%
	Vehicle manoeuvre	0.096	98.2%	0.087	100.0%
	Road location of vehicle	0.086	88.4%	0.072	82.3%
	Junction location of vehicle	0.094	96.8%	0.067	76.6%
	Skidding and overturning	0.067	68.4%	0.069	79.9%
Environmental	Light conditions	0.341	93.9%	0.422	100.0%
	Weather	0.363	100.0%	0.344	81.6%
	Road surface condition	0.296	81.6%	0.234	55.5%

vehicle (road type). This leads us to conclude that there are certain features of the infrastructure which are risky for all the cyclists. However, the level of risk that each attribute of the infrastructure poses is dependent upon the gender of the rider. In environmental variable model, the most important variable for males (normalized importance); are meteorological (100%), lighting conditions (94%), and road surface condition (82%), whereas for females; light conditions (100%), meteorological (82%), and road surface condition (56%). These results lead us to conclude that the environmental conditions have a significantly different impact on the gender of the trip maker.

Therefore, from the variable importance, we can conclude that the spatial and environmental variables are significantly different for males and female. Although few of the findings from this study have been reported in the literature, however, these have not been proved mathematically nor been quantified or predicted.

4 Conclusion

In this paper, six accurate cyclist crash prediction models are constructed for the personal attribute of the trip maker, with 21 different types of variables considered for each crash which broadly fall into i) Spatial, i) Infrastructure and iii) Environment categories. Through modelling, it has been established that the personal attributes of the rider affect its safe interactions within the natural road environment. The effectiveness of deep learning neural nets is demonstrated for modelling road safety with an overall accuracy of 84%, which is significantly higher than the available models in the literature. There have been very few works which have attempted to predict the crashes for the cyclist with such accuracy. The following deductions are made: i) The safety of the cyclists varies with personal attributes both temporally as well as spatially, ii) The safety of the cyclists is directly affected by spatial and infrastructural variables, whereas environmental conditions in combination with other variables affect the safety based upon the personal attribute of the rider, iii) For spatial variations, behavioural variable, i.e., journey purpose is critical for males, whereas, for the females, externalities such as traffic flow regime and lighting conditions are the most significant variables affecting their safety, iii) The infrastructural hazards present different level of risk to the cyclist based upon its gender, v) Lighting conditions have a significant effect on the safety of the cyclists, however, they have a more pronounced impact in females.

It is hoped that the proposed model can help in better designing of cyclist network, design, and planning. This work will contribute to a future integrated cyclist smart transportation system. We would like to thank Northumbria University for sponsoring the research (through Research Development Fund) and Gateshead city council for providing access to crash data.

References

1. Dolan, C.O., Stewart, K., Tricker, R., Dolan, C.O., Stewart, K.: Transport Research Institute, Edinburgh Napier University, and Reggie Tricker, Edinburgh City Council 1 (2014)
2. Gordon-Larsen, P., Boone-Heinonen, J.E., Sidney, S., Sternfeld, B., Jacobs Jr., D.R., Lewis, C.E.: Active and cardiovascular disease risk: the CARDIA study. Arch. Intern. Med. **169**, 1216–1223 (2009). https://doi.org/10.1001/archinternmed.2009.163.Active

3. Bell, M.C., Galatioto, F., Fryer, C., Wang, R.: The role of cycling in delivering sustainable travel by 2050. In: 48th Universities Transport Studies Group Annual Conference, Bristol, United Kingdom (2016)
4. Elvik, R.: The non-linearity of risk and the promotion of environmentally sustainable transport. Accid. Anal. Prev. **41**, 849–855 (2009). https://doi.org/10.1016/j.aap.2009.04.009
5. Transport Research Lab: Infrastructure and Cyclist Safety: Research Findings TRL Report PPR 580, Berks, UK (2011)
6. Bill, E., Rowe, D., Ferguson, N.: Does experience affect perceived risk of cycling hazards? In: Scottish Transport Applications Research Conference, pp. 1–19 (2015)
7. Beecham, R.: Exploring gendered cycling behaviours within a large, attribute-rich, transactional dataset. In: 45th Universities Transport Studies Group Annual Conference, Oxford (2013)
8. Bíl, M., Bílová, M., Müller, I.: Critical factors in fatal collisions of adult cyclists with automobiles. Accid. Anal. Prev. **42**, 1632–1636 (2010). https://doi.org/10.1016/j.aap.2010.04.001
9. Rodgers, G.B.: Bicyclist deaths and fatality risk patterns. Accid. Anal. Prev. **27**, 215–223 (1995). https://doi.org/10.1016/0001-4575(94)00063-R
10. Mindell, J.S., Leslie, D., Wardlaw, M.: Exposure-based, "like-for-like" assessment of road safety by travel mode using routine health data. PLoS ONE **7**, 1–10 (2012). https://doi.org/10.1371/journal.pone.0050606
11. Welander, G., Ekman, R., Svanström, L., Schelp, L., Karlsson, A.: Bicycle injuries in Western Sweden: a comparison between counties. Accid. Anal. Prev. **31**, 13–19 (1999). https://doi.org/10.1016/S0001-4575(98)00040-2
12. Aldred, R., Watson, T., Lovelace, R., Woodcock, J.: Barriers to investing in cycling: stakeholder views from England. Transp. Res. Part A Policy Pract. **128**, 149–159 (2018). https://doi.org/10.1016/j.tra.2017.11.003
13. Gettman, D., Pu, L., Sayed, T., Shelgy, S.: Surrogate safety assessment model and validation : final report, Report No. FHWA–HRT-08-051 (2008)
14. IBM: IBM SPSS Neural Networks 25, New York, USA (2017)
15. Araujo, P., Astray, G., Ferrerio-Lage, J.A., Mejuto, J.C., Rodriguez-Suarez, J.A., Soto, B.: Multilayer perceptron neural network for flow prediction. J. Environ. Monit. **13**, 35–41 (2011). https://doi.org/10.1039/b718582k
16. Haykin, S.: Neural Networks: A Comprehensive Foundation. Pearson Education (Singapore) Pte. Ltd (2005). https://doi.org/10.1142/s0129065794000372
17. Hajian-Tilaki, K.: Receiver operating characteristic (ROC) curve analysis for medical diagnostic test evaluation. Casp. J. Intern. Med. **4**, 627–635 (2013)

Author Index